ADVANCES IN POLYOLEFINS

The World's Most Widely Used Polymers

ADVANCES IN POLYOLEFINS

The World's Most Widely Used Polymers

Edited by

Raymond B. Seymour
University of Southern Mississippi
Hattiesburg, Mississippi

and

Tai Cheng
Advanced Polymer Systems
Redwood City, California

SPRINGER SCIENCE+BUSINESS MEDIA, LLC

Library of Congress Cataloging in Publication Data

Advances in polyolefins.

Bibliography: p.
Includes index.
1. Polyolefins. I. Seymour, Raymond Benedict, 1912– II. Cheng, Tai.
TP1180.P67A35 1987 668.4′234 87-7193
ISBN 978-1-4757-9097-9 ISBN 978-1-4757-9095-5 (eBook)
DOI 10.1007/978-1-4757-9095-5

Proceedings of the ACS International Symposium on Recent Advances in Polyolefins, held September 8–13, 1985, in Chicago, Illinois

Originally published by Plenum Press, New York in 1987
Softcover reprint of the hardcover 1st edition 1987

DEDICATION

Polyolefins, which are the world's most widely used polymers, are the product of Modern Polymer Science Research. Unlike other widely used polymers, such as polystyrene, polyvinyl chloride, the acrylics and condensation polymers of formaldehyde with plenolor utea, these polymers were invented after Staudinger, Mark and Carothers had demonstrated the existence of macromolecules.

It should be an inspiration to researchers to note that significant advances are being made in this mature branch of polymer science. Some of these advances which are described by international experts in the field, in this book, are based on theories championed by Dr. Paul J. Flory who died while preparing his report for the national meeting of the American Chemical Society at Chicago in September 1985 which was the site of the symposium on Advances in Polyeolins.

The delegates to the macromolecular secretariat at this meeting, voted unanimously to dedicate these proceedings on Advances in Polyolefins to the memory of our esteemed colleague, Professor Paul John Flory.

Raymond B. Seymour
Tai Cheng

PREFACE

Because of the many important new developments in other branches of science, some scientists fail to recognize that the volume of polyolefins produced annually is greater than that of all metals. Hence, the American Chemical Society sponsored symposia on the History of Polyolefins at its national meeting at Miami Beach in the Spring of 1985 and a Macromolecular Secretariat on Advances in Polyolefins at its national meeting in Chicago in the fall of that year.

The books on the proceedings of these landmark symposia and another book entitled "The Chain Straighteners" by Dr. F. M. McMillan will provide the scientist with background information which is essential for researchers in this important phase of polymer science.

The presentations at these international symposia and the publications of the reports presented, would not be possible without the dedicated efforts of our assistant editors and publisher. The list of contributors to ADVANCES IN POLYOLEFINS includes most of the leaders in this field, such as Dr. Mark, Mandelkern, Bruzzone, Hsieh, Kaneda, Chien, Tait, Karol, Kaminisky, Scott, Cook, Mirabella, Samuels, Kanamoto and Vigo.

These reports covered many phases of polyolefin science and technology, ranging from elastomers, single crystals, film and fibers to characterizations by modern instrumentation and many new innovations in catalysis which have brought about a revolution in polyolefin production.

We as editors are grateful for the unselfish assistance of those who made this publication possible. We are particularly appreciative of our appointment as cochairs for the Macromolecular Secretariat on ADVANCES IN POLYOLEFINS by the Macromolecular Secretariat Committee. Hopefully, this treatise will set the stage for subsequent publications describing future advances in polyolefins.

Raymond B. Seymour
Tai Cheng

CONTENTS

MAGNESIUM CHLORIDE SUPPORTED CATALYSTS

LINEAR LOW DENSITY POLYOLEFINS

CHARACTERIZATION OF POLYOLEFINS

POLYPROPYLENE FILM

INTRODUCTION

In spite of the fact that polyolefins were one of the last types of polymers to be commercialized, they have become the most widely used polymers throughout the world. Low density polyethylene (LDPE), which was the pioneer commercial polyolefin, is a highly branched polymer, produced at extremely high pressures, in a capital-intensive process. Nevertheless, over 2 million tons of this product is produced annually in the U.S. and an equal volume of this polymer is produced in other developed nations.

In spite of its higher density, which means higher cost per unit volume, the use of high density polyethylene (HDPE) continues to increase. Over 3 million tons of this linear polymer is produced annually in the U.S. and an equal volume is produced in other developed countries. This versatile product may be produced as a tough ultra high molecular weight polymer, it may be swollen in solvents and chain extended to produce filaments with unusually high strengths and may be used as a mica-filled high temperature resistant sheet, film and blow molded containers.

Sizable quantities (625 thousand tons) of the copolymer of ethylene and vinyl acetate (EVA) are produced annually in the USA but the polymer with the greatest potential is a copolymer of ethylene and alpha-olefins, such as 1-butene. These linear low density polyethylenes (LLDPE) which are characterized by toughness and flexibility, are now produced at an annual rate of 1.5 million tons in the U.S. The Unipol, Dowlex and related processes used to produce these superior products are more efficient, more versatile and less capital intensive than the traditional processes used to produce LDPE and HDPE.

The existence of several types of polyethylene and blends of these polymers provide the designer and molder with unusual versatility in resin specifications. Thus, polyethylene technology has progressed from its dependence on one low density polymer (LDPE) to many linear polymers and copolymers and blends which will extend the use of polyethylene to many heretofore unacceptable applications.

The story of polypropylene is also one of versatility and unusual growth potential. In spite of its greater susceptibility to degradation from outdoor exposure, over 2.5 million tons of polypropylene is produced annually in the USA. The use of about 700 thousand tons of polypropylene fibers and filaments is indicative of the versatility of this polymer. It is also used for blow molding, extruded film and pipe and injection molded packaging and battery cases.

Copolymers of ethylene and propylene (EPDM) are also used as light colored vulcanizable elastomers. These and other applications will increase as the result of the so called revolution in polyolefin catalysts which includes magnesium chloride-supported catalysts. Since this book is a chronicle of events leading to the polyolefin catalyst revolution, it has been possible to record these events by worldwide authorities in a single volume. However, because of the rapid and continued growth in this field, it will not be possible to describe subsequent events in a single volume.

Accordingly, we, as organizers of this international macromolecular secretariat on "Advances in Polyolefins" recognize that it has been a distinct privilege to edit and publish these important contributions at this time.

Raymond B. Seymour
Tai Cheng

HISTORY OF POLYOLEFINS

Raymond B. Seymour

Department of Polymer Science
University of Southern Mississippi
Hattiesburg, MS 39406

INTRODUCTION

A Brief History of Polyolefins

Polyolefins are, by far, the world's most widely used synthetic polymers. As shown in subsequent chapters, these ubiquitous macromolecules are used as coatings, fibers, elastomers and plastics. Yet, commercial polyolefins were essentially nonexistent prior to World War II. The history of the genesis of these versatile polymers illustrates the inherent genius and good as well as misfortunes of the pioneers in the science and technology of polyolefins.

The term polyolefin is confusing since this is the name given to polymers of olefins, which of course, are neither olefinic (unsaturated hydrocarbons) or alkenes. Actually, polyolefins are polyalkanes with the empirical formula of $H(CH_2)_nH$. This is also the formula for solid alkanes, such as paraffin wax. The latter is an oligomer with 26 to 30 repeat methylene (CH_2) units. Microcrystalline wax has a higher molecular weight and higher melting point (90°C) than paraffin wax. The naturally occurring elaterite is a polyolefin-like polymer.

In a lecture to the Chemical Society in 1863, Berthelot[1] reused the word "polymeric" which had been coined by Berzelius in 1832. Later, he coined the term polyethylene for a hexadecane produced by heating ethylene with alkali. He also obtained a higher molecular weight product i.e. tetramer and pentamer by heating propylene with sulfuric acid.[2] Yet, Goryainov and Butlerov were unsuccessful in their attempts to polymerize olefins by heating the sulfuric acid,[2] but Berthelot and Gaudechon were successful in their attempts to produce oligomers by the photopolymerization of ethylene in 1910.[3]

POLYMETHYLENE FROM DIAZOALKANES

Von Pechman[4] and Bamberger and Tschirner[5] obtained solid polymethylene by the decomposition of diazomethane in 1889 and 1900 respectively. A half century later, Buckley, Cross and Ray produced polymethylene $(CH_2)_n$ polyethylidene $(CH(CH_3))_n$ and polypropylidene $(CH(C_2H_5))$ by the copper-catalyzed decomposition of diazomethane, diazoethane, and diazopropane, respectively.[6] Kantor and Osthoff produced high molecular weight, semicrystalline, polymethylene by the decomposition of diazomethane in the presence of diethyl ether-boron trifluoride complex at 0°C.[7] Polymethylene was also prepared by Bawn and Rhodes by the copper, copper stearate and boron trifluoride decomposition of diazomethane.[8]

J. Hertzog, R. Schonbach, H. Merwein, W. Burnleit and W. Weile also produced polymethylene by the decomposition of diazomethane in the 1920's and 1930's.[9] Leitch, Gagnon, and Cambron also used this type reaction to produce a deuterated polymethylene.[10] Buckley published infrared spectograms showing different bands in polymers produced from diazoalkanes[11] and Oakes and Richards determined the average molecular weights of polymethylene using cryoscopy and intrinsic viscosity.[12] Polymethylene was patented in Germany, but was not produced commercially from diazomethane.[13]

These polymethylenes were characterized by crystalline melting point (137°C) which is in accord with thermodynamic predictions of Mark.[14] The properties of polyolefins, produced by the decomposition of diazoalkane have been summarized[15] and considerable information on these polyolefins has been recorded.[16] According to Nasini and Trossarelli, small amounts of crystalline polymethylene are produced by the decomposition of diazomethane in the presence of gold, presumably by a heterogeneous reaction but larger amounts of amorphous polymethylene are also produced by a homogeneous reaction.[17]

POLYETHYLENE BY CONDENSATION REACTIONS

Taylor and Jones produced oligoethylene by the thermal decomposition of diethylmercury, in the presence of ethylene.[18] Carothers produced polymethylenes with as many as 100 methylene groups by the condensation of decamethylene bromide.[19] He separated these polymers by fractional crystallization and recorded the melting points of specific polymers, such as $(CH_2)_{50}$, (92°C), and $(CH_2)_{70}$, (105°C).

H. Staudinger produced a crosslinked polythylene by the reduction of polyvinyl bromide[20] and W. Hahn and W. Muller obtained linear polyethylene by the lithium hydride reduction of polyvinyl chloride in the 1950's.[21]

FISCHER-TROPSCH REACTION

Koch and Ibing obtained linear crystalline polymethylene, with an average molecular weight of 200 and a melting point of 117°C, by the Fischer-Tropsch reduction of carbon monoxide with hydrogen.[22] Pichler and Buffleb used a ruthenium catalyst in the Fischer-Tropsch reaction to obtain polymethylenes with average molecular weights as high as 9,000[23] and this technique for the production of polymethylenes, using a metal tungstite catalyst, was patented by duPont in 1955.[24]

POLYISOBUTYLENE

Oligomers of isobutylene were produced in 1873 by the boron trifluoride-catalyzed polymerization of isobutylene.[25,26,27] This polymerization was reinvestigated by Hoffman[28] and Otto[29] who obtained and patented this polymer.[30] In 1930, Lebedev, Staudinger, and Bruner obtained oligomers and high molecular weight polymers of isobutylene by polymerization in the presence of Floriden clay at -80°C and Waterman[31] polymerized isobutylene in the presence of aluminum fluoride at -78°C in 1934. After a formal cooperative agreement was signed, Otto was invited to the Standard Oil Co. Laboratories at Bayway, NJ where he remained from 1933-1935.[32]

Bannon of Standard Oil patented the polymerization of isobutylene in ethylene using a cold, boron trifluoride - slurry technique, which is still in use for the production of polyisobutylene and butyl rubber.[33] Otto patented an endless belt technique for the polymerization of isobutylene.[34] Later, a solution of aluminum chloride in ethyl chloride was substituted for the boron trifluoride catalyst in the production of polyisobutylene.[35] Polyisobutylene was produced in Germany under the trademark of Oppanol and in the U.S. under the trademark of Vistanex.

Thomas and Sparks patented the production of a vulcanizable elastomer (butyl rubber) which was a copolymer of isobutylene, with a small amount of isoprene.[36] Some of the early problems in the commercial production of butyl rubber have been recorded.[37] Earlier studies by Hoffman and Otto[38] and subsequent studies by Polanyi[39,40] and others[41] showed that cocatalysts were essential for these cationic polymerizations. Over 600 thousand tons of isobutylene-based polymers are produced annually worldwide.

PRODUCTION OF LINEAR POLYETHYLENE

Goryainov and Butlerov were unsuccessful in their attempts to polymerize ethylene in the presence of sulphuric acid,[42] but Ipatiev obtained oligomers of ethylene by polymerization in the presence of zinc chloride and aluminum chloride in 1913.[43] As discussed previously in this report, several investigators have prepared polymethylene by the decomposition of diazomethane. These investigations were repeated by Sailors and Hogan[44] and Hobart and Ziegler[45] in order to characterize these polymers. These polymers had a density of 0.97 grams/cc, a melting point of 135°C, a tensile strength of 3900 psi, a relative viscosity in xylene (0.125%) at 85°C of 1.2 and were 91% crystalline.

LOW DENSITY POLYETHYLENE

The discovery of low density polyethylene (LDPE) was based on a series of fortuitous events which have been described by Swallow,[46] Ballard,[47] and McMillan.[48] This polymer, which was originally called high pressure polyethylene (HPPE), was, in fact, the result of an interest in ultrahigh pressure reactions by Nobel Laureate Robert Robinson, a consultant for ICI, who was aware of the pioneering efforts of P. W. Bridgman and A. Michels. Imperial Chemicals Industries (ICI) was formed in 1927 through the amalgamation of British Dyestuffs Company, Brunner Mond, United Alkali and Nobel Explosives Industries in an attempt to compete with I. G. Farben Industrie.

In 1933, E. W. Fawcett and R. O. Gibson attempted to combine ethylene and benzaldehyde at 170°C and 1400 atmospheres. These investigators, who received little recognition, recovered about 1 gram of residue from the reaction kettle. Since they recognized that this residue was polyethylene, they repeated the experiment without benzaldehyde, but the reactor exploded.

Without consent of ICI, Fawcett reported the production of polyethylene at the Faraday Discussion on Polymer Chemistry in 1935. H. Staudinger labeled Fawcett's statement as nonsense,[49] but, Fawcett was transferred to other work in ICI. A larger quantity (8g) of LDPE was produced in 1935, by M. W. Perrin, who accidentally used traces of oxygen which was present in ethylene, as a catalyst. It is of interest to note that Fawcett, Gibson and Perrin, who is now Sir Michael Perrin, all left ICI before the end of

World War II and were not involved in further production of LDPE. However, they were recognized as co-patentees in the ICI patent application.[50]

The LDPE produced in the 1930's had the following properties; melting point 115°C, density 0.91 grams/cc, could be drawn into filaments, was partially crystalline[51] and had a Staudinger molecular weight of 2000 to 24,000. Fox and Martin used infrared spectroscopy to show a larger than expected number of methyl groups than should be present in a linear polymer and concluded that LDPE was a branched polymer.[52]

A pilot plant was built, under the direction of J. C. Swallow, and the full scale production facility for LDPE came on stream in September 1939, on the same day that Hitler's troops invaded Poland. LDPE which was used for the insulation of radar cable was also produced by duPont, in the U.S., in 1943. The production of LDPE exceeded that of all other polymers until 1982 when some of this polymer was displaced by linear low density polyethylene (LLDPE) which is produced at moderate pressures.[53] Several reviews on the history of LDPE are available.[54,55,56,57,58]
Over 1700 tons of LDPE was produced in the U.S. in 1983.

LINEAR POLYETHYLENE

In 1930, Frederick and Marvel produced polyethylene, by the accidental polymerization of ethylene in the presence of alkyllithium and tetraethylarsenium bromide.[59] Marvel then passed ethylene through a solution of butyllithium, in hot mineral oil, and obtained linear polyethylene (HDPE). duPont examined this polymer but declined to consider its commercialization.[60]

Nevertheless, Krase[61] and Larcher and Pease[62] patented HDPE, produced at high pressure with peroxide initiators. These polymers had a density of 0.96 grams/cc, a relative viscosity of 1.2 (0.125% in xylene) at 85°C and a melting point of 127°C.

In 1940, Mayo also produced HDPE, by the free radical polymerization of ethylene, but his employer, U.S. Rubber Co., was not interested in this polymer.[63] M. Fischer patented the production of HDPE by the polymerization of ethylene in the presence of titanium tetrachloride and aluminum chloride.[64]

Karl Ziegler had used alkyllithium to polymerize butadiene in 1928.[65] Gellert used the more soluble lithium aluminum hydride in place of the insoluble lithium hydride.[66] Ziegler and his associates found that lithium aluminum hydride and triethylaluminum added ethylene readily to form 1- butene. After it was discovered that a trace of nickel from the stainless steel reactor was responsible for termination of the addition of ethylene, a search for the effect of other metals was initiated.

Breil obtained negative results with cobalt, copper, iron, gold, platinum, and silver but produced HDPE with zirconium on October 26, 1953.[67] The preferred catalyst system, which was titanium tetrachloride and triethylaluminum was patented by Ziegler.[68] This process was licensed to Petrochemicals Ltd. in England, Montecatini in Italy, Farbwerke Hoechst in Germany, and Hercules in the U.S. It is important to note that Ziegler acted as his own patent attorney and restricted his patent application to HDPE. Ellis also patented the polymerization of ethylene in the presence of alkyllithium supported on silicon dioxide.[69]

Breil noted that HDPE softened at 130-150°C, which was higher than the softening point of LDPE and used an IR spectrograph to show the presence of only a few methyl groups and, hence, he concluded that HDPE was a linear polymer. The average molecular weight was reported as about 200 thousand.

In 1943, G. Bailey and E. Reid[70] obtained an oligomer of ethylene by use of a nickel oxide catalyst on an alumina/silica support. Later, Hogan and Banks used a chromium salt, supported on alumina/silica and plugged the reactor pipe with solid HDPE. Shell chemists had also used chromium salts as catalysts for the dimerization of ethylene but discontinued the research when "waxy solids" were produced.[71] Hogan and Banks filed patent applications in the U.S., Australia, and Belgium in 1953, 1954, and 1955, respectively.

Considerable production and characterization information is given in a U.S. Hogan and Bank's patent which issued in 1958.[72] The solid HDPE had the following properties; density, 0.95-0.97 grams/cc, intrinsic viscosity 0.2-10 (corresponding to a weight average molecular weight of 5,000 to 250,000) shore D hardness 60-65, tensile strength about 4,000 psi, elongation of 10-40%, softening temperature about 200°F and crystallinity of about 80%. The amorphous portion was soluble in solvents, such as n-heptane but the crystalline portion was insoluble.

The melting point or actually the freezing point of HDPE was determined from a cooling curve of temperature vs. time. Infrared spectrograms showing similarity to Ziegler's HDPE and differences from LDPE were included in this patent application. The history of Phillips' HDPE has been reported in several articles.[73,74,75]

Zletz also patented the production of HDPE using a catalyst consisting of molybdenum oxide or cobalt molybdate, supported on alumina.[76] Plants using the Zletz process were built by Furukawa in Japan, in Sardinia, and by Amoco (formerly Standard Oil) in Texas City, Texas. The latter has been scrapped and Amoco is now producing HDPE, under a license from Solvay-USI. The Phillips' process is now the most widely used process for making HDPE. About 10 million tons of HDPE is produced annually, worldwide.

POLYPROPYLENE

The word "polypropylen" was first used by Buterlov,[77] who obtained tetramers and pentamers by the addition of sulfuric acid to propylene.[78,79] Berthelot also produced oligomers by this technique.[80] Oligomers of propylene were prepared, using copper pyrophosphate, and silica/alumina as initiators. Low molecular weight, amorphous polypropylene, obtained by the use of boron trifluoride with a hydrogen fluoride activator, was used as a viscosity index (VI) improver and as a lubricating oil during World War II.

Much of the early investigations were updated in the post war period, by Fontana at Mobil Oil Co. He obtained polypropylene with a number average molecular weight as high as 830,000 by the cationic polymerization of propylene at -44°C.[81,82,83,84] Fontana recorded a glass transition temperature (T_g) of -30°C for his amorphous polypropylene and stated this polymer had no tendency to crystallize. That this amorphous polymer was soluble in alkanes was evident by its end use as a viscosity improver in lubricating oils.

CRYSTALLINE POLYPROPYLENE

Breil attempted to polymerize propylene in Ziegler's laboratory but since the pressure did not decrease as dramatically as it did when ethylene was used, he concluded that no reaction took place. Hence, he added ethylene to the reactor and obtained a quantitative yield, which was probably a mixture of polyethylene and polypropylene. He recorded "Propylene cannot be converted into high molecular polypropylene" in his notebook and Ziegler concluded "Es geht nicht" and did not include propylene in his patent application.

Ziegler's next attempt to polymerize propylene was not made until May, 1954, when Martin obtained polypropylene in a new reactor. In the interim, Wright of Petrochemicals, a Ziegler licensee, produced polypropylene in a pilot plant when ethylene was in short supply. Since he mistakenly believed that the Ziegler also covered polypropylene he missed an opportunity to make a patent application.

In contrast, Natta, who had been asked by Ziegler to restrict his investigations to HDPE, directed P. Chini to make polypropylene. After the crystalline polypropylene had been obtained, Natta wrote the following in his laboratory notebook; "Today (March 11, 1954), we made polypropylene." Natta applied for a patent and published his findings in 1955.[85] Ziegler also applied for a patent for polypropylene but was late by 10 days. A patent was issued to Natta, in 1972, after many years of litigation.[86]

Since Natta was a crystallographer, he used x-ray diffraction techniques to obtain a repeat period of 6.50 Å for polypropylene and stated that each portion of the principal chain included in the elementary cell corresponded to 3 monomer units. He also reported that because the asymmetric carbon atoms have the same configuration and because of steric requirements of the chain substituents "a spiralization of the principal chain takes place." He followed his wife's suggestion and used the term isotactic to designate the stereoregularity of these polymers.[87]

That stereoregularity of polymers could be influenced by polymerization condition had been suggested by Huggins[88] and demonstrated by Schildknecht when he polymerized vinyl isobutyl ether cationally at low temperature.[89] Flory suggested an isotactic structure for Schildnecht's polymer[90] and Natta verified this configuration in 1955.[91]

POLYPROPYLENE PATENTS

Since Hogan and Banks and other investigators had produced crystalline polypropylene before Natta, an interference proceeding (No. 89,634) was declared by the U.S. Patent Office in 1958.[92] The contestants were Phillips, duPont, Montecatini (now called Montedison), Standard Oil of Indiana, and Hercules. Hercules was dropped from these proceedings because of late discovery dates and Montedison was awarded the basic composition of matter patent for crystalline polypropylene.[93]

An appeal by the losing parties resulted in a Civil Court trial in the Federal Court of Delaware in 1977-78. The Court awarded a patent to Hogan and Banks in 1980.[94] The losing parties appealed in 1982, but the Federal Appeals Court upheld the decision of the lower court and the U.S. Supreme Court refused to review the decision of the lower courts. Hence, a patent was issued to Hogan and Banks in 1983 with a single claim, "Normally solid propylene, consisting of recurring propylene units, having a substantial crystalline polypropylene content."[95] The losing parties have appealed the latest court's decision.

As stated previously, Vanderburg of Hercules, claimed to have produced polypropylene but was unable to show a sufficiently early date for his research. Baxter, Merkling, Robinson, and Stamatoff of duPont claimed to have produced crystalline polypropylene in 1954, but failed to show utility requirements of actual reduction to practice. Zletz, Peters, and Carmody of Standard Oil of Indiana, produced a product which was shown to be a copolymer of ethylene and 2-butene, resulting from a rearrangement of propylene to ethylene and 2-butene in the presence of a molybdenum catalyst.[96]

The decision of the court to issue the patent for polypropylene to Hogan and Banks was based on the following: Proof of reduction to practice using a chromic oxide - alumina/silica catalyst system in 1951 and 1952, recognition of the product as polypropylene, recognition of a specific practical use as a molded plastic and constructive reduction to practice by filing a patent application describing how the polymer was produced and stating a practical use for polypropylene.

There were actually 4 patents issued to Hogan and Banks, based on subject matter first disclosed in a patent application dated January 27, 1953. These patents were for polymers, catalysts, poly 4-methyl-1-pentene and polypropylene.[97]

As pointed out by Morawetz, the Zletz patent application claimed a rubbery polymer,[98] the Hogan and Banks application[99] claimed a solid fraction, insoluble in pentane but neither patent mentioned crystallinity.[100] However, Hogan and Banks did indicate the transition of molten polypropylene to solid polymer as a melting point and showed a distinct plateau in their cooling curve which is consistent with crystallinity. A noncrystalline polymer does not exhibit a sharp melting point.[101] Hogan and Banks reported a melting point of 144-150°C and Natta reported a melting point of 160°C. Fontana reported a glass transition temperature of -30°C for amorphous polypropylene which, as expected was much lower than the crystalline melting point.

PROPERTIES OF POLYPROPYLENE

Densities of 0.91-0.92 grams/cc and 0.92 grams/cc were reported by Hogan and Banks and Natta, respectively. The relationship between density and degree of crystallinity had been reported for cellulose in 1937.[102] Boyer has shown that the ratio of difference between the density of the sample less the density of the amorphous polymer and the difference between the density of the crystalline polymer less that of the amorphous polymer to be equal to the degree of crystallinity.[103]

Hogan and Banks had reported intrinsic viscosities in the range of 0.2 to 1.0 which would correspond to Staudinger molecular weights of 5,000 to 25,000. It is now agreed that crystalline polypropylene is a monoclinic crystal with a 3 fold helix having an identity period of 6.50Å and with low energy gauche and high energy transconformations in adjacent chains.[104]

Polypropylene, which is the fastest growing polymer, is produced at an annual rate of 2.2 million tons in the U.S. and at an annual rate of 6.5 million tons worldwide. It is used in the blow-molding of containers, extrusion of filaments, and fiber, and for the injection molding of containers, appliances, housewares, and battery cases.

CROSSLINKED POLYETHYLENE

In 1948, Dole discovered that LDPE may be crosslinked by irradiation. It may also be crosslinked by peroxide initiators.[105]

LINEAR LOW DENSITY POLYETHYLENE

In 1958, Hogan reduced the density of HDPE by producing a copolymer with 1-hexene. This copolymer was described in 1969[106] and patented in 1972.[107] Kaminsky has shown that linear low density polyethylene (LLDPE) with a density of 0.89 grams/cc can be produced by the addition of 9 mole percent of 1-hexene to ethylene before polymerization.[108] In the late 1980's, chemists at Union Carbide Company and Dow Chemical Company produced LLDPE by the copolymerization of ethylene with C_4-C_8 alpha olefins. The low pressure slurry (Unipol) process used 1-butene, 1-pentene, and 1-hexene and the solution process (Dowlex) used 1-octene, 4-methylpentene -1 and 1-hexene as comonomers for the production of LLDPE. Efficient Ziegler-Natta catalysts are used in these processes.[109] About 1 million tons of LLDPE is produced annually in the U.S.

duPont has been producing LLDPE in Canada since 1960. The first generation catalyst was used in the low pressure slurry process in 1970.[110] This was replaced by a second generation catalyst in 1976 and by a third generation catalyst in 1980. According to Tait, the first generation Ziegler-Natta catalysts are ball milled α- titanium trichloride and diethylaluminum chloride.[111] Ether-treated titanium trichloride - based catalysts have a 5 fold increase in activity over the first generation catalysts.[112]

The addition of Lewis acids also improve Ziegler-Natta catalyst activity.[113] Magnesium chloride, ball milled with ethyl benzoate has been used as a support to enhance Ziegler-Natta catalysts.[114]

G. Natta produced polymers of 2- methybutene -1, 4,4- dimethypentene -1, 1-hexene, 1-butene, and methypentene. The latter two low density polymers are in commercial production.

ETHYLENE COPOLYMERS

In addition to the previously mentioned copolymers of isobutylene and isoprene (butyl rubber), several other olefin copolymers are produced commercially. While both linear polyethylene and isotactic polypropylene are crystalline solids, the copolymer of these two olefins is an amorphous elastomer. The copolymer of ethylene and propylene (EPM) was produced commercially by Exxon in 1960.[115] A sulfur-vulcanizable copolymer was produced by the addition of 1,4-hexadiene, 5-methylene or ethylidene -2-norbornene, or dicyclopentadiene to the two monoolefins before polymerization. The first unsaturated terpolymer (EPDM) was produced commercially by duPont in 1963.[116,117] EPM and EPDM are produced by at least 11 companies in plants around the world. The annual production of these copolymers, world-wide, is about 500 thousand tons.

Copolymers of ethylene and methacrylic acid (ionomers) were produced in the laboratory in the 1940's and commercialized by duPont in the 1950's.[118] These ionomers, under the trade name of Surlyn, were described and patented in 1966.[119,120] Other commercial ethylene copolymers developed in the 1950's, were the heat meltable copolymer with vinyl acetate and the copolymer with ethyl acrylate. The ionomer is more resilient and has a higher modulus than the vinyl acetate or ethyl acrylate copolymer.

The poor resistance of polyolefins to gaseous diffusion has been overcome by the production of copolymers of ethylene and vinyl alcohol (EVAL). These copolymers are obtained by the hydrolysis of copolymers of ethylene and vinyl acetate (EVAC). Many other copolymers of ethylene have been synthesized and these will be commercialized as end uses are encountered which require these copolymers.

REFERENCES

1. M. Berthelot, Bull Soc Chem France, 114, 2 (1869).
2. B. Goryainov and A. Butlerov, Ann, 169, 146 (1873).
3. D. Berthelot and H. Gaudechon, Compt Rendu, 150, 1169 (1900).
4. H. von Pechman, Ber, 31, 2440 (1899).
5. E. Bamberger and F. Tschirner, Ber, 33, 955 (1900).
6. G. D. Buckley, L. H. Cross, and N. H. Ray, J. Chem Soc, 2714 (1950).
7. S. W. Kantor and R. C. Osthoff, J. Am. Chem. Soc., 75, 931 (1953).
8. C. E. Bawn and T. B. Rhodes, Trans. Faraday Soc., 50, 934 (1954).
9. R. B. Seymour, Polymer News, 10, 376 (1985).
10. L. C. Leitch, P. E. Gagnon, and A. Cambron, Can J. Research, B28, 256 (1950).
11. G. D. Buckley, L. H. Cross, and N. H. Ray, J. Chem Soc, 2715 (1950).
12. W.G. Oakes and R. B. Richards, J. Chem Soc, 2929 (1949).
13. R. C. Osthoff and S. W. Kantor, Ger Pat. Appl., B25,064 (1955), assigned to G.E.
14. H. F. Mark, J. Appl Phys, 12, 41 (1941).
15. J. Felzin, A. J. Restaino, and R. B. Mesrobian, J. Am. Chem Soc, 77, 206 (1955).
16. R. A. V. Raff and J. B. Allison, "Chap. 1 Historical Development" in: "Polyethylene," Interscience Publishers, New York (1956).
17. A. G. Nasini and L. Trossarelli, in: "Macromolecular Chemistry," J. Polym Sci, C4, M. Mgat, Ed. Interscience, New York (1964).
18. H. S. Taylor and W. H. Jones, J. Am Chem Soc, 52, 1111 (1930).
19. W. H. Carothers, J. W. Hill, J. E. Kirby, and R. A. Jacobson, J. Am. Chem Soc, 52, 5279 (1930).
20. H. Staudinger, Angew Chem, 64, 155 (1952).
21. W. Hahn, Makrolmol Chem, 16, 71 (1955).
22. H. Koch and G. Ibing, Brennstoff-Chem, 16, 141 (1935).
23. H. Pichler and H. Buffleb, Brennstoff-Chem, 21, 257 (1940).
24. H. R. Arnold and E. C. Herrick U.S. Pat 2, 726, 231 (1953) assigned to duPont.
25. H. A. Butlerov and B. Goryainov, Ber, 6, 561 (1873).
26. B. Goryainov and H. A. Butlerov, Ann, 169, 146 (1973).
27. H. A. Butlerov, Ann, 180, 247 (1976), 189, 44 (1877).
28. E. G. M. Tornqvist in: "Polymer Chemistry of Synthetic Elastomers," Part 1 J. P. Kennedy and E. G. M. Tornqvist eds, Interscience - Wiley, New York (1968).
29. M. Otto, Brennst-Chem, 8, 321 (1927).
30. M. Otto and M. Mueller - Conradi U. S. Pat 2, 130, 507 (1938) assigned to I. G. Farben Industrie.
31. W. L. Dunkel, R. F. Neu, and R. L. Zapp, Chapt. 12 in: "Introduction to Rubber Technology," M. Morton, ed., Reinhold, New York (1959).
32. E. G. M. Tornqvist, Chapt. 9 in: "History of Polyolefins," R. B. Seymour, T. Cheng, eds. D. Reidel Publishing Co., Dordrecht, Holland (1985).
33. L. A. Bannon, U. S. Pat 2,317,878 (1943) assigned to Joint American Study Co. (JASCO).
34. M. Otto, H. Gueterbork, and A. Hellemans, U. S. Pat 2,311,567 (1943) assigned to JASCO.
35. R. M. Thomas and O. C. Slotterbeck, U. S. Pat. 2,243,658 (1941) assigned to JASCO.
36. R. M. Thomas and W. J. Sparks, U. S. Pat 2,356,128 (1944) assigned to JASCO.
37. R. M. Thomas, I. E. Lightbown, and W. V. Sparks, Ind Eng Chem, 32, 1283 (1940).

38. F. Hoffman and M. Otto, Ger. Pat 504,730 (1933).
39. M. Polanyi, Nature, 158, 94 (1946).
40. A. G. Evans and M. Polany, J. Chem Soc, 252 (1947).
41. H. Morawetz, Chapt. 20 in: "Polymers: Their Origin and Growth of a Science," Wiley-Interscience, New York (1985).
42. V. Goryainov and A. Butleroy, Ann, 169, 146 (1873).
43. V. N. Ipatiev, Ber, 46, 1748 (1913).
44. H. R. Sailors and J. P. Hogan, Chapt. 17 "History of Polymer Science and Technology," R. B. Seymour, ed., Marcel Dekker, New York (1982).
45. H. Hoberg and K. Ziegler, Brennst Chem, 39, 302 (1958).
46. J. C. Swallow in: "Polyethylene," A. Renfew, P. Morgan, eds. Interscience, New York (1960).
47. D. G. H. Ballard, Chapt. 2 in: "History of Polyolefins," R. B. Seymour, T. Cheng, eds. D. Reidel Publishing Co., Dordrecht, Holland (1985).
48. F. M. McMillan, "The Chain Straighteners," Chapt. 2, The MacMillan Publishing Co., London (1979).
49. E. W. Fawcett, "Engineering," 136 32 (1933).
50. E. W. Fawcett, R. O. Gibson, M. W. Perrin, J. G. Patton, and E. G. Williams, Brit. Pat. 471,590 (1937) assigned to ICI.
51. C. W. Bunn, Trans Faraday Soc, 35, 482 (1939).
52. J. J. Fox and A. E. Martin, Trans Faraday Soc, 36, 897 (1940).
53. R. B. Seymour, Pop Plastics, 29 (7) 18 (1984).
54. J. C. Swallow, Endeavor, 3, 26 (1944).
55. P. C. Allen, Plastics (London), 9, 68 (1945).
56. F. A. Freeth, Brit Plastics, 18, 441 (1946).
57. A. Renfew, Plastics Inst (London) Trans, 19 (35) 5 (1951).
58. M. W. Perrin, Research, 6, 111 (1953).
59. M. E. P. Fredrick and C. S. Marvel, J. Am Chem Soc, 52, 376 (1930).
60. H. F. Mark, Preface in: "History of Polyolefins," R. B. Seymour, T. Cheng, eds. D. Reidel Publishing Co., Dordrecht, Holland, 1985.
61. N. W. Krase, U. S. Pat 2,396,791 (1946) assigned to duPont.
62. A. W. Larcher and D. C. Pease, U. S. Pat 2,816,883 (1957) assigned to duPont.
63. F. M. McMillan, Chapt. 2 in: "The Chain Straighteners," The MacMillan Press, London (1979).
64. M. Fischer, Ger. Pat 874,215 (1953) assigned to I. G. Farben Industrie.
65. K. Ziegler and K. Bahr, Ber, 61, 253 (1928).
66. K. Ziegler and H. G. Gellert, Ann, 567, 179 (1950).
67. K. Ziegler, E. Holzkamp, H. Breil, and H. Martin, Angew Chem, 67, 541 (1955).
68. K. Ziegler, H. Breil, E. Holzkamp, and H. Martin Ger Pat 973,626 (1960) assigned to K. Ziegler.
69. C. Ellis, U. S. Pat 2,212,155 (1941) assigned to C. Ellis.
70. G. C. Bailey and J. A. Reid, U. S. Pat 2,381,198 (1943) 2,581,228 (1952) 2,606,940 (1955) assigned to Phillips Petroleum Co.
71. F. M. McMillan, Chapt. 6 in: "The Chain Straighteners," The MacMillan Press, London (1979).
72. J. P. Hogan, R. L. Banks, U. S. Pat 2,825,721 (1958) assigned to Phillips Petroleum Company.
73. A. Clark, J. P. Hogan, R. L. Banks, and W. C. Lanning, Ind Eng Chem, 48, 1152 (1956).
74. R. V. Jones and P. J. Boeke, Ind Eng Chem, 48, 1155 (1956).
75. D. C. Smith, Ind Eng Chem, 48, 1161 (1956).
76. A. Zletz, U. S. Patent 2,692,257 (1954) assigned to Standard oil of Indiana.
77. H. F. Mark, Preface in: "History of Polyolefins," R. B. Seymour, T. Cheng, eds. Reidel Publishing Co., Dordrecht, Holland (1985).

78. A. M. Butlerov and V. Goryainov, J. Russ Chem Soc, 5, 302 (1873).
79. A. M. Butlerov and V. Goryainov, J. Russ Chem Soc, 9, 38 (1877).
80. M. Berthelot, Bull Soc Chim, 11, 4 (1969).
81. C. M. Fontana and G. A. Kidder, J. Am Chem Soc, 70, 3745 (1948).
82. C. M. Fontana, G. A. Kidder, and R. J. Harold, Ind Eng Chem, 44, 1688 (1952).
83. C. M. Fontana, R. J. Herold, E. J. Kinney, and R. C. Miller, Ind Eng Chem, 44, 2955 (1952).
84. C. M. Fontana, J. Phys Chem, 63 (1959).
85. G. Natta, J. Am Chem Soc, 77, 1708 (1955).
86. G. Natta, U. S. Pat. 3,715,344 (1972) assigned to Montecatini Corp.
87. P. Pino, "Giulio Natta, Present Significance of his Scientific Contributions," Editrice di Chinica, Milano (1982).
88. M. L. Huggins, J. Am Chem Soc, 66, 1991 (1944).
89. C. E. Schildknecht, S. T. Gross, H. R. Davidson, J. M. Lambert, and O. A. Zoss, Ind Eng Chem, 48, 2104 (1948).
90. P. J. Flory, "Principles of Polymer Chemistry," Cornell University Press, Ithaca, New York (1953).
91. G. Natta, I. W. Bassi, and P. Carradini, Makromol Chem, 18-19, 455 (1955).
92. J. P. Hogan and R. L. Banks, Chapt. 6 in: "History of Polyolefins," R. B. Seymour, T. Cheng, eds. D. Reidel Publishing Co., Dordrecht, Holland (1985).
93. G. Natta, U. S. Pat, 3,715,344 (1972) assigned to Montedison.
94. U. S. Patent Quarterly, P.676 Civil Action Suit 4319, January 11, 1980.
95. J. P. Hogan and R. L. Banks, U. S. Pat 4,376,851 (1983) assigned to Phillips Petroleum Co.
96. W. Bailey, Chapt. 18, "History of Polyolefins," R. B. Seymour, T. Cheng, Eds. D. Reidel Publishing Co., Dordrecht, Holland (1985).
97. J. P. Hogan and R. L. Banks, U. S. Pat. 2,825,721 (1958), 2,951,826 (1960), 4,342,854 (1982), 4,376,851 (1983).
98. A. Zletz, U. S. Pat. 2,692,257 (1954) assigned to Standard Oil of Indiana.
99. J. P. Hogan and R. L. Banks, Belg. Pat. 530,617 (1955) assigned to Phillips Petroleum Co.
100. H. Morawetz, "Polymers: The Origins and Growth of a Science," P.190 Wiley-Interscience, New York (1985).
101. B. Golding, "Polymers and Resins," Van Nostrand Co., Princeton, New Jersey (1959).
102. K. H. Meyer and L. Misch, Helv. Chim Acta, 20, 232 (1937).
103. R. Boyer, R. S. Spencer, and R. M. Wiley, J Polym Sci, 1, 249 (1946).
104. G. Natta and F. Danusso, "Stereoregular Polymers," Pergamon Press, New York (1967).
105. M. Dole, Chapt. 4 in: "History of Polyolefins," R. B. Seymour, T. Cheng, eds. D. Reidel Publishing Co., Dordrecht, Holland (1985).
106. P. Hogan, Chem Eng News, 47, (17) 15 (1969).
107. P. Hogan, U. S. Pat 3,666,766 (1972) assigned to Phillips Petroleum Co.
108. W. Kaminsky, Chapt. 15 in: History of Polyolefins," R. B. Seymour, T. Cheng, eds. D. Reidel Publishing Co., Dordrecht, Holland (1985).
109. R. B. Seymour, Chapt. 18 "History of Polyolefins," R. B. Seymour, T. Cheng, eds. D. Reidel Publishing Co., Dordrecht, Holland (1985).
110. F. J. Karol, Chapt. 12 in: "History of Polyolefins," R. B. Seymour, T. Cheng, eds. R. Reidel Publishing Co., Dordrecht, Holland (1985).
111. P. J. T. Tait, Chapt. 13 in: "History of Polyolefins," R. B. Seymour, T. Cheng, eds. D. Reidel Publishing Co., Dordrecht, Holland (1985).
112. J. P. Hermans and P. Henrioulle, U. S. Pat. 3,769,233 (1973) 4,210,738 (1980) assigned to Solvay et cie.

113. P. J. T. Tate in: "Developments in Polymerization," Vol. 2 R. N. Howard, ed. Applied Science Publication, London (1979).
114. B. L. Goodall and N. Kasawa in: "Transition Metal Catalyzed Polymerizations," Alkenes and Dienes Part A, P.355, 379, R. P. Quirk, ed. Howard Academic Press, New York (1983).
115. E. G. M. Tornqvist, Chapt. 9 in: "History of Polyolefins," R. B. Seymour, T. Cheng, coed. D. Reidel Publishing Co., Dordrecht, Holland (1985).
116. A. W. Anderson and N. G. Merckling, U. S. Pat 2,721,198 (1955) assigned to duPont.
117. A. W. Anderson, N. G. Merckling, and P. H. Settlage, U. S. Pat 2,799,688 (1957) assigned to duPont.
118. R. J. Statz, Chapt. 11 in: "History of Polyolefins," R. B. Seymour, T. Cheng, eds. D. Reidel Publishing Co., Dordrecht, Holland (1985).
119. R. W. Rees, Modern Plastics, 41, (9) 209 (1964).
120. R. W. Rees, U. S. Pat 3,264,272 (1964) assigned to duPont.

INTERNAL POLYOLEFINS AND A FEW HIGHLY SUBSTITUTED POLYVINYLS

H. F. Mark

Polytechnic Institute of New York
Brooklyn, New York

INTERNAL POLYOLEFINS

Internal olefins are molecules in which the ethylenic double bond is fully substituted by hydrocarbon moities, e.g.

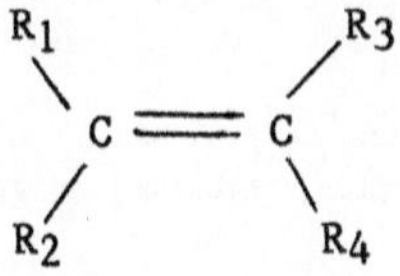

The simplest internal olefin is tetramethylethylene

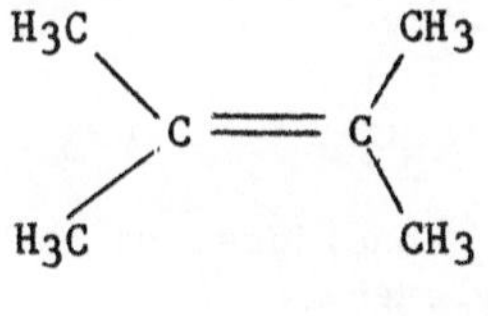

which may be also called 2,3-dimethyl-2-butene; its homopolymer is shown below;

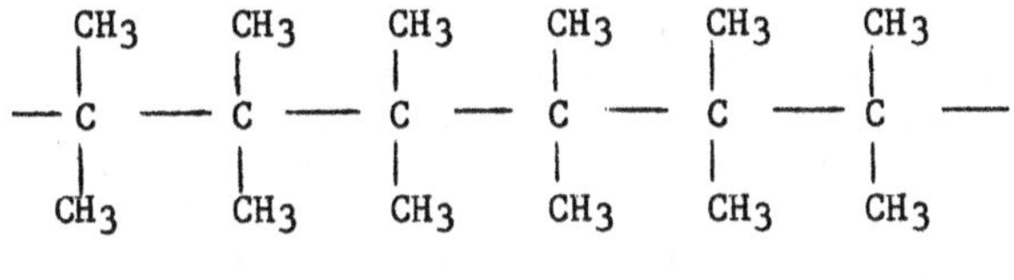

(1)

which could also be called poly-dimethyl-methylene. It has never been prepared but certain steps for its synthesis have recently been taken.

Before their enumeration, let us speculate about the properties of this polymer. The backbone chain has negligible hindrance for adopting different conformations. It is "flexible" like the chain of polyethylene although the methyl group substituents are "bulky" and any rotation about a C - C bond of the backbone chain is going to be "hindered" by the repulsive Van der Waals' forces of the CH_3 groups. By use of Fischer-Hirschfelder models, one arrives at the qualitative conclusion that the most favorable conformation of this fully methylated polymethylene will be a flat helix with a large radius. As a result, its long polymer chain will be relatively rigid, not as a consequence of cyclic units in the backbone but because of steric hindrance due to the "bulky" substituents.

HEAD TO HEAD CHAINS

The first step in the direction of exploring the properties of such polyolefins was carried out through the systematic studies of O. Vogl on head to head addition polymers, specifically polyisobutylene. As shown below, head to tail (normal) polyisobutylene is a soft rubbery material

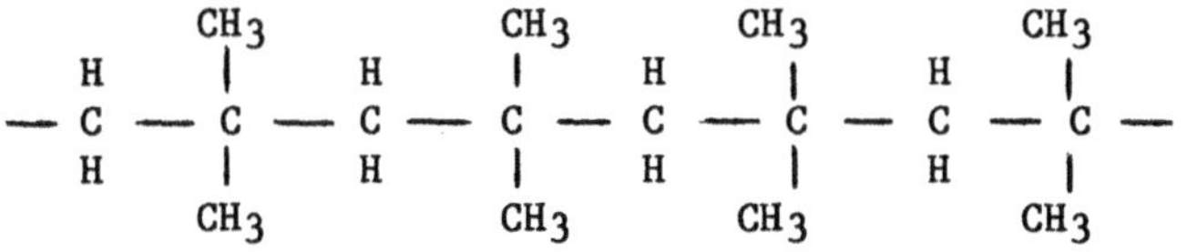

(2)

with a T_m of 5°C and a T_g of -61°C.[1]

In contrast, head to head polyisobutylene,

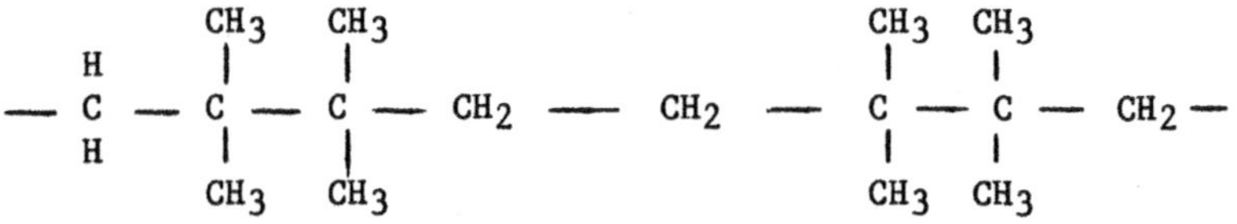

(3)

is a rigid resin with a T_m of 190°C and a T_g of 89°C. This clearly shows the stiffening action of the ordered segments with this chain as shown below;

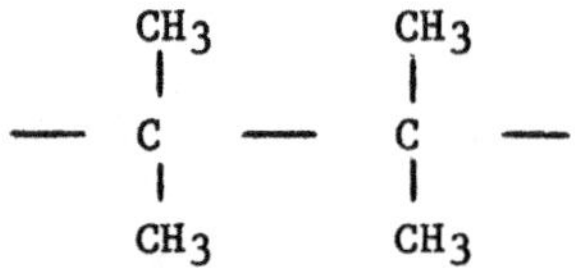

In this chain, which might be considered as an ordered copolymer of ethylene and 2,3-dimethylbutylene-2.

HIGH PRESSURE SYNTHESIS

Anderson, Hoover and Vogl[2] attempted to synthesize poly (2,3-dimethyl-2-butene) directly from its monomer (tetramethylethylene) and obtained a viscous oil by polymerization in the absence of an initiator at 300°C under a pressure of 6.56 Pa (65,000 bar). The molecular weight of the product was 3900 which corresponds to a DP of about 40. Infrared and NMR spectroscopy indicated that the product was identical with or very similar to formula (1). At present, there exists no more details on the products obtained by this high pressure technique.

THE USES OF NITRENES AND CARBENES

The first linear polymethylenes were prepared by Pechmann, Bamberger, and Tschirner at the beginning of this century by the decomposition of diazomethane

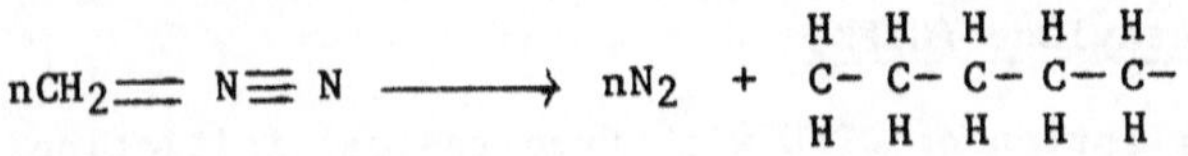

Repetition of this reaction showed that good yields and relatively pure products can be obtained. It could therefore be expected that diazoisopropane

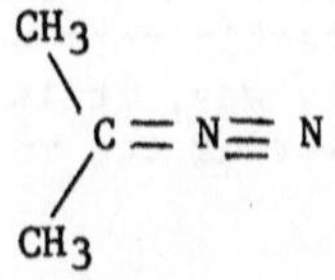

might decompose into nitrogen and an internal polyolefin of the formula (1). However, no such reaction has yet been described but Nasini and Saini[3] polymerized diazoethane

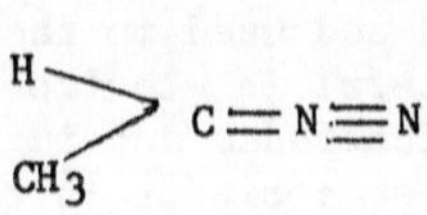

and obtained a solid polymer with the following formula:

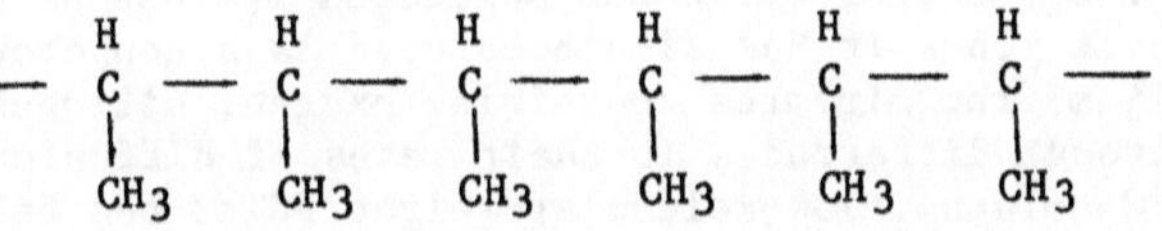 (4)

or

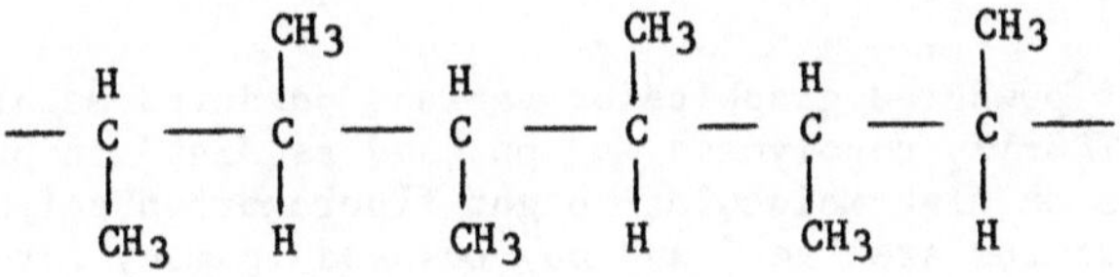 (5)

This substance is a polybutene which has also been obtained by other routes.

Finally, it might be expected that dimethylketene

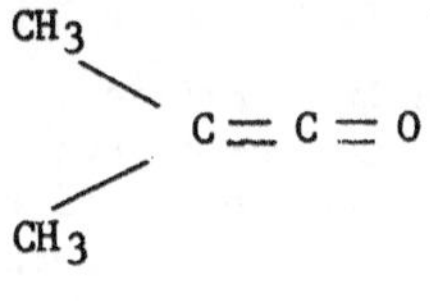

which is a well known stable molecule, could be polymerized with the elimination of CO to produce a substance with the formula shown in Figure 1.

A SHORT REVIEW ON A FEW HIGHLY SUBSTITUTED VINYLS

The chains of internal polyolefins are characterized by the presence of two substituents on each carbon atom. Such "crowding" of substituents was already known in highly halogenated polyvinyls, particularly in polychlorotrifluoroethylene (CTFE), tetrafluoroethylene (PTFE) and hexafluoropropylene (PHFP). It may, therefore, be of practical use to review the preparation, properties and uses of these polymers.

Chlorotrifluoroethylene (CTFE)

The polymerization of CTFE with free radical initiation, including photopolymerization was thoroughly studied over many years in bulk and solution;[4] the most widely used technique in the practical polymerization of CTFE as well as many other fluoroolefins is the colloidal dispersion method. In early efforts with this method, water-soluble initiators, such as alkali persulfates, were used in combination with alkali bisulfites in aqueous media. Later, investigators added silver salts, as accelerators, to these mixtures. In this way, it is possible to accelerate the polymerization without lowering the molecular weight of the resulting material.

The uses of high molecular weight PCTFE are many and diverse. It is used as a protective coating in the fabrication of many articles in which thermal and chemical stability are desirable. It can be compounded with metals or other inorganic materials in the preparation of self-lubricating bearings. It can be sintered and used in the preparation of alundum-like cells, which are extremely useful in electrochemistry. It has been used in the actual construction of electrodes and in components of storage batteries. The use of PCTFE as a binder in the preparation of gas generating propellants and pyrotechnic compositions is firmly established.

The chemical inertness of PCTFE has suggested its use as a food container. In recent years it has also been used as a nonporous membrane for separating helium from mixtures containing oxygen, nitrogen, carbon monoxide, etc., through differences in their rates of diffusion. In the construction of VPC columns, low-molecular-weight PCTFE has been used as a stationary phase material and high-molecular-weight polymers have been employed as a support medium in place of celite, firebrick, or diatomaceous earth.

Suspensions of powdered graphite or various powdered metals in low-molecular-weight CTFE-VF_2 copolymers may be used as "ink" in printing electrical circuits on high-molecular-weight fluorocarbon polymers. After "baking," these circuits are "set" and may be used in many forms of electrical apparatus. If organic dyes are substituted for the graphite or powdered metal, inks may be obtained which are useful in making both fluorocarbon and hydrocarbon polymeric materials.

Tetrafluoroethylene (PTFE)

TFE has been polymerized in a manner similar to that used for CTFE. Most of the systems that produce useful PCTFE will produce PTFE of comparable quality as well as copolymers of the two, with compositions varying with the mole ratios of the two monomers. The polymerization of TFE may be inadvertently photo-initiated when the monomer is stored in glass containers. It has also been known to polymerize spontaneously in dark metal containers. The polymerization reaction is strongly exothermic and occasionally occurs with great force.

The monomer is extremely unsafe to handle at room temperature unless it is inhibited with 1-2% by weight of limonene or other radical scavengers. Even though, it is not considered safe for shipment. Companies, that once supplied this material, now recommend that it be prepared *in situ* from 1,2-dibromotetrafluoroethane. It is desirable that reactions involving the transfer of pure uninhibited TFE be carried out in the laboratory at temperatures as low as possible. The refined persulfate systems are excellent for the preparation of PTFE if chain transfer agents are vigorously excluded.

The uses of PTFE are many and varied. Because of its thermal and chemical stability, it has been used for lining cooking utensils. Many patents have been issued in which it is compounded with other plastics or with metals in the preparation of self-lubricating bearings or compounded with aluminum silicate, glass, asbestos, or other nonmetallic fibrous materials in the construction of energy transmitting devices, such as clutch plates.

Low molecular weight PTFE is used in the preparation of high-temperature lubricants and adhesives. A conducting polymer is produced by mixing finely-divided PTFE with graphite, granulating the mixture, and coating the resultant material, first with molybdenum sulfide and then with silver, gold, or nickel.

In a very unusual process, powdered PTFE was added to monomeric lactam. Upon polymerization of this suspension, using sodium hydride catalyst and a diisocyanate initiator, a material was produced which possessed greater flexibility than PTFE and was more chemically stable than nylon.

Because of its thermal stability, electrical inertness, and nonflammability, PTFE is considered the finest insulation for electric wires used in space environments. PTFE may be copper plated by use of recently developed electrochemical techniques. Properly treated PTFE may be used as a support material in gas-liquid chromatography, in the manufacture of sintered plastic articles, in the preparation of gaseous fuel cells and as a substitute for leather.

Like PCTFE, it can be used in the preparation of membranes for separating helium from gaseous mixtures. A dropping mercury electrode composed of PTFE has identical polarographic properties with one made of glass. Its use in the fabrication of laboratory apparatus, tubing, pipes, gaskets, etc., is well known and firmly established.

Vinylidene Fluoride and Hexafluoropropylene

VF_2 and HFP have both been homopolymerized by previously discussed methods, but the homopolymers have very little commercial significance and only limited academic interest. The most important polymer involving either of these monomers is probably the copolymer of VF_2 and HFP. One material of this type, known commercially as Viton rubber, is a tough, flexible substitute for natural rubber. Copolymers with various percentages of these two monomers may be prepared by using the peroxide and persulfate initiators. In addition to general copolymerization studies, many investigations have been undertaken in which the copolymerization was directed toward specific objectives.

Specific uses of VF_2-HFP copolymers are numerous. In general they are used as sealants and caulking materials and in place of ordinary rubber when chemical and thermal stability are at a premium. They are excellent partitioning "fluids" for use as the stationary phase in gas-liquid chromatography and for the preparation of pyrotechnic mixtures. Thanks to extensive studies of effective solvent mixtures, many materials, especially porous materials, may be coated with these copolymers. Such materials include firebrick, celite, fabrics, paper, felt, wood and concrete.

POLYMERIZATION OF HALOGENATED BUTADIENES

The influence of halogenic substituents in butadiene on the polymerization and polymer properties was studied by W. H. Carothers. G. S.

Whitby has commented on these investigations as follows: Carothers' studies afford much valuable information as to the influence of substitution in the butadienoid system on the readiness with which polymerization occurs and the degree to which it proceeds. Considering, first halogen substitution: chloroprene, with chlorine in the 2-position, polymerizes about 700 times as fast as isoprene and yields a product of excellent rubber-like properties; the analogous bromo derivation polymerizes somewhat more rapidly, but the product, although having good extensibility and strength, shows less complete recovery from extension than does polychloroprene, and the lower limit of the temperature range within which it is elastic is higher than for polychloroprene 2-Iodo-1, 3-butadiene polymerizes at a markedly higher rate than either of the preceding, but the product (containing 70 percent of iodine), is rubber-like only under certain conditions.

The introduction of a second chlorine atom to one of the interior carbon atoms of the butadienoid system yields a compound (CH_2:CCl.CCl:CH_2), which polymerizes even more rapidly than chloroprene (CH_2:CCl.CH:CH_2). Either because of the high degree to which polymerization has proceeded or because of its high chlorine content, the product is a hard tough mass which lacks rubber like properties.

The high speed of polymerization of CH_2:CCl.CH:CH_2 is perhaps somewhat unexpected, if one were to judge by analogy with the effect of chlorine substitution in the ethenoid system, whereas the introduction of one atom of chlorine to ethylene renders the system highly polar, and produces vinyl chloride which is highly polymerizable. However, the introduction of a second atom of chlorine, symmetrically disposed with respect to the first, produces sym.-dichloroethylene which does not polymerize readily. It should be noted that unsubstituted terminal :CH_2 groups are present in 2,3-dichloro-1,3-butadiene and that their presence may be a condition for chlorine to exert an "activating" effect on polymerizability.

The substitution of chlorine in the terminal positions of butadiene has no such striking effect on polymerization as its substitution in the interior positions. 1-Chloro-1,3-butadiene, (ClCH:-CH.CH:CH_2) polymerizes at the same rate as isoprene and the product has inferior rubber-like properties. Similarly, 1,2,3-trichloro-1,3-butadiene, (ClCH:CCl.CCl:CH_2) polymerizes much more slowly than 2,3-dichloro-1, 3-butadiene, (CH_2:CCl.CCl:CH_2).

Observations on a number of chlorine compounds prepared from divinylacetylene, by the additions and the removal of hydrogen chloride, provide further data on the influence of chlorine substitution on the polymerization of conjugated systems. The formulas for these conjugated systems are as follows:

1. CH_2:C:CCl.CCl:C:CH_2

2. CH_2:C:C:CCl.CH:CH_2

3. ClCH_2.CH:C:CCl.CH_2:CH_2

4. ClCH_2.CH:CClCCl:C:CH_2

5. ClCH_2.CH:CCl.CCl:CH.CH_2Cl

Of these, 1 polymerized to a hard resin, when allowed to stand for 24 hours; 2 and 3 polymerized only slowly (3 changed to a viscous syrup on standing for three months); 4 and 5; in which one end or both ends of the molecule carries chlorine and in which a terminal vinyl group is lacking, showed no tendency at all to polymerize.

The effect of introducing other groups into the 2-chloro-butadienoid system, especially their introduction into the terminal positions, is to depress the polymerizability. 3-Methyl-2-chloro-1,3-butadiene, CH_2:CCl.C(CH_3):CH_2, polymerizes somewhat more slowly than chloroprene, and the polymer is somewhat inferior to polychloroprene in rubber-like properties: although tough and elastic, it has a rather low extensibility. The introduction to the last mentioned butadiene of a second, terminal methyl group has a more profound effect, 3-4-dimethylchloroprene, CH_2:CCl.C(CH_3):CH.CH_3, polymerizing not much more rapidly than isoprene and yielding a product which, although elastic, recovers only sluggishly from deformation. As contrasted with the relatively slight effect of a 3-methyl group on the polymerization of chloroprene; a terminal methyl group, as in 1-methylchloroprene, CH_3CH:CCl.CH:CH_2 reduces the speed of polymerization greatly and detracts markedly from the rubber-likeness of the polymeric product. Increase in the magnitude of the terminal alkyl group magnifies the effect in the same direction, n-heptylchloroprene, C_7H_{15}.CH:CCl.CH:CH_2, yielding on polymerization only a sticky solid with very slight elasticity. Such observations are in accord with previous findings that the introduction of terminal alkyl groups to butadiene reduces polymerizability and that, the larger the alkyl group, the greater is the reduction.

In papers of the Acetylene Polymers series published subsequent to Carothers' association with it, the influence of certain other substituents on the polymerizability of chloroprene is recorded. The introduction of a dialkylmethylamino group in a terminal position depresses the polymerization of the chloroprene system very greatly, the 1-(dialkylmethylamino)-chloroprenes, R_2N.CH_2CH:CCl.CH:CH_2, polymerizing very slowly indeed and yielding no rubber-like product. Alkoxy-methyl or alkoxy-(methyl)methyl groups in the terminal position have some but a far smaller effect on the polymerizability of chloroprene, CH_3O.CH_2-CH:CCl.CH:CH_2 polymerizing at one-sixth the rate of chloroprene and yielding a product with somewhat rubber-like properties.

In view of the profound effect on polymerization of the presence of chlorine in the 2-position of the butadienoid system, the effect of other substituents in this position was examined. As compared with isoprene, which has a methyl group in the 2-position, 2-heptyl-1,3-butadiene was found to polymerize rather less readily and to yield a weaker and softer rubber. 2-Phenyl-1,3-butadiene was found to polymerize considerably more rapidly than isoprene but to yield under ordinary conditions only a dimeric product. In continuation of Carothers' studies, the effect of introducing (a) alkoxy groups, (b) aliphatic acid residues with the 2-position has been examined. 2-Ethoxy-1,3-butadiene, CH_2:C(OC_2H_5).CH:CH_2, was found to polymerize at about one-third the rate of isoprene and to give a product of inferior rubber-like properties. 2-Acetoxy-1,3-butadiene, CH_2C(O.$COCH_3$).-CH:CH_2, and analogous compounds on the other hand were found to polymerize at a rate not vastly different from that of chloroprene and to yield rubber-like products.

REFERENCES

1. O. Vogl et al., Macromolecules, 10, 1331 (1977); M. Malanga and O. Vogl et al., Polymer Bulletin, 8, 9 (1982); M. Malanga and O. Vogl, Polymer Bulletin, 9, 236 (1983).
2. R. C. Anderson, C. L. Hoover, and O. Vogl, Macromolecules, 2, 686 (1969); 12, 222 (1978).

3. H. V. Pechmann, Berichte der Deutsche Chemischer Gesellschaft, 32, (II) 2301 (III) 3681 (1899); Bamberger and Tschirner, Berichte der Deutsche Chemischer Gesellschaft, 33, 955 (1900); A. Nasini and G. Saini, Atti Acad. Scien. Torino, 90, 586 (1955); W. Lwovski, "Nitrenes," John Wiley (1970).
4. "Fluoropolymers," A. Wall ed., Wiley-Interscience, New York (1972).
5. G. S. Whitby and H. Mark, "Collected papers of Wallace H. Carothers," Interscience Publishers, New York (1940).

ESTIMATING POLY(1-ALKENE) PROPERTIES FROM

OLIGOMER AND REPEAT UNIT DIMENSIONS

Charles H. Fisher

Roanoke College
Salem, VA 24153

The first of the two methods used in the present work to estimate polyolefin properties consisted in using oligomer (n-alkane) properties to calculate limiting properties (P_{∞}, properties of an infinite-length n-alkane); such limiting properties resemble the corresponding properties of amorphous, linear polyethylene (or amorphous polymethylene, APM).

The second method, applicable to the higher polyolefins (polypropylene, polybutylene, etc.), consisted in correlating polyolefin properties (P) with the dimensions, e.g., formula weight or carbons of the repeat units.

These two approaches, used to some extent by previous investigators,[1-16] were explored further in the present work and extended to additional properties.

N-ALKANES AND POLYETHYLENE

The properties of n-alkanes increase in a regular manner with increase in chain length (L), measured conveniently by formula weight (M) or number of carbons (C). The changes in properties per CH_2 increment become smaller with increasing L and approach zero as L approaches infinity. The limiting properties (P_{∞}) are those that are reached eventually for M_{∞} and C_{∞}.

The published properties (P) of the oligomers (n-alkanes) leading to polyethylene are relatively abundant, and hence there are many opportunities for calculating limiting properties (P_{∞}) by equations such as $P = P_{\infty} + m/(L + k)$, where m is slope and k is an adjustable parameter.

Homolog properties, in addition to n-alkane properties, can be used to estimate P_{∞} because all homologs $[H(CH_2)_xG]$, regardless of the nature of the functional group, G, lead to the same structure $[(CH_2)_{\infty}]$ and properties as M and C approach infinity.[6]

Broadhurst,[7] used n-alkane data to predict the thermodynamic properties of polyethylene, mentioned the advantage that n-alkanes can be prepared and used in highly pure states. Wu[8] favors the use of n-alkane data to estimate limiting surface tensions. However, the use of homolog to predict polyethylene properties has the disadvantage that some homolog properties (e.g., boiling points) are irrelevant for this purpose. Moreover, the homolog properties needed to predict some important polyethylene properties (e.g., tensile strength and elongation) are lacking. The homology approach presumably has limited promise for estimating the properties of crystalline polymers.

The principal purposes of the present work concerned with polyethylene were to review previous relevant work and compare limiting properties with the corresponding properties of amorphous polymethylene.

POLYPROPYLENE AND SIMILAR HIGHER POLYOLEFINS

It is not feasible to estimate the higher polyolefin properties from their oligomer properties (as done for n-alkanes and polyethylene) because of the scarcity of the required oligomer data. An alternative consists in correlating the higher polyolefin properties with the dimensions (formula weight or number of carbons) of the repeat units. In this approach, the main polymer chain may be considered the functional group and the pendant n-alkyl groups may be treated as the homolog chain for correlation purposes. Rehberg[9] used this method about 40 years ago to correlate properties of the polymeric n-alkyl acrylates and methacrylates with repeat unit dimensions. Others[10-13] have correlated polymer properties with repeat units. Somewhat similar correlations[14-16] are based on additive group contributions.

The principal purposes of the work related to the higher polyolefins (polypropylene, etc.) were to develop correlations between polyolefin properties and repeat unit dimensions, compare the calculated with the corresponding experimental values, and explore further this general method of estimating polymer properties.

Data Sources and Equations

Isotactic poly(1-alkene) data[17-32] were used in most of the calculations. The work was hindered by the lack of data for some properties and lack of information in some instances about degrees of crystallinity and branching. Selecting the P_{∞} values and polymer properties was frequently difficult because of disagreement among the published data. The selected data, however, are reasonably consistent with each other and appear suitable for correlation purposes. In some instances molar formulas, e.g., molar refraction and parachor, were used to calculate properties. Squared refractive indices were used as dielectric constants.[14]

Reciprocal equations (Equations 1-4) are attractive for correlating properties (P) with M and C because of their simplicity, versatility, relation to P_{∞}, previous successful use,[1-6] facile interconvertibility and other advantages (b, intercept; m, slope; and k, a, and h, adjustable parameters).

$$P = P_{\infty} + m/(L + k) \quad (1)$$

$$P(L + k) = b + P_{\infty} L \quad (2)$$

$$(L + a)/P = b + L/P_{\infty} \quad (3)$$

$$L/(P - h) = b + L/(P_{\infty} - h) \quad (4)$$

Equations 1-4 have the advantage of being related to Equations 5 and 6, which are widely used to correlate some homolog properties and functions with the number of carbons.

$$MP = b + mC \qquad (5)$$

$$M/P = b' + m'C \qquad (6)$$

A useful method for estimating P_∞ consists in developing equations (such as 1-6) in which P_∞ (or a function of P_∞) is a parameter. The expression $P_\infty = (C' - C)/(C'/P' - C/P)$ can be used to calculate P_∞ where appropriate data (preferably high C' and C values) are available. Methylene (CH_2) increments (m and m' in Equations 5 and 6), published for many properties,[14,15,33-35] can be used to calculate P_∞: $P_\infty = m/14.027$ and $14.027/m'$.

Comparison of Limiting and Polyethylene Properties

Limiting properties (P_∞) are shown in Table I and compared with the corresponding APM properties; the agreement between P_∞ and APM properties is reasonably good. The relatively high polyethylene dielectric constant, compared with the limiting value, might have been caused by impurities.

Van Krevelen[14] reported polyethylene refractive indices at 25°: Experimental, 1.49; calculated by the Lorentz-Lorenz, Gladstone-Dale, and Vogel molar refractions: 1.479; 1.478; and 1.469, respectively. The calculated values resemble the limiting refractive index (1.4752 at 25°) more than the reported experimental value.

The C' and C values used for calculating P_∞ in Table I [$P_\infty = (C' - C)/(C'/P' - C/P)$] were C_{40} and C_{30} with the following exceptions: d^{150}, $C_{36} - C_{28}$; γ, $C_{20} - C_{14}$; and solpars, $C_{12} - C_8$. When the C_{100} and C_{82} melting points were used, the calculated $T_{m\infty}$ was 412.4K. When the C_{100} and C_{90} boiling points (760 torr) were used, the calculated $T_{b\infty}$ was 1173 K.

The limiting properties obtained with methyl n-alkanoate data are given in Table II. The density and refractivity at 20°C agree with those estimated from n-alkane data (Table I).

Somayajulu[35] published methylene increments (m in Table III) for several thermodynamic and other properties at 25°C. The limiting properties calculated from Somayajulu's methylene increments also are given in Table III. The limiting densities calculated from Somayajulu's CH_2 increments agree with those in Table I.

Some limiting and polyethylene properties are approximately zero, e.g., dipole moments, vapor pressure, fluidity, iodine number, saponification value, acid strength, functional groups, and water solubility.

Correlating Properties of the Higher Poly(1-Alkenes) with Repeat Unit Dimensions

The polypropylene and higher polyolefin properties were correlated with repeat unit dimensions with the following useful assumptions: (1) the poly(1-alkenes) may be treated as homologs in which the pendant n-alkyl groups are the principal polymer chain for correlation purposes; (2) as the formula weight (M) and number of carbons (C) in the pendant groups approach infinity, the polyolefin properties (P_∞) resemble those of an infinite-

Table I. Comparison of Limiting Properties (P_{∞}) with Amorphous Polymethylene (APM) Properties[a,b,c]

Property	P_{∞} (ref.)	P_{∞} (calc'd)	APM (ref.)
Melting point, T_m K	414.3[36]	411.9	414.6[32]
Density, d_4^{20}	0.8541[2]	0.8547	0.855[18]
Density, d_4^{25}	0.8510[4]	0.8514	0.8519[32]
Density, d_4^{40}	0.8423[2]	0.8421	
Density, $d^{141.4}$			0.7835[32]
Density, d^{150}	0.779[26]	0.7818	0.7791[26]
Refractivity, N_D^{20}	0.4771[2]	0.4773	0.49[23]
Refractivity, N_D^{25}	0.4752[4]	0.4753	0.476[28]
Refractivity, N_D^{40}	0.4701[2]	0.4690	
Surface tension, /, 20°C	35.3[8]	36.2	35.7[8]
Dielectric constant, c, 20°C	2.182[4]	2.181	2.3[23]
Solpar, o, 25°C	8.2[27]	8.57	8.0[23]

[a]n-Alkane data used except for d_4^{40} and N_D^{40}.

[b]Albert[28] and Barton[29] give 0.476 and 0.483, respectively, as the refractivity of AP.

[c] ${}_D^tN = n_D^t - 1$.

Table II. Methyl n-Alkanoates: Limiting Properties (P_{∞})[a]

	Limiting Properties, P_{∞}	
	20°C	40°C
Density, d_4^t	0.85407	0.84225
Refractive index, n_D^t	1.47704	1.47010
Dispersion, Gladstone-Dale	0.00876	0.00858
Dielectric constant	2.15	2.14
Ultrasonic sound velocity, m/sec	1552	1489

[a]From Reference 2.

Table III. Methylene (CH_2) Values (m)[a] and Limiting Properties (P_{∞})[b] at 25°C

	Property	State	$m^{a,b}$	m^{c}	P_{oo}
1	$fH°/kJmol^{-1}$	g	-20.668	-20.61	-1.473
2	$fH°/kJmol^{-1}$	1	-25.627		-1.827
3	$vH°/kJ/mol^{-1}$	1	4.959	4.974	0.3535
4	$S°/JK^{-1}\ mol^{-1}$	g	39.205	38.95	2.795
5	$S°/JK^{-1}\ mol^{-1}$	1	32.219		2.297
6	$C_p°/JK^{-1}\ mol^{-1}$	g	22.624		1.613
7	$C_p°/JK^{-1}\ mol^{-1}$	1	31.733	30.4	2.262
8	$V/cm^3\ mol^{-1}$ (20°C)	1	16.428		0.8539
9	$V/cm^3\ mol^{-1}$	1	16.486	16.45	0.8508
10[d]	$R/cm^3\ mol^{-1}$ (20°C)	1	4.638		0.3307
11[d]	$R/cm^3\ mol^{-1}$	1	4.640	4.65	0.3308
12[e]	$-x/10^{-6}$ emu mol^{-1}	1	11.330	11.35	0.8077
13	P (parachor)	1	40.007	40.0	2.852
14	$Vc/cm^3\ mol^{-1}$	1	59.622	57.4	0.2353

[a]From Reference 35.
[b]P_{oo} = m/14.027, except for V = 14.027/m.
[c]From References 14 and 33.
[d]Molar refraction, Lorentz-Lorenz.
[e]Magnetic susceptibility.

length n-alkane; and (3) the expressions (Equations 1-6) found useful for n-alkanes and other homologs are useful also for polymer homologous pendant groups.

Wunderlich[32] referred to publications on several series of homologous polymers: n-alkyl acrylates and methacrylates, n-alkyl styrenes, fluoroalkyl acrylates, vinyl esters, and N-n-alkyl acrylamides; the poly(1-alkenes) were not included. The first publication listed by Wunderlich[32] appeared in 1944; the authors correlated brittle points and parallel plate plasticities of the polymeric n-alkyl acrylates and methacrylates with the number of carbons in the n-alkyl groups.

In the present work, approximately straight lines were obtained by plotting xT(K) of the poly(n-alkyl methacrylates) against x (where T represents brittle points and glass transition temperatures and x represents carbons plus oxygens in the repeat units); the xT values of the lower members are on one straight line, whereas the upper members are on another.

The polyolefin Tg values resemble those of the poly(methacrylates) in that different equations are needed for the lower and upper members. The functions M/Tg,K and CTg,K of the polyolefin glass transition temperatures are approximately linear with repeat unit M or C over substantial carbon spans (Table IV and Equations 7-11). The agreement between the calculated and literature values is reasonably good (except for C_7) for the C_3 - C_{16} polyolefins. The limiting Tg values suggested by Equations 9, 10, and 11 are 272, 268 and 269, respectively.

$$M/Tg(K) = -0.037813 + 0.066399C \quad (7)$$
(Corr. coeff., 0.99936)

$$CTg(K) = 156.35 + 207.83C \quad (8)$$
(Corr. coeff., 0.999425)

$$M/Tg(K) = 0.088296 + 0.051654C \quad (9)$$
(Corr. coeff., 0.999768)

$$CTg(K) = -360.84 + 267.84C \quad (10)$$
(Corr. coeff., 0.999833)

$$Tg(K) = 269.38 - 381.80/C \quad (11)$$
(Corr. coeff., -0.989371)

The melting points (T_m,K) of the C_{10}-C_{18} polyolefins are compatible with the reported 414K melting point of amorphous polymethylene.[32] Presumably Equation 12 can be used to calculate the melting points of all the higher polyolefins (Table V). Two-parameter expressions (Equations 13-15) can be used to correlate polyolefin melting points with repeat unit dimensions.

The melting points are linear with Tg,K over a considerable carbon span (Equations 16 and 17). According to Equation 17, the Tg,K value corresponding to 414K is 283K.

Table IV. Polyolefin Glass Transition Temperatures (Tg,K)[a,b]

C[c]	Tg,K	M/Tg	CTg	Equations			
				7	8	9	11
2	(237)[d]	.1184		(295)	(286)		
3	265	0.1588	795	261	260		
4	249	0.2253	996	246	247		
5	233	0.3010	1165	238	239		
6	233	0.3612	1398	233	234		
7	(242)[d]			230	230		
8	228	0.4922	1824	227	227		
9	226	0.5586	2034	226	225		227
10	232	0.6046	2320			232	231
11	236	0.6538	2596			235	235
12	237	0.7102	2844			238	238
14	241	0.8148	3374			242	242
16	246	0.9123	3936			245	246
18	(328)[d]					(248)	(248)

[a]Data from References 14, 17, and 22.
[b]Equations 9 and 10 give the same values.
[c]Carbons in repeat units.
[d]Not used in calculations.

$$T_m(K) = 414 - 1661.8/(C + 6.043) \quad (12)$$
($C_{10} - C_{18}$; corr. coeff., -0.985658)

$$T_m, K = 383.20 - 720.86/C \quad (13)$$
($C_8 - C_{18}$; corr. coeff., -0.992944

$$M/Tm, K = 0.094356 + 0.035575C \quad (14)$$
(Corr. coeff., 0.999583)

$$CTm, K = -742.02 + 384.84C \quad (15)$$
(Corr. coeff., 0.999740)

$$Tm, K = -143.73 + 1.9670\ Tg, K \quad (16)$$
(Corr. coeff., 0.999303)

$$Tg, K = 73.304 + 0.50769\ Tm, K \quad (17)$$
(Corr. coeff., 0.999303)

The molar volumes (V = M/d) of the amorphous polyolefins are approximately linear with the number of carbons (C) in the repeat units (Equation 18):

$$V = -0.11700 + 16.467C \quad (18)$$
(Corr. coeff., 0.999890; d_∞ = 0.8518)

The indicated limiting density (0.8518) is similar to the reported[32] APM density (0.8519 at 25°). Equation 18 divided by Equation 19, gives a useful expression (Equation 20) for calculating amorphous polyolefin densities (Table VI):

$$M = 14.027C \quad (19)$$

$$C/d = -0.008341 + 1.1740C \quad (20)$$

Table V. Polyolefin Melting Points (Tm,K)[a]

C	Lit.	Equations				
		12	14	15	13	16
7	290		286			
8	293		296	292	293	
9	(292)[b]		305	302	303	
10	313	310	312	311	311	313
11				317	318	320
12	322	322	323	323	323	322
13					328	
14	330	331	332	332	332	330
15	332	335	335	335	335	
16	340.5	339	338	338	338	340
18	345	345	344	344	343	
20		350	348	348	347	
22	(365)[b]	355	352	351	350	
∞		414	394	385	383	

[a]Melting points from Reference 32.
[b]Not included in calculations.

Table VI. Polyolefin Molar Volumes and Densities (25°C)[a]

	Repeat unit carbons, C						
	2	3	4	5	6	10	∞
M	28.054	42.081	56.108	70.135	84.162	140.27	14.027
V=M/d	32.93	49.34	65.3	82.5			
18[b]	32.82	49.28	65.75	82.22	98.69	164.6	16.467
20[b]	0.8548	0.8538	0.8533	0.8530	0.8528	0.8524	0.8518
25[b]	0.8526	0.8525	0.8594	0.8515	0.8519	0.8521	0.8522

[a]Wunderlich (32) molar volumes (isotactic polyolefins).
[b]Equation.

The densities of crystalline polyolefins at 25°C in Table VII are much higher than the amorphous densities in Tables VI and VII. Two-parameter Equation 21, which suggests the low d_{∞} value of 0.8845, presumably is not suitable for predicting d_c for the higher polyolefins. The adjustable parameter in Equation 22 was selected to give a perfect fit for C_2 - C_4. The adjustable parameter of 7.40 in Equation 23 was selected to give the intercept (1.003) of crystalline polyethylene. The densities from Reference 37 and Equation 24 (Table VII) are higher than the amorphous and lower than the crystalline densities. Equation 25, which is based on the d_a values in Table VII, has a limiting value (0.8522) that resembles the APM density (0.8519 at 25°C). The Equation 25 densities are compared with the amorphous densities in Table VI.

$$V_c = -3.6430 + 15.858C \quad (C_2\text{-}C_5;\ \text{corr. coeff., } 0.999582;\ d_{\infty} = 0.8845) \qquad (21)$$

$$d_c = 0.95518 - 0.0076989/(C-2.161) \quad (C_2\text{-}C_4;\ \text{corr. coeff., } -0.99999;\ d_{\infty} = 0.9552) \qquad (22)$$

$$d_c = 1.0030 - 0.59280/(C + 7.40) \quad (C_3\text{-}C_4;\ \text{corr. coeff., } -1.0000;\ d_{c\infty} = 1.0030) \qquad (23)$$

$$d = 0.87558 + 0.16429/C \quad (C_4\text{-}C_{16};\ \text{corr. coeff., } 0.999922;\ d_{\infty} = 0.8756) \qquad (24)$$

$$d_a = 0.85225 - 0.00071078/(C-4.1) \quad (C_2\text{-}C_5;\ \text{corr. coeff., } -0.899912) \qquad (25)$$

The Van der Waals' volumes published by Van Krevelen[14] (Table VIII) are defined by Equation 26:

$$V_w = -0.01000 + 10.230C \qquad (26)$$

The Lorentz-Lorenz (R_{LL}) and Gladstone-Dale (R_{GD}) molar refractions were used with molar volumes to calculate the polyolefin refractivities ($N_D = n_D-1$). The Gladstone-Dale group values (CH_2 = 7.831; CH_3 = 8.82; and CH = 6.80) were used to construct Equation 27. Equation 27 was divided by Equation 18 to get an expression (Equation 28), which correlates amorphous polyolefin refractivities with repeat unit carbons:

$$R_{GD} = -0.04200 + 7.831C \qquad (27)$$

Table VII. Densities of Crystalline (c) and Amorphous (a) Polyolefins[a,b]

C^c	d_c	V_c	d_c^d	d_c^e	d_c^f	d^g	d_a
2	1.003	27.97	0.9993	1.003	0.9399	0.8762	0.8519
3	0.946	44.48	0.9579	0.9460	0.9460	0.8933	0.8554
4	0.951	59.00	0.9384	0.9510	0.9510	0.9166	0.859
5	0.923	75.99	0.9271	0.9525	0.9552	0.9084	0.850
6	(0.726)		0.9198	0.9532	0.9588	0.9030	
8			0.9107	0.9539	0.9645	0.8961	
10			0.9053	0.9542	0.9689	0.8920	
∞			0.8845	0.9552	1.003	0.8756	

[a]Data from References 20, 32 and 37.
[b]Packing densities are given in References 20 and 32.
[c]Carbons in repeat units.
[d]Equation 21.
[e]Equation 22.
[f]Equation 23.
[g]The C_2 and C_3 densities are from Reference 37, the others from Equation 24.

$$V = -0.11700 + 16.467C \tag{18}$$

$$N_D(C - 0.00711) = -0.0025506 + 0.47556C \tag{28}$$

The polyolefin refractivities calculated by Equation 28 are shown in Table IX.

The R_{LL} expression (Equation 29) was developed with the group values: CH_3 = 5.644; CH_2 = 4.649; and CH = 3.616. Equation 29 was divided by the molar volume Equation 18 to get Equation 30, where Q is R_{LL}/V. Equation 31 was used to calculate $(n_D)^2$, which is also the dielectric constant (ε). The calculated refractivities (Table IX), which are approximately constant over a wide carbon range, resemble the limiting refractivity of $N^{25}_{D\infty}$ = 0.475.

$$R_{LL} = -0.038000 + 4.6469C \tag{29}$$

$$V = -0.11700 + 16.467C \tag{18}$$

$$Q(C-0.00711) = -0.0023076 + 0.28232C \tag{30}$$

$$n^2(1 - Q) = 1 + 2Q \tag{31}$$

Equation 32, developed with Reference 37 data, gives refractivities higher than the amorphous and lower than the crystalline refractivities (Table IX).

Table VIII. Literature and Calculated Van der Waals' Volumes

	Repeat unit carbons, C			
	2	3	4	5
Literature	20.46	30.68	40.91	51.14
Equation 26	20.45	30.68	40.91	51.15

Table IX. Calculated Refractivities ($N_D = n_D - 1$)

Equation	Repeat unit carbons, C				
	2	4	8	10	14
28	0.4760	0.4758	0.4757	0.4756	0.4756
31	0.4762	0.4764	0.4764	0.4765	0.4765
31	2.179	2.180	2.180	2.180	2.180
32		0.5105	0.5007	0.4987	0.4964

$$N_D = 0.49080 + 0.078839/C \quad (32)$$

(C_4–C_{16}; corr. coeff., 0.999911)

The high literature refractivities of the C_2, C_4, and C_{18} polyolefins were used to construct Equation 33. The calculated refractivities (Table X) may resemble those of the largely-crystalline polyolefins.

$$N_D = 0.50530 + 0.029285/C \quad (33)$$

Parachors (Equation 34) were used with polyolefin molar volumes to calculate polyolefin surface tensions. Equation 34, divided by Equation 18, gave an expression (Equation 35) that was used to calculate surface tensions: C_2, 35.32; C_4, 35.06; and C_8, 34.94. The polyethylene surface tension at 20°C has been reported[8] as 35.7 dynes/cm. The simple two-parameter Equation 36 can be used instead of Equation 35.

$$P_p = 40.00C \quad (34)$$

$$V = -0.11700 + 16.467C \quad (18)$$

$$\gamma^{1/4}(C - 0.00711) = 2.4291C \quad (35)$$

$$\gamma = 34.812 + 1.0147/C \quad (36)$$

The Small molar attraction constants published by Seymour[24] were used to develop Equation 37. Equation 37 was divided by Equation 18 to get Equation 38, which was used to calculate polyolefin solubility parameters (Hildebrands) (Table XI).

$$E = -24.000 + 133.00C \quad (37)$$

$$V = -0.11700 + 16.467C \quad (18)$$

$$S_p(C - 0.00711) = -1.4575 + 8.0768C \quad (38)$$

$$S_pC = -1.4575 + 8.0768C \quad (39)$$

Equation 40, constructed with Hoy attraction constants (CH_3 = 303; CH_2 = 269; and CH = 176), was divided by Equation 18 to get Equation 41.

Table X. Polyolefin Refractivities[a]

Equation	Repeat unit carbons, C					
	2	3	4	6	18	∞
	0.52		0.5125		0.507	
33	0.5199	0.5151	0.5126	0.5102	0.5069	0.5053

[a]Data from References 14 and 17.

Table XI. Calculated Polyolefin Solubility Parameters (S_p)[a]

Equation	Repeat unit carbons, C						
	2	3	4	6	10	18	oo
39[b]	7.35	7.59	7.71	7.83	7.93	8.00	8.08
39[c]	15.04	15.53	15.77	16.02	16.22	16.37	16.53
42[c]	14.54	15.14	15.44	15.74	15.98	16.14	16.34

[a]Hildebrands multiplied by 2.046 to get S.I. units.
[b]Hildebrands.
[c]S.I. units.

The adjustable parameter is omitted in Equation 42.

$$E = -59.000 + 269.00C \quad (40)$$

$$S_p(C - 0.0071) = -3.5829 + 16.336 \quad (41)$$

$$S_pC = -3.5829 + 16.336C \quad (42)$$

Published solpar values (29) for polyethylene range from 15.9 to 17.1 in S.I. units.

ACKNOWLEDGMENT

Vernon R. and Vernon C. Miller are thanked for the excellent computer program used in the work.

REFERENCES

1. J. Smittenberg and D. Mulder, Rec. trav. Chim., 67, 813 (1948).
2. T. H. Gouw and J. C. Vlugter, J. Am. Oil Chem. Soc., 41, 426 (1964).
3. M. L. Fein and C. H. Fisher, J. Am. Chem. Soc., 70, 52 (1948).
4. C. H. Fisher, Chem. Eng., Sept. 20, 111 (1982).
5. M. G. Broadhurst, J. Chem. Phys., 36, 2578 (1962).
6. C. H. Fisher and B. P. Huddle, Jr., Chem. Eng., Feb. 8, 99 (1982).
7. M. G. Broadhurst, J. Res. Nat'l. Bur. Stds., 67A, 233 (1963).
8. Souheng Wu, "Polymer Interface and Adhesion," Marcel Dekker, Inc., New York (1982).
9. C. E. Rehberg and C. H. Fisher, Ind. Eng. Chem., 40, 1429 (1948).
10. C. Overberger, J. Am. Chem. Soc., 75, 3326 (1953).
11. E. F. Jordan, Jr., J. Polymer Sci., 10, 3347 (1972).
12. R. A. Rhein, and D. D. Lawson, Chemtech, Feb., 122 (1971).
13. F. W. Billmeyer, Jr., "Textbook of Polymer Science," 3rd Ed., J. Wiley and Sons, Inc., New York (1984).
14. D. W. Van Krevelen and P. J. Hoftyzer, "Properties of Polymers," Elsevier Scientific Publishing Co., New York (1976).
15. A. Bondi, "Physical Properties of Molecular Crystals, Liquids and Gases," John Wiley and Sons, New York (1968).
16. D. H. Kaelble, Preprint, Am. Chem. Soc. Div. of Organic Coatings and Plastics, Chem., March, 46, 241 (1982).
17. J. Brandrup and E. H. Immergut, "Polymer Handbook," 2nd Ed., John Wiley and Sons, Inc., New York (1975).
18. R. C. Ferguson, private communication.

19. U. Gaur, S. Lau, B. B. Wunderlich, and B. Wunderlich, J. Phys. Chem. Ref. Data, 11, 1065 (1982).
20. U. Gaur, B. B. Wunderlich, and B. Wunderlich, B., J. Phys. Chem. Ref. Data, 12, 29 (1983).
21. Kirk-Othmer Encyclopedia of Chemical Technology, 3rd ed., John Wiley and Sons, New York, Vols. 15 and 16.
22. O. G. Lewis, "Physical Constants of Linear Homopolymers," Springer-Verlag, Inc., New York (1968).
23. R. B. Seymour and C. E. Carraher, "Structure-Property Relationships in Polymers," Plenum Press, New York (1984).
24. R. B. Seymour, "Handbook of Chemistry and Physics," CRC Press, Inc., Boca Raton, FL, C687 (1980-81).
25. E. A. Turi, Editor "Thermal Characterization of Polymeric Materials," Academic Press, New York (1981).
26. R. A. Orwoll and P. J. Flory, J. Am. Chem. Soc., 89, 6814 (1967).
27. C. H. Fisher, J. Coatings Technology, 57, 69 (1985).
28. R. Albert and W. M. Malone, J. Macromol. Sci. - Chem., A6, 347 (1972).
29. A. F. M. Barton, "Handbook of Solubility Parameters and Other Cohesion Parameters," CRC Press, Inc., Boca Raton, FL (1983).
30. R. A. V. Raff and J. B. Allison, "Polyethylene," Interscience Publishers, Inc., New York (1956).
31. R. A. V. Raff and K. W. Doak, "Crystalline Olefin Polymers, Part II," Interscience Publishers, New York (1964).
32. B. Wunderlich, "Macromolecular Physics," Academic Press, Inc., New York, Vol. 3 (1980).
33. D. R. Stull, E. F. Westram, Jr., and G. C. Sinke, "The Chemical Thermodynamics of Organic Compounds," John Wiley and Sons, Inc., New York (1969).
34. P. Sellers, et al., J. Chem. Eng. Data, 23, 250 (1978).
35. G. R. Somayajulu, Bulletin, Thermodynamic Research Center - Chemical Engineering Department, Sept. '84 - Mar. '85, Texas A&M University, College Station, TX 77843.
36. M. G. Broadhurst, J. Res. Nat'l Bur. Stds, 70A, 481 (1966).
37. R. M. Joshi, B. J. Zwolinski, and C. W. Hayes, Macromolecules, 1, 30 (1968).

RELATIONS BETWEEN MOLECULAR STRUCTURE - MORPHOLOGICAL STRUCTURE AND PROPERTIES OF POLYETHYLENES

Leo Mandelkern

Department of Chemistry
and Institute of Molecular Biophysics
Florida State University
Tallahassee, Florida 32306-3015

Studies of the crystallization of polymers has matured sufficiently to enable a detailed and specific quantitative analysis to be made of the relation between structure and properties. In this endeavor a careful distinction needs to be made between two different kinds of structures. On one hand there is the influence of molecular or chain structure. Here, the principal variables involved are the chain length for fractions, the polydispersity for whole polymers, and the structural regularity of the chain. In addition, the independent structural variables which define the crystalline state also need to be taken into account and in fact require major attention. There is a synergistic effect between these two different types of variables, the crystallization conditions and the properties of crystalline polymers.

A major advance has been made, in recent years, in identifying and in measuring the important independent structural variables which either individually, or in particular combination, contribute to, and control a specific property. Of interest are a diversity of thermodynamic, spectroscopic and mechanical properties. The structural variables which have been specifically identified are: the degree of crystallinity; the crystalline thickness distribution; the structure and relative amount of the interfacial region; the crystalline structure; the supermolecular structure; and the structure of the residual non-crystalline region.

The polyethylenes, homopolymers and copolymers, are most amenable for quantitative control and study, following the concepts that have been outlined. Each of these variables can be isolated and studied separately in these systems. Extensive studies have now been made and some selected typical results will be reported here with this set of polymers.[1] However, sufficient work has been carried out with other polymers to indicate the basic generality of the results obtained and conclusions reached.[1,2]

We will first describe the nature of the independent structural variables and show how they can be quantitatively measured. The concept of the degree of crystallinity was introduced very early in the study of crystalline polymers. Until recently, however, it was held in disrepute and was apparently banished to obscurity. However, from recent studies of properties by a variety of techniques, including electron microscopy,[3] it has been made abundantly clear that the lamellar crystallites are connected to one another by chain units which have a disordered liquid-like structure.

The concept of the degree of crystallinity has now been quantitatively reestablished. For homopolymers, depending on chemical type, molecular weight and crystallization conditions, the degree of crystallinity can range from 90-95% at one extreme to about 25-30% at the other. These changes in the level of crystallinity are systematic with molecular weight and crystallization conditions and are easily controlled. For copolymers and structurally irregular chains, this value can be reduced even further.

It has been demonstrated by a variety of experimental methods,[4] neutron scattering,[5] and theoretically,[6,7] that the interfacial structure characteristic of the basal plane is not comprised of regularly folded chains. Furthermore, adjacent reentry of a chain is severely restricted and is not nearly as prevalent as had been popularized. The interfacial content is relatively small for linear polyethylene, except for very high molecular weights. Here it is about 10-15%.[8] It increases significantly with the introduction of structural irregularities into the chain. Values as high as 20-25% have been reached.[1,8]

Recent detailed electron microscopic studies of the characteristics of the lamellar crystallites have shown that they are very dependent on the molecular weight and crystallization conditions.[3] At lower molecular weights and higher crystallization temperatures the lamellae are relatively flat and extend for several microns in the lateral direction. With increased molecular weight and lower crystallization temperature they become more curved, their lateral extent decreases and they eventually become highly segmented. The inclination of the chain axis to the basal plane is also a function of the crystallization temperature.

The organization of the lamellar crystallites into supermolecular structures has also been shown to depend systematically on molecular weight, polydispersity, chain regularity and crystallization conditions.[2,9-12] Among the major findings is the fact that spherulites are not the only type superstructure that can be observed. Under certain circumstances, supermolecular structures are not formed at all. Most surprisingly, and quite important, it has been found that the supermolecular structure has little, if any, direct influence on most properties.

A variable which is very important, and which has received relatively little attention, to date, is the detailed structure of the liquid-like region. It will be demonstrated that this region, particularly the "entanglement" structure, is the controlling factor in determining the level of crystallinity, and is a major factor, irrespective of the level of crystallinity, in determining mechanical and ultimate properties. Another important structural factor is the partitioning of sequences between those which return to the crystallite of origin and those which connect crystallites.

There are now available a large number of studies as to how these variables influence specific properties.[1,4] The alpha and beta relaxation transitions may be analyzed in detail. The important role of crystalline thickness in the former[13] and interfacial content in the latter are apparent.[13,14] The average correlation times that are involved may be determined by carbon-13 NMR.[15] It may be demonstrated that ultimate properties depend primarily on molecular weight and chain structure.[16]

From the point of view of basic principles, a general strategy has been developed which enables one to assess the value of any property of a crystalline polymer and to establish control.

REFERENCES

1. L. Mandelkern, Polymer J., 17, 337 (1985).
2. R. C. Allen and L. Mandelkern, J. Polym. Sci., Polym. Phys. Ed., 20, 1465 (1982).
3. I. G. Voigt-Martin and L. Mandelkern, J. Polym. Sci., Polym. Phys. Ed., 22, 1901 (1984).
4. L. Mandelkern, Farad. Soc. Disc., 68, 310 (1979).
5. D. Y. Yoon and P. J. Flory, Polymer, 18, 509 (1977); Farad. Soc. Disc., 68, 288 (1979).
6. P. J. Flory, D. Y. Yoon, and K. A. Dill, Macromolecules, 17, 862 (1984).
7. D. Y. Yoon and P. J. Flory, Macromolecules, 17, 868 (1984).
8. M. Glotin and L. Mandelkern, Coll. Polym. Sci., 260, 182 (1982).
9. J. Maxfield and L. Mandelkern, Macromolecules, 10, 1141 (1971).
10. J. Maxfield and L. Mandelkern, J. Polym. Sci., Polym. Phys. Ed., 17, 1913 (1979).
11. L. Mandelkern, M. Glotin, and R. A. Benson, Macromolecules, 14, 22 (1981).
12. M. Glotin and L. Mandelkern, Macromolecules, 14, 1394 (1981).
13. R. Popli, M. Glotin, L. Mandelkern, and R. A. Benson, J. Polym. Sci., Polym. Phys. Ed., 22, 407 (1984).
14. R. Popli and L. Mandelkern, Polym. Bull., 9, 260 (1983).
15. A. Dekmezian, D. E. Axelson, J. J. Dechter, B. Borah, and L. Mandelkern, J. Polym. Sci., Polym. Phys. Ed., 23, 367 (1985).
16. R. Popli and L. Mandelkern, in preparation.

LIVING POLYMERIZATION OF ISOBUTYLENE

R. Faust and J. P. Kennedy

Institute of Polymer Science
The University of Akron
Akron, Ohio 44325

INTRODUCTION

Truly living polymerizations, i.e., polymerizations that proceed in the absence of termination and chain transfer, are a most desirable objective of the synthetic polymer chemist. Living polymerizations are not only of great scientific interest but also of considerable commercial consequence and several modern industrial processes are based on living systems, e.g., cis-1,4-polybutadiene,[1] triblock polymers of styrene/butadiene/styrene,[2] polytetrahydrofuran;[3] however these processes are not carbocationic in nature.

Since the discovery of anionic living systems,[4,5] great efforts have been made to find conditions under which living carbocationic polymerization could be achieved but success remains limited to this date.

In the cationic polymerization of vinyl monomers the propagating carbocations are usually rather unstable and therefore tend to undergo chain transfer to monomer and termination reactions.[15] In view of these facts, in 1975, a distinguished practitioner of cationic polymerizations rather pessimistically concluded that it is "...unlikely that any cationic polymerization will display living characteristics...".[6]

The first break, in this bleak prediction, occurred only very recently when Higashimura, Sawamoto et al.[7,8,9] discovered truly living polymerizations of vinyl ethers[7] and other cationically highly reactive monomers such as N-vinylcarbazole[8] and p-methoxystyrene[9] with the HI/I_2 initiating systems. According to the Japanese authors[7] the mechanism of these living polymerizations of vinyl ethers involves monomer insertion into an iodine activated $\sim\underset{\text{OR}}{\text{CH}}\text{-I}$ bond.

A different recent approach is the "quasiliving" technique which has been used with conventional monomers as well i.e., styrene and its derivatives, isobutylene, vinyl ethers.[10,11,12] Under certain rather restrictive conditions (i.e., low temperatures, continuous slow monomer addition) quasiliving carbocationic polymerizations approach those of truly living systems; even though termination can be completely suppressed ($R_t = 0$), the rate of chain transfer to monomer approaches zero.

Recently, we have discovered a family of initiating systems that yield truly living polymerizations of simple olefins such as isobutylene, and copolymerizations of olefins with conjugated diolefins, such as isobutylene/isoprene pairs. This paper describes experiments that demonstrate the living polymerization of isobutylene by tertiary ester·BCL_3 initiating systems and provides information essential for the understanding of the fundamentals of this new initiating concept.[13,14]

A Hypothesis of Living Carbocationic Polymerizations: Initiating Complexes of Organic Esters with Lewis Acids (BCL_3)

The growing species in carbocationic polymerization is conventionally visualized as a carbenium ion accompanied by a counteranion arising from the coinitiator, e.g., $R\text{-}Cl + BCL_3 \rightleftharpoons R^{\oplus}BCl_4^{\ominus} \xrightarrow{+M} R\sim\sim\sim M^{\oplus}\ BCl_4^{\ominus}$.[15] Hence, the counteranion will strongly influence every elementary event of a polymerization, i.e., ion generation, cationation, propagation, all kinds of chain transfer steps, and termination. The counteranion will affect both the monomolecular and bimolecular chain transfer to monomer steps[15] (termed respectively, counteranion assisted chain transfer or chain transfer to the counteranion), and spontaneous or direct chain transfer to monomer[15] the latter, for example, by sterically shielding the propagating cation. Similarly the nature of the counteranion will determine the outcome and rate of termination.[15]

We hypothesized that living carbocationic polymerizations may arise in the presence of counteranions that would facilitate initiation and propagation but would not participate in chain transfer and termination events, and therefore set out to search for counteranions that promote initiation and propagation, but do not cause or assist proton elimination (i.e., chain transfer to monomer), and are sufficiently stable not to cause termination. A search for such living conditions has been in progress in our laboratories for many years. With the discovery of the living polymerization of isobutylene by initiating systems comprising tertiary esters as initiators in conjunction with BCl_3 as the coinitiator, this elusive target may have finally been attained.

A few comments on ester/BCl_3 complexes are in order. While well-characterized complexes of esters of carboxylic acids and Lewis acids i.e., BCl_3, are known,[16] they have not yet been used as initiating systems for cationic polymerization. Following a report on the preparation of $EtOAc/TiCl_4$,[16] a large number of ester/Lewis acid complexes have been prepared and characterized, mainly by Lappert et al.[17] According to these authors EtOAc readily forms 1:1 complexes with BF_3, BCl_3, $AlCl_3$, $AlBr_4$, $GaCl_3$, and $TiCl_4$ whereas with $SnCl_4$, $SnBr_4$ and $FeCl_3$ 1:2 complexes are obtained. The structures were determined by IR spectroscopy and other techniques and it was found that the Lewis acids were exclusively coordinated to the acyl oxygen, e.g.,

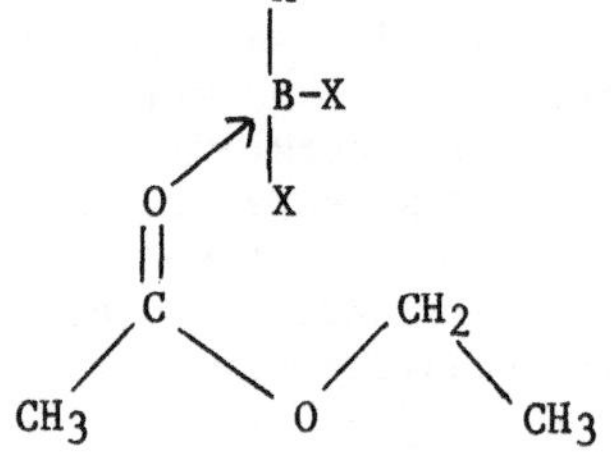

Since the C=O $\longrightarrow MX_n$ angle is $\sim 120°$, the possibility for cistrans isomerism arises due to the restricted rotation around the C=O bond. However, Lappert et al.[17] could not find a difference by NMR spectroscopy among the protons in the $(CH_3)_2CO/BF_3$ complex even at -80°C. To explain these findings the authors postulated a rapid exchange of BF_3 among the donor molecules, a process that has been demonstrated with alcohol adducts.

Based mainly on infrared absorption shifts of the carbonyl frequency of ethyl acetate and its complexes with various Lewis acids, Lappert[17] proposed the following order of relative acceptor strengths of these halides: B > Ga > Al > In,Sn > Ga > Si and Ti > Zr. Among the boron halides the acceptor strength order was $BBr_3 > BCl_3 > BF_3$.

The chemical behavior of ester/BCl_3 complexes was studied by Gerrard et al.[18] These complexes decompose thermally by cleavage of the acyl-oxygen bond (primary alkyl esters) or by the alkyl-oxygen bond (secondary or tertiary alkyl esters). The decomposition temperature decreases in the order primary > secondary > tertiary. The decomposition of the sec.-butyl acetate/-BCl_3 complex yielded sec.-butyl chloride, acetyl chloride and a boron-containing residue. The complex of 2 tert. butyl acetate/-3BCl_3 started to decompose at -25°C and yielded t butyl chloride, acetyl chloride and dichloro diacetyl-diborate.[18] The thermal decomposition (98°C) of BCL_3 complexes of neopentyl acetate and isobutyl acetate yielded mainly t-amyl chloride and t butyl chloride, respectively,[18] which we regard to be cationic processes. The decomposition of the optically active sec.-butyl- and 1-methylheptyl acetate·BCl_3 complexes afforded the inverted sec.-butyl and 1-methylheptyl chlorides, by an S_N1 process which may or may not be a carbocationic process.

We have hypothesized that the living carbocationic polymerization of isobutylene could be initiated, for example, by the t butyl acetate/BCL_3 complex and that initiation will proceed via the t $Bu^{\oplus}$ $BCl_3Ac^{\ominus}$ ion pair i.e., by the highly active tertiary cation and the highly stable $BCl_3OAc^{\ominus}$ counteranion.

This paper describes the living polymerization of isobutylene by various tertiary ester/BCl_3 complexes, a brief examination of the effect of solvents on these polymerizations and various supporting studies. The structure of the end groups of PIBs prepared under various conditions in living polymerizations induced by t ester/BCl_3 complexes has also been investigated. An attempt has been made to explain the data in terms of a mechanism, particularly in regards to the initiation and propagation steps.

Experimental

Materials. Cumyl acetate (CuOAc): Acetyl chloride (18.8g 0.24m) was slowly added to a solution of 27.2g (0.2m) 2-phenyl-2-propanol and 31.4g (0.26m) freshly distilled N,N-di-methyl-aniline in 110 ml anhydrous diethyl ether, at such a rate as to maintain gentle reflux. After refluxing it for another 2 days the mixture was cooled to room temperature, treated with water and the organic layer separated. The ether solution was washed with 10% H_2SO_4 aq., water, and dried over $MgSO_4$. The vacuum distilled product (80% 2-phenyl-2-propyl acetate, 20% unreacted 2-phenyl-2-propanol) was chromatographed on neutral Al_2O_3 (activity III) and afforded upon elution with n-pentane pure 2-phenyl-2-propyl acetate in 67% yield. 2,4,4-Trimethyl-2-pentyl acetate (TMPOAc): 2,4,4-Trimethyl-2-pentanol was prepared from 2,4,4-trimethylpentene-1 according to H. C. Brown et al.[19] The vacuum distilled pure 2,4,4-trimethyl-2-pentanol was esterified with acetyl chloride following the procedure described for the preparation of 2-phenyl-2-propanol acetate. The vacuum distilled product was chromatographed on neutral Al_2O_3 (activity III) and afforded upon elution with n-pentane pure 2,4,4-trimethylpentyl acetate in 40% yield. t-Butyl acetate (tBuOAc) (Aldrich) was freshly distilled from CaH_2. Methyl chloride was dried by passing the gas through a column packed with BaO and condensing it under nitrogen atmosphere. Methylene chloride was dried over molecular sieves, distilled from CaH_2, refluxed with triethylaluminum under nitrogen atmosphere overnight, and distilled on the day of experiment. Ethyl chloride (Fluka) was refluxed and distilled from CaH_2 under nitrogen atmosphere. n-Hexane was refluxed with fuming sulfuric acid, washed neutral with distilled water, dried on molecular sieves, refluxed and distilled from CaH_2 under nitrogen atmosphere. Isobutylene (IB) was dried by passing the gas through a column packed with BaO, and condensing it under nitrogen atmosphere. Boron trichloride was used as received.

Procedures. Polymerization experiments were carried out in a stainless steel enclosure (dry box), under a dry nitrogen atmosphere. The polymerization charges were quenched by the addition of prechilled MeOH. The solvents were evaporated and the polymer was redissolved in n-hexane. The n-hexane solution was decanted or filtered off the inorganic boron compounds, and the polymer was recovered by evaporating the n-hexane. Minute losses of the polymer were unavoidable.

In termination studies the polymerization charges and the TMPOAc·BCl_3 model compound was quenched by the addition of different nucleophiles, prechilled MeOH, Et_3N, pyridine, and CH_3OH/NaOH. The solvents were evaporated, the polymer dissolved in n-hexane, washed with 5% aq. sulfuric acid, followed by distilled water until neutral, dried over $CaCl_2$, filtered and the polymer (or the model compound) was recovered by evaporating the n-hexane.

The infrared spectra were recorded by a Perkin Elmer Model 521 spectrophotometer. A pair of Wilmad matched 1.0 mm liquid cells with NaCl windows were used. Only the C=O stretching frequency was followed due to warm up and decomposition caused by the IR beam.

Molecular weights were determined using a Waters high-pressure GPC instrument (Model 6000A Pump, a series of five μ-Styragel columns (10^5, 10^4, 10^3, 500 Å, 100 Å). Differential Refractometer 2401 and UV Absorbance Detector Model 440 and a calibration curve made by well fractionated polyisobutylene standards.

^{1}H NMR spectra were taken by a Varian T-60 Spectrometer using appropriate solvent solutions and TMS standards.

Results and Discussion

Preliminary Experiments. Preliminary experiments have shown that various tertiary esters, i.e., t-butyl acetate (tBuOAc) 2,4,4-trimethylpentyl-2-acetate (TMPOAc) and cumyl acetate (CuOAc) in conjunction with BCl_3 in polar diluents (i.e., CH_3Cl, CH_2Cl_2) readily initiate the polymerization of isobutylene (IB) and yield high conversions over the temperature range from -10 to -50°C. Initiation occurs only in tricomponent systems, i.e., organic acetate/BCl_3/IB, regardless of the introduction sequence of the chemicals. "Control" experiments i.e., experiments carried out by mixing IB with BCl_3 solutions in CH_3Cl yield insignificant conversions (<7%) and relatively higher molecular weights under essentially identical conditions.

Control experiments carried out under the same conditions by the use of CH_2Cl_2 diluent yield high conversions; however, these results are in line with previous experience[20] which showed that initiation of isobutylene polymerization in the presence of BCl_3 coinitiator in CH_2Cl_2 is due to traces of protic impurities (most likely moisture) which[21] are extremely hard to remove. Nonetheless even in these control experiments, we obtained much higher molecular weights and the conversions were noticeably below 100%; in contrast, in the presence of the various acetates conversions were invariably complete and the molecular weights were determined by the $[M]/[CH_3COOR^{t}]$ ratio.

Table I shows the results of some representative experiments. Polymerizations were very fast upon BCl_3 addition. The onset of polymerization was indicated by the appearance of haziness in the reactors. In several experiments we have ascertained that conversions reached 100% one minute after BCl_3 addition (indicated by triple asterisks in Table 1). Nonetheless, to be sure that the polymerizations always reached completion, we have terminated the reactions arbitrarily 30 min after BCl_3 introduction.

In all experiments, carried out in the presence of acetates, conversions were invariably 100% (except in one case in which a very low t-BuOAc concentration was used, see line 5 in Table 1). The I_{eff}'s obtained with t-BuOAc are much lower than 100%, most likely because initiation i.e., ion generation, is slow with this ester. I_{eff}'s were invariably $\sim$100% with the more efficient carbocation precursors TMPOAc and CuOAc.

The molecular weights were independent of the temperature and solvent at a given ester concentration, $[CH_3COOR^{t}]$, and in a few cases $\bar{M}_w/\bar{M}_n$ were below 2.0. Significantly, proton NMR analysis of the low molecular weight products obtained even at -10°C showed the absence of terminal unsaturation in PIB, i.e., indicating the absence of proton elimination which is an unmistakable sign for chain transfer to monomer.

A closer analysis of the data also showed that the molecular weights ($\bar{M}_n$) were inversely proportional to $[CH_3COOR^{t}]$; indeed with TMPOAc and CuOAc $\bar{M}_n$ was proportional to $[M_o]/[CH_3COOR^{t}]$. The $[BCl_3]$ was deemed to be of lesser importance: an excess was used to insure complete CH_3COOR^{t} utilization. In view of the controls, "free" BCl_3 was viewed to be of diminished significance.

A closer examination of the molecular weight data obtained in experiments using CH_2Cl_2 and CH_3Cl indicate that the effect of moisture impurities in $CH_3COOR^{t}\cdot BCl_3$ initiated IB polymerizations is most likely unimportant. According to the data the molecular weights obtained in the CH_2Cl_2 and CH_3Cl systems are similar although the level of moisture in these diluents is quite different (note the high conversions obtained in CH_2Cl_2 in contrast to the insignificant level of conversions found in CH_3Cl; cf. Table I). The concentration of the CH_3COOR^{t} initiators and moisture in CH_2Cl_2 are probably of the same order of magnitude ($\sim 10^{-3}$M). This implies that the effect of the $H_2O\cdot BCl_3$ on the polymerization of IB induced by $CH_3COOR^{t}\cdot BCl_3$ complexes is negligible. We theorize that the $CH_3COOR^{t}\cdot BCl_3$ complexes give rise to the $R^{t\oplus}$. $BCl_3CH_3COO^{\ominus}$ initiating species in the presence of which the equilibrium

$$H_2O\cdot BCl_3 \rightleftharpoons H^{\oplus}BCl_3OH^{\ominus} \qquad (1)$$

is strongly displaced toward the unionized species so that initiation by H_2O cannot occur. This proposition is directly supported by the observation that polymerization of IB is absent in CH_2Cl_2 in the presence of $(C_6H_5)_3C^{\oplus}BCl_3OAc^{\ominus}$.[22] The trityl cation is too stable to initiate (cationate) isobutylene polymerization[15] and, evidently ion generation by eq.

Table 1. Demonstration of Initiating Activity of Various tert.-Acetate BCl_3 Complexes under Various Conditions*

CH_3COOR^t R^t conc	Diluent	Temp °C	Conv. %	$\bar{M}_n$ g/mole	$\bar{M}_w$ g/mole	$\bar{M}_w/\bar{M}_n$ -	I_{eff} %
- -	CH_3Cl	-50	6.6	116400	230000	2.0	-
t-Bu $2.8x10^{-2}$	CH_3Cl	-50	100**	7370	15000	2.0	28
t-Bu $5.6x10^{-2}$	CH_3Cl	-50	100	4240	7210	1.7	24
- -	CH_3Cl	-30	60	36000	72000	2.0	-
t-Bu $5.6x10^{-3}$	CH_3Cl	-30	13.5	31000	48200	1.5	32
t-Bu $2.8x10^{-2}$	CH_3Cl	-30	100	4650	8930	1.9	44
t-Bu $5.6x10^{-2}$	CH_3Cl	-30	100	2950	5590	1.9	35
- -	CH_2Cl_2	-30	75	49800	103300	2.1	-
t-Bu $5.6x10^{-3}$	CH_2Cl_2	-30	100	27500	66800	2.4	36
t-Bu $2.8x10^{-2}$	CH_2Cl_2	-30	100	4230	10420	2.4	48
t-Bu $5.6x10^{-2}$	CH_2Cl_2	-30	100	2470	5950	2.4	42
- -	CH_2Cl_2	-10	68	28600	80000	2.8	-
t-Bu $2.8x10^{-2}$	CH_2Cl_2	-10	100	4370	9200	2.1	47
t-Bu $5.6x10^{-2}$	CH_2Cl_2	-10	100	2040	5090	2.5	51
- -	CH_3Cl	-50	6.6	116400	23000	2.0	-
TMP $5.6x10^{-3}$	CH_3Cl	-50	100	10500	21500	2.0	97
TMP $2.8x10^{-2}$	CH_3Cl	-50	100	1800	6300	3.6	121
TMP $5.6x10^{-2}$	CH_3Cl	-50	100	1400	3400	2.4	80
- -	CH_3Cl	-30	6.0	36000	72000	2.0	-
TMP $5.6x10^{-3}$	CH_3Cl	-30	100	10770	15150	1.5	94
TMP $2.8x10^{-2}$	CH_3Cl	-30	100	2090	5320	2.5	103
TMP $5.6x10^{-2}$	CH_3Cl	-30	100	1290	2800	2.2	88
- -	CH_3Cl	-30	6.0	36000	72000	2.0	-
Cu $5.6x10^{-3}$	CH_3Cl	-30	100***	8500	13600	1.6	115
Cu $2.8x10^{-2}$	CH_3Cl	-30	100	2000	3400	1.7	104
Cu $5.6x10^{-2}$	CH_3Cl	-30	100	1160	2820	2.4	96
- -	C_2H_5Cl	-30	2.0	-	-	-	-
Cu $5.6x10^{-3}$	C_2H_5Cl	-30	100***	9500	14100	1.5	107
Cu $2.8x10^{-2}$	C_2H_5Cl	-30	100	2390	3980	1.7	90
Cu $5.6x10^{-2}$	C H Cl	-30	100	1260	2050	1.6	91

*$[BCl_3] = 2.6x10^{-1}$ mol/ℓ, [M] = mol/ℓ, time = 30 min.
**Pzn. time was 1 hour
***Conv. was 100% after 1 min.

1 from $H_2O \cdot BCl_3$ is suppressed on account of the presence of the $(C_6H_5)_3C^{\oplus} \cdot BCl_3{-}OAc^{\ominus}$ ion pair.

In line with these observations and considerations the living polymerization of isobutylene induced by organic ester·BCl_3 complexes can be carried out reproducibly under conventional laboratory conditions and the use of super-dry solvents is unnecessary.

Encouraged by these observations we proceeded to examine in greater detail the possibly living nature of these systems.

Experimental Proof for the Living Polymerization of Isobutylene in the Presence of CH_3COOR^t/BCl_3 Complexes

A diagnostic proof of living polymerizations is linear $\bar{M}_n$ versus conversion plots starting from the origin. If initiation is instantaneous and faster than propagation ($R_i > R_p$) the molecular weight distributions will approach the Poisson distribution (i.e., $\bar{M}_w/\bar{M}_n \sim 1.0$). To insure rapid ion generation initially we chose to work with the TMPOAc·BCl_3 and CuOAc·BCl_3 systems since the expected cations from these initiators $(CH_3)_3CCH_2C^{\oplus}(CH_3)_2(TMP^{\oplus})$ and $C_6H_5C^{\oplus}(CH_3)_2(Cu^{\oplus})$ respectively have been shown to be efficient initiating entities for isobutylene polymerizations (see also Table I).[23,24] The use of $TMP^{\oplus}$ as the initiating entity is particularly advantageous since the structure of this cation and that of the propagating polyisobutylene carbenium ion $PIB^{\oplus}$ are for all practical purposes identical. Indeed the addition of BCl_3 to TMPOAc/IB or CuOAc/IB charges dissolved in CH_2Cl_2 caused immediate polymerizations at -30°C and the reactions proceeded to complete conversions within a few seconds. Essentially similar observations have been made by the use of CH_3Cl diluent.

Since the polymerizations were too fast for kinetic investigations, i.e., for the withdrawing of a series of samples from a single reactor for conversion/$\bar{M}_n$ determinations, we carried out experiments by the use of a series of reactors. Thus 5-6 large test tubes (or three-neck flasks) were charged with CuOAc/IB or TMPOAc/IB in CH_2Cl_2 or CH_3Cl diluents and at zero time an appropriate amount of BCl_3 coinitiator was added to all the charges at -30°C. The systems were strongly mixed (turbomix) and replaced in a constant temperature bath at -30°C. After an arbitrary 30 mins the polymerization was quenched in the first reactor (addition of a few ml prechilled methanol) whereas an additional quantity of IB was added to all the remaining reactors. After 30 min, the reaction in the second reactor was quenched and a further quantity of IB was added to the rest of the reactors. This procedure was continued until all the charges in the series have been quenched. Subsequently, the amounts of polymers formed (conversions) were determined by gravimetry, and $\bar{M}_n$ and $\bar{M}_w/\bar{M}_n$ by GPC.

Figure 1 shows a representative $\bar{M}_n$ versus g polymer formed plot obtained in a series of experiments carried out with TMPOAc in CH_3Cl and CH_2Cl_2 diluents at -30°C. The different symbols indicate replicate experiments. Similarly, Figures 2 and 3 show the results obtained with the CuOAc/BCl_3 system using CH_2Cl_2 and CH_3Cl diluents at -30°C, respectively. The inserts in Figures 1-3 show the number of PIB chains in these systems N as a function of polymer formed. The N was obtained from g/M_n. The linear $\bar{M}_n$ and horizontal N versus g PIB plots as shown in Figures 1-3 strongly suggest living polymerizations. The fact that the $\bar{M}_w/\bar{M}_n$ values decrease with increasing monomer addition suggests that the growing ability of the living species persists throughout the experiment.

The molecular weight dispersities ($\bar{M}_w/\bar{M}_n$ values) of PIBs obtained with the TMPOAc/BCl_3 system are broader than those produced by the CuOAc/BCl_3 system. Evidently the rate of initiation with the TMPOAc/BCl_3 combination

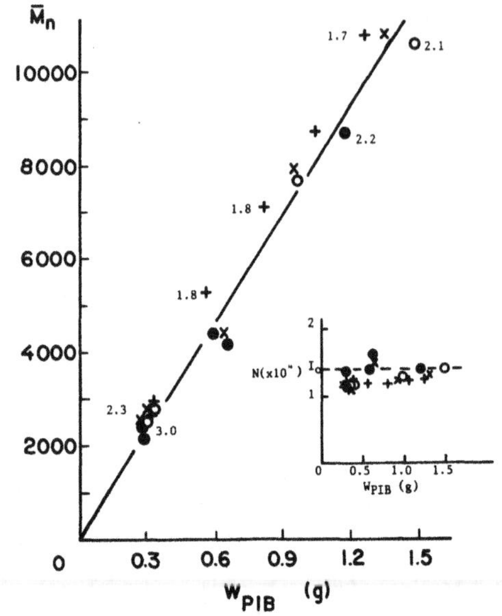

Fig. 1. $\bar{M}_n$ and N, the number of PIB chains (insert), versus the weight of PIB formed W_{PIB} in the TMPOAc/BCl_3/IB polymerization system using the incremental monomer addition (IMA) technique at -30°C: CH_2Cl_2 (O,●), and CH_3Cl (+,X), $[I_o]$ = 5.6 x 10_3M, $[BCl_3]$ = 2.8 x 10^{-1}M. Numbers indicate $\bar{M}_w/\bar{M}_n$ values.

is lower than that with the CuOAc/BCl_3 system. Indeed the $\bar{M}_w/\bar{M}_n$ of the initially formed PIB was appreciably lower when the individual components were premixed and the monomer was added to the preformed TMPOAc·BCl_3 complex (e.g., 2.3 versus 3.7 in CH_3Cl). The molecular weight distributions decreased further with each monomer addition in this series of experiments.

The fact that the $\bar{M}_n$ of the first formed PIB's in which BCl_3 was added to TMPOAc/BCl_3 charges, and in which the monomer was added to the preformed TMPOAc·BCl_3 complexes are equal within experimental error, indicates that the rate of formation of the initiating TMPOAc·BCl_3 complexes is extremely high and that initiation follows very rapidly. Moreover, experiments have been carried out in which TMPOAc·BCl_3 complexes were prepared (mixing TMPOAc plus BCl_3 in CH_3Cl or CH_2Cl_2 at -30°C) and after 30 mins of aging various amounts of isobutylene were added; immediate polymerization to 100% conversion ensued and the $\overline{DP}_n$ of the product obeyed [IB]/[TMPOAc·BCl_3]. These experiments suggest that the TMPOAc·BCl_3 complex is active "living PIB dimer" and that the structure of this dimer and that of the propagating end of the living PIB polymer are essentially identical.

According to the data shown in Figures 1-3 the molecular weight distributions gradually decrease with each monomer addition in both the CuOAc/BCl_3 and TMPOAc/BCl_3 systems in spite of the heterogeneous nature of the polymerizations.

The Effect of Diluent

The data indicate that the living polymerization of isobutylene can be induced by the TMPOAc/BCl_3 and CuOAc/BCl_3 complexes in CH_2Cl_2 and CH_3Cl diluents (cf. Figures 1-3). Since PIB with $\bar{M}_n$ higher than about 3000 g/mole is insoluble in CH_2Cl_2 and CH_3Cl these polymerizations tend to be

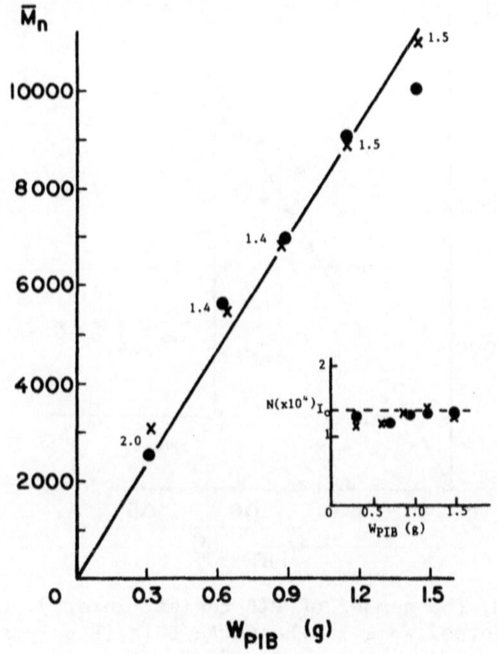

Fig. 2. $\overline{M}_n$ and N, the number of PIB chains (insert), versus the weight of PIB formed W_{PIB} in the $CuOAc/BCl_3/IB/CH_2Cl_2$ polymerization system using the IMA technique at -30°C: $[I_o] = 5.6 \times 10^{-3}$ M, $[BCl_3] = 2.8 \times 10^{-1}$ M. Numbers indicate $\overline{M}_w\overline{M}_n$ values.

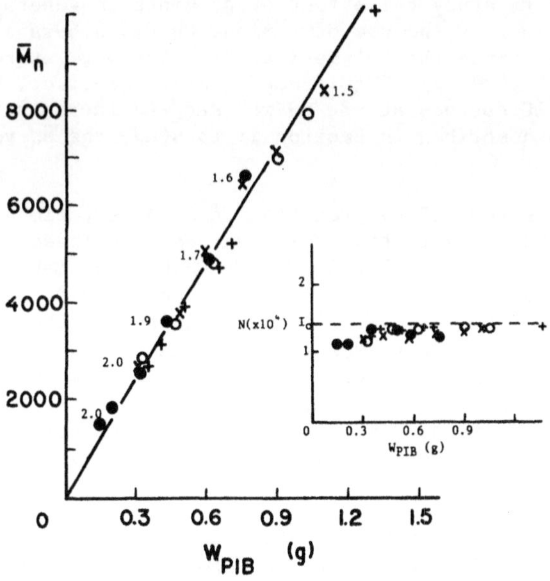

Fig. 3. $\overline{M}_n$ and N, the number of PIB chains (insert), versus the weight of PIB formed W_{PIB} in the $CuOAc/BCl_3/IB/CH_3Cl$ polymerization system using the IMA technique at -30°C: $[I_o] = 5.6 \times 10^{-3}$ M, $[BCl_3] = 2.8 \times 10^{-1}$ M. Numbers indicate $\overline{M}_w/\overline{M}_n$ values.

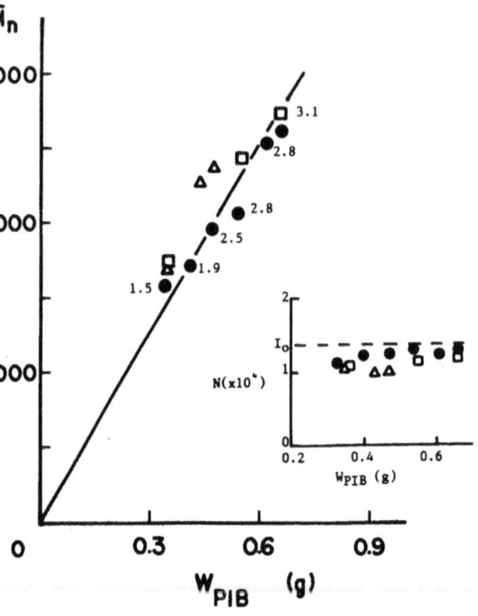

Fig. 4. $\bar{M}_n$ and N, the number of PIB chains (insert), versus the weight of PIB formed W_{PIB} in the $CuOAc/BCl_3/IB$ polymerization system using the IMA technique at -30°C with different solvents: C_2H_5Cl (●), CH_2Cl_2/n-hexane 80v/20v (□), CH_2Cl_2/n-hexane 60v, 40v (△) $[I_o] = 5.6 \times 10^{-3}$M, $[BCl_3] = 2.8 \times 10^{-1}$M. Numbers indicate $\bar{M}_w\bar{M}_n$ values.

heterogeneous. This circumstance and the very rapid propagation (relative to initiation) in these diluents results in somewhat broader than Poisson molecular weight distributions. To overcome problems caused by system heterogeneity and to study the effect of solvents in general, experiments have been carried out by the use of various CH_2Cl_2/n-hexane mixtures, C_2H_5Cl and $HCCl_3$. Since the polymerization rates were extremely high even by the use of 60/40 v/v $CH_3Cl/n-C_6H_{14}$ mixtures (i.e., polymerizations were complete in 60 seconds at -30°C) we used the incremental monomer addition technique as described in Section II to study the $\bar{M}_n$ versus g PIB formed profile.

Figure 4 shows some of the results. All the experiments were homogeneous in nature and the initial $\bar{M}_w/\bar{M}_n$ values were indeed smaller than those obtained in the more polar CH_2Cl_2 or CH_3Cl systems. However, in the later course of the experiment the molecular weight distributions broadened and finally became bimodal. The findings obtained in the experiment with C_2H_5Cl are shown in Figures 4 ($\bar{M}_w/\bar{M}_n$ values) and 5. This broadening of the distribution was even more pronounced in experiments conducted with 80/20 and 60/40 v/v $CH_3Cl/n-C_6H_{14}$ mixtures. The broadening of molecular weight distributions with decreasing solvent polarity is conceivably due to the presence and accumulation of a second (most likely less dissociated or less active) propagating species.

Broad distributions which ultimately become bimodal suggest an increasingly slower equilibrium between a dormant and active growing species.[29] Importantly, however, this equilibrium can be controlled, i.e., the equilibrium is reached slowly, as indicated by the narrow distributions obtained in the experiments with C_2H_5Cl in which all the monomer was present at initiation (cf. last three entries in Table I). In these instances theoretical molecular weights ($\overline{DP}_n = [M_o]/[CH_3COOR^t]$) and narrow

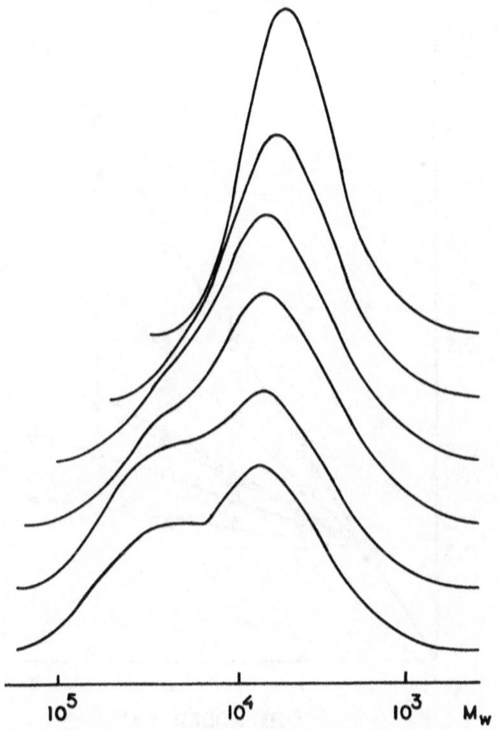

Fig. 5. MWD of PIBs obtained by the use of the CuOAc/BCl_3/IB/C_2H_5Cl system using the IMA technique at -30°C; $[I_o]$ = 5.6 x 10^{-3}M, $[BCl_3]$ = 2.8 x 10^{-1}M.

molecular weight distributions were achieved at 100% conversions. Evidently, then, theoretical $\bar{M}_n$, conversion, and $\bar{M}_w/\bar{M}_n$ values can be obtained not only in polar diluents but under select conditions even in less polar media.

A series of experiments have also been carried out by the use of $HCCL_3$, however, the conversions were so low that experimentation with this solvent was discontinued (High or 100% conversions were obtained only in the presence of relatively high CuOAc concentrations).

It was of interest to determine the amount of PIB formed as a function of IB added in various solvents and solvent mixtures by the use of the incremental monomer addition technique described above. Figure 6 shows the results obtained with the CuOAC/BCl_3/IB system but similar results have also been obtained with TMPOAc/BCl_3. Since the time elapsed between individual monomer additions was always the same, arbitrarily 30 mins, the g PIB formed reflects the relative polymerization rates. The results obtained in quadruplicate experiments by the use of CH_2Cl_2 diluent are theoretical, within what is considered to be experimental error (i.e., unavoidable minute losses of PIB during work up; see Experimental).

Similarly, the amount of PIB formed in the first reactor of the series of reactors, i.e., in the charge that contained all the monomer at the moment the BCl_3 was added, is theoretical (100% conversions) irrespective of the particular solvent used in the series. From this point on the amount of PIB that arises (i.e., the overall rate of polymerization) at each incremental monomer addition is mainly determined by the nature of the solvent. It appears that the rate of polymerization is directly related to

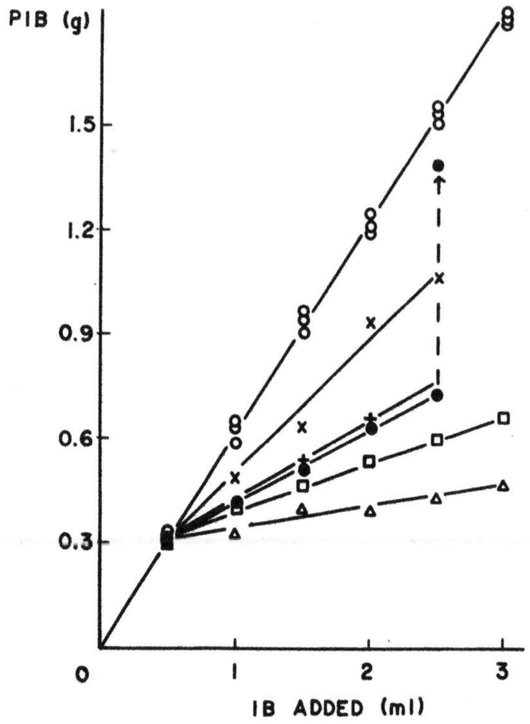

Fig. 6. The weight of PIB versus cumulative amount of IB added by the use of various solvent systems at -30°C. CH_2Cl_2 (O), CH_3Cl (●), CH_2Cl_2/n-hexane 80v/20v (+), C_2H_5Cl (□), CH_2Cl_2/n-hexane 60v/40v (△) - time between monomer additions 30 min, CH_3Cl (X) - time between monomer addition 60 min. $[I_o] = 5.6 \times 10^{-3}$ M, $[BCl_3] = 2.8 \times 10^{-1}$ M.

the overall polarity of the system. By this measure CH_2Cl_2 is more polar than CH_3Cl which is in turn more polar than C_2H_5Cl.

The absolute amount of g PIB formed is a function of experimental conditions, e.g., time of experiment, temperature, concentrations, heterogeneity of the system. Specifically, the less than theoretical quantities (<100% conversions) of PIB formed in CH_3Cl are most likely due to polymer precipitation which must retard propagation (decreased diffusion). Indeed in an experiment in which the precipitated system was not quenched after 30 mins but was allowed to "live" for an additional 180 mins (see dotted line in Figure 6), the yields gradually increased and finally reached a close to theoretical value. Else, in C_2H_5Cl and CH_2Cl_2/n-C_6H_{14} mixtures the limited quantities formed (lower rates) are most likely due to the formation of less-dissociated (dormant?) species. Interestingly, close to theoretical values have been obtained by the use of CH_2Cl_2 although the system appears to be heterogeneous in this diluent.

Insight into the Mechanism of Chain Transfer to Monomer: An Operational Definition of Living Polymerization. After having demonstrated that living isobutylene polymerization can be readily achieved with the TMPOAc/BCl_3 and CuOAc/BCl_3 initiating systems by the use of CH_2Cl_2 and CH_3Cl diluents up to -30°C (cf. Figures 1-3) and that the t-BuOAc/BCl_3 combination also yields encouraging results up to -10°C (cf. Table I), we have decided to explore in more detail the efficient CuOAc/BCl_3/IB/CH_2Cl_2 system.

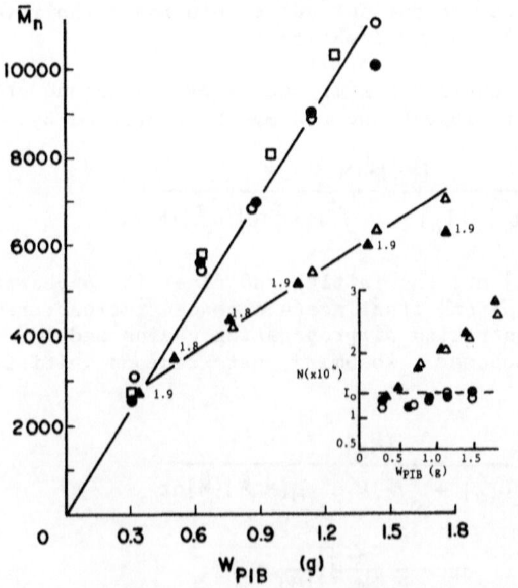

Fig. 7. $\bar{M}_n$ and N, the number of PIB chains (insert), versus the weight of PIB formed W_{PIB} in the $CuOAc/BCl_3/IB/CH_2Cl_2$ polymerization system using the IMA technique at -30°C (○,●), -10°C (△, ▲), and with all the monomer present at the addition of BCl_3, (□). $[I_o] = 5.6 \times 10^{-3}$ M, $[BCl_3] = 2.8 \times 10^{-1}$ M.

Figure 7 shows the $\bar{M}_n$ versus g PIB plots obtained with this system at -30°C and -10°C but otherwise under identical conditions. The linear plot characteristic of the -30°C experiment together with the insert, showing that the number of growing chains remains constant throughout the series, indicates a living system at this temperature (cf. also Figure 2). In contrast the experiment at -10°C yielded a kinked $\bar{M}_n$ versus g PIB plot according to the insert in Figure 7 the number of PIB molecules increases monotonically which indicates significant chain transfer to monomer.

These findings and an inquiry into the detailed nature of chain transfer in these systems led to some interesting conclusions. Chain transfer to monomer is visualized to occur by β-proton elimination either by a second-order bimolecular or first-order monomolecular process. Schematically:

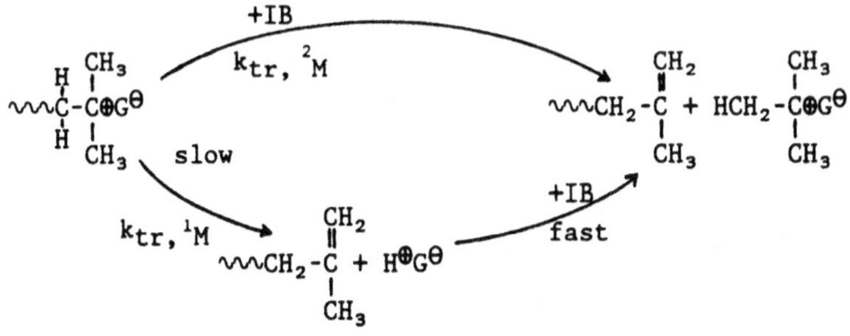

where $k_{tr,}$ ^2M and $k_{tr,}$ ^1M are the rate constants of the second and first order chain transfer to monomer, respectively. In line with previous investigations[15] first we assumed that the more important pathway is by the second order process and set out to elucidate the rate constant ratio $k_{tr,}{}^{2}M/k_p$ by the following treatment:

Assuming that chain transfer occurs by the second order route, and since termination is absent the $\overline{DP}_n$ may be expressed by:

$$\overline{DP}_n = \frac{[M_o]-[M_t]}{[I_o]-[I_t] + \int_o^t k_{tr,}{}^{2}M[M_n^{\oplus}][M]dt} \qquad (2)$$

where $[M_o]$ and $[M_t]$ are the initial and final IB concentrations, $[I_o]$ and $[I_t]$ are the initial and final acetate concentrations, respectively, $[M_n^{\oplus}]$ is the concentration of propagating chains and $[M]$ is the momentary concentration of monomer. Assuming instantaneous initiation and 100% conversion:

$$\overline{DP}_n = \frac{[M_o]}{[I_o] + \int_o^t k_{tr,}{}^{2}M[M_n^{\oplus}][M]dt}$$

Since

$$dt = \frac{d[M]}{k_p[M_n^{\oplus}][M]} \qquad (3)$$

we get

$$\overline{DP}_n = \frac{[M_o]}{[I_o] + \int_{M_o}^{o} \frac{k_{tr,}{}^{2}M}{k_p} \cdot d[M]} \qquad (4)$$

thus

$$\overline{DP}_n = \frac{[M_o]}{[I_o] + \frac{k_{tr,}{}^{2}M}{k_p} \cdot [M_o]} \qquad (5)$$

and

$$\overline{DP}_n^{-1} = \frac{[I_o]}{[M_o]} + \frac{k_{tr,}{}^{2}M}{k_p} \qquad (6)$$

A series of experiments have been designed to test this expression. Thus isobutylene was polymerized at -10°C by the CuOAc/-BCl_3 system with varying $[M_o]$ at $[I_o]$ = constant. Table II shows the results and Figure 8 the corresponding $1/\overline{DP}_n$ versus $1/[M_o]$ plot. Obviously the straight line can be smoothly back extrapolated to the origin and there is no evidence for an intercept. The slope of the plot yields $[I_o]$ within experimental error. According to this evidence $k_{tr,}{}^{2}M/k_p \approx 0$, i.e., bimolecular chain transfer to monomer is absent, and by inference, chain transfer at -10°C, as indicated by the data in Figure 7, is most likely due to the monomolecular process. This is the first time that one is able to distinguish experimentally with some measure of confidence between monomolecular and bimolecular chain transfer processes.

Significantly, the first data point (duplicates) of the -10°C experiment in Figure 7 also fall on the theoretical line indicating the absence of chain transfer even at this unusually high temperature. These first data were obtained in the experiment in which the BCl_3 was added last to monomer/CuOAc charges whereas the subsequent data were collected in experiments in which increasing amounts of monomer were added to polymerized living CuOAc/BCl_3/IB charges. Thus, the question arises as to why is chain transfer absent when BCl_3 is added to CH_3COOR^t/IB charges but is present when additional monomer is added to active CH_3COOR^t/BCl_3 charges i.e., to charges 2,3, etc. in the incremental monomer addition series?

Table II. Polymerization of Isobutylene at Various Monomer Concentrations*

$\underline{M}$(mol/ℓ)	conv.%	$\overline{M}_n$	$\overline{M}_w$	$\overline{M}_w/\overline{M}_n$	I_{eff}
0.225	100	2640	5000	1.9	92%
0.45	100	5800	3900	1.7	81%
0.676	100	8200	13900	1.7	85%
0.9	100	10400	16320	1.6	88%

*Cumyl acetate initiator [I] = $5.6 \cdot 10^{-1}\underline{M}$, [$BCl_3$] = $2.6 \cdot 10^{-1}\underline{M}$, CH_2Cl_2, -10°C.

The answer to this puzzle becomes apparent by considering the long delay that elapses between incremental monomer additions. Evidently, the polymerization is living in nature even at -10°C as long as sufficient monomer is present (i.e., when BCl_3 is added last) because the rate of the only operational chain transfer step, i.e., the monomolecular process, is negligible as compared to that of propagation.

Due to the very high polymerization rate, the monomer disappears in a few seconds by living polymerization; however given a sufficient length of time, in the absence of monomer, the living chain ends will slowly undergo proton elimination to the counteranion and thus the stage is set for monomolecular chain transfer to monomer. Thus, it is not too surprising that living polymerizations more readily occur at -30°C because monomolecular chain transfer is presumably "frozen out" at this lower temperature level. However, living systems may still be obtained at -10°C or even at higher temperatures provided sufficient monomer is initially available in the system and the polymerization is over before the much slower monomolecular chain transfer to monomer process becomes noticeable.

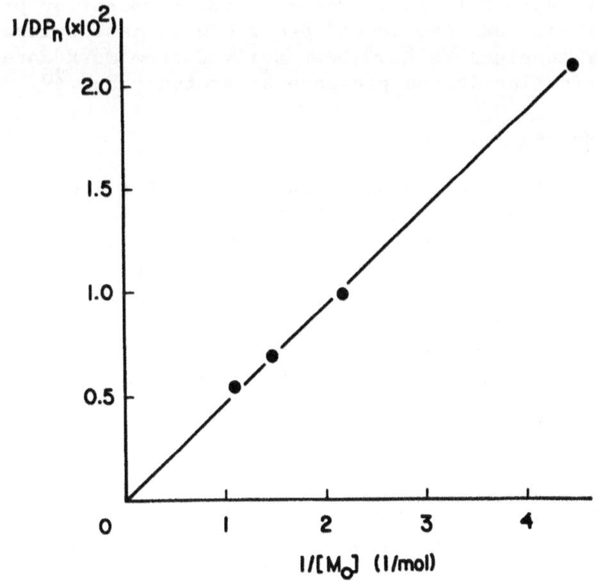

Fig. 8. $1/\overline{DP}_n$ versus $1/M_o$ plot for the CuOAc/BCl_3/IB/CH_2Cl_2 system at -10°C. [I_o] - $5.6 \times 10^{-3}\underline{M}$, [$BCl_3$] = $2.8 \times 10^{-1}\underline{M}$.

A direct experiment indicates that this is indeed the case, and living polymerization can be obtained even at -10°C, provided the charge contains all the monomer when initiation occurs. In this experiment, we added BCl_3 to a series of IB/CuOAc charges in CH_2Cl_2 at -10°C. The conversions were complete and as shown by the plot in Figure 7 the $\bar{M}_n$ versus g monomer added plot is close to theoretical. This important lead has been followed up and recently we are routinely carrying out various living cationic polymerizations by batch and continuous techniques at rather high temperatures.

Following this train of thought the incremental monomer addition technique developed in the course of these investigations is a more sensitive method to demonstrate the existence of living systems than the conventional intermittent monomer withdrawal technique because only by the use of the former method can the monomolecular chain transfer process be diagnosed.

These conclusions have a bearing on the interpretation of data obtained by the popular Mayo analysis.[25] In line with the above considerations the Mayo equation $1/\overline{DP}_n = k_{tr,M}/k_p + (k_t/k_p)1/[M]$ should be expanded as follows:

$$\frac{1}{\overline{DP}_n} = \frac{k_{tr,{}^1M}[M_n^{\oplus}] + k_{tr,{}^2M}[M_n^{\oplus}][M]}{k_p[M_n^{\oplus}][M]} + \frac{k_t[M_n^{\oplus}]}{k_p[M_n^{\oplus}][M]}$$

and thus

$$\frac{1}{\overline{DP}_n} = \frac{k_{tr,{}^2M}}{k_p} + \frac{k_{tr,{}^1M} + k_t}{k_p} \cdot \frac{1}{[M]}$$

In other words, the absence of an intercept of such an expanded Mayo plot would indicate the absence of only the second-order chain transfer to monomer process, however, the slope of the plot may still contain the first-order component, or, the presence of an intercept would suggest not only the operational presence of the second-order process but also that the first-order process is buried in the intercept. Since the first-order process is operational even when the second-order process is absent (see above data), it appears that, in general, the first-order process is probably more important than the second-order one in isobutylene polymerizations. Similar conclusions have been derived from work done with α-methylstyrene polymerization in the presence of proton traps.[26]

Termination Studies

While the living species may remain active for a considerable length of time, if termination is desired (e.g. for terminal functionalization) it could be brought about either by increasing the temperature above the decomposition temperature of the propagating ester/BCl_3 complex or by the addition of strong nucleophiles, e.g., alcohols, amines. We have examined the structure of the end groups of PIBs obtained in living systems after the active ester/BCl_3 complex was decomposed by heating to room temperature or by the addition of various amines, CH_3OH, etc. We have also carried out model experiments with the TMPOAc/BCl_3 complex to corroborate the results obtained in polymerization experiments (see Experimental for details).

Surprisingly all the polymerization experiments irrespective of the particular killing method employed yielded t-chloro end groups, i.e., where R = $(CH_3)_3C$-, $(CH_3)_3CCH_2C(CH_3)_2$-, or $C_6H_5(CH_3)_2C$- depending on whether the initiator was t-BuOAc, TMPOAc, or CuOAc, respectively. The analysis of the end groups has been carried out by ^1H NMR spectroscopy by the use of well-established methods.[27]

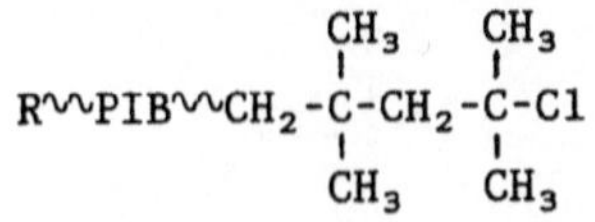

The results of these polymerization experiments have been fully corroborated by model experiments. These studies were carried out by the TMPOAc$\cdot BCl_3$ complex which was decomposed ("killed") in the same manner the polymerization experiments were. According to ^1H NMR analysis the product was invariably TMP-Cl:

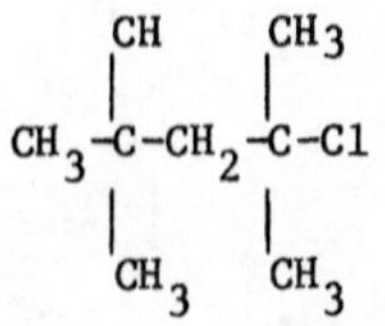

Additional model thermal decomposition experiments have been carried out with t-BuOAc/BCl_3 and TMPOAc/BCl_3 complexes mimicking the active PIB chain end. Thus we have followed the disappearance of the carbonyl stretching frequency at ν C=O= 1570 cm^{-1} characteristic of acetate$\cdot BCl_3$ complexes and the appearance of the free C=O bond in acetyl chloride at ν C=O = 1800 cm^{-1} (acetyl chloride is one of the decomposition products of the above complexes) in CH_2Cl_2 at different temperatures. The complexes did not decompose at -10°C after 1 hour, however, decomposition occurred at room temperature and after 1 hour above 50% of the complexes decomposed. The decomposition products were acetyl chloride and the corresponding alkyl chloride, i.e., t-BuCl and TMP-Cl, respectively. These findings are in line with the results of Gerrard et al.[18]

Conclusions; Mechanistic Speculations

Evidence has been presented that demonstrates, for the first time, the truly living polymerization of isobutylene induced by TMPOAc/BCl_3 and CuOAc/BCl_3 complexes in CH_2Cl_2 and CH_3Cl diluents. The PIBs so obtained do not contain terminal unsaturation which is evidence for the absence of chain transfer to monomer. PIB molecular weights are controlled by the relative ratio of monomer over initiator (i.e., CH_3COOR^t) concentrations, i.e., $\overline{DP}_n = [M_o]/[I_o]$. The molecular weight distributions are somewhat broader, $\overline{M}_w/\overline{M}_n$ = 1.3-2.0, most likely due to rapid propagation relative to initiation and various physical effects i.e., system heterogeneity, insufficient stirring (heat effect) of very rapid polymerizations. Experiments designed to give narrower molecular weight distributions (to be reported later) have yielded $\overline{M}_w\overline{M}_n$ = 1.17-1.3.

Protic impurities (moisture) have very little effect on these living polymerizations and the experiments can be carried out under conventional laboratory conditions by the use of conventionally dried solvents under a blanket of dry nitrogen, for example.

An attempt is made to rationalize these results by the use of information gleaned from the literature in regard to the nature of ester/BCl_3 complexes.[17,18] Thus the following mechanism may explain some of the features of living isobutylene polymerization by, for example, the TMPOAc/BCl_3 system (R = $-CH_2-C(CH_3)_3$):

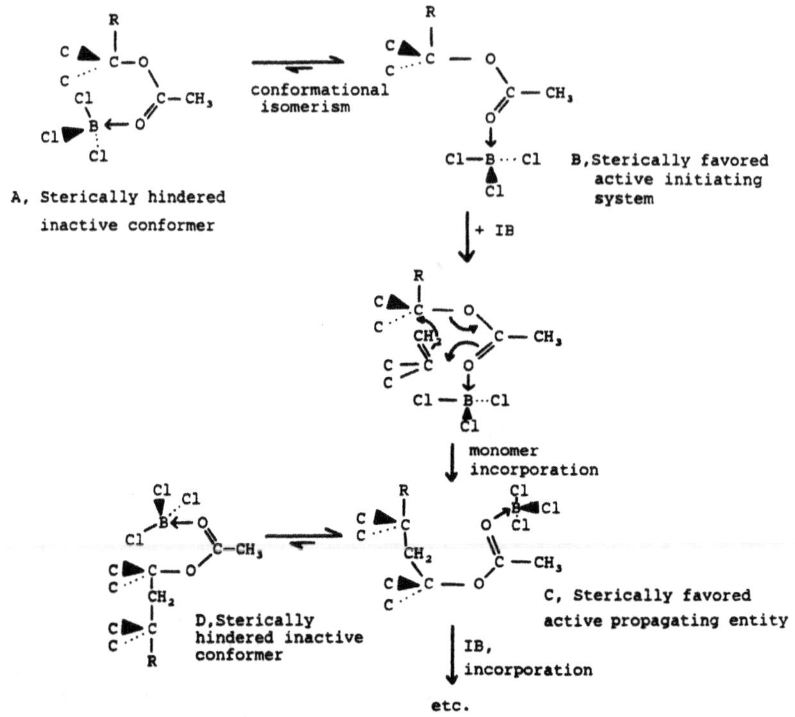

According to previous workers and by our own data the rate of complex formation between organic acetates and BCl_3 is very rapid. The site of complexation is the acyl oxygen[17] and equilibration between the two possible conformational isomers A and B is most likely also very rapid. Inspection of molecular models would indicate that conformer B is sterically more favored to incorporate incoming monomer. Significantly, monomer incorporation would yield a tertiary acetate·BCl_3 complex structurally identical to the initiating complex. In this sense this system is akin to a group transfer polymerization.[28]

Whether incorporation is by monomer insertion or by an ionic process is obscure. The facts that chain transfer to monomer and termination are absent certainly indicate a non-conventional active species.

The large rate enhancing effect of polar diluents (CH_2Cl_2, CH_3Cl, C_2H_5Cl) may be due to the presence of a dormant (largely covalent) and an active (largely ionic) species connected by a mobile equilibrium. Assuming that the polar diluent accelerates equilibration and/or accentuates the ionic character of the active species:

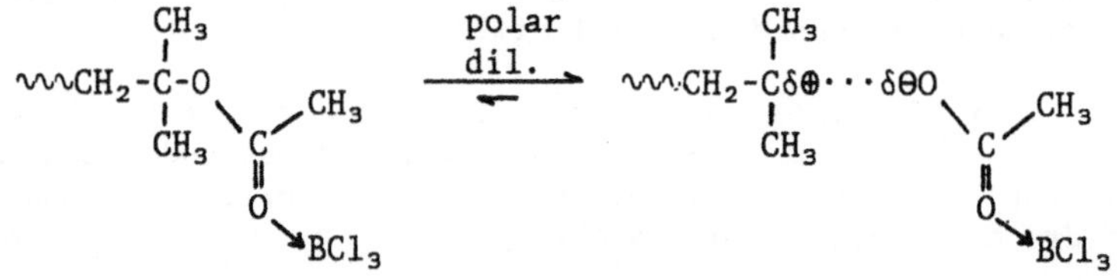

Extensive studies are in progress to elucidate the nature of the propagating entity.

Inspection of molecular models provided some insight into the mechanism of forced termination. We speculate that termination could occur thermally or by strong nucleophiles by the following intramolecular S_{Ni} process:

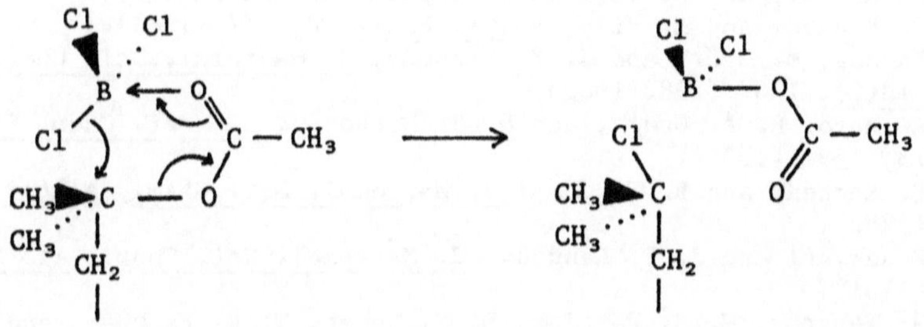

One of the Cl's of the BCl_3 is in the proximity to the propagating tertiary carbon and either heating or nucleophilic attack on the complexed B center may induce the above rearrangement. Previous work on the decomposition of acetate·BCl_3 complexes,[18] and our polymer end group and model studies indicate the formation of the expected tertiary alkyl chlorides together with the boron-containing species. The synthesis of terminal tertiary chlorines by this simple route is a valuable new technique for the preparation of end-functional polymers.

ACKNOWLEDGEMENT

Support by the Polymer Program of the NSF (Grant DMR-8418617) is gratefully acknowledged.

REFERENCES

1. W. M. Saltman, Encylopedia of Polymer Science and Technology, Vol. 2, John Wiley and Sons, New York (1965).
2. L. J. Fetters, Elastoplast, 4, 32 (1971).
3. S. Penczek, P. Kubisa, and K. Matyjaszewski, Cationic Ring Opening Polymerization, Advances in Polymer Science, 68/69 (1985).
4. M. Szwarc, Nature, 178, 1168 (1956).
5. M. Szwarc, M. Levy, and R. Milkovich, J. Am. Chem. Soc., 78, 2656 (1956).
6. D. C. Pepper, J. Polym. Sci., Symp. No. 50.
7. T. Higashimura, M. Miyamoto, and M. Sawamoto, Macromoles, 18, 611 (1985).
8. T. Higashimura, H. Teranishi, and M. Sawamoto, Polymer J., 12, 393 (1980).
9. T. Higashimura and O. Kishiro, Polymer J., 9, 87 (1977).
10. R. Faust, A. Fehérvári, and J. P. Kennedy, J. Macromol. Sci. Chem., A18(9), 1209 (1982).
11. J. Puskás, G. Kaszás, J. P. Kennedy, T. Kelen, and F. Tüdős, ibid., A18(9), 1229 (1982).
12. M. Sawamoto and J. P. Kennedy, ibid., A18(9), 1275 (1982).
13. R. Faust and J. P. Kennedy, Abstract, ACS Meeting, Chicago (1985).
14. R. Faust and J. P. Kennedy, Abstract, ACS Meeting, New York (1986).
15. J. P. Kennedy and E. Marechal, Carbocationic Polymerization, John Wiley-Interscience Publishers, New York (1982).
16. Demarcay, Bull. Soc. Chim. France, 20, 127 (1873).
17. M. F. Lappert, J. Chem. Soc., ibid. (1961); ibid., 542 (1962).
18. W. Gerrard and M. A. Wheelans, J. Chem. Soc., 4196 (1956).
19. H. C. Brown and P. J. Geoghegan, J. Org. Chem., 35, 1844 (1970).

20. J. P. Kennedy, S. Y. Huang, and S. C. Feinberg, J. Polym. Sci., 18, 2801 (1977).
21. F. Y. Chen, Ph.D. Dissertation, The University of Akron (1986).
22. J. P. Kennedy and A. Fehérvári, unpublished data (1985).
23. J. P. Kennedy and M. Hiza, Polym. Bull., 8, 557 (1982).
24. G. Kaszás, M. Győr, and J. P. Kennedy, J. Macromol. Sci. Chem., A18(9), 1367 (1982-1983).
25. F. A. Mayo, R. A. Gregg, and M. S. Matheson, J. Amer. Chem. Soc., 73, 1691 (1951).
26. J. P. Kennedy and R. T. Chou, J. Macromol. Sci. Chem., A18(1), 17 (1982).
27. A. Fehérvári and J. P. Kennedy, J. Macromol. Sci. Chem., A15(2), 215 (1981).
28. O. W. Webster, W. R. Hertler, D. Y. Sogah, W. B. Farnham, and T. V. Rajan-Babu, J. Macromol. Sci. Chem., A21(8/9), 819.
29. M. Szwarc, Carbanions, Living Polymers and Electron Transfer Processes, Wiley-Interscience, New York (1968).

THERMOPLASTIC ELASTOMERS BASED ON BLOCK COPOLYMERS OF ETHYLENE AND PROPYLENE

George A. Lock

Hercules Incorporated

INTRODUCTION

Polyolefins having the melt processability of thermoplastics but the solid state properties of elastomers are clearly attractive products for a broad range of applications. Commercially available olefin-based thermoplastic elastomers fall into three major categories: simple alloys of crystalline polypropylene with ethylene-propylene rubber, EPR; similar alloys in which the rubbery phase, EPDM, ethylene-propylene-diene rubber, is partially crosslinked; and those in which the rubber is completely crosslinked. Many of the simple polypropylene-EPR alloys are in situ blends made by sequential feeds to the polymerization reactor; others are mechanical blends of separately polymerized polypropylene and EPR. In the latter two cases, crosslinking is induced during melt processing of the blend. Because of the crosslinking in these materials, recycle-processing, if possible, is usually difficult. Simple alloys, on the other hand, while easier to reprocess, are often inferior in tensile properties.

Several approaches have been undertaken to develop olefin-based thermoplastic elastomers. One approach appeared particularly attractive: i.e., the development of technology which would enable the production of poly(propylene-b-(ethylene-co-propylene)-b-propylene) ABA block copolymers. The direct polymerization of block copolymers was believed to be the least costly method of producing an olefinic thermoplastic elastomer relative to processes based on post-polymerization coupling, grafting, or crosslinking reactions.

A true block copolymer should impart the better mechanical properties expected of a thermoplastic elastomer when small crystalline homopolymer domains are well dispersed in an incompatible second phase of a rubbery copolymer matrix and when the blocks are actually bonded through the covalent bonds in the polymer backbone. An illustration of this hypothesis is depicted in Figure 1. Melting of the crystalline phase would permit thermoplastic processing. Subsequent cooling would lead to recrystallization and reestablishment of the crosslinks and elastomeric properties. The existence of physical, rather than chemical, crosslinks would permit recycle processing without a sacrifice in tensile properties. Thus, our objective was to develop a process for preparing such copolymers and to optimize their properties by controlling the relative block sizes and their composition.

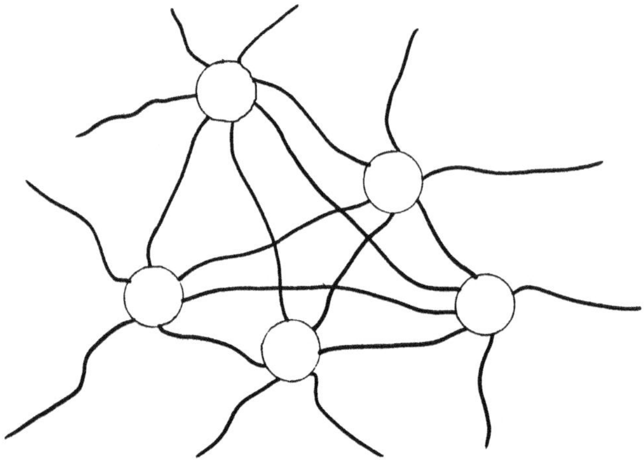

Fig. 1. Morphological Model of the Physical Network

POLYMERIZATION KINETICS

In general, Ziegler-Natta polymerization of olefins is the result of successive monomer insertions into a transition metal-alkyl carbon bond. The initial active site is created on a small fraction of a metal chloride catalyst surface by exchange of a chloride for an alkyl group from an activator such as diethylaluminum chloride. The actual lifetime of the subsequently growing polymer chain has been the subject of much discussion for many years. Chain lifetimes ranging from seconds to several minutes are commonly reported, with longest lifetimes usually being obtained at low temperatures.[1]

This typical behavior is illustrated in Figures 2 through 4 for propylene polymerization catalyzed by titanium trichloride and activated with

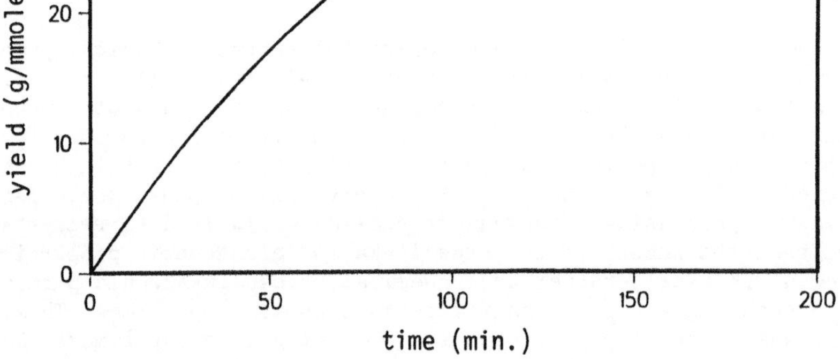

Fig. 2. Propylene Polymerization Polymer Yield vs Time

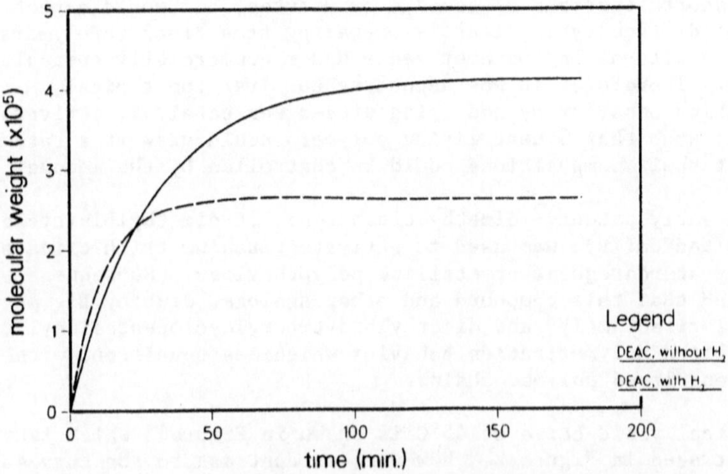

Fig. 3. Propylene Polymerization
Polymer Molecular Weight vs Time

diethylaluminum chloride (DEAC) at 60°C. While polymer yield increases with time (Figure 2), the increase in polymer molecular weight is much less rapid and ultimately reaches a maximum (Figure 3), even in the absence of a chain transfer agent, such as hydrogen. Thus the number of polymer chains must increase as induced or spontaneous chain transfer reactions occur (Figure 4).

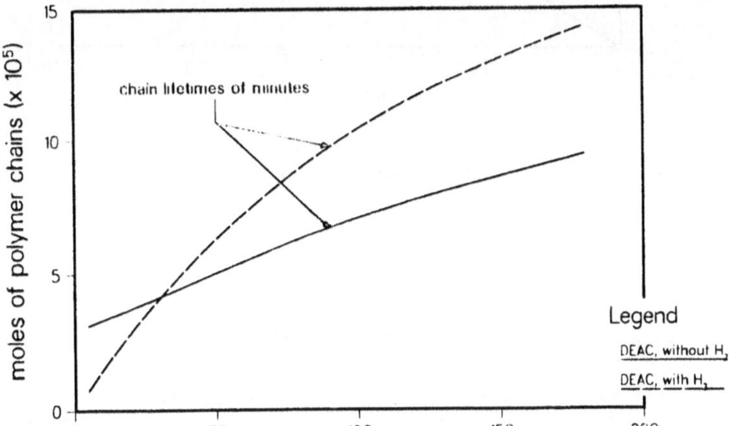

Fig. 4. Propylene Polymerization
Moles of Polymer vs Time

With short lifetimes of seconds or minutes, one would expect significant difficulty in directly preparing true block copolymers by sequentially alternating monomer feeds under commercially operable conditions. Therefore, it was necessary to alter the typical polymerization behavior by modifying either the catalyst, activator, or conditions, such that linear living polymers could grow at a rate slow enough that their compositions could be controlled by the monomer feed.

In an early patent,[2] dimethyltitanocene, or dimethylbis(cyclopentadienyl)titanium(IV), was used to activate titanium trichloride and produce highly stereoregular crystalline polypropylene. Subsequent work[3,4] demonstrated that this compound and other analogs, dimethylbis(methylcyclopentadienyl)titanium(IV) and dimethylbis(t-butylcyclopentadienyl)titanium (IV), exhibited polymerization behavior which was consistent with the existence of long-lived polymer chains.

A typical yield curve at 45°C is shown in Figure 5 which is similar to that illustrated in Figure 2. However, in contrast to the curves in Figure 3, the molecular weight of the polymer in Figure 6 continues to increase throughout the polymerization. As a result, as shown in Figure 7, the number of polymer chains remains nearly constant.

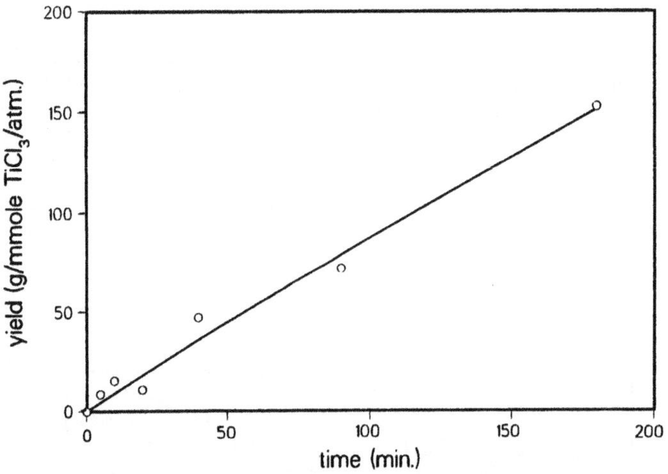

Fig. 5. $(\eta^5\text{-}CH_3C_6H_4)_2Ti(CH_3)_2$ and $TiCl_3$
Polymer Yield vs Time

To further demonstrate the ability of the catalyst - activator system to minimize chain transfer reactions, a polymerization was carried out as before, but after twenty minutes, the monomer was removed from the system and replaced by argon. The inert system was left in this condition for 17 hours after which the monomer was replaced at its original concentration and polymerization resumed. The reaction was quenched after three hours in the presence of monomer. The results are illustrated in Figure 8. As in the previous experiment, the number of polymer chains remains constant throughout the entire twenty-hour period.

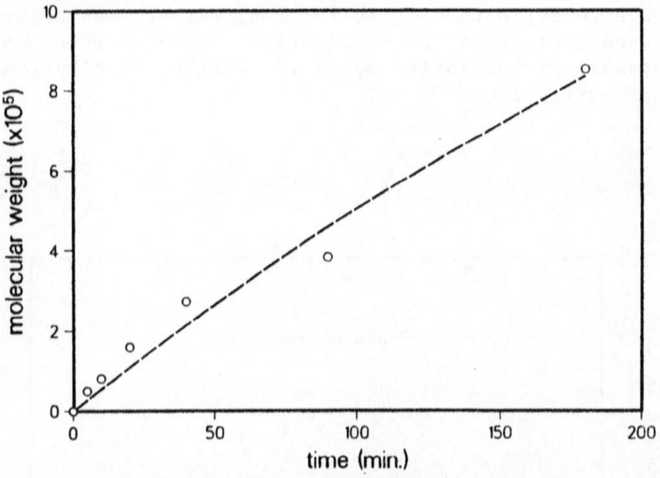

Fig. 6. $(\eta^5\text{-}CH_3C_6H_4)_2Ti(CH_3)_2$ and $TiCl_3$
Polymer Molecular Weight vs Time

Unlike the living ionic polymerizations, we find that molecular weight distributions are in the range of 4 to 5. However, the distribution remains essentially unchanged throughout the polymerization. A constant distribution, markedly narrower than those typically observed in Ziegler-Natta polymerizations, is probably a reflection of a distribution of active sites

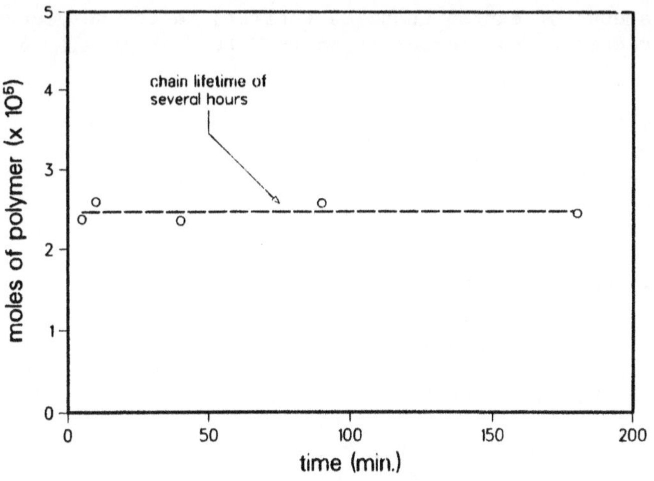

Fig. 7. $(\eta^5\text{-}CH_3C_6H_4)_2Ti(CH_3)_2$ and $TiCl_3$
Moles of Polymer vs Time

with a consequent distribution of activities, rather than significant chain transfer, delayed activation, or termination. Each of which would be expected to produce an increasing molecular weight distribution over the duration of polymerization.

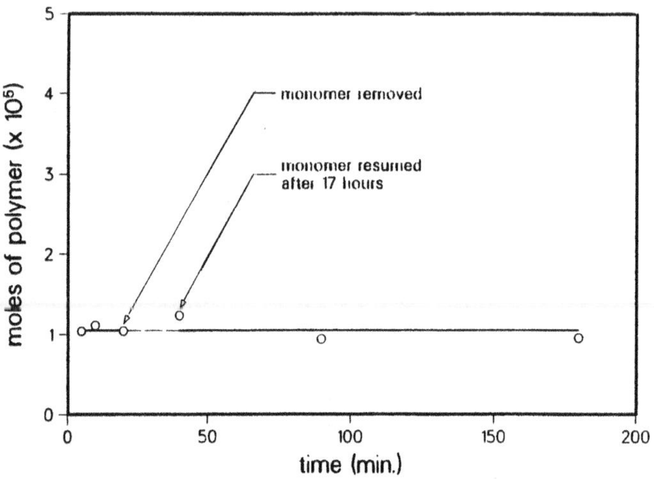

Fig. 8. (η^5-$CH_3C_6H_4$)$_2$Ti(CH_3)$_2$ and $TiCl_3$
Moles of Polymer vs Time

Finally, polymerization catalyzed by titanium trichloride activated with dimethylbis(methylcyclopentadienyl)titanium(IV) were quenched with radiochemically labelled carbon monoxide at several different polymerization mileages.[5] Radioactivity counting techniques applied to the polymer product indicated that the number of chains agreed in all cases with the constant number of chains calculated from yield and molecular weight data, and that the amount of carbon monoxide retained in the polymer after work-up related directly to the number of active sites originally present in the system.

These combined, yet independent results, are consistent with our requirements for direct polymerization of block copolymers: i.e., 1) all growth sites are initiated simultaneously; 2) chain transfer is minimal for significant periods of time.

BLOCK COPOLYMERIZATION PROCESS

The use of a living catalyst system should permit the direct preparation of block copolymers in which the blocks are covalently bonded to one another through the chain itself. The size of the blocks and their composition should be easily regulated by careful control of the monomer feed(s) to the polymerization vessel. Semi-batch and plug flow reactors would be amenable to block copolymer production; the development of this process was conducted in semi-batch reactors.

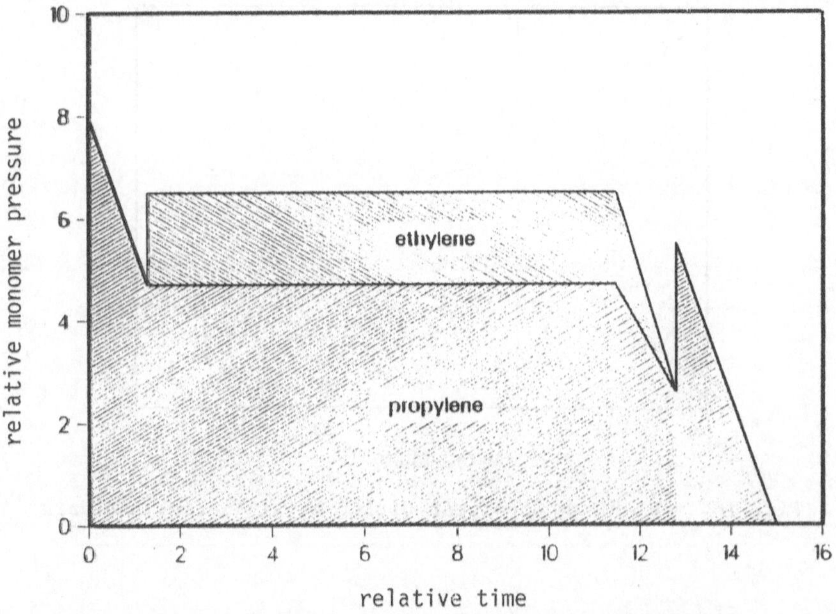

Fig. 9. Block Copolymer Process Monomer Feed Profile

A simple schematic of the process is illustrated in Figure 9. The reactor, containing hydrocarbon diluent and the activator, is filled with enough propylene to make an A block of polypropylene and to provide a "pad" of monomer for the steady-state B block copolymerization of ethylene and propylene. Polymerization is initiated by the injection of titanium trichloride catalyst. When sufficient propylene has been consumed (as indicated by a pressure drop of a predetermined amount), ethylene is added as rapidly as possible.

At this point, it is intended that the relative concentrations of ethylene and propylene be that required for preparation of a random copolymer containing 65 mole (55 weight) percent ethylene. The relative monomer concentrations were calculated according to the copolymer equation.[6] Reactivity ratios were determined independently and found to range generally from 12 to 15 and .06 to .14 for ethylene and propylene, respectively, when copolymerized with this catalyst and activator combination.[7] Variations in reactivity ratios have been attributed to batch-to-batch and source-to-source differences in the titanium trichloride catalyst.

After the requisite amount of ethylene has been charged, a mixed feed of ethylene and propylene containing 65 mole percent ethylene is fed at a rate sufficient to maintain a constant pressure until a B block of the desired size has been made. Monomer feeds are then discontinued and the reaction continues until the ethylene is essentially consumed; of course, significant amounts of propylene remain. Enough additional propylene is added to make the final A block of the desired size when the reaction is complete.

POLYMER CHARACTERIZATION

The copolymer products were characterized by a variety of methods. A sample differential thermal analysis trace is shown in Figure 10. The random copolymer glass transition can be observed at ~ -60°C. The broad endo-

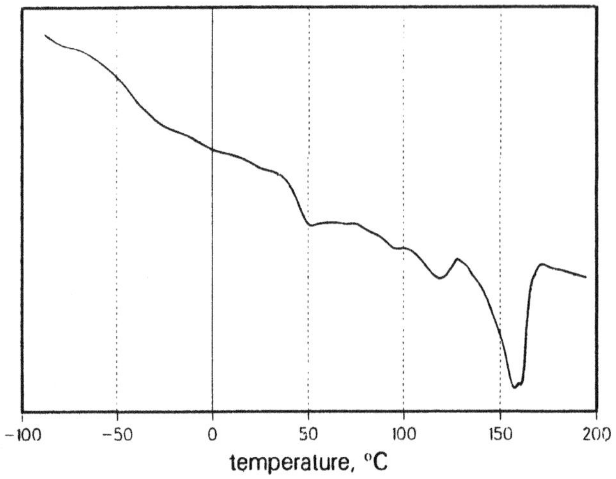

Fig. 10. ABA Block Copolymer Differential Thermal Analysis

therm is due to a low melting form of polypropylene, small amounts of crystalline polyethylene, and crystalline polypropylene. The low melting polypropylene and crystalline polyethylene are formed during the B block copolymerization and their relative heats and thus their relative concentration were correlated with the B block composition. Such compositions were also consistent with the x-ray diffraction patterns exhibited by these copolymers, such as the one shown in Figure 11. The relative heat of the

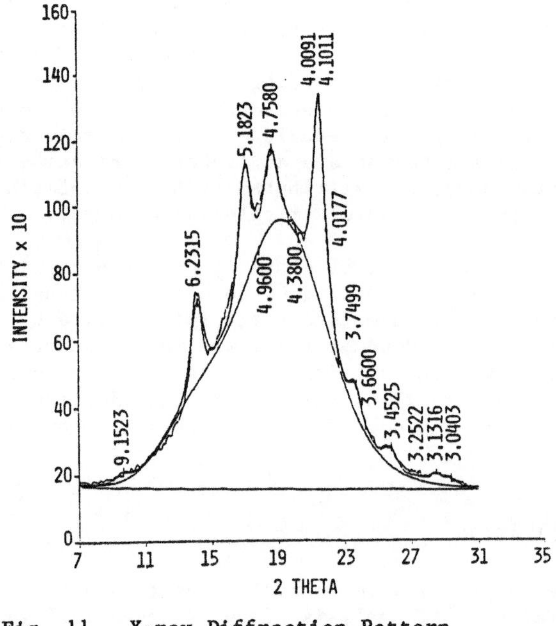

Fig. 11. X-ray Diffraction Pattern
19.3% PP, 2.9% PE, 77.8% Amorphous

crystalline polypropylene endotherm was found to correlate directly with the concentration of polypropylene measured by x-ray diffraction.

This catalyst system was also used to prepare random copolymers of ethylene and propylene at mileages equivalent to these used for the B block in the block copolymer. When such random copolymers were produced with 65 mole percent ethylene, minimum concentrations of low melting polypropylene and crystalline polyethylene were observed; no crystalline polypropylene was found. As a consequence, the observed concentration of crystalline polypropylene in the block copolymers was found to be a good indicator of A block size and molecular weight.

Molecular weights were calculated from intrinsic viscosity data. Mark-Houwink parameters varied according to the weight fraction of ethylene incorporated into the copolymer. These values were determined by the mathematical method of Ogawa and Inaba.[8] The total ethylene content of the copolymers, and thus the ethylene content of the B block, was determined by the infra red spectrum in the region from 775 to 675 reciprocal centimeters.[9]

Compositions, including actual block sizes, and molecular weights were consistently in agreement with material balance calculations and molecular weight predictions. Since such predictions are based on the number of previously determined concentration of active catalyst sites and the assumption that the number of growing chains remains constant throughout the polymerization, good agreement over a wide range of compositions and molecular weights is highly supportive of a living type of polymer chain growth.

After analyses of the types described above, polymer samples were milled on a two-roll mill and compression molded. Some fewer number of samples were injection molded as well. Elastomeric properties were measured primarily on compression molded samples. Rubber tensile properties were determined at room temperature and occasionally at 70°F. The dynamic tensile properties of complex modulus and loss angle were measured as a function of temperature. Hardness values are reported in Shore A units. Permanent sets were measured ten minutes after release from 100 percent strain. Compression sets were determined after the samples had undergone 22 hours of compression to 75 percent of their original thickness.

In order to examine the integrity of the individual blocks in the copolymer and the extent to which the A and B blocks are actually bonded, samples were fractionated in heptane. The procedure involves dissolution of 1.5 g of copolymer in 150 ml of heptane at 135°C followed by cooling slowly to room temperature and isolating the soluble and insoluble fractions. Earlier experiments with polyolefin blends demonstrated that this procedure results in fractionation on the basis of crystallinity rather than molecular weight.[10] If bonding does not exist between the crystalline A and the rubbery B phases, one would anticipate fractions of insoluble polypropylene and soluble EPR. The amount of insoluble material above that observed for crystalline polypropylene or polyethylene by DTA or x-ray methods should represent the amount of EPR actually bound to the insoluble A block. This hypothesis assumes a clean block copolymer of crystalline A and amorphous B.

In order to test the effectiveness of the fractionation method, an alloy of 15 percent polypropylene and 85 percent EPDM, i.e., non-bonded A and B phases, was prepared and fractionated. After correcting the insoluble fraction for the crystalline polyethylene that was present in the EPDM, fractions of 18 percent insoluble and 81 percent soluble material were found. A DSC thermogram showed the insoluble fraction to be essentially polypropylene. Thus, the method appears to be effective in separat-

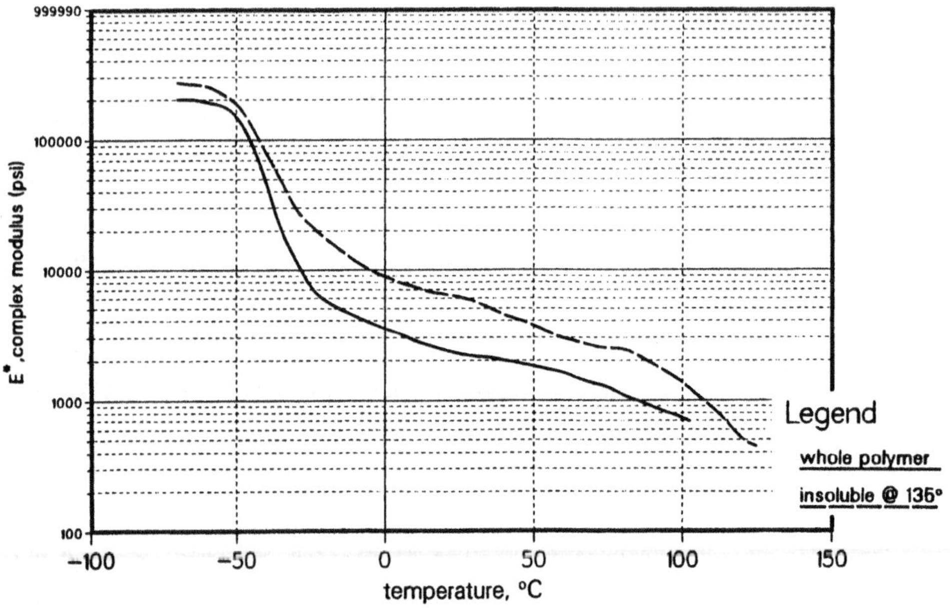

Fig. 12. ABA Block Copolymer
Complex Modulus vs Temperature

ing non-bonded phases of polypropylene and EPR.

The extent of block character of an ABA copolymer containing 14 weight percent A block and 86 weight percent B block, as measured by DTA, was examined by this method. A total of 71 weight percent of the copolymer was found to be insoluble, which implies that 71 - 14 = 57 percent of the material is B blocks and that 57/86 = 66 percent of the B block chains are chemically bonded to A block chains. The dynamic tensile properties of the insoluble fraction (Figures 12 and 13) indicate that the B segment glass transition and loss angle peak are still clearly present in the insoluble fraction. However, a higher modulus across the temperature range is observed as would be expected if unblocked B chains were extracted. Additional characterization by DSC showed that all of the crystalline phase was present in the insoluble fraction, and the intrinsic viscosity of both fractions was essentially equivalent to that of the whole polymer. These observations imply that the majority of the B chains are bonded to A segments; however, there are no data which truly distinguish between ABA triblock and AB diblock sequences.

The absence of a perfectly flat plateau in the rubbery region of the dynamic mechanical spectra, and the presence of an albeit small but soluble fraction in 135°C heptane indicate that the copolymer is not perfectly blocked and/or that the phases are not perfectly separated. Such conclusions are consistent with the molecular weight distributions of 4 to 5. The complex modulus plot of a commercial styrene-butadiene block copolymer is illustrated in Figure 14 for comparison to the ethylene-propylene block copolymers reported here. A plot for a commercial blend of polypropylene and crosslinked EPDM is shown as well. The flatter rubbery plateau of the styrene-butadiene copolymer is consistent with the narrower molecular weight distribution characteristic of the ionic, living polymerization used for its production. Furthermore, one might expect a greater degree of phase separation in a system where neither monomer is common to both

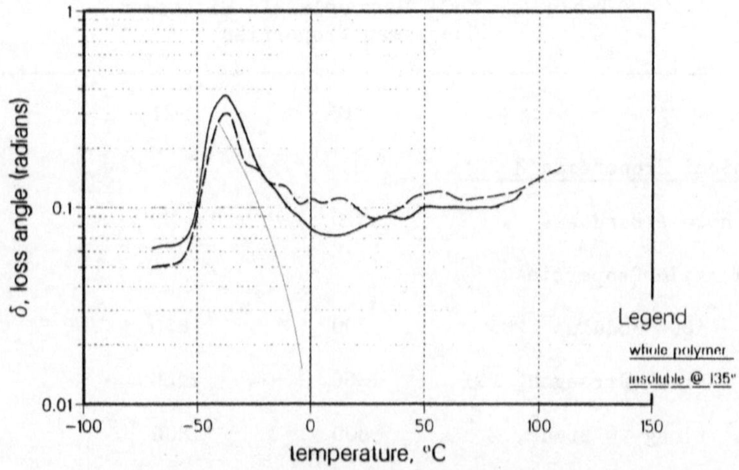

Fig. 13. ABA Block Copolymer
Loss Angle vs Temperature

phases. The ethylene-propylene block copolymers compare quite favorably with the polypropylene-EPDM blend and have done so without the need for crosslinking.

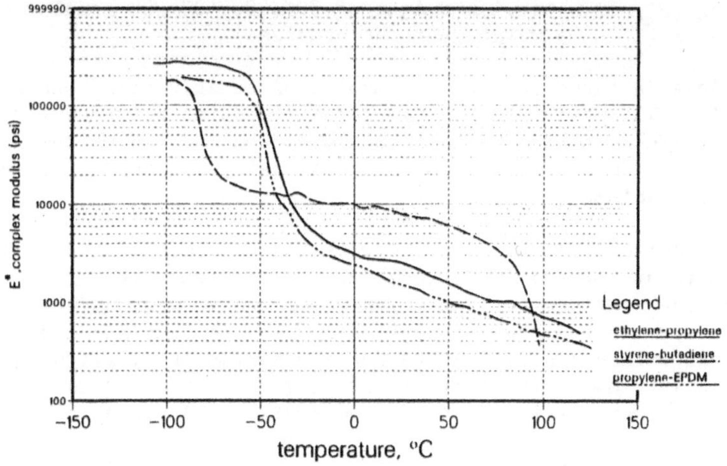

Fig. 14. ABA Block Copolymer
Complex Modulus vs Temperature

Table 1. C_2C_3 Thermoplastic Elastomer Copolymer Properties

Wgt % A	12	21	29
Mechanical Properties @ 23°			
Shore A Hardness	70	80	90
Tensile Properties			
100% Modulus, PSI	300	650	1000
Break Strength, PSI	1800	2200	2800
Elong. @ Break, %	800	800	700
Delayed Permanent Set, %,			
After 100% Extension	12	15	20
Compression Set, %	-	25	-
Tear Strength, lb/in	165	200	-
Hysteresis, % (1st, 3rd)		53, 28	

POLYMER PROPERTIES

The optimum composition of these copolymers, in terms of their physical properties, has been found to be within the following ranges:

1) A blocks: 6 to 14 weight percent of the total
intrinsic viscosity of 0.4 to 1.1 dl/g
molecular weight from 32,000-96,000 g/mole

2) B block: 71 to 88 weight percent of the total
intrinsic viscosity of 3.3 dl/g
molecular weight of 300,000-400,000 g/mole
55 weight percent ethylene

Materials in these ranges of composition exhibit stiff but processible melts, similar in appearance to commercial blends of polypropylene and lightly crosslinked EPDM. Melt viscosity can be reduced by the addition of extending oils. Some orientation is occasionally observed after processing by injection molding.

The mechanical properties of compression moldings of these thermoplastic elastomers are listed in Tables 1 and 2. Moderate strengths and high elongations to break are evident. The rubber tensile properties of these copolymers vary, as one might expect, according to specific composition,. but a typical stress - strain curve for a material of 21 weight percent A block and an intrinsic viscosity of 4.0 dl/g is shown in Figure 15. The behavior is characteristic of an elastomer with good strength.

Increasing the temperature causes a decline in mechanical properties. The decrease is probably related to the small amounts of polyethylene and low melting polypropylene made during B block copolymerization which begin

Table 2. C_2C_3 Thermoplastic Elastomer Copolymer Properties

Wgt % A	12	21	29
Mechanical Properties @ 70°			
Shore A Hardness	57	64	82
Delayed Permanent Set, %			
After 100%, Extension	16	20	25
Compression Set, %	-	55	-
Tear Strength, lb/in		60	

to reduce the strength of the hard phase at elevated temperatures.

A hardness range of 70 to 90 Shore A is readily achieved. A lower hardness can be obtained by oil extension. High hardness has been achieved by using reinforcing fillers.

Hysteresis is fair to good and seems to depend heavily on the block character of an individual sample.

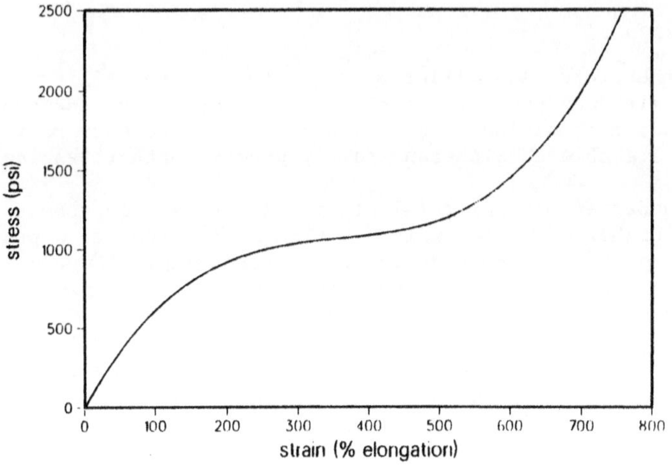

Fig. 15. ABA Block Copolymer Rubber Tensile Propertie
Stress vs Strain (23°C)

The creep behavior of an ethylene-propylene thermoplastic elastomer of high block character is shown in Figure 16. Resistance to creep is better than that for the styrene-butadiene block copolymer and polypropylene-EPDM blend thermoplastic elastomers tested.

The solvent resistance of these thermoplastic elastomers is quite good compared to the styrenic based thermoplastic elastomers and better in some solvents than polypropylene-EPDM blends.[11] Excellent resistance was observed in ASTM #3 oil, isooctane, acetone, tetrachloroethylene, dioctyl phthalate, 30 percent aqueous sodium hydroxide and 20 percent aqueous sulfuric acid. This behavior is probably a reflection of the crystalline block in the ethylene-propylene block copolymers.

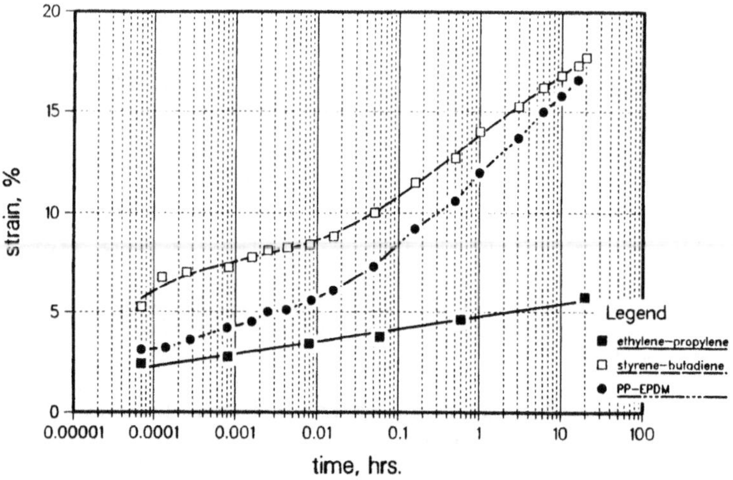

Fig. 16. ABA Block Copolymer Creep at 23°C, 50 psi

Given the proper stabilizer package, the weatherability should be quite good, although specific testing of thermoplastic elastomers in the range of these compositions has not been tested. However most stabilized polyolefins are able to withstand fairly severe weather testing conditions.

The boundaries of commercial utility for these elastomers are shown in Figure 17. Higher molecular weight B blocks will lead to a product with a melt viscosity that is too high for good processing. Lower molecular weight A blocks will reduce the concentration of crosslinks and a deleterious effect on product tensile properties will result. Reducing the molecular weight of the B block and increasing the molecular weight of the A block will eventually lead to a phase inversion; indeed, such a material may, at some point, function as a high impact plastic with rubbery domains dispersed in and bonded to a continuous phase of crystalline polypropylene.

CONCLUSION

Block copolymers of the type propylene-b-(ethylene-co-propylene)-b-propylene have been prepared in which relatively small, crystalline polypropylene blocks serve as physically crosslinked domains in a rubbery EPR matrix. A common Ziegler-Natta catalyst, titanium trichloride, when activated by a novel co-catalyst, dimethylbis(alkylcyclopentadienyl)titanium (IV), permits polymerizations with low rates of chain transfer and long chain lifetimes. Thus polymer chains with a high degree of block character can be produced in semi-batch copolymerizations by control of the monomer

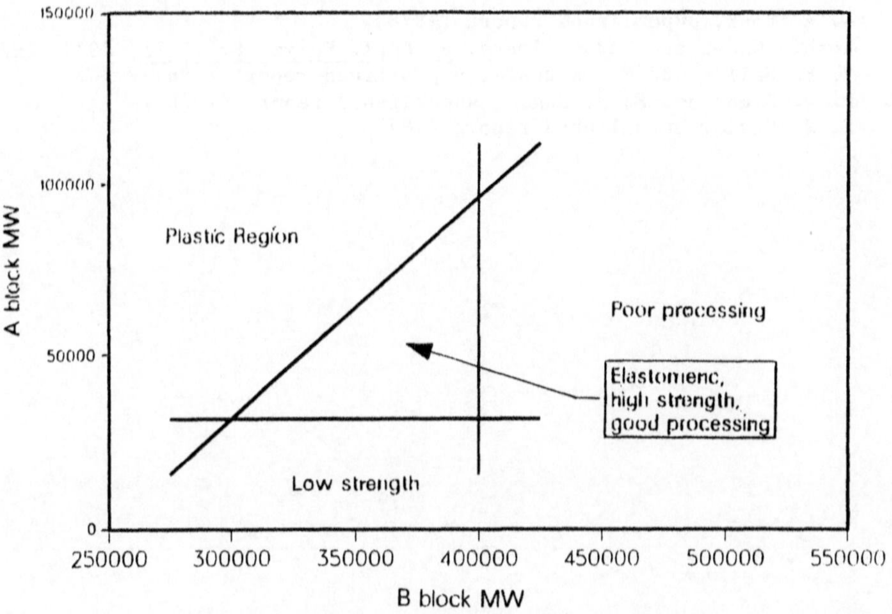

Fig. 17. Effect of Block MW on Properties

feed. The products exhibit high strength, low set, thermal reversibility, and excellent solvent resistance over a hardness range of 70 to 90 Shore A. The specific combination of properties can be controlled by the relative size of the copolymer blocks.

The relatively high degree of block character apparent from the fractionation experiments, the competitive behavior of the complex modulus with increasing temperature, the ability to achieve such behavior without chemically crosslinking, the use of melt processible recyclable crosslinks and the expected ability to resist attack by many solvents which are deleterious to the properties of styrenic based copolymers are encouraging for the utility of these block copolymers.

ACKNOWLEDGEMENTS

The author is indebted to H. W. Blunt for development of much of the process, suggestions, and for many years of patient guidance; E. J. Engle for much of the characterization and property work; and W. P. Long for contributions to development of the activator and methods of kinetic analysis.

REFERENCES

1. J. Boor, Jr., Ziegler-Natta Catalysts and Polymerizations, p. 501-ff, Academic Press (1979) and references therein.
2. E. H. deButts, U. S. Patent 2,992,212 (1961).
3. G. A. Lock, unpublished report (1976).
4. W. P. Long, unpublished report (1975).
5. R. C. Heaton and R. E. Stapleford, unpublished report (1978).
6. P. J. Flory, Principles of Polymer Chemistry, pp. 179-ff, Cornell University Press (1953).

7. G. A. Lock, unpublished report (1978).
8. Toshio Ogawa and Tadami Inaba, J. Appl. Polym. Sci., 21, 2979 (1977).
9. J. E. Gailey and M. Margosis, unpublished report (1964).
10. H. W. Blunt and E. J. Engle, unpublished report (1976).
11. E. J. Engle, unpublished report (1977).

NEW EP ELASTOMERS

G. Foschini and F. Milani

Centro Ricerche G. Natta
44100 Ferrara, Italy

INTRODUCTION

EPM elastomers, discovered around the middle of the 1950's by Natta and co-workers, quickly became a subject of research in many laboratories throughout the world. Thus, the research efforts were started, directed towards ever improved knowledge of the properties of these products, the development of more effective catalyst systems suitable for better performance and in improvement in the production processes.[1] This report deals with the possibility of producing elastomeric ethylene-propylene copolymers with special high yield catalysts:

ZIEGLER-NATTA HIGH YIELD CATALYST PERFORMANCE

The titanium salt-based high yield catalysts are employed to produce thermoplastic poly-alpha-olefins and the common opinion still prevails that these catalysts are not useful for the synthesis of elastomeric EP copolymers. Nevertheless, the research efforts at ME/DUTRAL S.p.A. have identified catalysts capable of yielding amorphous ethylene-propylene copolymers. The elastomers produced are similar to, and better than those produced with vanadium-based catalysts.

Three families of titanium catalysts have been studied which will be called A, B and C. The copolymers obtained with these catalysts, as demonstrated by ^{13}C-NMR analysis, are free from inversions and have $r_1 \times r_2 \gg 1$ for type A and B catalysts and $r_1 \times r_2 \gtrsim 1$ for the type C catalyst, whereas the vanadium-based catalyst contains a considerable amount of inversions (about 15%) and has $r_1 \times r_2 < 1$. This proves that there is an increasing reduction of propylenic and ethylenic type sequence blocking from catalyst A, B, C to V, respectively and consequently the degree of crystallinity changes according to the type of supported catalyst used (Figure 1). The most interesting catalysts are B and C. Figure 2 reports molecular weight distribution (MWD) of products obtained with these two catalysts in comparison with a similar product obtained with a vanadium catalyst: all high yield products have broader MWD.

NEW SIMPLIFIED HIGH YIELD PROCESS

Most EPM/EPDM elastomer manufacturers are currently employing solution

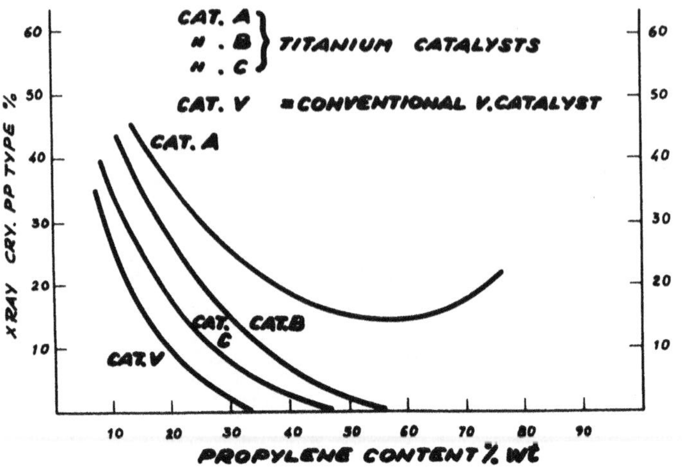

Fig. 1. Polyethylene or Polypropylene Crystallinity Content (From X-rays Analysis)

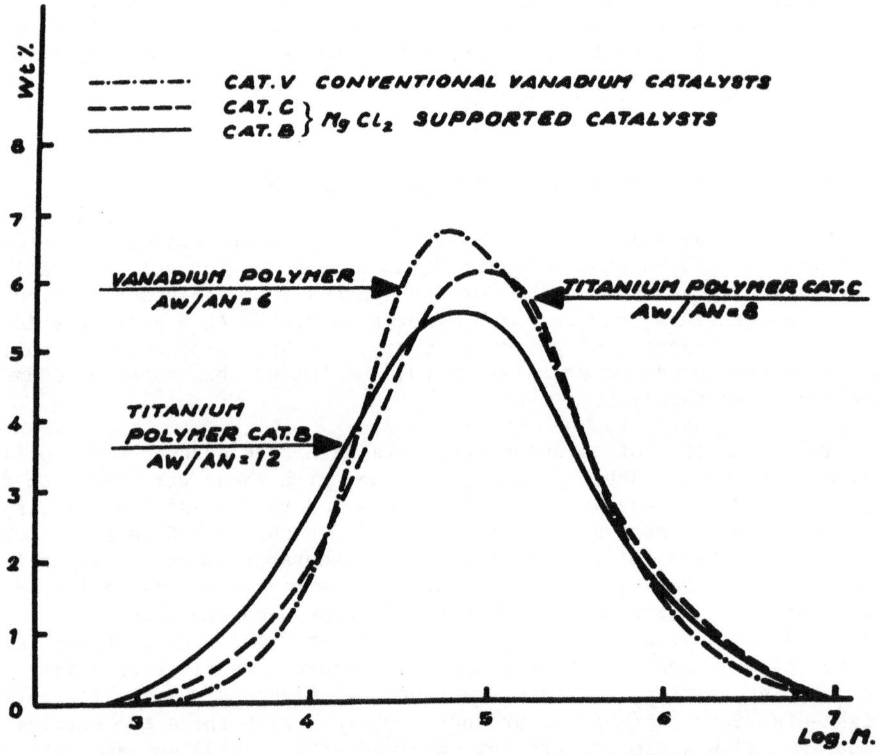

Fig. 2. Molecular Weight Distribution

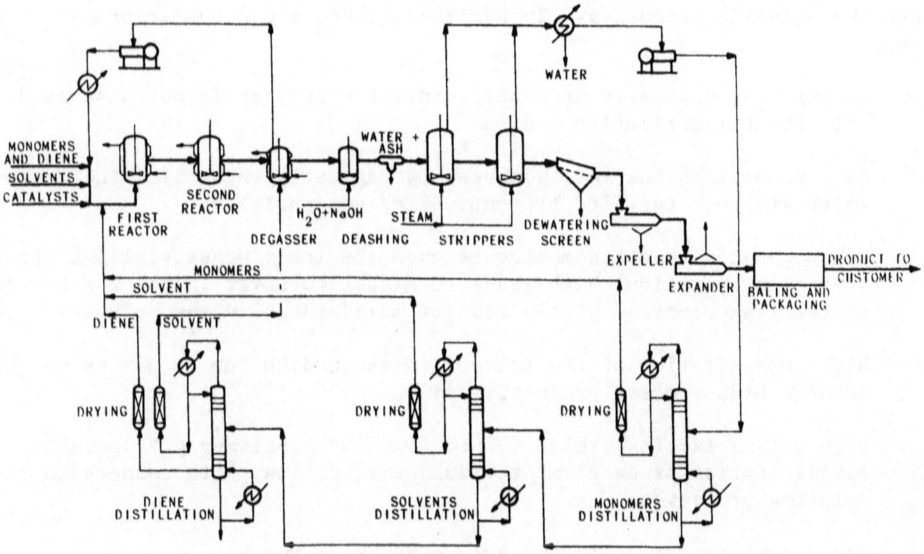

Fig. 3. Solution Process

processes. The reaction medium, in such processes, is a solvent to which ethylene, propylene and the catalyst are added. The end-product is then separated from the solution. The traditional flow sheet of this process is shown in Figure 3. ME/DUTRAL S.p.A. has, from the beginning, preferred a liquid propylene reaction medium to which ethylene and the catalyst are added. The polymer is produced as fine particles suspended in the liquid mixture. The flow sheet in Figure 4 shows how this process is much simpler

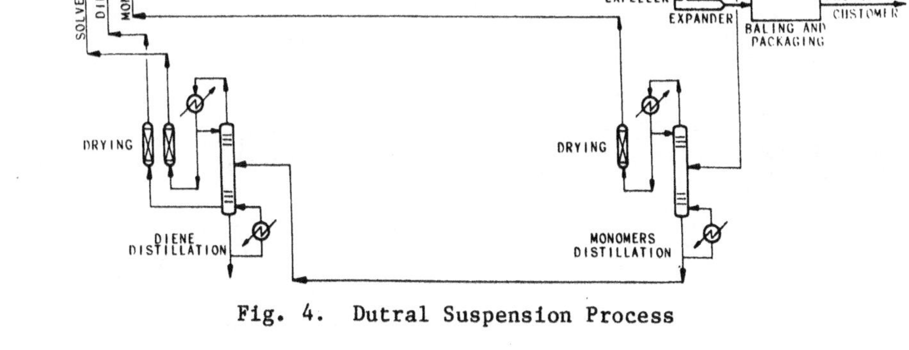

Fig. 4. Dutral Suspension Process

than the solution processes. In addition, there are a number of advantages:

a) no stirring viscosity problems, since the polymer is not dissolved into the polymerization medium;

b) easy removal of the reaction heat by partially vaporizing the monomers which are then recycled by compression-condensation;

c) a homogeneous reaction medium because the low viscosity allows for an intense evaporation which leads to a fast turnover in the reactor and an excellent control of the monomer molar ratio in the polymer;

d) high concentration of the polymer in the medium (up to 35% by weight) means a high process productivity;

e) high polymerization yields (20 to 50 x 10^3 g.polymer/g.V) permit simplification of catalyst residue purification, with respect to the solution processes;

f) the capability of producing very high MW polymers;

g) lower investment and production costs.

By employing high yield catalysts, DUTRAL S.p.A. has obtained dramatic simplification of its own suspension process. In fact, polymers can be obtained which do not need the usual stripping and drying operations which are indispensable in the current DUTRAL S.p.A. vanadium catalyst process. Since the catalyst polymerization yields (Figure 6) are about one percent higher than those obtained using vanadium catalysts, it follows that the ash and chlorine are about one percent lower.

One way of representing the new simplified process is illustrated in Figure 5.[2] The precess steps are:

1) suspension polymerization in liquid monomers or monomer mixtures with aliphatic hydrocarbons with a boiling point -10°C lower than (propane, butane, etc.);

2) flash off of the reaction medium with a consequent evaporation of monomers and diluents;

3) recovery of copolymer elastomers in powder, flakes or bales.

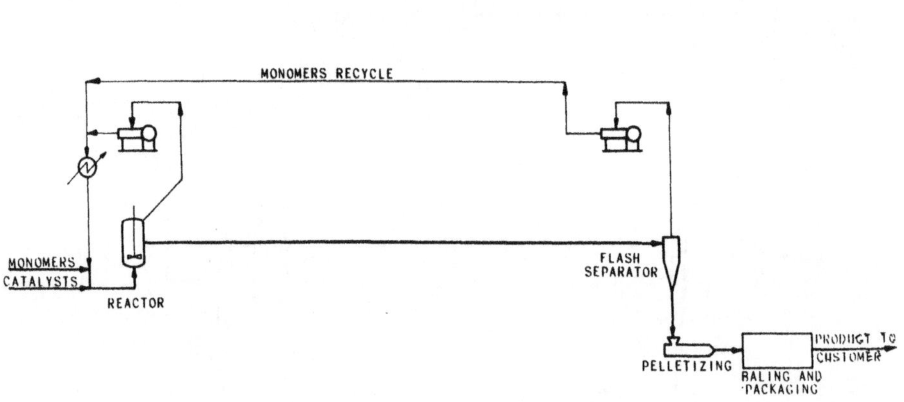

Fig. 5. New Dutral Suspension Process

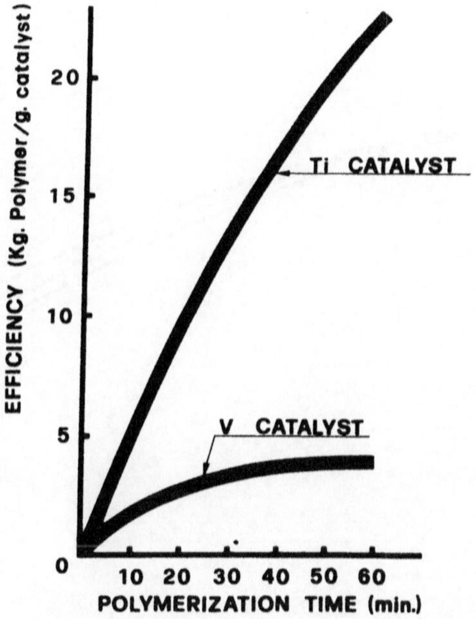

Fig. 6. Polymerization in Liquid Monomers Ethylene-Propylene Molar Ratio 10/100, 12 Bar, 20°C

Thanks to the remarkable simplification of the process, schematically involving polymerization and finishing alone, the only substances remaining in the finished polymer are the catalyst residues, basically: aluminum, titanium and magnesium compounds. Once the active catalyst has been killed, these compounds consist mainly of oxides. Again, since the catalyst polymerization yields (Figure 6) are about one order of magnitude higher than those obtained using vanadium acetylacetone, residual titanium and magnesium are about one tenth lower than the vanadium. If one then considers that, in place of diethylaluminumchloride, an aluminumtrialkyl is used, in a much lower concentration (ranging from about 1 to 0.3 g per liter), it is clear that the chlorine content in the finished product is also drastically reduced. In fact, the ash contents decrease to 200-300 ppm and the chlorine content to 10-20 ppm. Finally, if one considers that, apart from perhaps atmospheric humidity, the polymers never come into contact with, nor contain water, it is easy to deduce that, in addition to their chemical purity, transformation of this new generation of elastomeric copolymers is corrosion free.

POLYMER FEATURES

A. Rheology. Figure 7 shows the behavior of melt viscosity as a function of the shear rate. It can be seen that high yield products have a greater pseudoplasticity, probably due to their broader MWD. Figure 8 shows the behavior of the newtonian viscosity, determined with a Weissemberg rheogoniometer,[3] as a function of the average viscosimetric molecular weight (Mv).[4] Log η_o/log Mv correlations for the three families of products are practically the same. The broader MWD of this new class of elastomers in respect to the corresponding vanadium elastomers leads to the following advantages:

- absence of cold flow even at temperatures considerably higher than normal, so that products having very low mooney viscosities

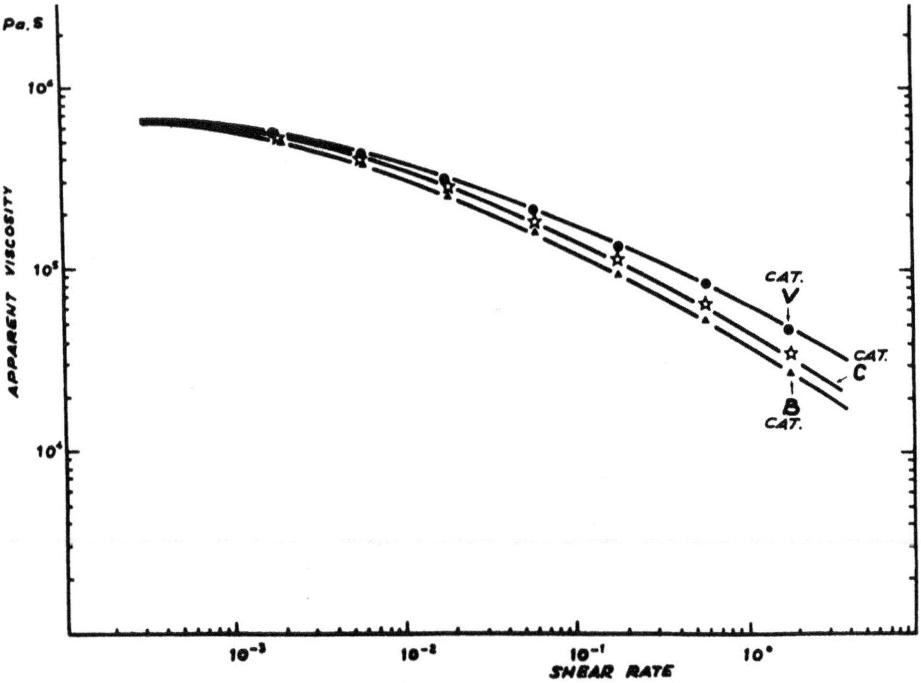

Fig. 7. Apparent Viscosity versus Shear Rate. T=150°C

(viscosity of ML(1+4)100°C = 10 can be reached) with an intrinsic viscosity of approximately 1 can easily be produced and handled;

- greater processability in roller mixers.

B. Mechanical Characteristics. The mechanical characteristics of two non-vulcanized copolymers with different molecular weights can be seen in the stress-strain diagrams of Figures 9 and 10. When these curves are similar to those of cross-linked compounds and the polymer is essentially amorphous, it is our opinion that the trend of the curve is due both to the broader MWD and the greater monomer blockiness in the macromolecules. In this respect, it is worth noting the characteristics of those polymers with the highest green strength, i.e. with the lowest propylene content. Figure 11 points out that the stress-strain curve of titanium copolymers is more influenced than that of vanadium copolymers since the former has a more pronounced ethylenic-type blockiness. As a matter of fact, at propylene contents of about 35% by wt., the titanium copolymers have much higher green strength values than those of vanadium copolymers.

C. Mechanical Characteristics of Vulcanizates. The mechanical characteristics of the peroxide-vulcanized titanium elastomers are generally similar to those of the traditional copolymers. Figure 12 shows that the tensile strength increases with a decrease in the propylene content at the same molecular weight. This curve fits well with the data of vanadium polymers (DUTRAL commercial grade C0059 and C0038 having an average propylene content of 45% and 30% respectively). On the other hand, the tension set curves of titanium vulcanized copolymers differ slightly toward the high ethylene contents even though they approach the elastic behavior of vanadium elastomers at high propylene contents (see Figure 13).

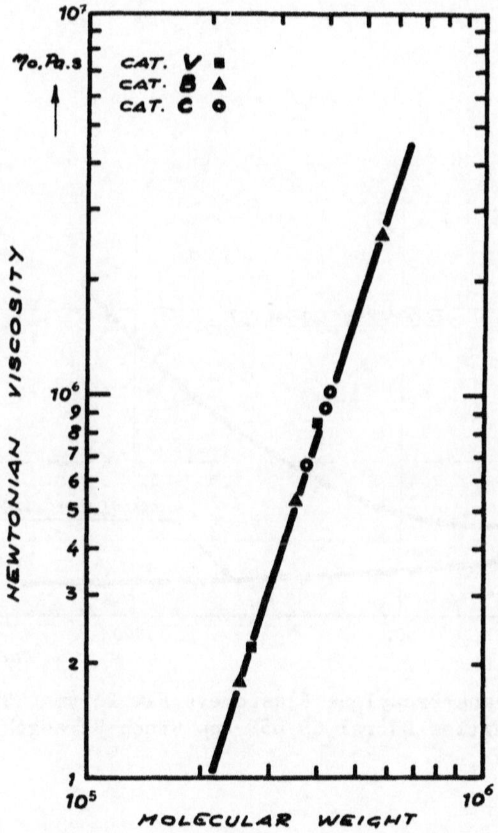

Fig. 8. Newtonian Viscosity at 150°C as a Function of Molecular Weight
Log. η o = 12,9 + 3,36 Log. $\bar{M}v$

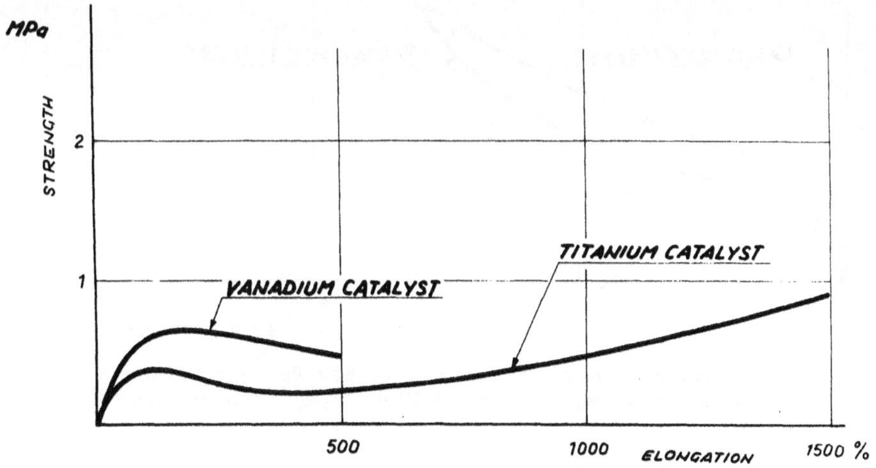

Fig. 9. Ethylene-Propylene Elastomers Raw Polymer Stress-Strain Properties Dutral CO 054-Low Green Strength Grade

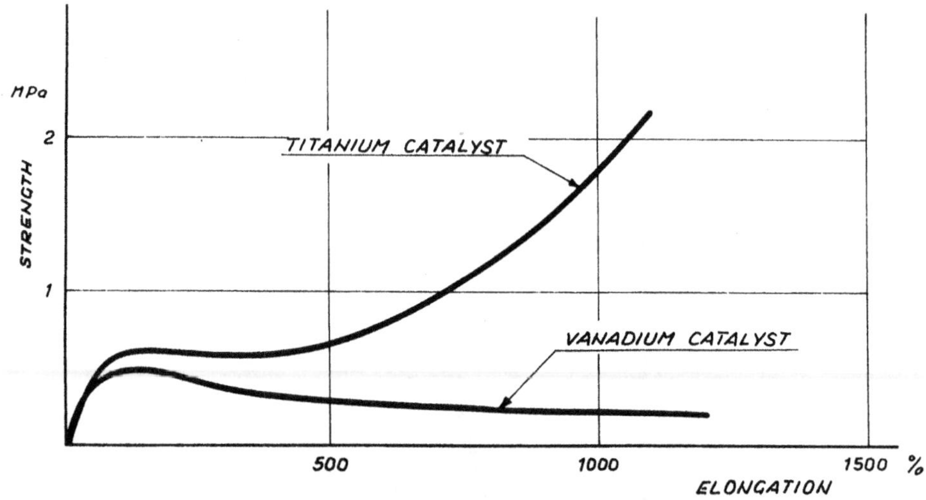

Fig. 10. Ethylene-Propylene Elastomers Raw Polymer Stress-Strain Properties Dutral CO 059-Low Green Strength Grade

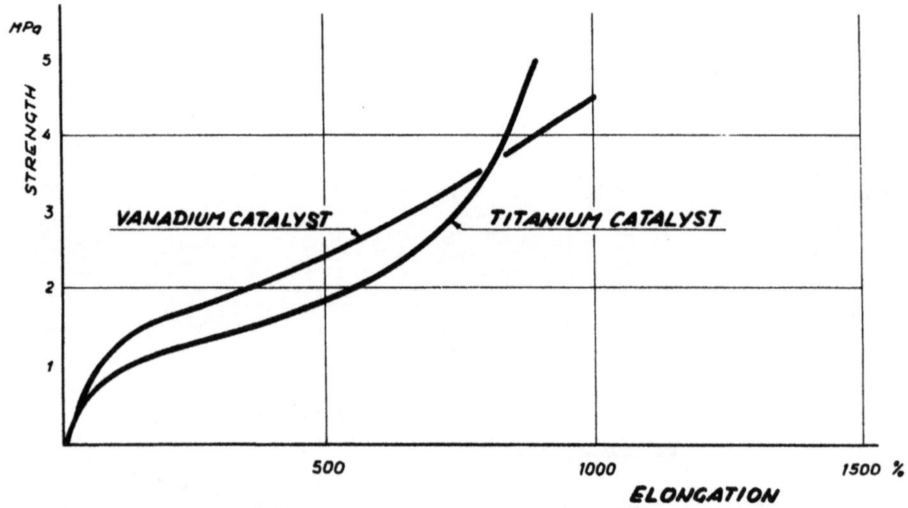

Fig. 11. Ethylene-Propylene Elastomers Raw Polymer Stress-Strain Properties Dutral CO 038-High Green Strength Grade

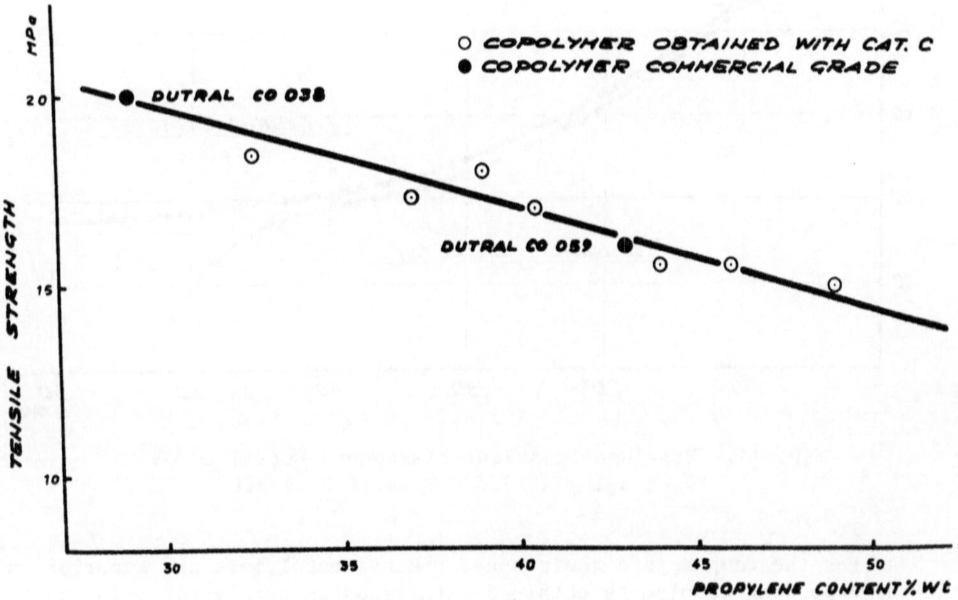

Fig. 12. Cured Compound Tensile Strength

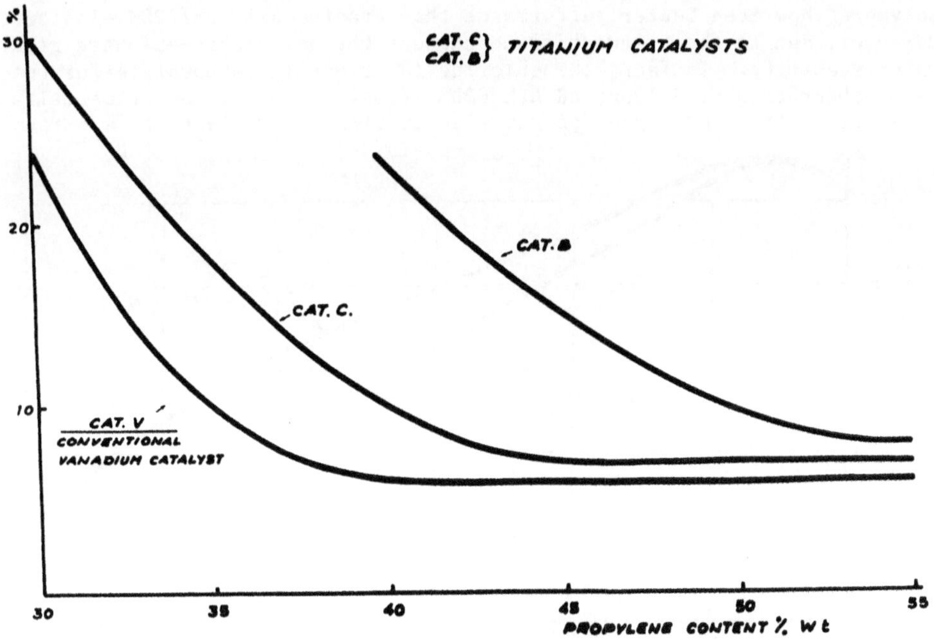

Fig. 13: Cured Compound-Room Temperature Tension Set; 200% Elongation For 10 min Reading 10 min After Release

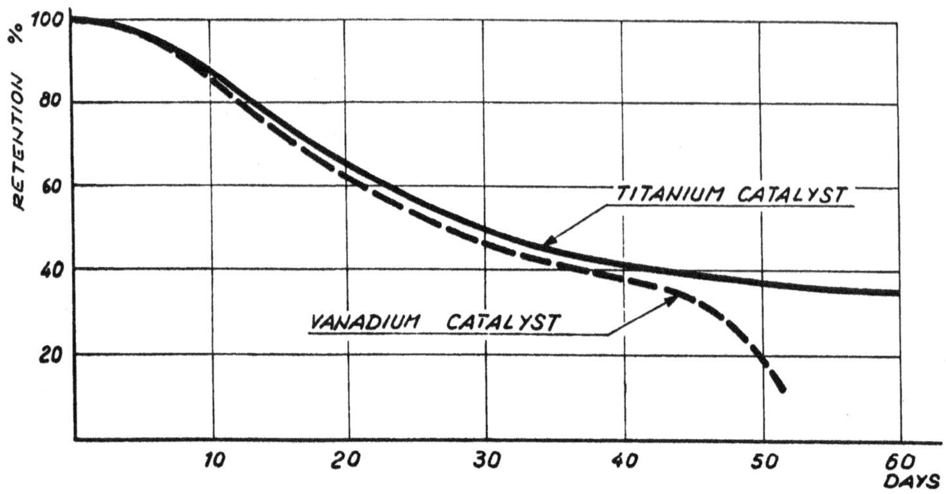

Fig. 14. Ethylene-Propylene Elastomers Dutral CO 038
Oven Aging at 150°C-Tensile Strength

As for the temperature resistance, the new copolymers are superior to the corresponding copolymers obtained with vanadium catalysts, owing to the high chemical purity of the titanium elastomers. This is shown by Figures 14 and 15 where the variations in tensile strength and elongation at break at 150°C are reported. The formulation used is given in Figure 16.

D. Dielectric Properties. Dielectric properties of titanium copolymers in comparison with the corresponding vanadium copolymers are listed in Figure 17. As might be expected, due to their high purity, the first copolymers show even better performance than traditional EPM/EPDM elastomers. Moreover, due to their low chlorine content the new elastomers have greater water stability. In fact, the chlorine is primarily responsible for the water absorption of vulcanized EPM/EPDM items. As proved experimentally (see Figure 18), the degree of water absorption in an elastomer was found

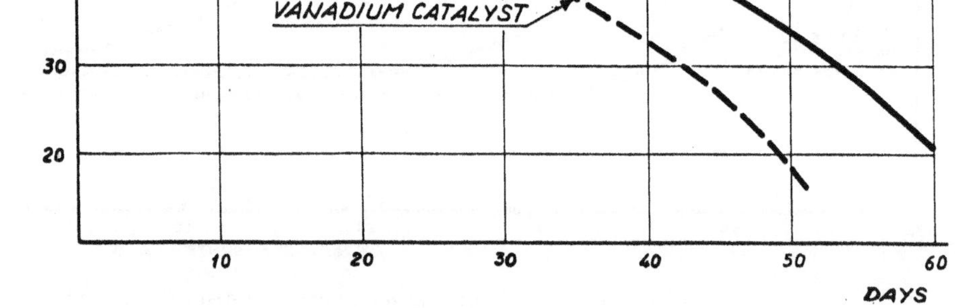

Fig. 15. Ethylene-Propylene Elastomers Dutral CO 038
Oven Aging at 150°C-Elongation at Break

POLYMER		100
ZINC		5
STEARIC ACID		1
FEF BLACK		80
PARAFFINIC OIL		30
MB		2
ANOX		0.5
SULPHUR		0.3
PEROXIMON F.40		4.5
CURING AT 165° FOR 40'		
MECHANICAL PROPERTIES OF CURED COMPOUND	TITANIUM CATALYST	VANADIUM CATALYST
TENSILE STRENGTH MPa	16.2	16.3
ELOGATION AT BREAK %	420	400

Fig. 16. Ethylene-Propylene Elastomers - Dutral CO 038 Pormulation for Aging Test

	TITANIUM CAYALYST	VANADIUM CATALYST
DIELECTRIC CONSTANT (1000 Hz)	1.8 + 1.9	1.9 + 2.0
DIELECTRIC BREAKDOWN KV/mm	25 + 29	24 + 28
VOLUME RESISTIVITY Ω x cm	$1.5+1.7 \times 10^{17}$	$1.3+1.6 \times 10^{17}$

Fig. 17. Ethylene-Propylene Elastomers Dielectric Characteristics

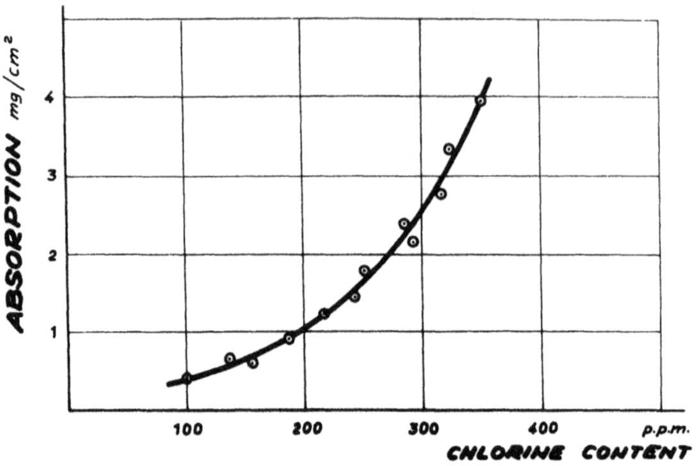

Fig. 18. Ethylene-Propylene Elastomers Water Absorption of Crosslinked Compound After Immersion in Water at 100°C

to be exponential, in respect to the amount of chlorine content. Finally, Figure 19 shows the water absorption, as a function of time, for both titanium and vanadium copolymers.

POSSIBLE DEVELOPMENTS

One can foresee interesting process and product developments with these new titanium-based high yield heterogeneous catalysts. In fact, the use of suitable catalysts makes it possible to exert great control over molecular structure (ethylenic and propylenic sequence length and composition) as well as the size and shape of polymer particles (the well-known replica factor). Thus the following can be achieved:

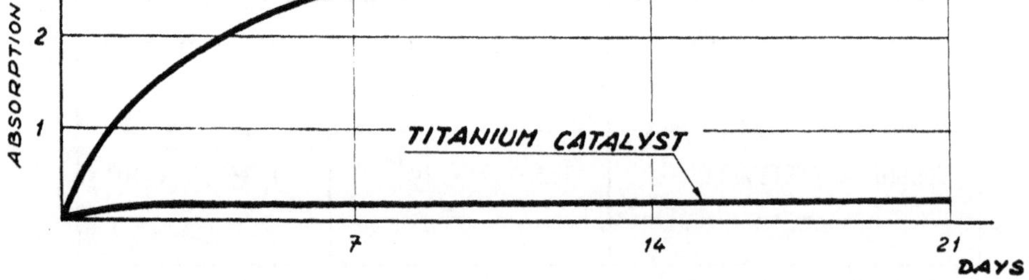

Fig. 19. Ethylene-Propylene Elastomers Water Absorption of Pure Crosslinked Compound After Immersion in Water at 100°C

- even higher reaction polymer concentrations in suspension processes without substantial polymer build up within the reactor;

- modification of certain product characteristics by introducing appropriate α-olefins (i.e. butene, hexene) in the chain without any significant fall in polymerization yield;

- improvement in copolymer adhesion and/or compatability with polar polymeric matrixes by grafting with polar monomers after synthesis.

CONCLUSIONS

The most recent important development in the field of Ziegler-Natta catalysts is the identification of high yield catalysts capable of synthesizing elastomeric products. The new $MgCl_2$-based catalysts are capable of doing a job which, up to now, was considered impossible. Furthermore, due to the characteristics of these catalysts, their marriage with the suspension process is a very convenient goal. The new products show some structural differences with respect to EP-elastomers produced by vanadium catalysts. Nevertheless, they have a range of properties very similar to the latter which allow them to successfully replace their vanadium counterparts in all fields currently employing EP-elastomers.

REFERENCES

1. R. H. Schwarr and C. H. Chien, Ethylene-Propylene CO and TER polymer Rubber, Report n.4B, Stanford Research Institute, Menlo Park, Calif. (1981).

2. P. Galli, F. Milani, and R. Simonazzi, Polymer J., 17, N. 1, 37 (1985).

3. J. R. Van Wazer, J. W. Lyons, K. Y. Kim, and R. E. Colweel, "Viscosity and flow measurement," Interscience Publishers, John Wiley & Sons, Inc., New York (1966).

4. F. P. Baldwin and G. Ver Strate, Rubber Chem. and Tech., 45, 709 (1977).

HYDROGENATED STAR BLOCK COPOLYMERS OF BUTYLSTYRENE - ISOPRENE AND DIVINYLBENZENE

J. M. Hoover, T. C. Ward, and J. E. McGrath

Polymer Materials and Interfaces Laboratory
Chemistry Department
Virginia Polytechnic Institute and State University
Blacksburg, VA 24061

INTRODUCTION

Anionically prepared triblock and star block copolymers of styrene with butadiene and isoprene have been widely synthesized and thoroughly studied by many investigators.[1] These systems have also provided a basis for many commercially important thermoplastic elastomers, including KRATON, SOLPRENE, etc. The thermal and mechanical properties of these materials are well established in their microphase-separated morphologies, which may be tailored, along with other polymer parameters, to achieve specific physical properties. Anionic polymerization allows the controlled synthesis of block polymers, and thus facilitates the precise engineering of mechanical properties, such as tensile strength and modulus by controlling parameters, such as block molecular weight, purity, and microstructure.

Relatively little attention[2,3] has been given to ring substituted styrenic monomers such as paramethylstyrene (PMS) and tertiary-butylstyrene (TBS).

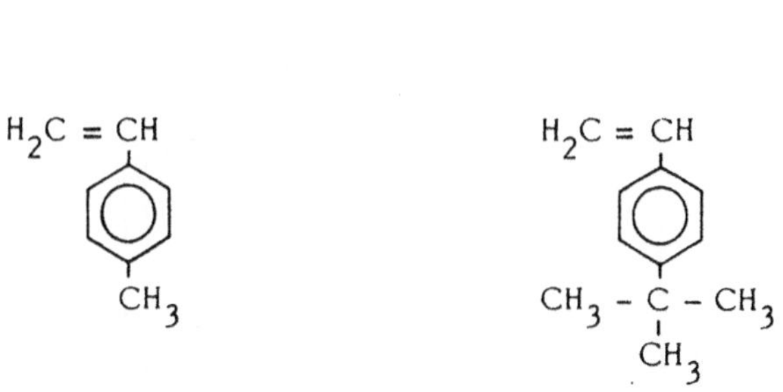

PMS

TBS

Figure 1. Chemical structures of paramethylstyrene (PMS) and tertiary butylstyrene (TBS)

These monomers may, in some cases provide the opportunity to improve the high temperature performance of conventional S-B-S and S-I-S systems. At the same time, other new systems permit fundamental studies on the role of microphase separation in the structure - property relations of well defined, thermoplastic, block copolymers. Some investigations of solution behavior have already been reported.[4] It has been previously demonstrated[3] that the high miscibility of triblock copolymers of poly(TBS) with polydienes limits the degree of phase separation that can be conventionally achieved in these block copolymers, and thus influences their physical properties. Since polyolefins are believed the exhibit lower solubility parameters than polydienes, we investigated hydrogenation as a method of altering the degree of miscibility. Fetters and coworkers[3] briefly noted that hydrogenated sample showed improved mechanical strength.

Catalytic hydrogenation has been employed as a tool for enhancing microphase separation in a variety of poly(TBS) - polyisoprene block copolymers. A description of the synthesis and catalytic hydrogenation of a poly(TBS) - polyisoprene star block elastomer follows.

EXPERIMENTAL

Radial or star shaped block copolymers, schematically represented in Figure 2b, containing various proportions of "hard" (glass - G) to "soft" (rubber - R) phases, were synthesized under nitrogen pressure and hydrogenated in modified Chemco[5] glass bowl/stainless steel reactors as follows.

Cyclohexane (99^+ mol %) obtained from Phillips Petroleum Company was used as the polymerization solvent. It was stored under a dry nitrogen at-

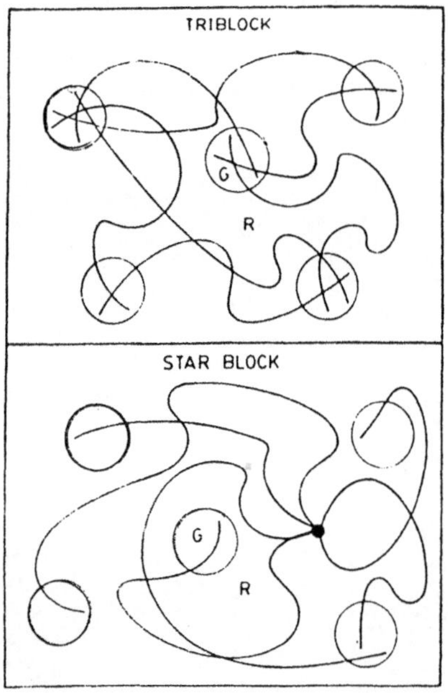

Fig. 2. Schematic Representation of (a) Triblock, and (b) Star Block Copolymer Topology

mosphere and purified by transfer through steel columns containing molecular sieves. A solvent charge of 436 ml (340 g) was measured in a calibrated holding tank and transferred to a 700 ml reactor. The solvent was heated to 050°C while stirring under a dry N_2 atmosphere of 10 psig.

Tertiary-butylstyrene (TBS) (95% para isomer), provided by the Dow Chemical Company, was purified by exposure to activated alumina, and stored at 5°C under a dry N_2 atmosphere. A charge of 18.2 ml (16 g) TBS was measured with a 20 ml syringe and transferred to the reactor.

Secondary butyllithium (1.37 M in hexane) provided by Lithium Corporation was added dropwise to the reactor via a 1 ml syringe. After the addition of 0.07 ml, a pale yellow color was observed to persist and an initiator charge of 0.47 ml (0.64 mmol) was added. The orange-colored polymerization was allowed to proceed for 60 min. before a 30 ml sample of "living" poly(TBS) was removed from the reactor and terminated. GPC characterization, using polystyrene standards, indicated a peak molecular weight of 22,700 g/mol, with a polydispersity of 1.15.

Isoprene, provided by ARCO Chemical Company, was purified by transfer through steel columns, containing activated alumina and molecular sieves, followed by distillation from dibutylmagnesium. A charge of 66 ml (45 g) was measured and transferred to the reactor via syringe. The resulting colorless polymerization was allowed to proceed for 2 hr. at 50°C before a 30 ml sample was removed and terminated. GPC analysis of this diblock copolymer indicated a peak molecular weight of 120,000 g/mol with a polydispersity of 1.10, based on polystyrene standards.

Practical grade divinylbenzene (DVB) was purified by stirring with powdered calcium hydride, followed by exposure to neutral alumina and vacuum distillation from dibutylmagnesium. Purified DVB was diluted with purified hexane to obtain a 0.72 M working solution. Next, 5.4 ml 0.72 M DVB (3.84 mmol) was syringed into the reactor, producing a red color within several minutes.

This linking reaction was allowed to proceed for 2 hrs. @ 60°C after which the "living" system was terminated with 0.5 ml HPLC grade, degassed methanol. GPC analysis of this polymer indicated a peak molecular weight greater than 10^6 g/mol with a polydispersity of 1.2, suggesting that the radial topology molecule contains at least 8 to 10 arms. FTIR and FT ^{1}H NMR analysis of this polymer indicated that the polydiene microstructure consisted of 72% cis-1,4, 22% trans-1,4 and 5% 3,4 polyisoprene, which was in agreement with expectations.

Catalytic hydrogenation of the above star block copolymer was accomplished in situ[6]. The polymer solution was cooled to 50°C and purged with hydrogen. A prepared catalyst was added via syringe and a 50 psig hydrogen atmosphere was maintained for 24 hours. The catalyst was prepared by adding 0.228 g nickel octoate (0.66 mmol or 0.1 mol % based on polyisoprene) and a magnetic stirring bar to a 50 ml glass serum bottle which was then sealed with a septum and purged well with N_2. Then, 15 ml purified cyclohexane was added to dissolve the $Ni(oct)_2$.

Triethylaluminum, 1.13 ml (1.65 mmol), was syringed into the bottle, dropwise, while stirring vigorously. An opaque, black colloidal suspension formed immediately and was allowed to stir for 10 minutes at room temperature. The catalyst was then added to the reactor at 50°C via syringe. FTIR and FT ^{1}H NMR were used to observe the rate of the hydrogenation reaction, and indicated greater than 90% saturation of the polydiene after 24 hours. Two examples of absorbance FTIR spectra, appropriate for hydrogenation quantification, are provided in Figure 3.

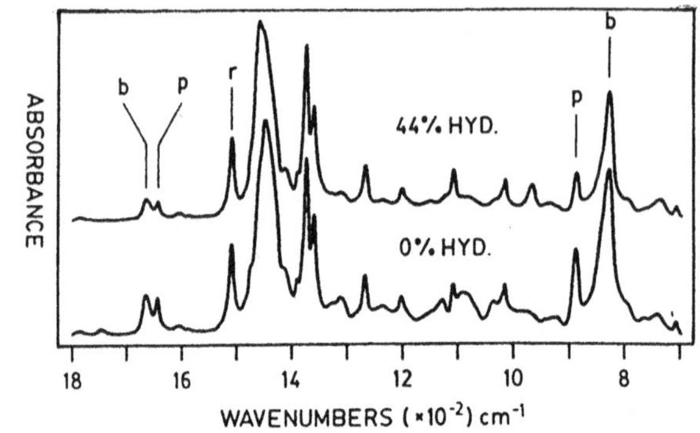

Fig. 3. FTIR Spectra of 25 TBS25K751*

A sample hydrogenated approximately 44% may be compared to its unhydrogenated precursor using the absorbance peaks labeled b (backbone unsaturation - 1,4) p (pendant unsaturation - 3,4) and r (reference - benzene ring deformation). When the absorbance peaks for unsaturation are normalized by their reference peaks, they may be ratioed to those in the unhydrogenated spectrum to determine the levels of residual unsaturation. The accuracy of this technique has been confirmed by FT ^{1}H NMR experiments.

RESULTS AND DISCUSSION

Five series of star block copolymers were synthesized and hydrogenated from 0 to 95%. Four of these series were of the same composition (25 wt. % poly(TBS)), and differed only in the molecular weight of the poly(TBS) blocks (ranging from 10,000 to 25,000 g/mol). These polymers were characterized by DSC, DMTA, and stress-strain experiments. The fifth series was of a non-elastomeric composition (60 wt. % poly(TBS)), and was characterized by DSC and DMTA experiments.

Dynamic mechanical thermal analysis (DMTA) and differential scanning calorimetry (DSC) have been employed to assess the thermal and mechanical behavior of two types of poly(TBS) - polyisoprene star block copolymers and their hydrogenated derivatives. Thermoplastic elastomers containing 15 to 25 wt. % poly(TBS), and rubber modified thermoplastics containing 60 to 75 wt. % poly(TBS) have been studied. An example of each type of material will be given.

Figure 4a is an example of DMTA data for a star block elastomer containing 25 wt. % poly(TBS) at 25,000 g/mol. Log_{10} flexural modulus and tangent delta are plotted as a function of temperature @ 1 Hz. The low room temperature modulus (0.66 MPa) and the broad low temperature relaxation are characteristic of a highly phase mixed block polymer. The polyisoprene and poly(TBS) phases have mixed sufficiently to cause "contamination" of the matrix rubber with glass, broadening the low temperature relaxation, and plasticization of the glassy "cross-linking" phase which sharply reduces the modulus at temperatures above the lower glass transition.

When this same star block polymer is hydrogenated approximately 90%, the low temperature relaxation sharpens greatly (Figure 4b), the room temperature modulus increases nearly an order of magnitude to 5.0 MPa, and a second glass transition is evidenced by the appearance of a high temperature damping maximum at 115°C. These changes suggest that demixing has

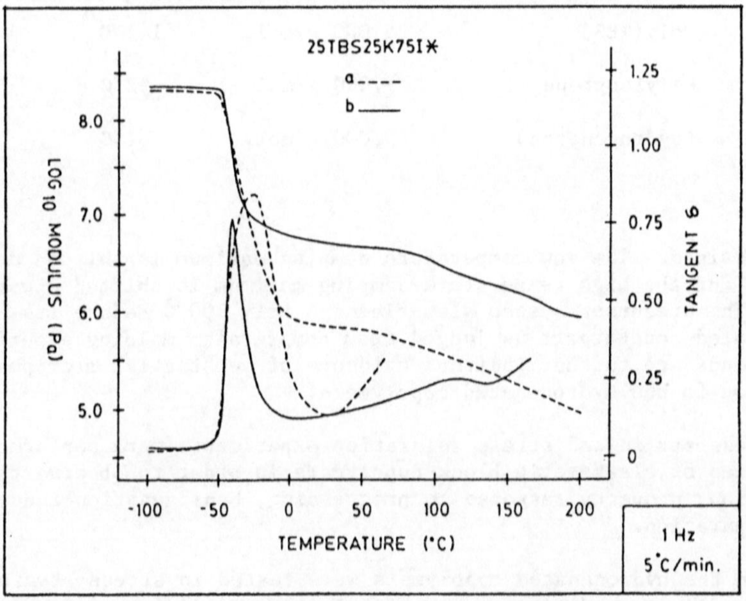

Fig. 4. DMTA Spectra of 25TBS25K75I*

resulted in more homogeneous or pure phases, with thermal properties more characteristic of the individual block components as shown in Table 1.

Additional star block copolymers were prepared in which the poly(TBS) block mol. wt. and % composition were increased to 60,000 g/mol and 60% (diblock mol. wt. of 100,000 g/mol). The DMTA data provided in Figure 5a were obtained for a representative sample. In this case, the broad low temperature damping and the relatively low (64°C) damping peak for the poly(TBS) block are again characteristic of a phase mixed system. When this block polymer was hydrogenated approximately 80%, the curve in Figure

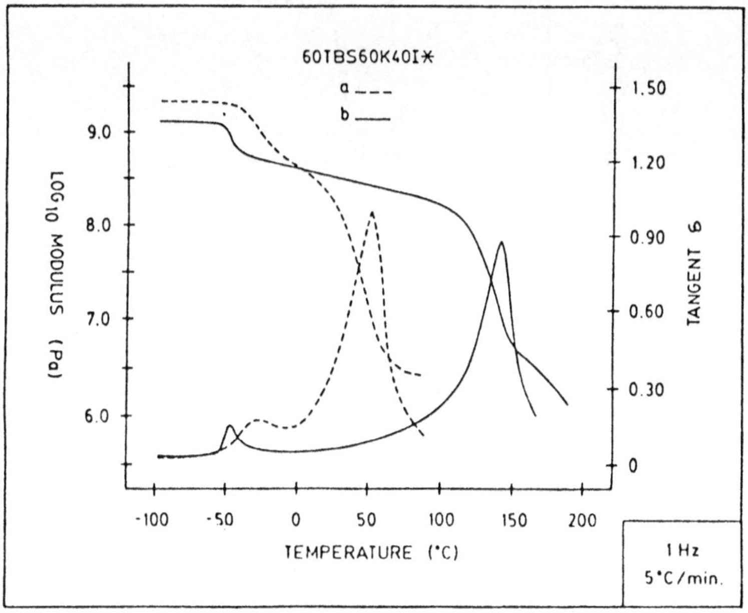

Fig. 5. DMTA Spectra of 60TBS60K40I*

Table 1. Glass Transitions of Homopolymers (DSC, 10°C/min.)

Poly(TBS)	25,000 g/mol.	141°C
Polyisoprene	75,000 g/mol.	-62°C
(hydrogenated)	75,000 g/mol.	-61°C

5b is obtained. The low temperature damping maximum is shifted down 22°C to -48°C and the high temperature damping maximum is shifted upward 78°C to 142°C. The precursor system also flowed nearly 100°C before its hydrogenated counterpart as judged from compression molding experiments. These trends are further indirect evidence of substantial microphase separation in the hydrogenated copolymers.

Stress-strain and stress relaxation experiments were performed on the four series of elastomeric block copolymers in order to observe their large deformation, property response to progressing, hydrogenation-induced microphase separation.

When the hydrogenated copolymers were tested in stress-strain experiments at ambient conditions, very low elastic moduli (ca. 0.5 MPa) and very high elongations (ca. 3000%) were observed. These polymers were also extremely tacky and proved to be interesting pressure sensitive adhesives.[7] Small to moderate levels of hydrogenation (ca. 50%) greatly reduced their tack and ultimate elongation, and increased their modulus and ultimate stress as shown in Figure 6. The change in solubility parameter that occurs as polyisoprene is hydrogenated to poly(ethylene-propylene), causes the rejection of poly(TBS) from the rubber phase to yield discrete domains of poly(TBS). As they develop, these glassy domains act as thermoplastic crosslinks to provide the observed improvements in elastomeric properties.

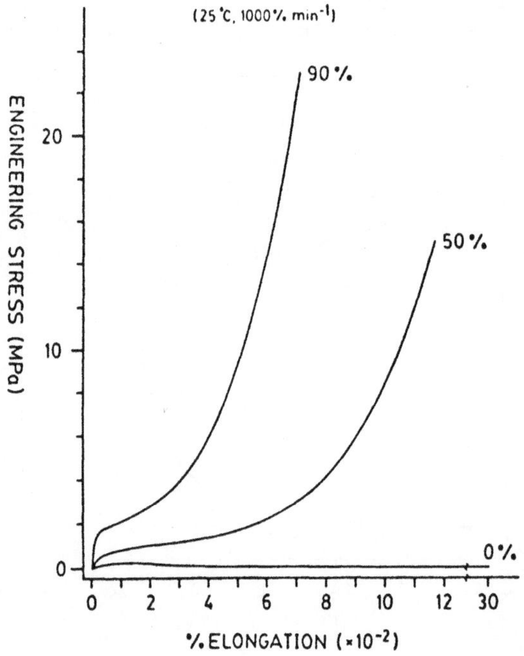

Fig. 6. Stress-Strain Relations for 25TBS20K751* Hydrogenated to the Levels Indicated

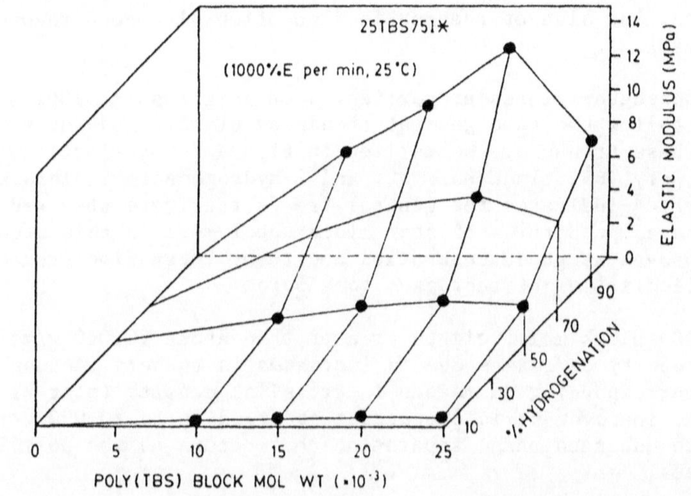

Fig. 7. Elastic Modulus for Star Block Elastomers Studied

As hydrogenation and phase separation proceed, the polymer perfects its phase separated character and is capable of supporting ultimate stresses in the range of 20 to 30 MPa, similar to previously mentioned, commercial, thermoplastic elastomers. It should be noted that these block copolymers contain only 10 mol% of the glassy tie-phase, in contrast to conventional triblock systems which typically contain 15 to 20 mol% or more of the glassy tie-phase. The 10 min. permanent set (after rupture) of the well hydrogenated samples is approximately 7 to 8 %. Thus, hydrogenation is able to produce poly(TBS) - polyisoprene copolymers of not only reason-

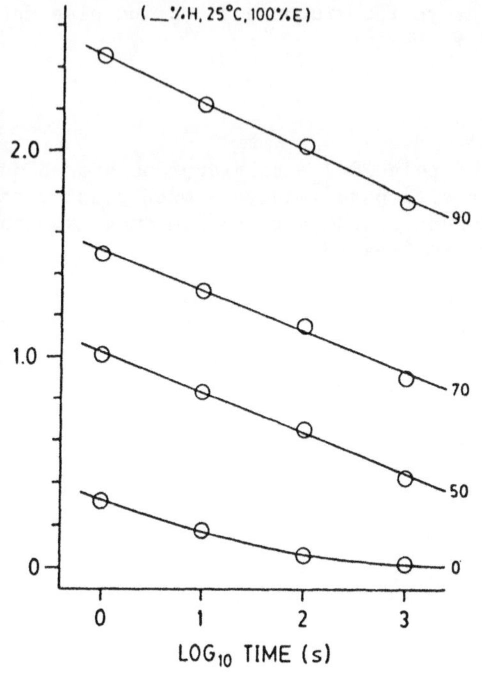

Fig. 8. Stress Relaxation for 25TBS20K75I* as a Function of Hydrogenation

able strength, but also of reasonable elongation with good recovery of large deformations.

Other strength-related properties, such as stress at 100% and 300% elongation, follow the same general trends as elastic modulus and tensile strength. These trends are summarized in Figure 7 for elastic modulus as a function of poly(TBS) block mol. wt. and % hydrogenation. This property response surface indicates the general trends that were observed for the strength related properties of star block copolymers in this study. The property improves as phase separation due to hydrogenation progresses, and as block molecular weight increases to a point.

Poly(TBS) block mol. weights greater than about 20,000 g/mol appear to result in property decreases due to increases in rubbery (network) chain length and corresponding decreases in crosslink density (star branch points). The improvement in properties observed up to 20,000 g/mol may be attributed to enhanced phase separation that occurs as the poly(TBS) block size increases.

Stress relaxation experiments were also performed on these polymers, as a function of poly(TBS) block size and % hydrogenation, to determine their ability to support a stress over extended periods of time. The series of curves shown in Figure 8 is typical of the response of these elastomers to strain and time. The shape and slope of these curves strongly suggests their ability to be horizontally shifted to form a time - degree of hydrogenation master curve. This master curve could then be used to predict the stress at extremely long times. Such a master curve would predict the total relaxation of stress (at 100% elongation) after $1 \times 10^{13.5}$ seconds or 1×10^{6} years, for this sample. Such a prediction is in good agreement with the time scale suggested for the relaxation of stress in glassy polymers by time - temperature superposition. Thus, such a prediction should also be appropriate for an elastomer held together by glassy domains, and strongly suggests the development of a structurally sound second phase that obeys the viscoelastic principles established for homogeneous glassy polymers.

CONCLUSIONS

In the types of poly(TBS) - polyisoprene star block polymers studied, hydrogenation appears to have induced a much greater degree of microphase separation as evidenced by their thermal-mechanical properties and greatly improved stress-strain behavior.

In samples where DSC transitions are very difficult to observe, DMTA has been found to be an especially valuable technique for assessing the relaxation characteristics of multiphase polymers which may possess a broad range of multiple damping phenomena. The poly(TBS) - polydiene block systems provide an excellent model for the investigation of hydrogenation influences on microphase separation phenomena.

ACKNOWLEDGEMENTS

The authors would like to thank the Army Research Office (ARO), the DOW Chemical Company, and the EXXON Foundation for support of this research.

REFERENCES

1. A. Noshay and J. E. McGrath, "Block Copolymers: Overview and Critical Survey," Academic Press (1977).
2. F. C. Schwab, Polymer Preprints, 25(2), 61 (1984); "Advances in Polymer Synthesis," B. M. Culbertson and J. E. McGrath, Editors, Plenum, in Press 1985.
3. L. J. Fetters, E. M. Firer, and M. Defauti, Macromolecules, 10, 1200 (1977).
4. P. K. Das, J. M. Hoover, T. C. Ward, and J. E. McGrath, Polymer Preprints, 25(2), 96 (1984); P. K. Das, Ph.D. Thesis, Virginia Polytechnic Institute and State University, December 1984.
5. Chemical Equipment Corporation, Tulsa, OK., J. M. Hoover et al., to be published.
6. R. J. A. Eckert, U.S. Patent 4,156,673, (to Shell), May 29, 1979.
7. M. M. Sheridan, J. M. Hoover, T. C. Ward, and J. E. McGrath, Polymer Preprints, 26(1), 186 (1985).

POLYALPHA-OLEFIN CATALYSTS

M. A. Bruzzone

R & D Enichem Elastomeri
Assago (Milan)
Italy

INTRODUCTION

A "polyethylene energy pie" is shown in Figure 1. This was extracted from a well known book of industrial energy analysis.[1] In this pie, the very large "slice" on the top is the energy content of the feedstock.

Ninety percent of this energy "slice" can be recovered at the end of the material life, just by burning it, as we would have done with the parent petroleum or natural gas. This portion of energy contained in the polyolefin is not "coercible" of course, since it is related to the hydrocarbon nature of the polymer, but in some way it is "recyclable", provided that the final object is disposed of in a correct way after use.

In contrast, the rest of the pie, which is related to the energy spent during the series of processes, going from petroleum to naphtha, from

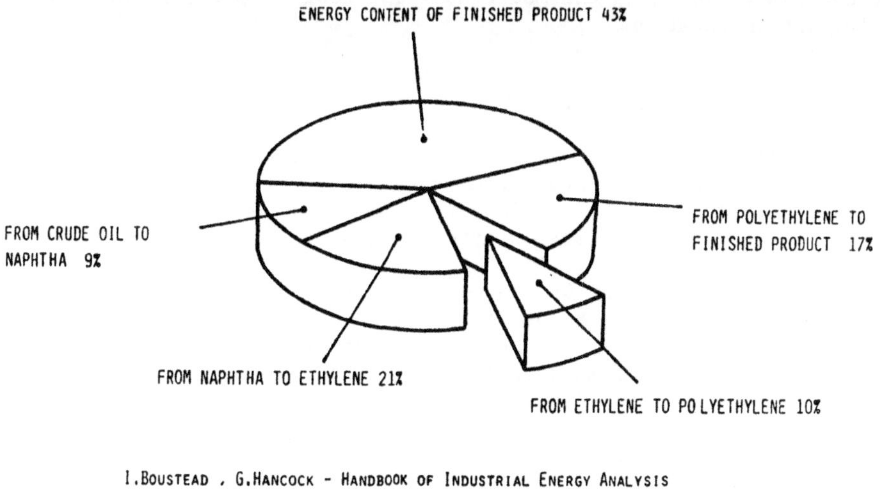

Fig. 1. From Crude Oil to Finished Product Energy Requirement

naphtha to ethylene, from ethylene to polyethylene and eventually from polyethylene to the final object, is in principle, susceptible to savings, by improving the related processes.

For a macromolecular scientist, the slices of interest are the ones related to polyethylene synthesis and the second one, a little bit neglected in the past, which is related to polyolefin transformation into the final object.

I admit that when I saw this energy pie I was amazed by the small size of the slice accounted for polyethylene synthesis, a mere 10-11% of the total.

I was then forced to ask myself: is a further improvement of the economy of the large scale polyolefin processes really necessary? Is this essential, in spite of the fact that monomer cost (which encompassed the feedstock plus the energy necessary for the monomer obtention, as shown in the energy pie) is already overwhelming the polymerization cost? Are our further efforts in reducing polymerization cost just a demonstration of "virtuosity" (as it has been called by my friend P. Galli) by scientists and technicians who have already done extremely well in bringing the polymerization processes close to their top level of efficiency?

My personal answer to this question is that, aside from the fact that a scientist should be motivated even by the simple joy of discovery, our future work in catalysis should be more addressed to product diversification than to process economy, unlike it has been done in the past years.

Also, the large slice of energy still spent in polymer transformation should be looked at, in order to find polymer structures, and hence catalysts, allowing easier polymer transformation conditions.

I trust that catalysis will be the key point for polyolefin diversification, as it has been the key point for the recent improvements in polyolefin synthesis.

According to these general observations, I shall consider two aspects of polyolefins: the first one will deal with some recent research on high yield Ziegler-Natta catalysts (both stereo and nonstereo specific ones), and the second one with olefin copolymers obtained therefrom. I believe that copolymerization is the most important tool at our disposal for polyolefins diversification.

ZIEGLER NATTA HIGH YIELD CATALYSTS

Also, the last generation Ziegler-Natta catalysts for polyolefins, patents and scientific literature embrace such a high number of chemical combinations that a general rationalization of the experimental data is impossible. The rationalization attempts generally follow two main routes.

The first one is that of synthesizing simpler, soluble catalysts based on combinations of Ti and Mg or Cr and Si compounds. The second one is that of studying directly the heterogenous catalysts, notwithstanding their complexity, by using the most up-to-date analytical techniques, for establishing the nature of the active centers.

The first requirement of a "high yield" or "high mileage" catalyst is of course its ability to produce, in a reasonable time, a high quantity of polymer in mild conditions of monomer concentration and temperature.

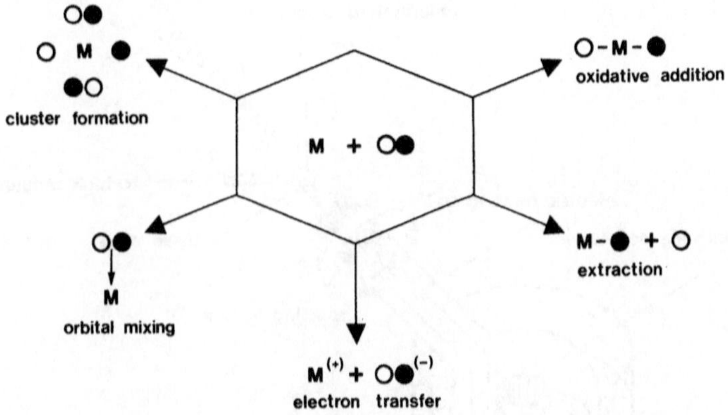

Fig. 2. Chemical Reactions by Metal Vaporization Technique

The main practical aim in reaching a high yield condition is not a catalyst saving, which is a minor factor in polymerization cost, but the elimination of the highly expensive polymer purification step. This is achieved when yields of polymer in excess of 200 kg./g of transition metal are attained.

Owing to the absence of any complication of regio- and stereoelectivity, ethylene is the monomer of choice for assessing the catalyst ability to provide high polymer yields. It is more fruitful to focus on recent results obtained in our laboratories and partly reported in patent literature.[2]

My aim in presenting our results in not that of adding another polyethylene catalyst to the above mentioned list, but that of showing you that well characterized, and relatively simple high yield catalyst can be obtained by a new straightforward technique.

The heterogeneous catalyst which is obtained contains Ti^{+3}, and only Ti^{+3}, into (or onto) disordered delta type crystals of $MgCl_2$. The only ligand is chlorine. The new technique is based on metal vaporization in high vacuum, followed by vapor quenching in a cold, reactive substrate.

This technique has already been applied to several chemical reactions, as shown in Figure 2, and is particularly interesting for synthesizing unstable organometallic compounds not achievable for conventional routes. However, it has never been used, to my knowledge, for preparing solid solutions suitable for use as polyolefin high yield catalysts. This technique is often referred to, as in a recent review of Chemical Engineering News,[3] as "metal atom synthesis".

The apparatus is shown in Figure 3. Mg metal is put in a crucible, heated by an electric coil, and vaporized in high vacuum. The rotating vessel wall is kept at low temperature by a cooling bath. The vessel contains a diluent (for example an aliphatic hydrocarbon) and a certain quantity of $TiCl_4$.

Unlike solid Mg metal, Mg vapor reacts instantaneously with the $TiCl_4$ solution according to the equation shown in Figure 1. A solid solution of Ti^{+3} and $MgCl_2$ is formed which precipitates in the diluent as a fine slurry.

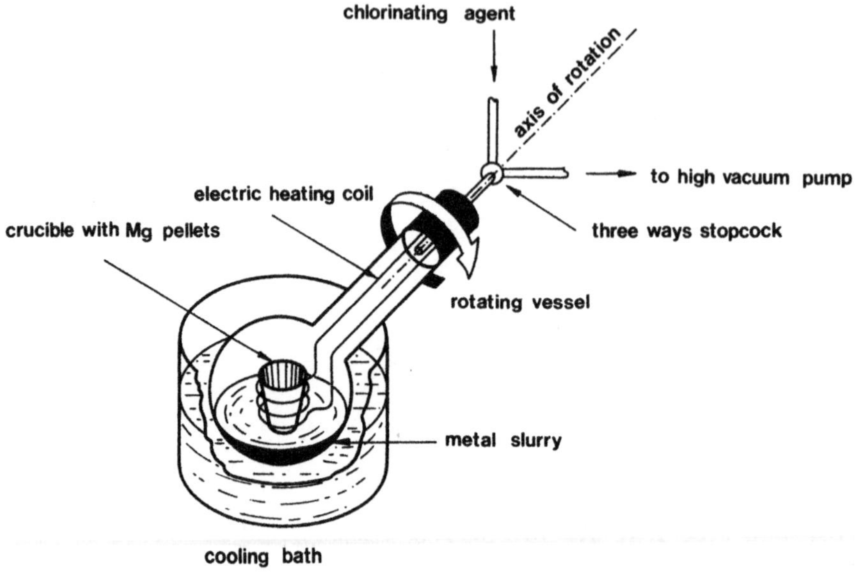

Fig. 3. High Yield Catalysts by Metal Vaporization Technique

This slurry, in conjunction with aluminum alkyls is an active catalyst in ethylene polymerization. However - and this is an important point - its catalyst activity can be tremendously improved just by "diluting" the solid solution with additional amounts of $MgCl_2$. In order to do this, the Mg vaporization is continued beyond the stoichiometry shown in Figure 4. The excess Mg condenses in the diluent as a very fine, black slurry, which blends, in the rotating vessel, with the former solid solution of Ti and Mg chlorides.

As soon as the desired Mg/Ti ratio is reached, the Mg vaporization is stopped and the Mg black slurry is transformed into $MgCl_2$ by admitting a chlorinating agent into the vessel. It can be either an organic or inorganic chlorinating agent which rules out the necessity of a Grignard formation as an intermediate step. In this way, solid solutions of Ti^{+3} into (or onto) highly disordered delta type crystals of $MgCl_2$, in any desired Mg/Ti ratio are obtained.

The improvement of polyethylene yield obtained by increasing Mg/Ti ratio in the solid solution is shown in Figure 5, in which polymer yield is reported on the left ordinate. In order to understand the reason of this increased catalyst yield, electron spin resonance (ESR) spectra of the solid solutions have been performed. These spectra show that, by increasing the Mg/Ti ratio, a progressive isolation of the Ti^{+3} sites present in the initial solid solution (with a Mg/Ti ratio of 0.5) takes place. This is detected by the ESR signal attributable to Ti^{+3} paramagnetic cations isolated in a diamagnetic environment.

$$Mg_{vap} + 2\ TiCl_4 \longrightarrow MgCl_2 \cdot 2TiCl_3$$

Fig. 4. Catalyst by Metal Vaporization
Formation of Mg/Ti Solid Solution

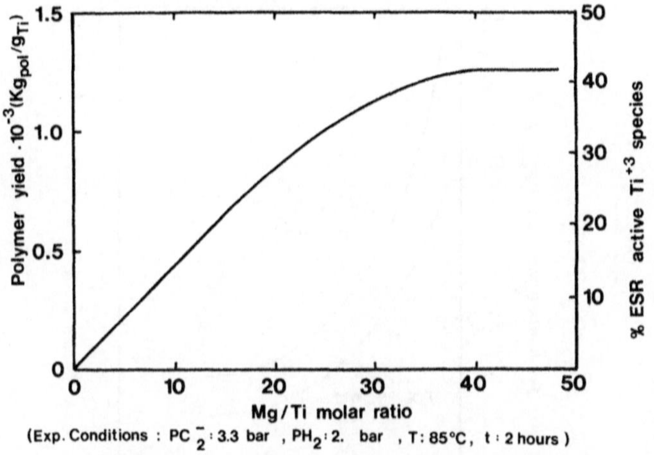

Fig. 5. Catalysts by Metal Vaporization: Effect of Mg/Ti on Polymer Yield and on ESR Active Ti^{+3} Species

The amount of Ti^{+3} isolated species obtained at high Mg/Ti ratios can be as high as 50% of the total Ti^{+3} present in the solid solution, as shown on the right ordinate in Figure 5. The concurrent increase of these species and of catalyst yield supports an intercorrelation of the two facts, or, in other words, that isolated Ti^{+3} species are the origin of the high catalyst activity.

Two other features of this catalyst deserve additional considerations. First, the catalyst is not stereospecific in propylene polymerization. Second, polyethylene produced by this catalyst has a rather narrow MWD, suggesting a possible correlation with the fact that it contains homogeneous Ti^{+3} sites in a similar electronic environment, unlike other, more complex catalysts.

A substantial research effort has been spent on high yield catalysis both with the aim of finding a way for total utilization of the transition metal sites and for understanding the nature of the activity and of the stereospecificity of the active centers.

Soluble Ziegler Natta catalysts based on V, Cr, Ti can be readily obtained by using proper ligands. However, neither their activity nor their stereospecificity matches that of heterogeneous catalysts. Also recent works on soluble catalysts based on bimetallic compositions containing Ti and Mg[4,5] have shown polymer yields on the low side, when compared to heterogeneous catalysts.

However, other transition metals, different from the customary Ti, Cr, V, are able to provide extremely high polymer yields. In fact, soluble Zr compounds, such as biscyclopentadienyl Zr dimethyl, when used in combination with aluminoxanes, which are cyclic compounds obtained by aluminum alkyls and water, is able to homo and copolymerize alpha olefins with extremely high yields[6] albeit without a stereoregulation. This result also points out the importance of the co-catalyst structure on activity.

Cyclic cocatalyst structures of aluminoxane type or cage structures, such as that of polyiminoalanes, obtained by reacting aluminum hydride and primary amines[7] are already well known in dienes polymerization for enhancing either the catalyst activity or its stereospecificity, or both.[8]

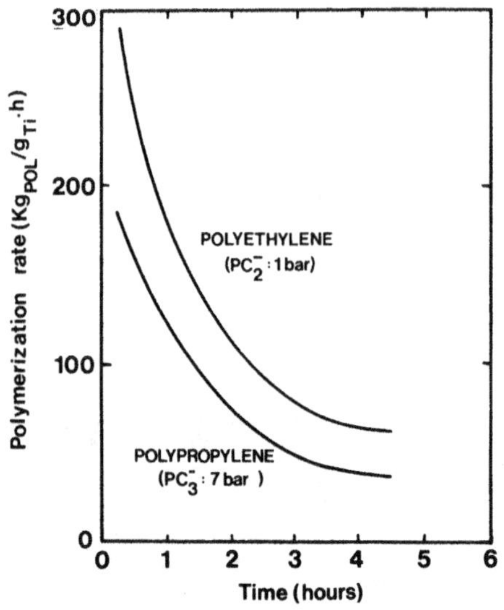

Fig. 6. Activity Decay of Supported Titanium High Yield Catalysts Polyethylene and Isotactic Polypropylene

An apparently general feature of high yield catalysts is the fact that they can display such a high yield notwithstanding their rapid loss of activity during the course of polymerization. A prolongation of the life span of the initial activity could have a dramatic effect on polymer yield. Typical activity decays as functions of time are shown in Figure 6 for slurry polymerization of ethylene and propylene with supported Ti catalysts.

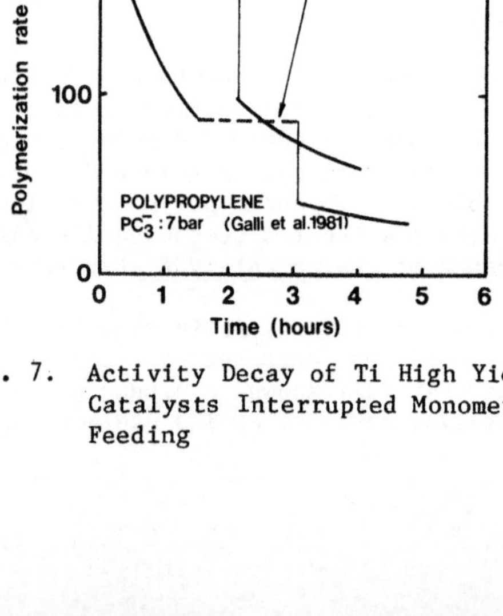

Fig. 7. Activity Decay of Ti High Yield Catalysts Interrupted Monomer Feeding

Two hypotheses have been put forward for explaining the catalyst decay. The first one explains it by a progressively increased difficulty of monomer to diffuse toward the active center, owing to the increased quantity of polymer surrounding the catalyst or clogging its pores. The second one explains the catalyst decay by a progressive chemical deactivation of catalyst sites.

The diffusional hypothesis is ruled out by some authors[9] on the basis of experiments shown in Figure 7 in which the polymerization is interrupted for a while, and then resumed. By these experiments, the activity decay seems chronologically dependent and not dependent on polymer build up, as expected from a diffusion dependent decay.

Another crucial point of high yield polymerization is whether the high catalyst activity is due to a higher number of active sites per transition metal atom, or to a higher propagation rate. A general agreement on the first hypothesis is found in literature, whereas the second hypothesis is still debated.

CATALYST STEREOSPECIFICITY

One of the most important recent successes in polyolefin catalysis has been the production of catalysts which, beside a high polymer yield, are highly stereospecific in propylene polymerization. In fact, high yield and stereospecificity seem in a certain way incompatible, as shown in Figure 8, based on a paper of P. Galli et al.[9]

For some experiments it seems that high yield supported Ti catalyst display at least two different highly active polymerization sites, one of which is stereospecific whereas the other is not. Of course paramount importance has the recognition of the structure of both sites, in order to find a way for increasing, or decreasing at will, the percentage of tactic polymer.

High yield stereospecific catalysts are generally obtained by a very complicated sequence of operations involving chemicals, labelled as modifiers, promoters, activators.

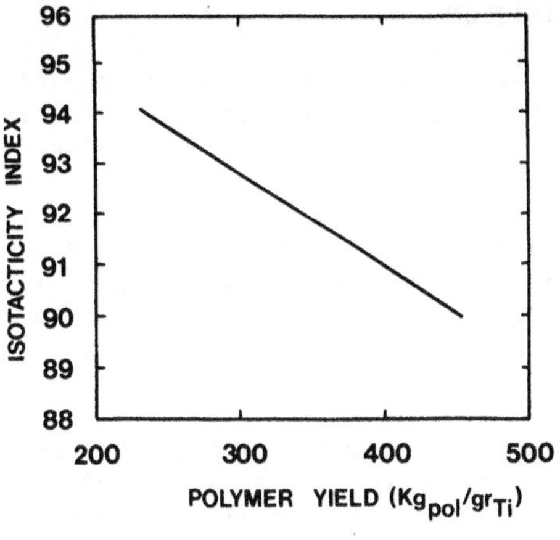

Fig. 8. Isotacticity Index and Polymer Yield: Polypropylene P. Galli, L. Luciani, G. Cecchin - Die Angew. Makromol. Chem. 94, 63 (1981)

Prof. Chien has recently tried to establish the structure of different active sites involved in a typical, Ti based stereospecific catalyst, capable of giving highly isotactic polypropylene in high yield.[10-14]

The main steps of catalyst preparation involve a support preparation, an implantation on the support of Ti sites, of which only 30% are in trivalent state whereas the others are tetravalent. Only 1/5 of Ti^{+3} is detectable by ESR, a small quantity in comparison with the catalyst obtained via metal atoms previously described.

The active catalyst, obtained by reacting the catalyst with $AlEt_3$ and methyl p-toluate, shows that most Ti^{+4} is reduced to a lower valence state, and that the structure of the Ti^{+3} complexes on the catalyst surface is modified giving rise to several ESR observable species. Just a few of these are believed to be active stereospecific sites whereas other sites are suspected to be the cause of atactic or block polymer formation. The electron donor function, used together with the cocatalyst, has been explained in different ways, but certainly has an important role.

Other tactic polymers, such as isotactic poly (1-butene), isotactic poly (3 methyl-1-pentene), isotactic polystyrene and syndiotactic polypropylene, should be investigated by taking advantage of the knowledge acquired on polypropylene high yield catalysts.

Another consideration on stereospecific catalysis is that in some polyolefins a stereoregulating effect is not desirable. In rubbery ethylene propylene copolymers (EPM), any crystallinity generated by both polyethylene or isotactic polypropylene sequencies should be avoided, in order to attain the best elastomeric properties and, in particular, a low permanent set.

The best catalyst for EPM has been the one based on V, whose yield can be improved either by using catalyst reactivators or by operating the polymerization in bulk.

Montedison, however, has recently proposed,[15] also for EPM, Ti based high yield catalysts which are apparently a spin off of the knowledge acquired on catalyst stereospecificity, but applied in the opposite direction, i.e. that of achieving the minimum polyethylene and polypropylene crystallinity.

POLYOLEFIN COPOLYMERIZATION

I believe that olefins copolymerization is one of the most important tools we have for polyolefins diversification. The trend toward copolymerization is not new. In fact, not only rubbery polyolefins but most plastics polyolefins are copolymers.

In the mid fifties, HDP and PP were the only homopolymers available and the trend was that of achieving the maximum polymer regularity (both structural and stereo). However, the concept "the purer the better," which generally holds for low molecular weight molecules, is hardly true for macromolecules.

Hence, this trend has been inverted and defects in the crystalline lattice of polyolefins are deliberately introduced. The "effect of defects," as it has been defined[16] has become one of the key points of polyolefins chemistry, physics and technology. It is a sort of a "vendetta" (retaliation) of the amorphous state, which has been disregarded in the past to the advantage of the crystalline state.

LDPE is a copolymer, from the structural point of view, with short and long side groups protruding from the main backbone. LLDPE is, of course, a copolymer. In comparison with LDPE, co-units are added deliberately. The experience gained so far on LLDPE indicates that technological properties are not just dependent on polymer density (or crystallinity) since LLDPE's behave differently depending not only on co-units amount, but also on their distribution, on the size of their side groups and even on the shape of side groups. Comonomers from 1-butene up to 1-octene, or branched comonomers such as 3-methyl-1-pentene are currently used.

HDPE often requires small amounts of 1-butene in order to improve the environmental stress cracking, the impact strength, or other properties. PP needs small amounts of ethylene as a co-unit, in a random or block arrangement, in order to decrease the brittleness at low temperature.

In isotactic poly (1-butene), a small quantity of a co-unit is required in order to increase the rate of the solid/solid transition, peculiar to this polymer, so to avoid the formation of frozen stresses in molded articles. In EPM and EPDM the co-unit is required in order to avoid any polyethylene crystallinity which would hamper their behavior as rubbers.

These examples tell us that the co-unit selection is a very important point in order to attain in the best product.

From the point of view of macromolecular physics, side groups were formerly considered to be expulsed by the crystalline lattice of polyethylene. The related theory has been developed by Flory in the early fifties[17] on the basis of phase equilibrium. Flory pointed out also the effect of the defect distribution along the chains, a fact unknown in small molecules crystallization.

Also, the crystallization case, in which an inclusion of the co-unit defects in the crystalline lattice takes place, has been theoretically considered. It has been experimentally found that methyl side groups can be included in substantial concentration within a polyethylene lattice as equilibrium requirement,[18] unlike more bulky side groups obtained by copolymerization with higher olefins.

The use of methyl groups in polyethylene lattice should cause a smaller decrease of crystallinity in respect to larger side groups, at equal molar concentration. However, when the amount and the distribution of methyl groups are enough to completely disrupt the polyethylene lattice, the polyethylene sequencies are so short that their interaction is unlikely even when the chains are extended.

The opposite situation holds for bulky side groups which are expulsed by the crystalline lattice of polyethylene which remains pure. In this way, the reticular forces hold together the crystalline lattice of plastics such as LLDPE, whereas the cohesion of a cured EPM relies merely on crosslinks.

Beside these rather well defined situations of isotropic polymers crystallization, other types of crystallization should be taken into account when the polymer, during its processing or during its use, experiences high deformations.

In fact, besides a "bulk" crystallization, in which a chain folded lamellar crystallization takes place, also an extended chain crystallization is noticed, when large strain or shear is applied to the polymer melt.[19]

This crystallization can take place also at temperatures higher than the isotropic melting point as pointed out earlier by Flory on strain crystallization of rubber.[20] Also a few fibrous crystals formed under shear in a polyethylene melt could have a substantial effect on the subsequent "bulk" crystallization of undeformed chains, since they can act as crystallization nuclei. The result of this dual type crystallization is the formation of "shish kebob" structures.

Coming back to polyolefins structure, it is quite reasonable that side groups of different bulkyness and shape would perform differently, depending on the mechanical history of the melt. However, a rational explanation of the different technological behavior among ethylene copolymers is still largely lacking.

Some useful indication on polyolefin behavior under stress can be gained by some experiments on cis 1,4-polybutadiene crystallization under strain. By taking advantage of the fact that the polymer was crosslinked, and must be kept in an elongated state for a sufficient time for performing the experiment, we have found a clear demonstration of the formation of a slow "bulk" crystallization after the rapid formation of fibrous crystals originated by strain.[21]

The copolymer of 1-hexene with 1-butene contains a small quantity of a diene (such as 5-methyl-1,4-hexadiene) in order to achieve sulphur vulcanization.[22]

The new polymer, besides the customary advantages of EPDM, shows good building tack, damping properties similar to butyl rubber and a so good flex life which suggests its use in demanding bio-applications (artificial hearts).

POSSIBLE FUTURE DEVELOPMENT OF POLYOLEFIN COPOLYMERS

One route could be that of copolymerizing ethylene (or propylene in isotactic configuration) with small amounts of dienes, in order to obtain crystalline polymers which can be upgraded by vulcanization. We have already shown the possibility of producing high density polyethylenes containing small amounts of butadiene in transconfiguration by high yield catalysts.[23,24]

After vulcanization, which can be performed in the solid state (i.e., below the polymer melting point) environmental stress cracking and impact resistance can be substantially improved. Or, vice versa, the initial molecular weight can be lowered so as to improve the polymer flow in manufacturing technologies (e.g. rotational molding) in which compacting forces applied to polymer pellets are small.

One important structural characteristic of these ethylene-butadiene copolymers is based on the fact that butadiene co-units are in the trans-configuration which is different than the other butadiene configurations, but interfere very little with potential energy calculations, with the crystallization of polyethylene sequencies.[25] In this way, the polymer rigidity (a property desirable in many applications) is very little affected, in contrast to conventional diene termonomers used for EPDM.

A second possible diversification route for polyolefins copolymers is that of functionalization. Functionalized polyethylene copolymers are currently produced by radical high pressure catalysts. The most important comonomers are methacrylic acid, acrylates and methacrylates, vinyl acetate, vinyl chloride and acrylonitrile.

However, some of these copolymers are difficult to produce because they require very high pressures, because the comonomers are corrosive, because the MW is not sufficiently high, or because of side reactions.

Therefore, some of these copolymers have been produced by an indirect route, i.e. by a radical copolymerization of butadiene with acrylonitrile, followed by a complete hydrogenation of the butadiene units.

This route, notwithstanding its complication and cost, has allowed one to assess the good properties of this product as an oil resistant rubber.

The copolymerization of monoolefins with acrylonitrile by transition metal catalysts is not possible with different diolefins.[26]

However, it has already been shown that it is possible to insert, in the chain of different polyolefins, very reactive conjugated double bonds systems, by copolymerization of alpha olefins with suitable trienes[27] which can act as precursors of functional groups, subsequently introduced by post reaction.

The triene can be a 1,3,5 triene, which is inserted by a 1,6 addition (likewise to a 1,4 addition of a conjugated diene) or a triene containing an isolated double bond, reactive in polymerization, and a conjugated double bonds system non-reactive in the polymerization step.

A conjugated double bonds system is reactive to catalytic post-reactions (radical, cationic etc.) and also to non-catalytic ones, such as a Diels Alder reaction.

Another possible important field of development for polyolefin copolymers is that of block copolymers. Block polyolefins copolymers are already well known. However, the block length control is still rudimentary in comparison to living anionic polymerization. Unfortunately the alpha olefins suitable for anionic polymerization are few (substantially styrene and its derivatives).

It is certainly difficult to conceive a living polymerization, when the rate of growth of a macromolecule is measured in fractions of a second, as in Ziegler-Natta polymerization, in contrast with the theoretical unlimited life of a living carbanion. However, Prof. Kennedy has recently announced a living cationic catalysis so that also living transition metal catalysis cannot be excluded "a priori".

In any case, ABA block copolymers of alpha olefins, such as styrene/ethylene, and 1-butene/styrene block copolymers are already being produced. However, also in this case likewise the polyolefins functionalization already discussed, the synthesis problem has been bypassed by synthesizing a styrene/butadiene/styrene block copolymer and then by hydrogenating it, in order to saturate the unsaturations of the central block (composed of 1,4 and 1,2-butadiene units).

Also these copolymers have already found interesting applications, which should stimulate our fantasy for finding simpler, more economic routes for obtaining the same result as those obtained by starting from alpha olefins.

ACKNOWLEDGEMENT

I have the pleasure to acknowledge the outstanding contributions in the field of polyalphaolefins of my associates of the Polymer Research

Laboratory of ENI (formerly ASSORENI), and in particular those of:
A. Balducci, S. Cesca, M. Corbellini, A. Magni, A. Greco, A. Priola (catalysts and polymer chemistry)
L. Gargani (polymer physics)
S. Gordini, M. Bazzotti (process engineering)
M. Cesari, C. Corno, C. Busetto, G. Del Piero (structure analysis of catalysts and polymers)

REFERENCES

1. I. Boustead and G. Hancock, Handbook of Industrial Energy Analysis, E. Horwood Ltd. (1979).
2. USP 4,194,992 to Snamprogetti, March 25, 1980; USP 4,203,866 to Snamprogetti, June 30, 1977; USP 4,217,245 to Snamprogetti, June 30, 1977.
3. R. Dagani, Chem. Eng. News, 63, 23 (April 8, 1985.
4. E. Albizzati, E. Gianetti, and U. Giannini, Chem. Rapid Commun., 5, 673 (1984).
5. A. Greco, G. Bertolini, and S. Cesca, J. Appl. Polym. Sci., 25, 2045 (1980).
6. W. Kaminsky, Polym. Bull., 9, 464 (1983).
7. S. Cucinella, Chim. Ind. (Milan), 59, 696 (1977).
8. A. Balducci, M. Bruzzone, S. Cucinella, and A. Mazzei, Rubb. Chem. Technol., 48, 736 (1975).
9. P. Galli, L. Luciani, and G. Cecchin, Die Angew. Makromol. Chemie, 94, 63 (1981).
10. J. Chien, J. Wu, and C. Kuo, J. Polym. Sci., Polymer Chem. Ed., 20, 2019 (1982).
11. J. Chien, J. Wu, and C. Kuo, ibid, 20, 2445 (1982).
12. J. Chien and J. Wu, ibid, 20, 2461 (1982).
13. J. Chien, J. Wu, and C. Kuo, ibid, 21, 725 (1983).
14. J. Chien, J. Wu, and C. Kuo, ibid, 21, 737 (1983).
15. P. Galli, F. Milani, and L. Corbelli, Scandinavian Rubber Conf. (Copenhagen), June 10-12 (1985).
16. C. Vonk, Polyethylenes 1933-1983, D 21.
17. P. Flory, J. Chem Phys., 17, 223 (1949).
18. M. Richardson, P. Flory, and J. Jackson, Polymer, 4, 221 (1963).
19. A. Pennings, J. Polym. Sci.C, 16, 1799 (1967).
20. P. Flory, J. Chem. Phys., 15, 397 (1947).
21. M. Cesari, G. Perego, A. Zazzetta, and L. Gargani, Makromol. Chem., 181, 1143 (1980).
22. J. Lal, P. Sandstrom, and M. Senyek, 127th Meeting ACS Rubber Division Los Angeles (April 23-26, 1985).
23. M. Bruzzone, A. Carbonaro, and C. Corno, Makromol. Chemie, 179, 2173 (1978).
24. USP 4,446,289 to Assoreni (July 29, 1981).
25. E. Cernia, G. Conte, L. D'llario, N. Pavel, and E. Giglio, J. Polym. Sci., Polym. Chem. Ed. 13, 125 (1975).
26. J. Furukawa and E. Kobayashi, Rubber Chem. Techn., 51, 600 (1978).
27. S. Cesca, A. Priola, and M. Bruzzone, Adv. Polym. Sci., 32, 1 (1979).

CHARACTERIZATION OF CHROMIUM-SILICA CATALYSTS

Alan Ellison,[1] J. A. Moulijn,[2] B. Scheffer,[2]
A. Brown,[3] B. Herbert,[1] P. Humphrey,[3] G. Diakun,[4]
P. Worthington,[5] F. E. Mabbs,[6] and D. Collison[6]

[1]Humberside College of Higher Education, Hull HU6 7RT England
[2]Laboratory for Chemical Technology, Amsterdam University 1018WV Amsterdam, The Netherlands
[3]University of Manchester Institute of Science and Technology, Manchester M60 1QD, England
[4]SRS Daresbury Laboratory, Warrington WA4 4AD, England
[5]Hull University, Hull HU6 7RT, England
[6]University of Manchester, Manchester M13 9PL, England

INTRODUCTION

The commercial importance of chromium-on-silica ethylene polymerization catalysts continues to attract the efforts of many researchers and to produce a wealth of related literature.[1]

Despite this high level of activity, the experimental ingenuity of individuals and the range and sophistication of modern techniques, controversy still surrounds characterization of the Cr species and the polymerization-active sites in this catalyst system.

Thus, Krauss and co-workers, in an extensive series of papers,[2] affirm that, after reduction in CO or H_2, isolated Cr(II) species are responsible for catalyst activity, without the induction period associated with Cr(VI)/SiO_2 catalysts. In contrast, isolated Cr(III) species are cited for their activity[1] and Rebenstorf et al[3] find that both Cr(II) and Cr(III) species are active towards polymerization.

Contrasting arguments simultaneously concern the state of dispersion of the active chromium. McDaniel et al[4] remain indecisive on the nature of the active species in CrO_3/SiO_2 catalysts. Their extensive and elegant experimentation apparently suggests that monodisperse chromate (VI) species constitute the majority species leading to activity, but always the authors reserve judgment on the participation of dichromate (VI) species. Nevertheless the similarly extensive work of Rebenstorf[5] leads to the conclusion that, independently of method of sample preparation, the active sites are dimeric, dinuclear Cr species.

Recent FABMS (fast atom bombardment mass spectrometry) and ESR (electron spin resonance) spectroscopic studies[6,7] have shown that extensive aggregation of supported Cr occurs even at relatively low Cr loadings. In this report, we present further evidence to elucidate the cluster structure of Cr in these catalysts and the predominance of these structures on the silica surface.

EXPERIMENTAL

Sample Preparation

The silica support used was a Gasil 35 micronized xero gel (Unilever Ltd) calcined at 525°C in dry air, surface area 350 m^2g^{-1} and pore volume 1.0 cm^3g^{-1}.[6,7] This support showed no adsorptive property with respect to CrO_3 and all CrO_3/SiO_2 samples were therefore prepared by immersion of the SiO_2 in an aqueous solution of CrO_3, subsequently filtering the slurry without water washing. Samples were dried in vacuum ($<10^{-4}$ torr) at room temperature to reduce contamination and decomposition of the supported Cr (VI). Small amounts of the dried solids were then calcined in dry, flowing air at temperatures between 100 and 600°C.

Samples dried at 25°C, and infrequently after calcination at 300°C, were analyzed by neutron activation analysis.

Samples are designated according to their chromium content (wt%), the calcination temperature in air (°C) and the support: thus, 2-300-Si contains approximately 2 wt% Cr on silica and was calcined in air at 300°C. When a decrease in chromium content was detected after calcination, due to the sublimation of excess deposited chromium oxide,[6] the new analytical value is included in the designation in parentheses, thus: 24-300-Si (21.8%).

In addition, 'standard' samples of chromium compounds and of chromium oxide, prepared by the air calcination of chromium (VI) oxide, are referred to in the subsequent text. These are designated by their formulae or as CrO_3 (290°C), to signify the product formed on heating CrO_3 in air at 290°C. These materials were of ANALAR grade and were dried in flowing air at 100°C before use.

Table I shows the actual chromium contents of CrO_3/SiO_2 samples and their designations.

Electron Spin Resonance (ESR). ESR spectra of the powder samples were obtained as first-derivatives at X-band (ca 9.4 GH_z) at the temperatures 298K, 77K and 4.2K, using a low value of modulation amplitude,[7] maximum

Table I. Results of Cr Analyses on CrO_3/SiO_2 Samples

Actual Cr content /%Cr	Sample designation (air dried at 25°C)
0.19	0.2-25-Si
0.82	0.8-25-Si
1.66	2-25-Si
4.98	5-25-Si
7.20	7-25-Si
10.70	11-25-Si
16.20	16-25-Si
23.50	24-25-Si

4.0 Gauss, and diphenylpicrylhydrazide, DPPH, as reference with a Varian E115 spectrometer.

Temperature Programmed Reduction (TPR)

The TPR apparatus has been described elsewhere.[8] A high-purity, 67% H_2/Ar, mixture was used as a reducing gas at a flow rate of 12×10^{-6} mol s^{-1} and at a pressure of approximately 1 bar. The sample size was chosen to maintain the amount of chromium in the experiment constant at about 50×10^{-6} mol. The conversion of hydrogen was therefore maintained at below 10%. A heating rate of 10K min^{-1} was used in all experiments, over the range of room temperature to 1400°C. Detection was achieved via a thermal conductivity detector which indicated, not only the consumption of H_2, but also the production of O_2 and CH_4, after water vapor, NH_3, CO_2 and organic impurities had been trapped by 3A and 5A molecular sieves. A flame ionization detector was also used to detect CH_4 alone, but CH_4 was completely absent during these experiments.

Catalysts were not pretreated, in situ, before TPR but were flushed at room temperature, in the reducing gas, to establish a stable baseline before the temperature program was begun. At very low Cr loading, the large sample size caused difficulties in the collection of accurate data, and these aspects are the subject of a continuing investigation.[9]

Fast-atom Bombardment Mass Spectrometry (FABMS)

FABMS spectra were obtained using a Vacuum Generators Analytical MM12-12S system. Argon, fast-atom bombardment at 10^{-5} torr was used as the primary beam of cross-sectional area 4 mm^2. Analyses were obtained at high-resolution as digital values of ion-fragment current at unit values of m/e for positive-bias spectra, using a high energy (2.0 - 3.0 keV) atom gun producing 1nA secondary-ion current. Samples were obtained as finely-ground, free powders and out-gassed in the SIMS-analysis chamber without in-situ heating to remove the adsorbed water vapor.

X-ray Absorption Spectra (EXAFS/XANES)

Extended X-ray absorption fine structure (EXAFS) and X-ray absorption near-edge structure (XANES) spectra were obtained in transmission, using the synchroton radiation source with a typical beam energy of 1.96 GeV and a circulating current of between 140 and 94 mA. The transmission of the reference ionization chamber was 90% and that of the signal chamber 30%. The comparatively low energy of the SRS and the ion-chamber settings throughout helped to reduce the harmonic contamination of the transmission measurement.

All samples were measured at room temperature as finely ground powders, particle size 1m, as 1mm thickness films entrapped in Mylar. To achieve an optimum maximum absorbance of approximately 2.2, samples were thoroughly mixed and diluted with boron nitride, whenever necessary.

Chromium metal foil, chromium (VI) oxide, chromium (III) oxide (α-Cr_2O_3), sodium chromate (VI) and sodium dichromate (VI) were used as reference materials and as calibrants for the energy of the Cr K-absorption edge at 5.989 keV (Si(III) crystal monochromator). The thermal decomposition products of CrO_3 were also used as reference compounds, the products CrO_3 (360°C) and CrO_3 (560°C) in air corresponding to chromium (IV) oxide[10] and chromium (III) oxide, respectively.

RESULTS AND DISCUSSION

Temperature Programmed Reduction (TPR)

Figure 1 shows the TPR patterns obtained for crystalline CrO_3, precalcined in dry air at 290°, 360° and 560°C. The CrO_3 (290°C) sample, figure 1(a), shows two distinct peaks at ca. 280°C and 320°C, for which the ratio of the peak areas is approximately 1:2. This behavior should be compared with that of pure, crystalline CrO_3 which is reduced to Cr_2O_3 in peaks at 270° and 320°C.

CrO_3 (360°C), figure 1(b), exhibits only one peak at ca 340°C together with a shoulder on the low temperature side at ca 290°C. A much larger sample was used in the TPR experiment for CrO_3 (560°C) to render the extremely small peaks visible. One peak at low temperature is seen together with a broad pattern at higher temperature, figure 1(c), both features caused by the reduction of impurities in the sample. Major reduction of chromium to lower oxidation states was not observed for this sample and reduction to chromium metal did not occur for any sample in the temperature range considered.

The total hydrogen consumption in the reduction of CrO_3 (290°C) and of pure CrO_3 corresponds to the reaction:

$$2\,CrO_3 + 3H_2 \longrightarrow Cr_2O_3 + 3H_2O$$

The ratio of the areas of the two peaks observed in figure 1(a) is in accordance with a mechanism proposed for the reduction of CrO_3 in vacuum:[11]

$$CrO_3 \longrightarrow CrO_{2.4}\ (Cr_2(CrO_4)_3)\ \text{at } 370°C \quad (1)$$

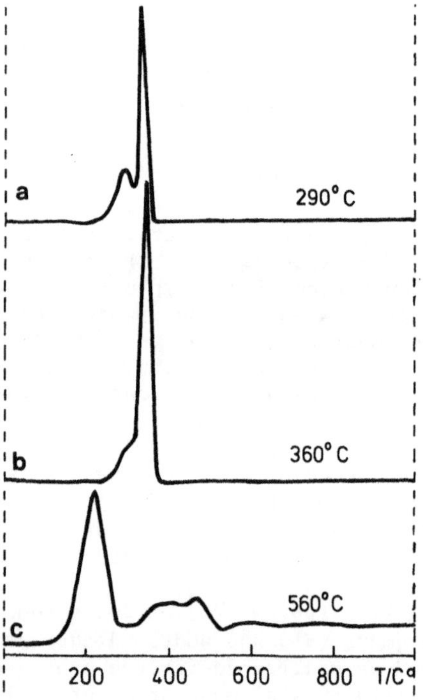

Fig. 1. TPR patterns obtained at a heating rate of $10 K min^{-1}$ for crystalline CrO_3 precalcined in dry air at the temperatures: a) 290°C, b) 360°C and c) 560°C.

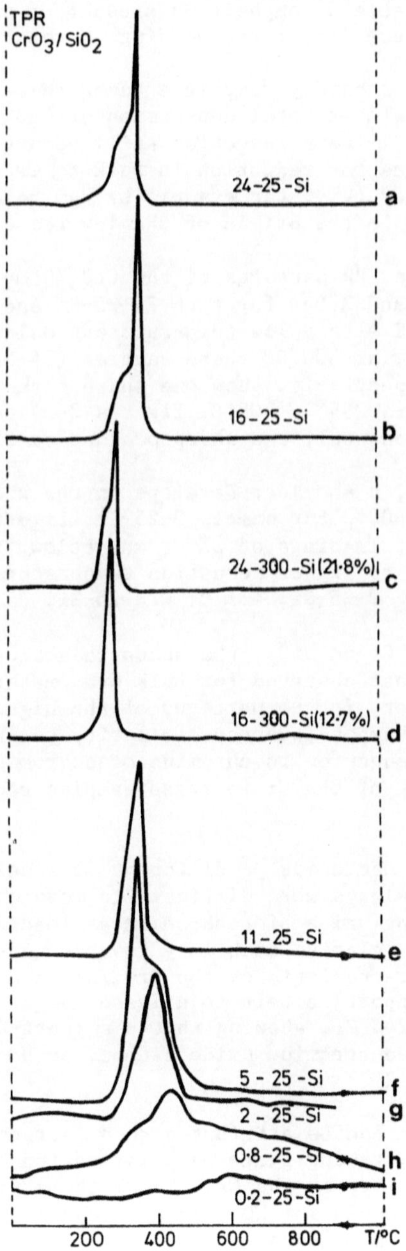

Fig. 2. TPR patterns obtained at a heating rate of $10Kmin^{-1}$ for the CrO_3/SiO_2 samples: a) 24-25-Si, b) 16-25-Si, c) 24-25-Si(21.8%), d) 16-25-Si(12.7%), e) 11-25-Si, f) 5-25-Si, g) 2-25-Si, h) 0.8-25-Si, i) 0.2-25-Si.

$$CrO_{2.4} \longrightarrow CrO_{1.56}\ (Cr_2O_3)\ \text{at}\ 495°C \qquad (2)$$

The reduction experienced in the H_2/Ar atmosphere of these experiments proceeds at lower temperatures than in vacuum but the reduction intermediate appears to be the same, that is $CrO_{2.4}$. In air, the 'reduction' temperature has been stated to be even higher[11] than in vacuum with reduction proceeding via the same proposed mechanism. The CrO_3 (360°C) product is itself a characteristic decomposition product of CrO_3 and displays only a single TPR peak, figure 1(b) corresponding to reaction.[2]

The CrO_3 (560°C) is hardly reducible under these conditions as the sample corresponds to almost total conversion of CrO_3 to crystalline Cr_2O_3. According to Mackenzie,[11] some reduction still occurs at high temperature but there is no evidence for reduction in bulk to Cr (II). Surface reoxidation of Cr (III) to Cr (VI), for example by H_2O before commencement of the TPR experiment may be the origin of the low temperature peak.

Figure 2 shows the TPR patterns of the CrO_3/SiO_2 samples. At high loading, figures 2(a) and 2(b), for both 24-25-Si and 16-25-Si a sharp peak is observed at ca 350°C with a low-temperature shoulder at ca 320°C. After precalcination, in air, at 300°C, these samples (24-300-Si (21.8%)) and 16-300-Si (12.7%), respectively, show one sharp peak which is shifted to lower temperature, at ca 295° - 305°C, figures 2(c) and 2(d). The sample 11-25-Si, meanwhile shows only one sharp peak at ca 345°C, figure 2(e).

At lower loadings, a shoulder develops on the high-temperature side of the sharp peak, at ca 400°C for sample 5-25-Si, figure 2(f), and becomes the dominant feature at loadings of 2% Cr and below, figures 2(g) to 2(i). This latter peak moves to higher reduction temperature as the Cr loading decreases, viz: 400°C, 2-25-Si; 435°C; 0.8-25-Si; ca 590°C, 0.2-25-Si.

Between 5 and 24% Cr on SiO_2, the sharp reduction peak occurs at a temperature close to that observed for bulk CrO_3 materials. Indeed, the low-temperature shoulder, in the patterns of the highest loading samples, 16-25-Si and 24-25-Si, emphasizes the similarity to 'bulk' behavior and indicates a step (1) reduction to chromium monochromate, $Cr_2^{III}(Cr^{VI}O_4)_3$. Presumably a major part of the Cr in these samples occurs as Cr oxide crystallites.

As the Cr loading decreases, a different Cr species becomes increasingly dominant and is always more difficult to reduce than bulk CrO_3 samples. The increasing value of peak-shift as loading decreases demonstrates the increasing relative stability towards reduction with Cr dispersion, particularly characteristic of the CrO_3/Al_2O_3 system.[12] This stabilizing effect of the support, albeit to a lesser degree, is evident at loadings of between 5 and 24% Cr, showing that a support-Cr interaction persists even for supported chromium oxide clumps, as has also been observed for WO_3[13] and MoO_3.[14]

Such stabilization can be attributed to a 'strong' interaction between the support and the Cr, taking place by polarization of the Cr-O bond and decreasing the reducibility of the Cr. Thus, the effect is smallest for CrO_3 and the most pronounced for WO_3, in accordance with the relative polarizabilities of the M-O bonds.

It is interesting and significant that the 24-25-Si and 16-25-Si samples, after precalcination in air at 300°C (i.e. 24-300-Si and 16-300-Si) lose Cr by sublimation, 23.5% → 21.8% and 16.2% → 12.7%, and thereafter, figures 1(c) and 1(d), show TPR patterns, which are closely similar to those of bulk CrO_3, but without the peak-temperature shifts shown by their uncalcined precursors. One is led to conclude that the 'sublimation' process is accompanied by redistribution of chromium oxide to produce more aggregated species which experience less stabilizing influence from the silica support. Such redistribution on calcination has also been observed for supported Re_2O_7.[15]

The lowest-loading samples, in addition to showing the least reducibilty, exhibit a breadth of peak-temperature and peak-shift which indicate both support and Cr-site heterogeneity.[13,14] At 0.2% and 0.8% Cr single-molecular-layer Cr species exist in which no 'crystalline' CrO_3 is present. The structure is 2-dimensional insuring maximum polarization interaction with the support.

Fast Atom Bombardment Mass Spectrometry (FABMS)

The raw data from the FABMS analyses consisted of digital values of ion-current at integral values of $^m/e$. In most instances, the mass spectrum fragmentation pattern of a particular Cr species overlapped with those of silica-based fragments and with other Cr species. This is, of course, due to the natural abundance of the isotopes of constituent atoms, within the fragments, and to the presence of proportions of OH or H, a consequence of the hydroxyl population on the support surface and of residual water in the samples. Thus, for example, CrO_3^+ is accompanied by CrO_3H^+, $CrO_3H_y^+$, $CrO_3H_2^+$ . . . etc.

A preliminary indication of a fragment-family in the data was easily ascertained by comparison of data values of $^m/e$ with values calculated from the formulae $Cr_mSi_nO_pH_q$ and Si_xO_y. Computer analysis was used to express an ion-count as being specific to a particular fragment and to demonstrate by least squares computer-fit that an ion-count-pattern within a particular range of $^m/e$ was specific to a fragment-family. Initially the program SIMS2[16] gave helpful results but proved to be inadequate for the analysis of this catalyst system, the maximum number of overlapping fragments being restricted by this program to 6 only. In consequence a PASCAL program was written[17] to analyze the mass spectrum isotope pattern resulting from, at present, up to 30 overlapping secondary-ion fragments.

The prediction of the primary isotope pattern for each atom in a fragment was based upon the combinatorial calculation used in the IPPKS program of Carberry and Dombek,[18] the overlapping sum of patterns for the atoms in a fragment ion, q, at a particular value of $^m/e$,p, being computed as x_{pq}.

If y_i is the observed count at the i'th value of $^m/e$, due to the sum of the proportions of contributing fragments, total q, then:

$$y_i = m_1x_{i1} + m_2x_{i2} + \cdots m_jx_{ij} \cdots + m_qx_{pq} \qquad (3)$$

with $1 \leqslant i < p$ and $1 \leqslant j < q$, where:

y_i is the observed ion count in the total profile of p $^m/e$ values;
m_j is the proportion of fragment j in the total profile;
x_{ij} is the calculated count at the i'th $^m/e$ value due to the fragment j.

The values of m_j were then estimated to minimize the error sum of squares between the estimated ion-counts $\hat{y}$ and the observed ion-counts, y, in the total profile, using the method of Kendal and Stuart.[19]

Using the matrix notation:

$$Y = \begin{pmatrix} y_1 \\ y_2 \\ \vdots \\ y_p \end{pmatrix} \qquad X = \begin{pmatrix} x_{11} & x_{12} & \cdots & x_{1q} \\ x_{21} & & & \vdots \\ \vdots & & & \vdots \\ x_{p1} & \cdots & \cdots & x_{pq} \end{pmatrix} \qquad M = \begin{pmatrix} m_1 \\ m_2 \\ \vdots \\ m_q \end{pmatrix}$$

then (3) becomes:

$$\hat{Y} = XM \quad (4)$$

The total error sum of squares is given by:

$$S = (Y-\hat{Y})^1(Y-\hat{Y})$$

where ^{1} denotes the matrix transpose, so that:

$$S = (Y-XM)^1(Y-XM) \quad (5)$$

For a minimum value of S we must have $\frac{\partial S}{\partial M} = 0$

i.e. $\frac{\partial S}{\partial m_1} = 0$, $\frac{\partial S}{\partial m_2} = 0$ ... etc., so that:

$$2X^1\,(Y-XM) = 0$$

and $$X^1XM = X^1Y \quad (6)$$

Since X is computed from the summed isotopic abundancies, Y is the observed data matrix and M contains the unknown m_j values to be optimized, equation (6) represents a set of simultaneous, linear equations. Solving, we obtain:

$$M = (X^1X)^{-1}X^1Y \quad (7)$$

The values m_j in matrix M are effectively the percentages of each constituent fragment, total q, in the profile.
From the values of m_j and y_i, it is then simple to calculate the contribution ion count due to the j'th fragment at the i'th value of $^m/e$. Using a Norsk Data Nord 560 main frame computer a 30 fragment problem, each fragment with the general formula $Cr_mSi_nO_pH_q$, gave apparently instantaneous analysis of fragment percentage and fragment-specific ion-count. In comparison, the SIMS2 program configured for the Apple II+ microcomputer, with which analyses of simple systems were compared, required up to 40 min execution time for a maximum of 6 fragments.

Figure 3 shows the excellent 'fit' achieved between the computed and observed ion-counts in the FABMS spectrum of 24-25-Si, for the overlapping fragment profile of CrO^+, $CrOH^+$, $CrOH_2^+$ with $Si_2O_2^+$, $Si_2O_2H^+$ and $Si_2OH_2^+$.

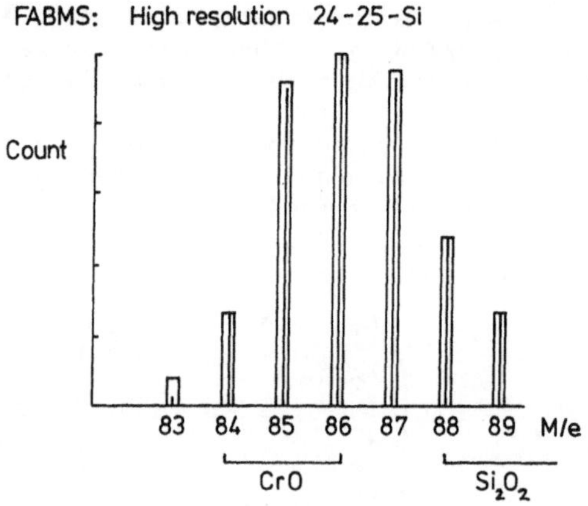

Fig. 3. FABMS spectrum, high resolution, for 24-25-Si
⊓ experimental data, | computed fit.

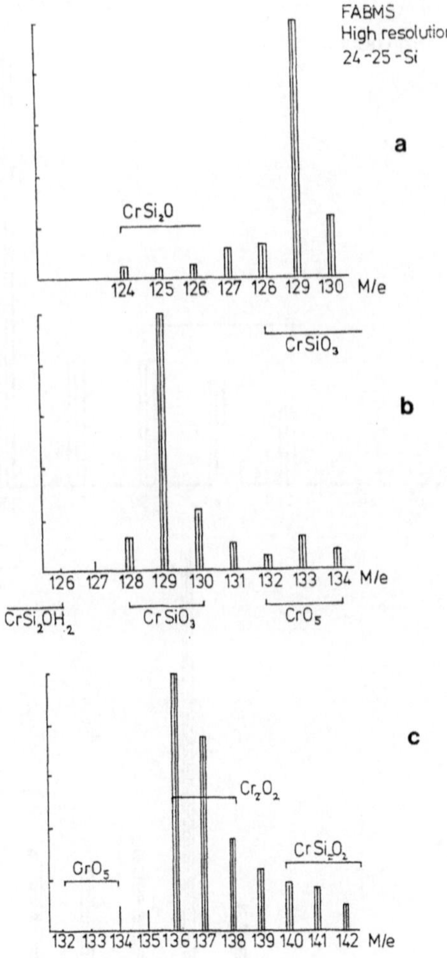

Fig. 4. FABMS spectra, high resolution, for 24-25-Si
⊓ experimental data, | computed fit.

Similarly, figure 4 shows FABMS computer-fitted spectra for 24-25-Si over different m/e ranges. The purpose of these analyses is demonstrated well in figure 4(a), where an accurate estimate of specific counts for the $CrSi_2O^+$ ... family was heavily obscured by the contributory counts from the more preponderant family of $CrSiO_3^+$ .... Likewise, as in figure 4(b), a confident assignment of specific count is now possible for the minority species CrO_5^+ ... in the presence of the $CrSiO_3^+$ ... family. Nevertheless the poor mismatch obtained at m/e 139, figure 4(c), reveals the presence of contaminating species which are usually hydrocarbon residues adsorbed on the sample. Thus, m/e 91 always shows a large, spurious ion current, presumably, due to the particularly stable tropllium cation, $C_7H_7^+$. At higher values of m/e, where the occurrence of fragments generates almost a continuum of overlapping patterns, the analyses yielded excellent computer fits, enabling useful comparisons of ion-yield for species such as CrO_6^+ ... and $Cr_2O_3^+$ ..., figure 4(d), $Cr_2SiO_2^+$ ... and $Cr_2O_4^+$ ..., figure 4(e), and $Cr_2Si_2O^+$ ... and $Cr_2SiO_3^+$ ... figure 4(f).

Figure 5 reproduces the extremes of comparative behavior of CrO_3/SiO_2 samples in the FABMS experiments. Each ion-count at a value of m/e is a specific count at that mass due to a specific fragment, all other contribu-

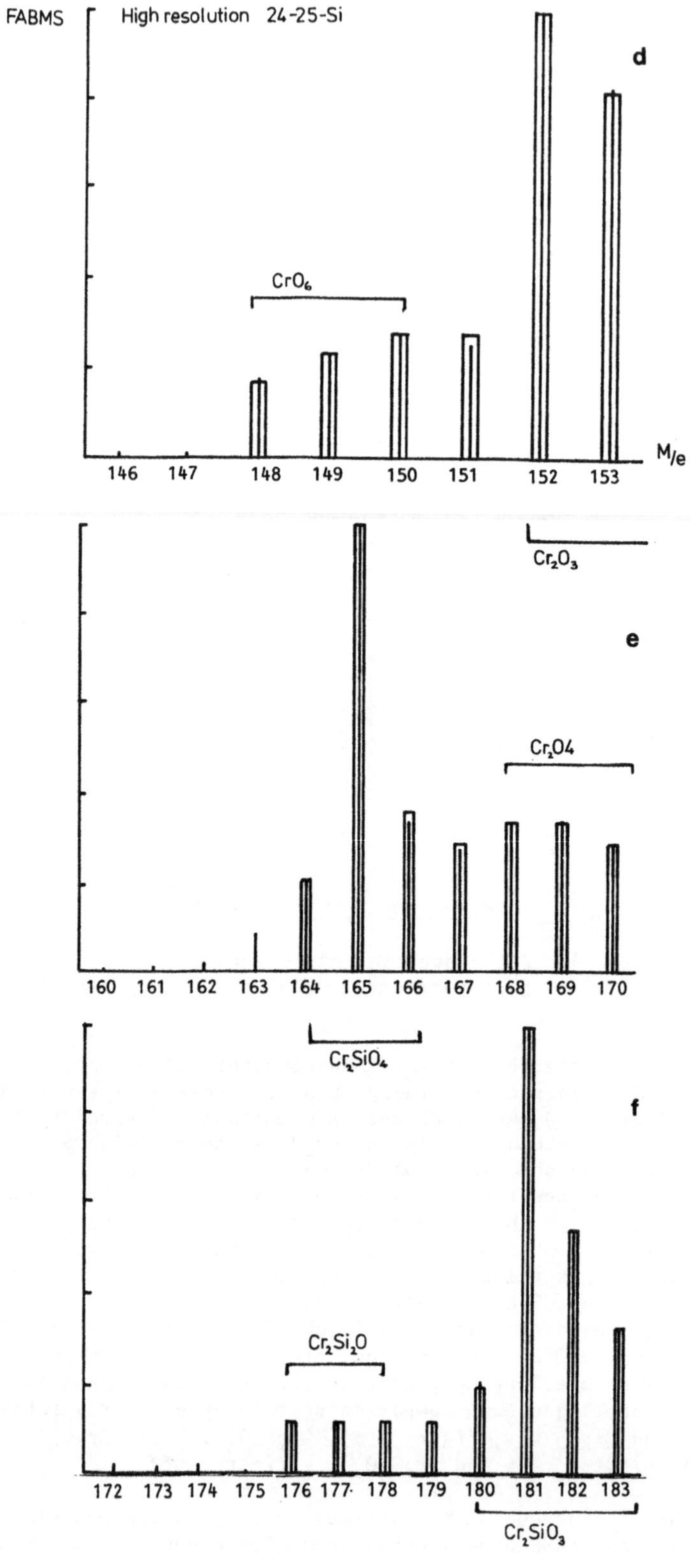
FABMS
High resolution 24-25-Si
d
CrO6
Cr2O3
M/e
146 147 148 149 150 151 152 153
e
Cr2O4
Cr2SiO4
160 161 162 163 164 165 166 167 168 169 170
f
Cr2Si2O
Cr2SiO3
172 173 174 175 176 177 178 179 180 181 182 183

Fig. 4 cont'd

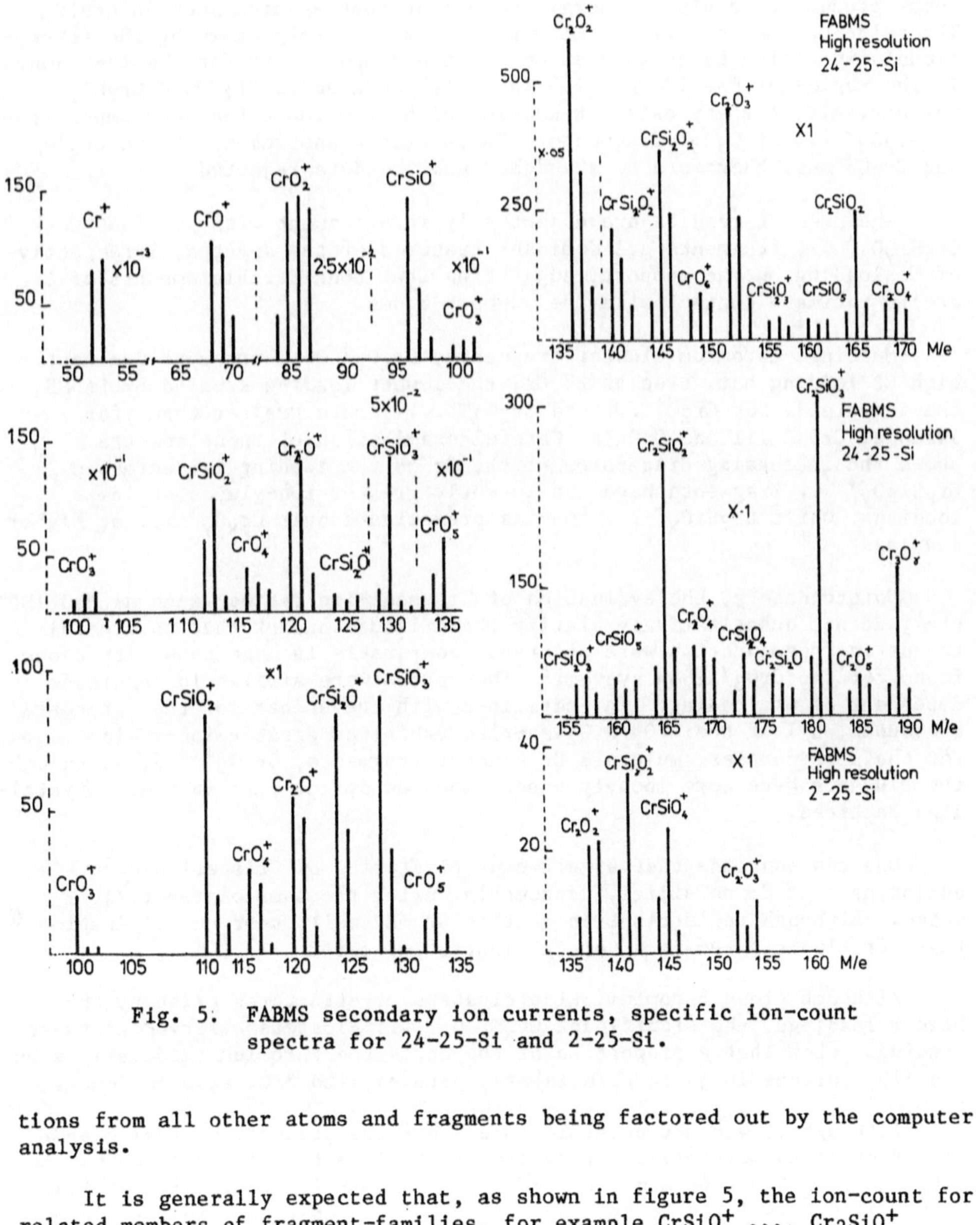

Fig. 5. FABMS secondary ion currents, specific ion-count spectra for 24-25-Si and 2-25-Si.

tions from all other atoms and fragments being factored out by the computer analysis.

It is generally expected that, as shown in figure 5, the ion-count for related members of fragment-families, for example $CrSiO^+$ ..., Cr_2SiO^+ ..., Cr_3SiO^+ ... or $CrSiO^+$ ..., $CrSiO_2^+$ ... $CrSiO_3^+$ ... usually decreases as the value of $^m/e$ increases. It is therefore significant that species such as $CrSiO^+$ ..., $CrSiO_2^+$ ... and $CrSiO_3^+$ ... have consistently greater ion-counts than do the essentially simpler species such as CrO_3^+ ..., CrO_4^+ ..., CrO_5^+ ... etc. Care should be exercised in expressing significance to these values in an absolute sense, since the sputtering coefficients of such fragments remain unknown.

The detection of a fragment does however reflect the fact that the atoms in that secondary-ion fragment must have been nearest-neighbors in the surface of the sample before the fast-atom bombardment. A high value of ion-yield reflects both the preponderance of the species and the stability of that species, in relative terms, in the spectrometer conditions. Thus, a species such as $CrSiO^+$ may have a separate existence both on the catalyst surface and in the secondary-ion beam as well as being a decay product of a higher energy, more transient species such as $CrSiO_3^+$. The value of the computer-fitting procedure was amply shown by the (exceptional) inability to achieve an acceptable computer-fit for the ion counts in the region of $^m/e$ 104 and 105 facilely assigned to Cr_2^+ and Cr_2H^+, respectively, for any catalyst sample, with or without the superimposition of $Si_2O_3^+$ ... $Si_2O_3H_3^+$ fragments. The previous assignment (6) to Cr_2^+ and Cr_2H^+ must therefore be abandoned and the data rejected.

The present results are essentially in agreement with previous work.[6] $Cr_xSi_yO_z^+$ ... fragments are dominant features of the spectra, irrespective of Cr loading on the support, suggesting that considerable amounts of Cr are in intimate contact with the support atoms.

Multiple-chromium cluster fragments, $Cr_xO_y^+$ ..., are more dominant at high Cr loading but, even at 2% Cr, the lowest loading studied by FABMS, the ion-yields for Cr_2O^+ ... and $Cr_2O_2^+$ ... remain greater than, for example, CrO_4^+ 111 and CrO_5^+. Careful examination of these spectra shows the increasing dispersion of the Cr as the loading is decreased. $Cr_xSi_2O_y^+$ ... fragments have consistently greater ion-yields at lower loadings, while $Cr_2SiO_x^+$ ... species predominate over $Cr_xO_y^+$ ... at higher loadings.

Unfortunately, the evaluation of fragment-ion ratios, such as M_2O^+/MO^+ etc., do not quantitatively clarify these issues except that no general trends in fragmentation were observed, comparable in magnitude with those found for the CrO_3/Al_2O_3 system.[6] The ratios were similar in magnitude, dependent of Cr loading. In comparison with the spectra of the 'standard' compounds, all of the CrO_3/SiO_2 samples exhibited greater ion-yields for the more complex, multiple Cr cluster fragments, $Cr_xO_y^+$ ..., as though the clusters were more loosely bonded when on silica than in their crystalline lattices.

One can conclude that significant clustering of Cr species occurs at as low as 1.6% Cr on silica, presumably during the sample preparation stage. Although redistribution of this Cr may well occur on calcination,[6] these Cr clusters remain after calcination at 525°C.

Although gross chromium oxide clusters or structures exist at the higher loadings, the significant $Cr_xSi_yO_z^+$ ... clusters observed at these loadings, show that a proportion of the deposited chromium oxide exists on the SiO_2 surface in quite thin layers, perhaps 1 to 3 Cr atom in depth.

Although it was not possible to express the proportion of single-Cr with respect to multiple-Cr fragments, it is important in this context to cite the results of Ohlmann et ak[20] for extremely low Cr loading silica. At 0.001% Cr, isolated Cr species were found to congregate in the vicinity of the contact areas of the primary silica particles. Such species will be the least accessible to the fast-atom sputtering beam and therefore essentially non-contributory to the secondary-ion yield and may well be enveloped within growing Cr clusters at higher loadings. The inter-particle voids as well as the pore-structure of the particles provide centres for metal oxide aggregation due to capillary action.[15]

X-ray Absorption Spectra (EXAFS/XANES)

The raw data was processed using the program EXCALIB, part of the SRS Daresbury suite of EXAFS programs, to convert the data based upon Si (III) monochromator angle into normalized spectra of absorption versus electron energy in eV.

Figure 6 shows the spectra produced by EXCALIB for both the standard samples and the highest loading CrO_3/SiO_2 sample, 24-25-Si. It is still possible to use these spectra to 'fingerprint' the chromium oxide phases.[21] Thus, for the standards featured in the spectra, figures 6(a) to 6(c), it is obvious that the Cr oxidation state (VI) is characterized by a pronounced pre-edge peak; this feature is absent for Cr (III). Immediately, for the catalyst samples for which the oxidation state is well defined and generally predictable, the same comments are tenable: a pre-edge feature for 24-25-Si, Cr (VI) oxide merely vacuum dried, which is absent for 24-525-Si, a green colored solid after air calcination at 525°C.

The spectrum for crystalline Cr_2O_3, figure 6(a), shows considerable XANES/EXAFS structure which is perceptible, without background reduction, to 600eV beyond the Cr-edge. The near-edge (XANES) structures of $Cr_2O_7^{2-}$, figure 6(b), and of CrO_3, figure 6(c), do show significant differences, as do the EXAFS features in the peak-shoulders within the energy range 6075 - 6150 eV. The CrO_3 (290°C) sample, figure 6(d), shows a somewhat intermediate pattern with the XANES region being similar to that of CrO_3, figure 6(c), while the EXAFS pattern, 6075 - 6150 eV, is more similar to that of $Cr_2O_7^{-2}$, figure 6(b). In fact the X-ray diffraction patterns of CrO_3 and of CrO_3 (290°C) are identical within experimental error suggesting that the latter product is merely an undecomposed distillate of CrO_3 (see TPR section). The product CrO_3 (360°C) however, figure 6(e), shows a distinguishable XANES structure which is quite unlike the other spectra in this series of 'standards'. The X-ray powder diffraction pattern of this material, while showing dissimilarities from that of CrO_3, is not identical to that claimed for CrO_2.[10]

The spectra of 24-525-Si, figure 6(f), and of CrO_3(560°C), are virtually superposable with the spectrum of crystalline Cr_2O_3, figure 6(a). The X-ray powder diffraction patterns of these samples, and of 11-525-Si, also have very similar d-spacings, within $\pm$ 0.02 Å for d > 2.18 Å and $\pm$ 0.004 Å for d < 2.18 Å. Even at 11% Cr on silica calcination at 525°C, in air, produces a majority Cr-clump structure similar to that of bulk, crystalline Cr_2O_3.[7]

The black-colored material, 24-300-Si, has a spectrum (figure 6(g)) which is intermediate between figures 6(c) and (d); the X-ray diffraction pattern revealed somewhat broad and diffuse reflections which had more similarity with CrO_3 (360°C) than with CrO_3. This material also has a well-defined β_2-phase ESR signal, whose characteristics suggest mixed-valency composition and double-exchange-ferromagnetic interactions.[7]

The spectrum of 24-25-Si, figure 6(f), closely resembles that in figure 6(b) of $Cr_2O_7^{2-}$ species, suggesting that a dichromate (VI) phase is deposited on the surface of the support rather than a phase of linked CrO_4^{2-} tetrahedra.

Brown et al[22] have shown that for dispersed manganese dioxide oxidation catalysts, the shift of the Mn-edge, relative to Mn-metal, can be related to the manganese oxidation state. In the case of chromium, the edge structure is particularly complex and the Cr-edge position should properly be computed via the first differential of the absorption along the edge.[22] Table II gives the values of Cr-edge shift, Δ E, plotted in figure 7, against the Cr oxidation number of the standard compounds Cr(0), Cr_2O_3, and CrO_3 and $Na_2Cr_2O_7$.

Although the Cr-edge positions have merely been estimated, the data do show that the Cr-edge shift is related to the Cr-oxidation state. While the Cr oxidation number for CrO_3 (290°C) is found to be probably (VI), that of CrO_3 (360°C) approximates to (IV), in accord with an intermediate mixed valence chromium oxide. Similarly, an oxidation number of (V) is

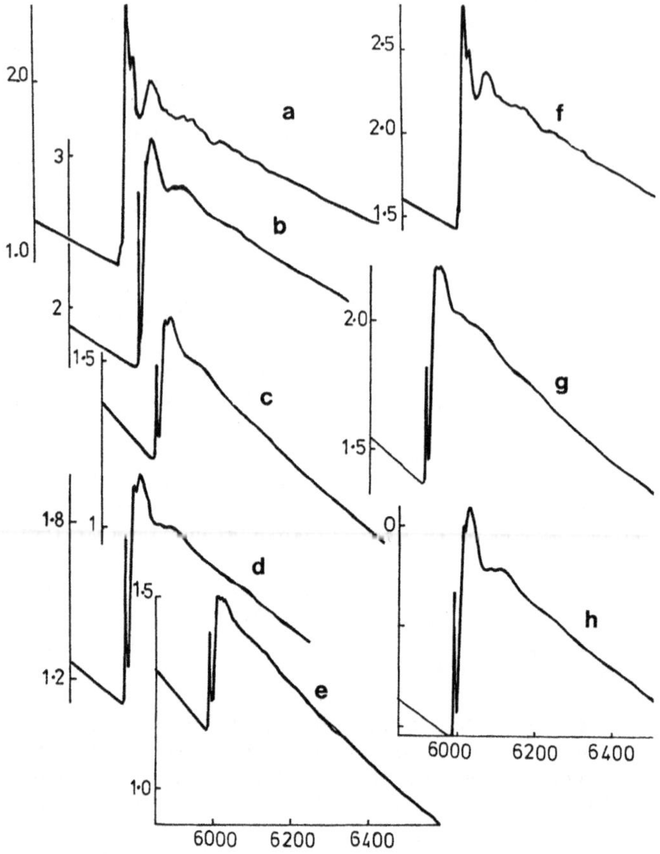

Fig. 6. Normalized X-ray absorbtion spectra of standard chromium compounds and CRO_3/SiO_2 samples.
a) Cr_2O_3, b) $Na_2Cr_2O_7$, c) CrO3, d) CrO_3(290°), e) CrO_3(360°C), f) 24-525-Si, g) 24-300-Si, h) 24-25-Si.

found for the 24-300-Si sample, suggesting that the supported chromium is in the form of a chromium (III) chromate (VI), $Cr_2^{III}(Cr^{VI}O_4)_3$, in agreement with the TPR conclusions. The continuing analysis[21] will relate the edge-shift ΔE with the nearest-neighbor Cr-O distances and with the effective atomic charge on the Cr in addition to the evaluation of metal ion cluster size and near-neighbor shell radii.[22]

Electron Spin Resonance (ESR)

The development of the so-called γ-resonance[23] with increasing calcination temperature is depicted in figure 8 for 1-25-Si and 18-25-Si. At loadings of 7% Cr or greater, an extremely narrow resonance was obtained for the samples immediately after drying at 25°C, $\bar{g} \sim 1.973$.[7] On calcination at higher temperatures in air, the samples exhibit a 'normal', anisotropic γ-resonance, although with a much smaller peak-to-peak width, ΔHpp, than usually reported, and whose intensity is linked to the formation of a β_2-mixed valence phase of chromium.[7]

The temperature-dependence of ΔHpp was found to be irreconcilable with that behavior expected on theoretical grounds for magnetically-dilute Cr(V) species and the spectra suggest the presence of more than one resonance component, coupled by magnetic exchange forces. Thus figure 9

Table II. X-ray Absorption Edge Positions, Edge Shifts and Related Data For The Catalyst Samples and Reference Materials

Material	Cr-edge, E/eV	edge shift ΔE/eV	oxidation number
Cr metal	5989.1	0	0
Cr_2O_3	5999.5	10.4	3
$Na_2Cr_2O_7$	6007.3	18.2	6
CrO_3	6008.0	18.9	6[a]
CrO_3(290°C)	6007.2	18.1	6[a]
CrO_3(360°C)	6004.4	15.3	4[a]
CrO_3(560°C)	6000.5	11.4	3[a]
24-25-Si	6007.5	18.4	6[a]
24-300-Si	6006.2	16.9	5[a]
24-525-Si	6003.2	11.2	3[a]

a values derived by interpolation in figure 7

shows the γ -resonances observed for 7-25-Si, 11-25-Si and 18-25-Si at 298K, for which the ESR parameters are given in Table III.

For 11-25-Si (298K) only one extremely narrow component is apparently observed, the width increasing markedly at 77K and marginally further at 4.2K, figure 10. The apparent $\bar{g}$-value also increases at 77K but decreases slightly at 4.2K. Although the spectra of 7-25-Si and 18-25-Si, figures 9(a) and (c), at 298K show greater width due to the presence of distinguishable compounds, the parameters of the spectra at 77K and 4.2K, figure 10, Table III show similar trends to those of 11-25-Si and have almost the same absolute values.

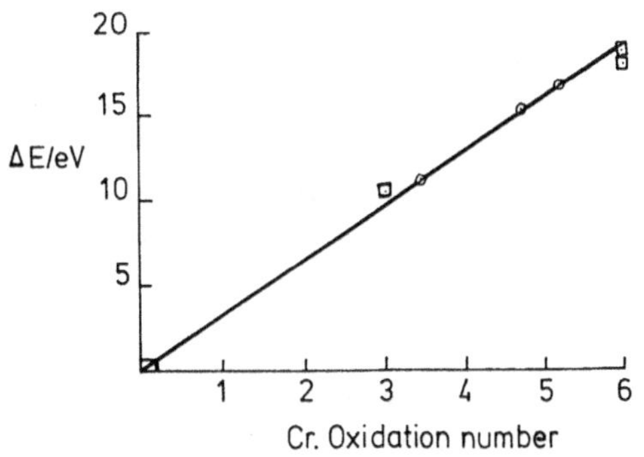

Fig. 7. Observed Cr-edge shift, ΔE, plotted against Cr oxidation number. □ standard Cr compounds, ○ interpolated points.

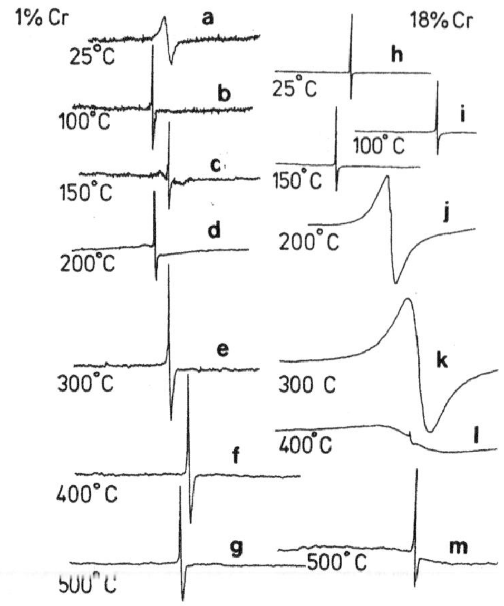

Fig. 8. ESR spectra of CrO_3/SiO_2 samples after calcination in air. Room temperature spectra, modulation amplitude 1.0 G, field sweep 5kG, X-band.
a) 1-25-Si, b) 1-100-Si, c) 1-150-Si, d) 1-200-Si, e) 1-300-Si, f) 1-400-Si, g) 1-500-Si, h) 18-25-Si, i) 18-100-Si, j) 18-150-Si, k) 18-200-Si, l)18-300-Si, m) 18-400-Si, n) 18-500-Si.
Reproduced with permission from Ref (7). Copyright 1984, Royal Society of Chemistry.

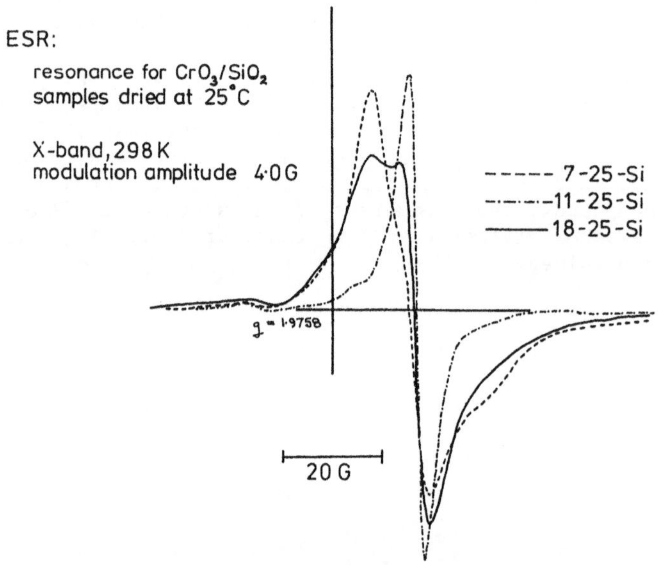

Fig. 9. γ-resonances observed in the ESR spectra of CrO_3/SiO_2 samples dried at 25°C. Room temperature spectra, modulation amplitude 4.0 G, field sweep 200 G.
- - - 7-25-Si; •-•-•-•11-25-Si; ——— 18-25-Si.
Reproduced with permission from Ref. (7). Copyright 1984, Royal Society of Chemistry.

The more recognizable, anisotropic γ-resonances of 2-200-Si and 11-100-Si, figure 11, also show the presence of more than one component and have unusual shape-variation with temperature.[7] Nevertheless, their $\bar{g}$-values clearly relate these resonances to 7-25-Si, 11-25-Si and 18-25-Si, Table III. The width and shape-variations with temperature were suggested[7] to arise from species experiencing exchange and dipole-dipole interactions, so that spin transitions occurred within a manifold of spin-coupled states whose population in zero magnetic field is temperature-dependent.[7]

The first-derivative resonances of 7-25-Si, 11-25-Si and 18-25-Si obtained over a 200 Gauss field scan, were numerically integrated to yield the resonance absorption envelopes. The computer program performed a trapezoidal integration for each integral value of field between 3246 and 3446 Gauss. Computer simulations of resonance line-shapes were then fitted to the experimental resonances using the following procedures:

1. Isotropic fit. The program assumed either Gaussian or Lorentzian line-shapes, although initial calculations quickly demonstrated that the experimental resonances were close to Lorentzian and the Gaussian function was abandoned.

Adopting the Lorentzian expression quoted by Gordy,[24] the shape function in terms of magnetic field, $f(H-H_o)$, has the form:

$$f(H-H_o) = \frac{1}{\pi} \times \frac{\Delta H_{1/2}}{(H-H_o)^2 + \Delta H_{1/2}^2} \qquad (8)$$

where $\Delta H_{1/2}$ is related to the effective relaxation time T_{eff} by:

$$\Delta H_{1/2} = \frac{h}{g\beta} \times \frac{1}{T_{eff}}$$

and to the peak-to-peak width, ΔH_{pp}, through:

$$\Delta H_{1/2} = \frac{\sqrt{3}}{2} \Delta H_{pp} \qquad (9)$$

$\Delta H_{1/2}$ is the half width of the resonance, measured at half-height, and for each experimental resonance, H_o is the 'turning-point' or the field-equivalent value of $\bar{g}$.

The computer program enabled any experimental signal to be 'best-fit' to the summed envelope of Lorentzian components, for which H_o, $\Delta H_{1/2}$ and f_{max} were freely varied, gradually increasing the number of components to achieve the best-fit. In all cases these calculations yielded components whose H_o values were identical between samples, within the enforced integration error of $\pm$ 1G. The constant widths, $\Delta H_{1/2}$, increased with measurement temperature for each sample. The distribution of component intensities varied with temperature and this effect was the major factor determining line-shape. It was intended to extend these analyses to γ-resonances with significant anisotropy, for example, $g_{\perp} = 1.9761$, $g_{/\!/} = 1.9544$, and therefore these precurser resonances were analyzed using a more general relationship, for which isotropy is a limiting case.

2. Anisotropy fit. The analysis of ESR signals derived from powder samples has been well described in Gordy,[24] who suggests that the equa-

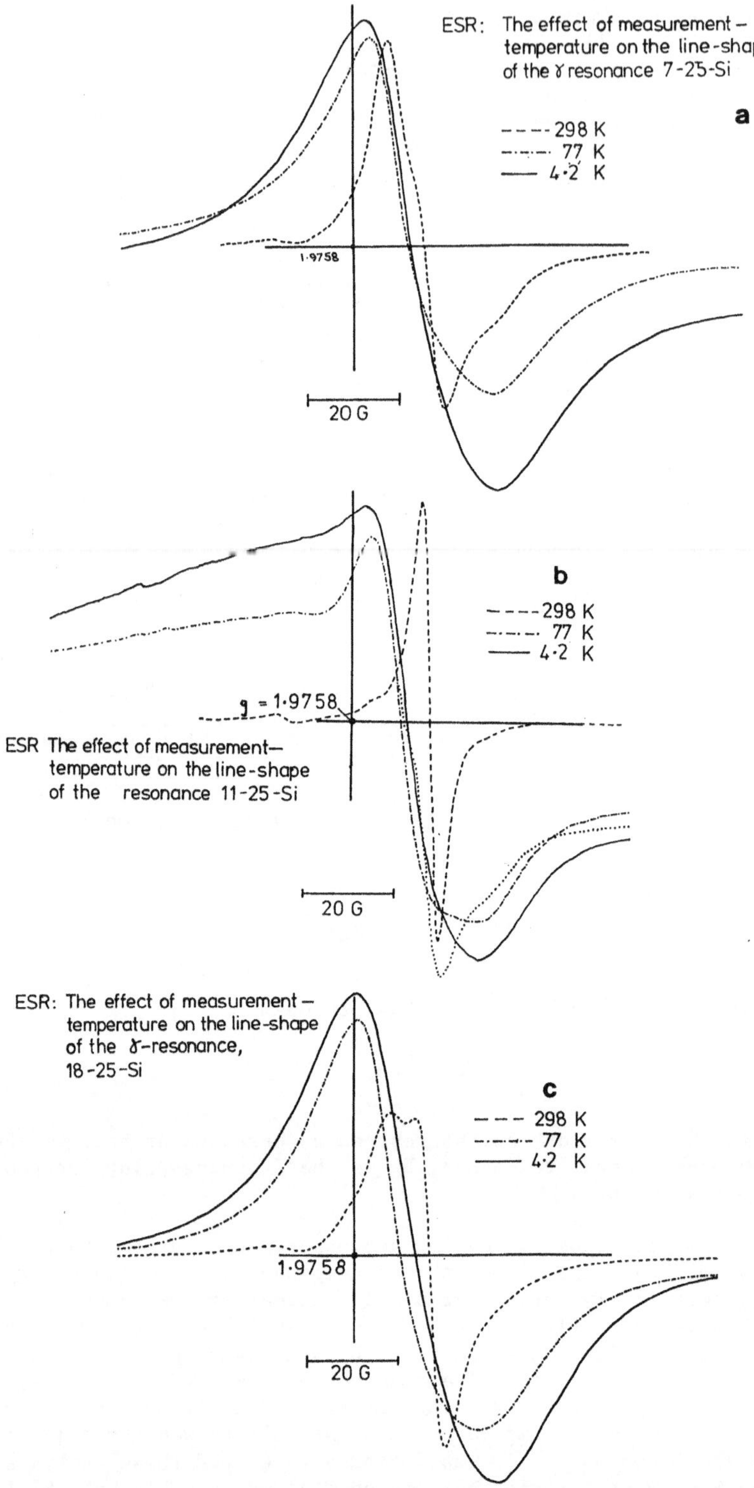

Fig. 10. Variation in γ-resonance shape with measurement temperature. Modulation amplitude 4.0 G, X-band spectra, field sweep 200G. a) 7-25-Si; - - - 298K, ·-·-·-·77K, ——— 4.2K. b) 11-25-Si, c) 18-25-Si. Reproduced with permission from Ref (7). Copyright 1984, Royal Society of Chemistry.

Table III. ESR Parameters for γ-Resonances of CrO_3/SiO_2 Samples

Sample	Temperature /K	$\bar{g}$	ΔHpp/G min	ΔHpp/G max
7-25-Si	298	1.9657	12.0	20.25
	77	1.9676	-	26.75
	4.2	1.9663	-	29.0
11-25-Si	298	1.9642	-	3.75
	77	1.9679	-	23.5
	4.2	1.9674	-	25.75
18-25-Si	298	1.9650	6.50	12.25
	77	1.9677	-	26.75
	4.2	1.9664	-	30.25
2-200-Si	298	1.9693	-	27.15
	77	1.9694	-	32.75
	4.2	1.9691	-	28.50
11-100-Si	298	1.9687	-	34.75
	77	1.9683	-	34.25
	4.2	1.9689	-	34.0

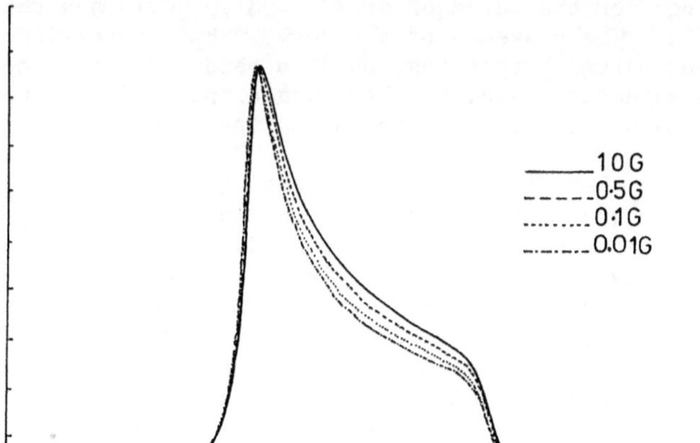

Fig. 11(a) Computer simulation of parameter Del X H,3326G
H 3406G
$\Delta H_{1/2}$ 3.75G

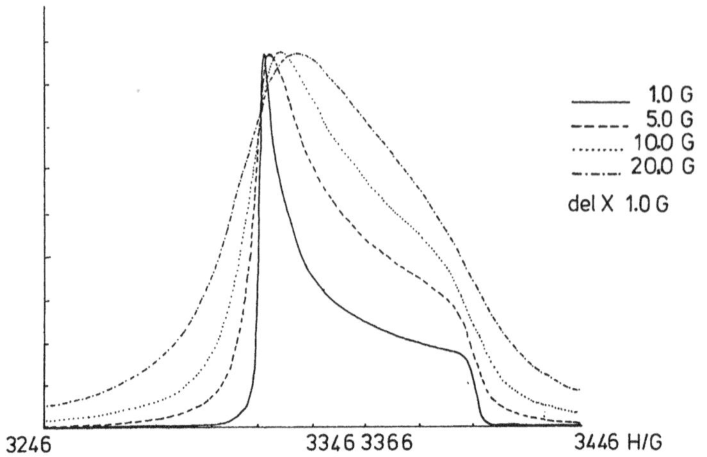

Fig. 11(b) Computer simulation effect of parameter $\Delta H_{1/2}$3326H ,$H_{/\!/}$3406G
Del X 1.0G

tion of Ibers and Swalen[25] is of particular value as it includes the variation of the absorption coefficient with g-value. The expression describes the intensity function with respect to magnetic field, I(H), for the cumulative envelope of all crystal orientation in the magnetic field, assuming an axially symmetric crystal field about the spin system i.e. $H_{/\!/} = H_z$ ($g_{/\!/} = g_z$) and $H_{\perp} = H_x = H_y$ ($g_{\perp} = g_x = g_y$)

with H $>$ H

Numerical integration between the limits $H_{\perp}$ and $H_{/\!/}$ was performed, again using the trapezoidal rule, H_r adopting each value in this range via the incremental parameter delX. As H_r approaches $H_{\perp}$, the integral rapidly approaches infinity so that H_r must never equal $H_{\perp}$ and the neglected integral part determined by ($H_r - H_{\perp}$) then constitutes numerical error depending upon the value of delX. $\Delta H_{1/2}$ again has the relationships implicit in (9). The presence of the term, $1/H_r^2$, determines that the function is not a true Lorentzian, as it depends severely upon the absolute values of the magnetic field, H. I(H) was computed for values of H between 3246 and 3446 Gauss, the field range used experimentally.

Figure 11(a) shows the results of testing the effect of delX on the approach to convergence of the envelope, minimizing the error (H_r-$H_{\perp}$), with ($H_{/\!/} - H_{\perp}$) = 80 Gauss. The value of delX is seen to be critical as convergence is only satisfactorily achieved with delX $\ll$ 0.01 Gauss. DelX also critically determines the absolute values of $H_{\perp}$ and $H_{/\!/}$ in the fittings procedure as the point corresponding to dI/dH = 0, and hence $H_{\perp}$ and $H_{/\!/}$, moves to lower field as delX increases. For comparison with Ibers and Swalen's calculation, figure 11(b) shows computed line shapes for different values of $\Delta H_{1/2}$. $H_{\perp}$ is easily identified as that point where all curves cross at H = 3326 Gauss but, as $\Delta H_{1/2}$ increases, the point dI/dH = 0 corresponding to $\bar{g}$ moves to increasing values of H and is of no real significance with respect to the experimental parameters $g_{/\!/}$ and $g_{\perp}$. Finally, figure 11(c) shows computed line-shapes (as a function of $\Delta H_{1/2}$) for the 'natural' values of $H_{/\!/}$ and $H_{\perp}$ which correspond to

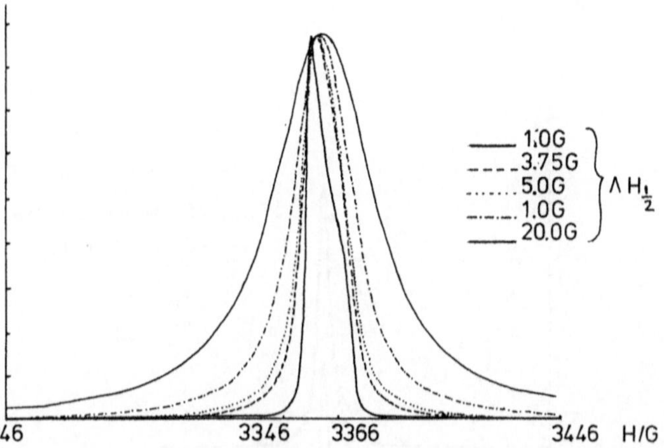

Fig. 11(c). Computer Simulation Effect of Parameter $H_{1/2}$H 3354G,H 3370G

interpretable g-values in the derivative spectrum of 11-25-Si (298K). Again $H_{\perp}$ is easily identified and there is an obvious shift in $\bar{g}$ as $\Delta H_{1/2}$ increases.

A computer program[26] was then used to achieve 'best fit' simulations by variations of the parameters $H_{\parallel}$, $H_{\perp}$, $\Delta H_{1/2}$ and peak intensity for an increasing number of components. During initial fitting delX was set to 0.1 Gauss, final fit being achieved at 0.01 Gauss. All parameters were freely varied between sensible limits, every combination of the parameter set being tested by a least squares criterion of best-fit between pre-set limits of magnetic field.

Figure 12(a) shows the experimental absorption spectrum for 11-25-Si at 298K, apparently a single, symmetric resonance at $\bar{g}$ = 1.9642 (H = 3365.35G) and ΔH_{pp} = 3.75 G, Table III, figure 9. Assuming one component with $H_{\parallel}$ = 3363.0 G, $H_{\perp}$ = 3361.25 G and a value of $\Delta H_{1/2}$ calculated from ΔH_{pp} as 3.249, figure 12(a) shows that the resonance is indeed predominantly a single Lorentzian component with a poor fit at low field. The best-fit curve, figure 12(b), uses the components and parameters in Table IV.

The experimental curve for 7-25-Si at 298K is compared, in figure 13(a), with single Lorentzians based on the maximum and minimum values of $\Delta H_{1/2}$ derived from the ΔH_{pp} values in Table III. $\Delta H_{1/2}$ max is an obviously poor fit. Although $\Delta H_{1/2}$ min appears to be a closer approximation, the poor fit around $\pm$ 20G of the peak maximum suggests the presence of at least one other component at lower field. Similar considerations apply for 7-25-Si at 77K, figure 13(b), and 7-25-Si at 4.2K, figure 13(c). At 77K, $\Delta H_{1/2}$ derived from ΔH_{pp} is 23.2 Gauss, a value yielding a very poor fit. Two components with $\Delta H_{1/2}$ at only 11.6 G improves the fit at $\pm$ 20G about the peak maximum, but any increase in $\Delta H_{1/2}$ above this value causes serious errors in the wings of the curve. A better fit is seen to be achieved for a single component with $H_{\parallel}$ at 3377 G and $H_{\perp}$ at 3351 G, but it is impossible to explain the increase in crystal field anisotropy with a decrease in temperature and spin-lattice relaxation theory must also cope with the increase in $\Delta H_{1/2}$. More rationally, the best fit curve is shown in Figure 13(b). A similar analysis with similar problems is encountered for the 18-25-Si series of samples for which a full list of parameters is given in Table IV.

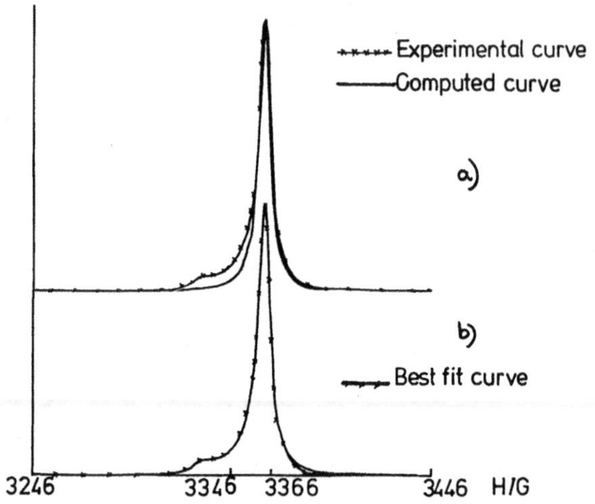

Fig. 12. ESR - Spectrum simulation for 11-25-Si(298 K)

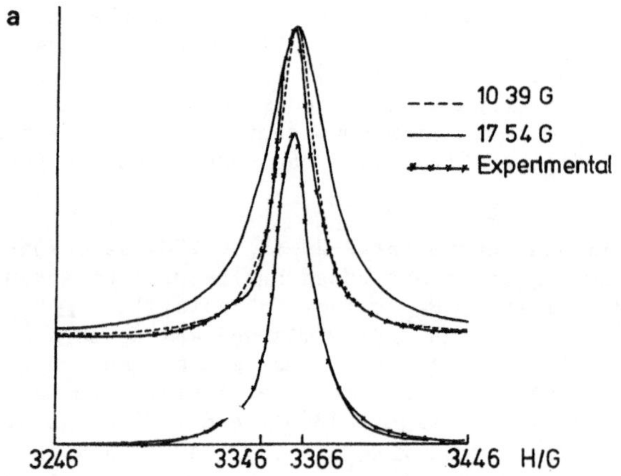

Fig. 13. Computer best-fit simulations for 7-25-Si at various temperatures. x x x x experimental curve, —— best fit curve. a) 298K: single components, —— $\Delta H_{1/2}$ 17.54G, - - - $H_{1/2}$ 10.39G.

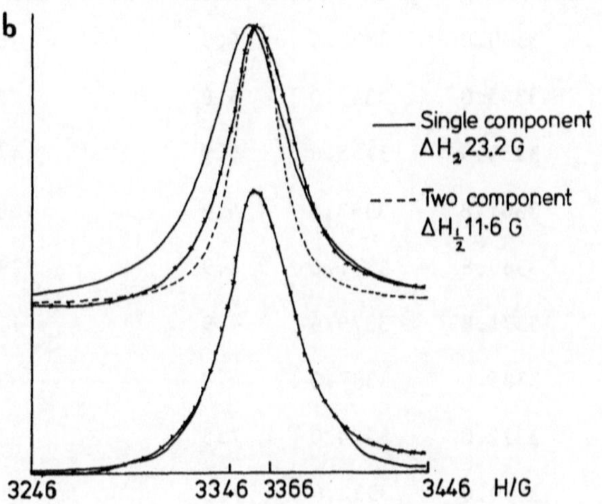

Fig. 13. b) 77K: ——— single component, $\Delta H_{1/2}$ 23.2G.
- - - two components, $\Delta H_{1/2}$ 11.6G

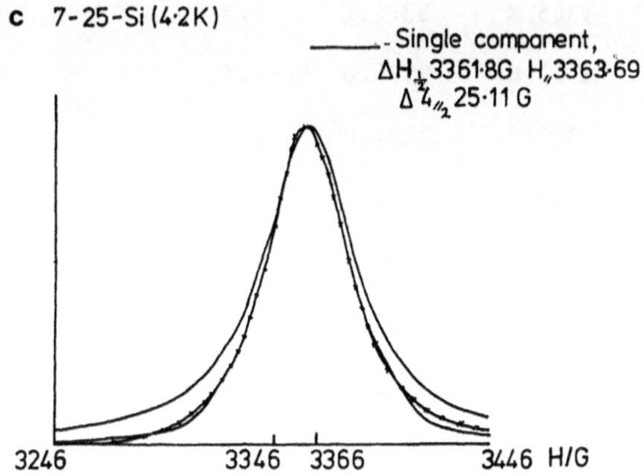

Fig. 13. c) 4.2K: ——— single component, $\Delta H_{1/2}$ 25.11G.

Table IV. Best-Fit ESR Parameters for the CrO_3/SiO_2 sample, 18-25-Si, at different measurement temperatures T/K.

T/K	peak	$H_\perp$ /G	$H_{//}$ /G	$\Delta H_{1/2}$/G	Intensity
298	2	3330.0	3332.0	4.9	33
	3	3345.0	3347.0	4.9	22
	4	3354.0	3355.6	4.9	478
	5	3361.8	3363.6	4.9	1000
	6	3369.8	3371.6	4.9	298
	7	3371.8	3379.6	4.9	113
	8	3385.0	3387.0	4.9	32
77	1	3322.0	3324.0	7.5	56
	2	3330.0	3332.0	7.5	113
	3	3337.8	3339.6	7.5	121
	4	3345.8	3347.6	7.5	600
	5	3353.8	3355.6	7.5	1000
	6	3361.8	3363.6	7.5	829
	7	3369.8	3371.6	7.5	561
	8	3377.8	3379.6	7.5	292
	9	3385.8	3381.2	7.5	147
	10	3393.8	3395.6	7.5	72
42	1	3322.0	3324.0	8.0	66
	2	3330.0	3332.0	8.0	127
	3	3337.8	3339.6	8.0	243
	4	3345.8	3347.6	8.0	535
	5	3353.8	3355.6	8.0	1000
	6	3361.8	3363.6	8.0	983
	7	3369.8	3371.6	8.0	751
	8	3377.8	3379.6	8.0	486
	9	3385.8	3381.2	8.0	321
	10	3393.8	3395.6	8.0	150

The spectrum of 11-25-Si at 298K is an exception, displaying only one major resonance at $H_{\parallel}$ = 3363.0 $\pm$1G and $H_{\perp}$ 3361.0 $\pm$1G with a narrow peak width, $\Delta H_{1/2}$ = 3.5 G. For all other analyzed spectra, computer simulation yields the same pattern of resonances, yielding the component peaks listed as the mean values $\bar{H}_{\parallel}$ and $\bar{H}_{\perp}$ in Table V, with error $\pm$0.5 Gauss. The overall mean anisotropy ($\bar{H}_{\parallel}$ - $\bar{H}_{\perp}$) is 1.8 $\pm$ 0.1 Gauss, the overall mean separation of parallel components is 8.0 $\pm$ 0.3 Gauss and the overall mean separation of perpendicular components is 8.0 $\pm$ 0.3 Gauss. At any value of measurement temperature, $\Delta H_{1/2}$ is constant for all components, $\pm$ 0.3 Gauss, but this value increases as the temperature decreases (see, for example, Table IV) as does the value of ΔH_{pp} for the experimental envelopes, Table III.

It is the combination of the changing $\Delta H_{1/2}$ values together with changes in the intensity distribution of components on decreasing the measurement temperature, which causes the distinctive changes in shape shown in figures 9 and 10.

These analyses confirm that the observed resonance envelopes arise from spin transitions within a manifold of spin-coupled states whose zero-field energies are determined by magnetic exchange interactions and whose population is temperature dependent. Studies are continuing[27] to evaluate the zero-field degeneracies and splittings of the spin states in correlation with the computed line separations and intensity distributions to determine the cluster size of neighboring Cr species.

The samples which exhibit these unusually narrow and anomalously temperature-dependent spectra are relatively high loading, heavily clustered CrO_3 based materials, in which a minimal yet unavoidable reduction to low oxidation states has occurred. If the odd electron or electrons associated with a lower oxidation state is delocalized over a number of neighboring chromium centers the motional exchange narrowing in the cluster would generate a narrowed envelope of overlapping component resonances subjected to increasing dipole-dipole broadening as the measurement temperature decreases.

Excellent support for these conclusions is provided by the recent work of Fricke and Ohlmann[28] on the related area of free - and supported - 12 - heteropolyacids. This work concerns ESR studies of air and vacuum dehydration and reduction of $H_3PMo_{12}O_{40}$ (PMo_{12} - HPA) and $H_4SiMo_{12}O_{40}$ ($SiMo_{12}$ - HPA) heteropolyacids, in both the free state and supported on silica and alumina. The principal feature of these materials is a cage structure of 12Mo(VI) ions in the so-called Keggin ion. Even low temperature dehydration of these compounds, leads to some reduction of molybdenum to Mo(V) which, having the isotopes ^{95}Mo and ^{97}Mo with 23.5% total abundance and I = $^5/2$, should show well-resolved and distinctive hyperfine structure in its ESR spectra.

We find a great variety of unusual lineshapes rendering analysis difficult and speculative, while most spectra are devoid of hyperfine structure. At low temperatures of dehydration up to 150°C, while the free-HPA is dissolved in its own water of crystallization, a single most symmetric and narrow resonance is observed, figure 14(a). This signal is well known and characteristic of strong electron delocalization of the unpaired electron in the mixed-valence Keggin ion structure,[29] so that fast hopping of the electron occurs between the 12Mo species, with a hopping rate in excess of $10^8 s^{-1}$.

Significantly, the same signal is observed for 20% $PMo_{1/2}/Al_2O_3$, figure 15(a), immediately after drying in air and after calcination at temperatures up to 100°C. Not only is the Keggin ion structure preserved at this

Table V. Mean Values of $H_{//}$ and $H_{\perp}$ for CrO_3/SiO_2 Samples Dried at 25°C

Peak	$\bar{H}_{//}$ /G	$\bar{H}_{\perp}$ /G
3	3339.5	3337.6
4	3347.4	3345.5
5	3355.6	3353.9
6	3363.3	3361.6
7	3371.5	3369.7
8	3379.5	3377.6
9	3387.4	3385.5

loading and heat treatment, but sufficient water remains at the catalyst surface to enable fast electron-hopping to proceed with motional exchange narrowing of the resulting cluster of lines. Prados and Pope[29] estimate that for Mo-HPA the envelope should consist of 64 unresolved and extremely narrow resonance components. The similarity with this CrO_3/SiO_2 system is obvious for high loading and low-temperature dried samples which yield similar spectra.

As the dehydration temperature is increased, anisotropic resonances are observed for both free and supported HPA, figures 14 and 15, but their

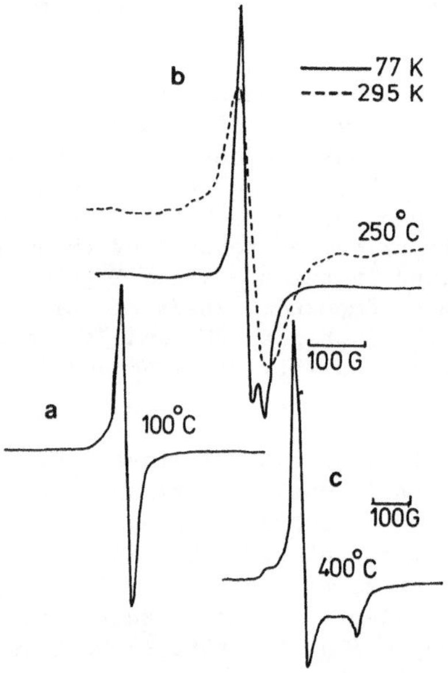

Fig. 14: ESR spectra of PMo_{12}-HPA dehydrated in air for 1 hour at various temperatures; X-band spectra. a) 100°C, b) 250°C, c) 400°C. - - - 295K, —— 77K. Reproduced with permission from Ref (28). Copyright 1985, Royal Society of Chemistry.

widths remain small and hyperfine structure is still not observed. The line shape behavior is not simple, figure 14(b) and figure 15, again suggesting correlations with the CrO_3/SiO_2 system. These signals are identified[28] with a dehydrated but structure-preserved Keggin ion, for which the axial symmetry indicates that the unpaired electron is more localized at Mo positions. The loss of water limits the mobility of the electron, drastically lowering the hopping rate.

The absence of hyperfine structure, the narrowness of the lines and their temperature behavior indicates the influence of exchange interactions coupling the electron spins. In the CrO_3/SiO_2 system, after calcination at $\geqslant$ 100°C,[7] narrow, anisotropic γ-resonances are observed. The strange shape-measurement temperature-calcination temperature relationships remain until relatively high (> 400°C) calcination has produced severe reduction of Cr(VI) and complete dehydration.

The strange shape behavior reported[7] for CrO_3/SiO_2 is reproduced in the spectra of mildly, hydrogen-reduced free HPA, figure 16, and of PMo_{12}/SiO_2 reduced in H_2, figure 17. Unresolved components, whose influence on line shape, arises from changes in their widths. Spin-lattice relaxation effects cannot explain the temperature-dependent spectra. The line width diminishes as the amount of reduction increases, presumably due to exchange narrowing.

These correlations help to confirm[30] that γ-resonance in CrO_3/SiO_2 arises from chromium clusters, in contrast to the more usual assignment towards magnetically isolated Cr(V) species. The PMo_{12}/SiO_2 spectra show that structure can be preserved on the surface of a support confirmed by the TPR, FABMS and EXAFS studies here. The mixed oxidation state species

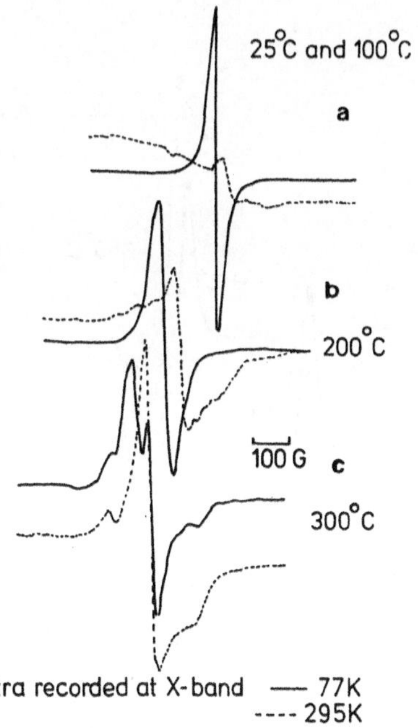

Fig. 15. ESR spectra of PMo_{12}/Al_2O_3 (20% HPA), dehydrated in air for 1 hour at various temperatures, X-band spectra. a) 25°C and 100°C, b) 200°C, c) 300°C. - - - - 295K, ——— 77K. Reproduced with permission from Ref (28). Copyright 1985, Royal Society of Chemistry.

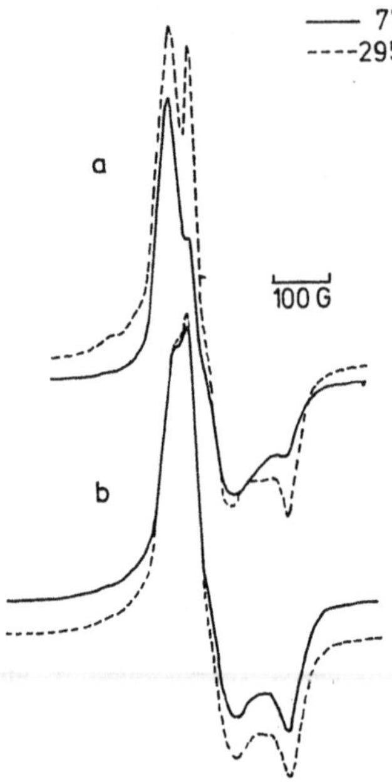

Fig. 16. ESR spectra of PMO_{12}-HPA reduced with hydrogen at 300°C for 2 hours, X-band spectra. a) without further treatment, b) further evacuated at 300°C for 1 hour. - - -- 295K, ——— 77K. Reproduced with permission from Ref (28). Copyright 1985, Royal Society of Chemistry.

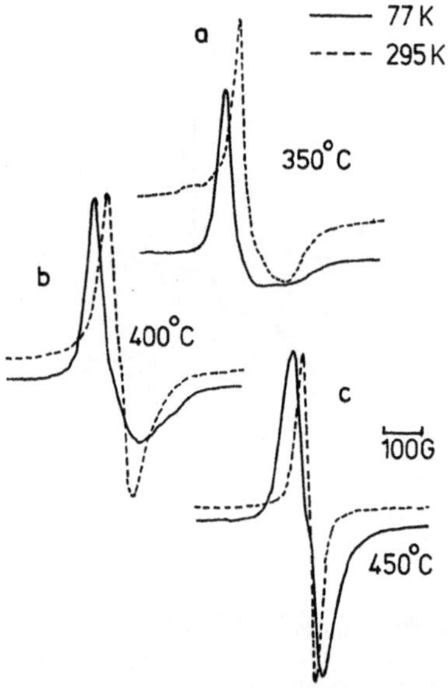

Fig. 17. ESR spectra of PMo_{12}/SiO_2 reduced in hydrogen for 1 hour at various temperatures, X-band spectra. a) 350°C, b) 400°C, c) 450°C. - - - - 295K, ——— 77K. Reproduced with permission from Ref (28). Copyright 1985, Royal Society of Chemistry.

involved in the CrO_3/SiO_2 require more evaluation, but at least up to calcination temperatures of 400°C, the evidence[6,7] suggests a Zener exchange-coupled phase of Cr^{3+} and Cr^{6+} species.

SUMMARY

The combination of techniques used[6,7] demonstrates that at loadings above 5 wt% Cr, gross structures of chromium oxide exist on the silica surface. After drying, they have either chromate or dichromate structure; after calcination between 200°C and 400°C, they have a Cr(III), Cr(VI) mixed valency structure; and above 500°C calcination temperature the structures are close to α-Cr_2O_3. At loadings between 0.2% and 2%, single molecular layer clusters of chromium species predominate on the surface.

This clustering behavior is to be expected from the known chromium redox chemistry[31] and its redox behavior in catalysis. Even in solution, the reduction of Cr(VI) is dominated by facile electron transfer between bi-nuclear intermediates. For example, the oxidation of alcohols requires the self-redox reaction: $Cr(IV) + Cr(VI) \rightarrow 2Cr(V)$, but each slow rate-determining step in the reactions requires an activated complex including 2 Cr atoms. Thus, although $Cr(V) + e \rightarrow Cr(IV)$ is found to be the slow step in the reduction of Cr(VI) and the oxidation of Cr(III),[31] a chemical equilibrium is required between Cr(III) and Cr(VI) to fit the observed kinetics, so that $Cr(V) \propto [Cr(III)]^{1/3}\ [Cr(VI)]^{2/3}$.

The cluster hypothesis is in agreement with the work of Rebenstorf and Larsson[32,33] who recognize the facile redox cycle of Cr which is associated with multi-Cr species. They suggest[33] that the dinuclear species offer an easy solution to the mechanistic problem which involves a 'missing' hydrogen atom in the polymerization initiation reaction. In the insertion of 1 ethylene molecule between the 2 Cr centers, no C-H bond needs to be broken!

Groeneveld et al[34,35] conclude that all active chromium complexes on silica contain at least 2 Cr cations, probably occurring as Cr(II) - Cr(III) dimers, even at 1% Cr. Contrastingly, the elegant work of McDaniel et al[4] is interpreted via the basic assumption of monodispersed surface (Cr(VI) in CrO_3/SiO_2. The assumption implies that surface hydroxyls are evenly and uniformly distributed over the heterogeneous silica surface so that at low Cr loading, random interaction between Cr species and surface hydroxyls gives rise to isolated Cr species.

In consequence, their method of expression of Cr loading, as Cr atom per unit surface area, is quite misleading. At approximately 2.5 Cr nm^{-2}, corresponding to approximately 6 wt% Cr, McDaniel finds[36] the maximum surface-isolated limit and the maximum amount of Cr established as Cr(VI),[37] but the calcined catalyst is observed to change from orange to brown and eventually to green.[38] The brown color is observed in these systems when Cr(III) and Cr(VI) occur together in a mixed valence species, while the green color signifies Cr(III) in a clumped condition - Cr(III) on silica at very low loadings in a more dispersed state, is pale blue or grey.

Above this loading McDaniel finds[37] α- Cr_2O_3 from X-ray powder diffraction, signifying that the clumps are now so extensive and structured, even at relatively low loadings, that a Cr(III) oxide lattice exists. It is inconceivable that there is a discontinuity at or about 6% Cr (or 2.5 $Crnm^{-2}$) in which isolated Cr species are transformed into gross macrostructure of Cr oxide by only a small extra loading of chromium. Similarly, even when chromium (III) acetate is used[39] to impregnate the silica to produce Cr(VI) and Cr(II), no cognizance is given to the fact that this material may exist as polymeric chains and may be expected to retain this structure on the silica surface. Certainly, the 20% PMo_{12}/SiO_2 corresponding to 12% Mo on the silica support retains its full Keggin ion, cage structure,[28].

ACKNOWLEDGEMENTS

We thank the Royal Society, Humberside College of Higher Education and the S.E.R.C. for financial help and support, and Professors C. Boelhouwer and L. L. van Reijen for valuable discussions.

REFERENCES

1. D. L. Myers and J. H. Lunsford, J. Catalysis, 92, 260 (1985).
2. P. Morys, U. Gorges, and H. L. Krauss, Z. Naturforsch, 39B, 458 (1984).
3. B. Rebenstorf and R. Larsson, J. Catalysis, 84, 240 (1983).
4. M. B. Welch and M. P. McDaniel, J. Catalysis, 82, 110 (1983).
5. B. Rebenstorf, Z. anorg. allg. Chem., 513, 103 (1984).
6. A. Ellison, J. Chem. Soc., Faraday Trans. 1, 80, 2567 (1984).
7. A. Ellison, J. Chem. Soc., Faraday Trans. 1, 80, 2581 (1984).
8. P. Arnoldy and J. A. Moulijn, J. Catalysis, 93, 38 (1985).
9. B. Scheffer, J. A. Moulijn, and A. Ellison, unpublished results.
10. R. S. Schwartz, I. Fankuchen, and R. Ward, J. Am. Chem. Soc., 74, 1676 (1952).
11. R. C. Mackenzie, Ed "Differential Thermal Analysis," Academic Press (1970).
12. A. Ellison, J. O. V. Oubridge, and K. S. W. Sing, Trans. Farad. Soc., 66, 1004 (1970).
13. R. Thomas, J. A. Moulijn, V. H. J. de Beer, and J. Medema, J. Mol. Catalysis, 8, 161 (1980).
14. P. Arnoldy, J. C. M. de Jonge, O. J. Wimmers, and J. A. Moulijn, accepted for publication, J. Catalysis (1985).
15. A. Ellison, A. K. Coverdale, and P. F. Dearing, J. Mol Catalysis, 28, 141 (1985).
16. K. Harrison, L. Hazell, Surface Reactivity and Catalysis Group Newsletter, Roy Soc. Chem. (Jan. 1984).
17. A. Ellison and R. Herbert, to be submitted to the Quantum Chem. Program Exchange, Indiana Univ. (1985).
18. E. Carberry and B. Dombek, private communication.
19. J. Kendal and P. Stuart, Advanced Theory of Statistics, 2, 79 (1973).
20. B. Parlitz, W. Hanke, R. Fricke, M. Richter, U. Roost, and G. Ohlmann, J. Catalysis, 94, 24 (1985).
21. G. Diakun, A. Ellison and P. Worthington, unpublished results.
22. N. M. D. Brown, J. B. McMonagle, and G. N. Greaves, J. Chem. Soc., Faraday Trans. 1, 80, 589 (1984).
23. D. E. O'Reilly and D. S. MacIver, J. Phys. Chem., 127, 1914 (1962).
24. W. Gordy, "Electron Spin Resonance," Wiley, New York (1980).
25. J. A. Ibers and J. B. Swalen, Phys. Rev., 127, 1914 (1962).
26. A. Ellison, to be submitted to the Quantum Chem. Program Exchange, Indiana Univ., Chapter 8 (1985).
27. B. Dean and A. Ellison, unpublished results.
28. R. Fricke and G. Ohlmann, J. Chem. Soc., Faraday Trans. 1, accepted for publication (1985), papers 4/1969 and 5/963.
29. R. A. Prados and M. T. Pope, Inorg. Chem., 15, 2574 (1976).
30. L. L. van Reijen, private communication.
31. J. K. Beattie and G. P. Haight, Progr. Inorg. Chem., 17, 93 (1972).
32. B. Rebenstorf and R. Larsson, Z. anorg. allg. Chem., 478, 119 (1981).

33. B. Rebenstorf and R. Larsson, J. Mol. Catalysis, 11, 247 (1981).
34. P. P. M. M. Wittgen, C. Groeneveld, P. J. C. J. M. Zwaans, H. J. B. Morgenstern, A. H. van Heughten, C. J. M. van Henmen, and G. C. A. Schuit, J. Catalysis, 77, 360 (1982).
35. C. Groeneveld, P. P. M. M. Wittgen, J. P. M. Lavrijsen, and G. C. A. Schuit, J. Catalysis, 82, 77 (1983).
36. M. P. McDaniel, J. Catalysis, 76, 29 (1982).
37. M. P. McDaniel, J. Catalysis, 76, 37 (1982).
38. M. P. McDaniel, J. Catalysis, 76, 17 (1982).
39. R. Merryfield, M. McDaniel, and G. Parks, J. Catalysis, 77, 348 (1982).

HIGHLY ACTIVE CHROMIUM CATALYSTS FOR ETHYLENE POLYMERIZATION

Kazuo Soga, Shian-Ing Chen, Y. Doi and T. Shiono

Research Laboratory of Resources Utilization
Tokyo Institute of Technology
4259 Nagatsuta
Yokohama 227, Japan

INTRODUCTION

The activity of Ziegler-Natta catalysts for olefin polymerization has been remarkably increased by supporting the titanium compound on a solid carrier.[1-23] A variety of metal halides have been reported as carriers, the most common being magnesium chloride ($MgCl_2$) which was originally observed by Petrochemicals Ltd.[24] and Shell.[25] The marked improvement of the activity by using $MgCl_2$ has been revealed to be attributed to a significant increase in the propagation rate as well as a considerable increase in the number of active species.[16-20] However, the precise reason why $MgCl_2$ accelerates the propagation rate is quite open to discussion.

We have recently investigated the effect of carriers on the rate of propylene polymerization by using the soluble $Ti(OBu)_4/AlEt_2Cl$ catalyst combined with a variety of metal chlorides (MCl_x). A clear correlation was found between the propagation rate constant and the electronegativity $X(M^{x+})$ of metal ion in MCl_x; metal chlorides having the electronegativities below $X(Ti^{3+})$; increased the propagation rate constant, whereas those above $X(Ti^{3+})$ decreased it.[26]

These results together with the XPS spectra of Cl(2p) in MCl_x and of Ti(2p) in $TiCl_4/MCl_x$ have led to the conclusion that MCl_x involved in the active species influences the electronic structure of the active Ti^{3+} by inductive effects.[27,28]

We have prepared Cr-based catalysts by modifying the Battelle-type catalyst[29] with a variety of MCl_x and investigated the effect of MCl_x on the rate of ethylene polymerization. This report summarizes the results obtained.[30,31]

EXPERIMENTAL

Research grade ethylene, 2-ethyl-1-hexanol(2-EHA) and toluene (from Takachiho Chemical Co.) were further purified according to the usual procedures. Ultra high purity nitrogen (from Nippon Sanso Co.) was purified by passing it through the molecular sieve 3A column. The other chemicals (research grade) were commercially obtained and used without further purification.

PREPARATION OF CATALYSTS

The following two types of catalysts based on $Cr(CH_3COO)\cdot(CH_3CO)_2O$ and $Cr(tBuO)_4$ were prepared. The $Cr(CH_3COO)_3\cdot H_2O$ was treated in a mixture of acetic anhydride and acetic acid at 120°C for 4h, and excess acetic anhydride and acetic acid were removed under reduced pressure to give a bright green find product, $Cr(CH_3COO)_3\cdot(CH_3CO)_2O$.

$Cr(tBuO)_4$ was prepared according to the method reported by Bradley et al[32]: $Cr(tBuO)_3$ was first obtained by addition of excess tBuOH to $Cr(NEt_2)_3$ which had been prepared by the reaction of $CrCl_3$ with $LiNEt_2$ in in hexane. The system was oxygenated and the diethylamine, solvent and excess of tBuOH were evaporated off, followed by vacuum distillation to obtain $Cr(tBuO)_4$. Since $Cr(CH_3COO)_3\cdot(CH_3CO)_2O$ was sparingly soluble in toluene, it was dissolved in 2-EHA(5×10^{-3}M)[Cat-A]. However, $Cr(tBuO)_4$ was soluble in toluene (5×10^{-4}M)[Cat-B]. Metal chlorides (MCl_x) were also soluble in 2-EHA(1.5×10^{-1}M).

POLYMERIZATION AND ANALYTICAL PROCEDURES

In a 300 cm^3 glass reactor, equipped with a magnetic stirrer, were placed measured amounts of Cat-A or Cat-B, MCl_x and toluene under nitrogen. The reactor was degassed, followed by the introduction of ethylene (1 atm.). After the mixture was agitated for 10 min., a specified amount of alkylaluminumium was added. The polymerization was terminated by adding a dilute hydrochloric acid solution in methanol. The precipitated polymer was washed with methanol, followed by drying in vacuum at room temperature. The molecular weight distribution of the polymer was measured at 145°C by GPC using 1,2,4-trichlorobenzene as the solvent. DSC measurements were made at a heating rate of 10K/min. The ESR spectra were recorded with a Varian E-12 ESR spectrometer at room temperature at a microwave frequency of 9.528 GHz.

RESULTS AND DISCUSSION

Polymerization of Ethylene with the Original Cat-A

The original Cat-A showed no activity for ethylene polymerization, by itself or in combination with $AlEt_3$ at a mole ratio Al/Cr = 2 $\sim$ 30. However, when combined with $AlEt_2Cl$, this catalyst was capable of catalyzing the polymerization ethylene (about 1 kg-polymer/g-Cr·h; see (Figure 1) to give linear polyethylene with a very broad polydispersity (Q $\doteq$ 35). Upon adding $AlEt_2Cl$ to the original Cat-A in toluene, a very fine, white precipitate was formed. Correlating with the polydispersity of polyethylene, it may be considered that this catalytic system is heterogeneous.

Cat-A showed an ESR signal with a g-factor of 1.993 which is attributed to Cr^{3+}. When it was mixed with $AlEt_2Cl$ with increasing concentration, the signal diminished gradually and completely disappeared (Cr^{2+}) at a mole ratio Al/Cr = 10. Reduction of Cr^3 to Cr^1 by $AlEt_3$ lead to a new signal with a hyperfine structure, which is typical of the Cr^{1+} complex. These results have lead to the conclusion that the oxidation state of Cr in the active catalyst is 2+.[33]

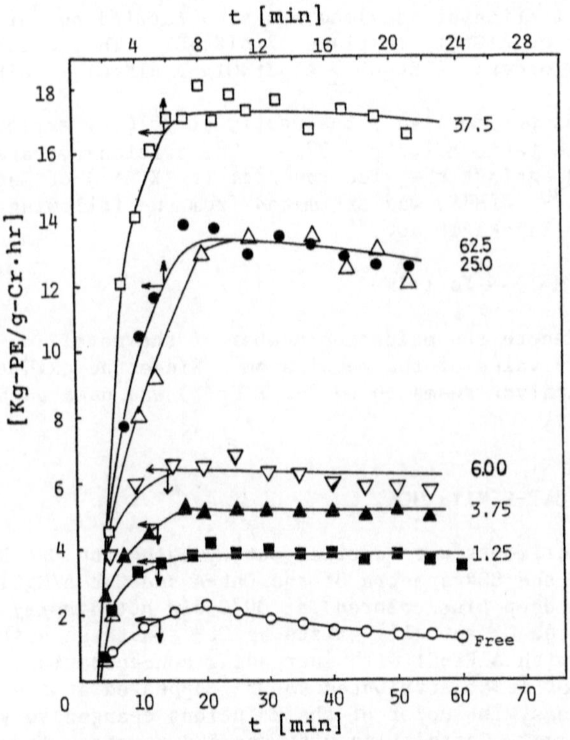

Fig. 1. Kinetic Curves for Ethylene Polymerization with the Cat-A/$AlEt_2Cl/MgCl_2$ catalyst system
Polymerization conditions: 30°C, ethylene pressure = 1 atm, toluene = 160 cm^3, $Cr(CH_3COO)_3 \cdot (CH_3CO)_2O$ = 1,6 x 10^{-2} mmol, $AlEt_2Cl$ = 14 mmol. Numbers in the Figure indicate the Mg/Cr mole ratio

Modification of Cat-A with MCl_x

Polymerization of ethylene was first carried out at 30°C with the Cat-A/$AlEt_2Cl$ system by using different amounts of $MgCl_2$. The rates of polymerization are shown in Figure 1. With an increase in the mole ratio of Mg/Cr, the rate of polymerization drastically increased at first and attained a maximum value, followed by a slight decrease. The rates of polymerization obtained in the presence of $MgCl_2$ seem to be almost unchanged during the course of polymerization, indicating that the active species, Cr^{2+}, is stabilized by forming a complex with $MgCl_2$.

Some analytical results of the polymers obtained with various mole ratio of Mg/Cr are shown in Table 1 from the IR and ^{13}C NMR measurements these polymers were identified to be linear polyethylenes. Both the molecular weight and polydispersity of polyethylene were changed by adding $MgCl_2$, clearly indicating that new highly active species are formed by the addition of $MgCl_2$. The narrow polydispersity of polyethylene strongly suggests that the catalytic system is homogeneous.

The polymerization of ethylene was then carried out at temperatures in the range of 20 - 60°C, under the optimum conditions [Mg/Cr = 37.5, $AlEt_2Cl$ = 7x10^{-3}M]. When the polymerization occurred at above 40°C, a significant deactivation was observed during the course of polymerization. Thus, a maximum activity was obtained at around 30°C.

The polymerization of ethylene was also carried out at 30°C by using $AlEt_3$, $Al_2Et_3Cl_3$ or $AlEtCl_2$ in place of $AlEt_2Cl$. The activity decreased in the following order: $AlEt_2Cl > Al_2Et_3Cl_3 > AlEtCl_2 > AlEt_3$ (= 0).

Ethylene was polymerized subsequently at 30°C by employing a variety of MCl_x at a mole ratio MCl_x/Cr - 37.5. The stationary rate of polymerization was plotted against the electronegativity $X(M^{x+})$ of metal ion in MCl_x (Figure 2).[30] $X(M^{x+})$ was estimated from the following empirical equation reported by Tanaka et at.,[34]

$$X(M^{x+}) = Xo(1+2x) \qquad (1)$$

where x and Xo denote the oxidation number of the metal ion and Pauling's electronegativity value of the metal atom. Since the oxidation state of Cr in the active catalyst seems to be 2+, $X(Cr^{2+})$ was used as a standard for electronegativity.

MODIFICATION OF CAT-B WITH MCL_x

Similar experiments were carried out by using the Cat-B in place of the Cat-A. Both the ESR spectra of the Cat-A and Cat-A/$MgCl_2$ solutions in toluene (both deep blue colored) at 30°C did not display any signal, indicating that the Cr actually exists as Cr^{4+} in these systems. When they were mixed with $AlEt_2Cl$ with increasing concentration, a weak signal with a g-factor of 1.986 attributed to Cr^{1+} appeared at a very slow rate. During this process, the color of the solutions changed very quickly from deep blue to orange. Correlating with the ESR spectra, it may be concluded that the Cr^{4+} species are instantaneously reduced to Cr^{2+} followed by a very slow reduction to Cr^{1+}. On the other hand, when $AlEt_3$ was added, a very strong signal with the same g-factor having a hyperfine structure appeared at a fast rate. The color of the solutions in these cases changed instantaneously from deep blue to orange followed by a quick change to black, indicating that the Cr^{4+} species are quickly reduced to Cr^{2+} and then to Cr^{1+}. The Cat-B/$MgCl_2$ in combination with $AlEt_2Cl$ showed a very high activity for ethylene polymerization, while the activity with $AlEt_3$ was very small. These results confirm the preceding conclusion that the oxidation state of Cr in the active catalyst is 2+.

Table 1. Analytical Results of Polyethylene Obtained With Various Mole Ratios of Mg/Cr[a]

Mg/Cr	Molecular Weight			Tm
	$\bar{M}n(x10^{-10})$	$\bar{M}w(x10^{-4})$	$Q(\bar{M}w/\bar{M}n)$	(°C)
0	1.7	61.6	35.5	140
6	124	317	2.6	145
25	96	303	3.2	145
37.5	153	330	2.2	144

[a]Polymerization conditions as in Figure 1.

Table 2. Analytical Results of Ethylene Polymerization With the $Cr(OBu^t)_4/AlEt_2Cl$ and $Cr(OBu^t)_4/AlEt_2Cl/MgCl_2$ Catalysts[a]

Run No.	Mg/Cr	Polym. Time (min)	Yield (g)	Activity (Kg/g-Cr·h)	$\bar{M}n$ ($x10^4$)	$\bar{M}w$ ($x10^4$)	Q ($\bar{M}w/\bar{M}n$)	N/Cr[b]
1	0	25	0.35	3.3	1.2	43.6	35.2	(5.6)
2	5	5	0.31	7.7	33.5	83.1	2.5	0.1
3	5	10	0.67	8.3	37.9	91.3	2.4	0.19
4	5	20	1.29	8.0	52.6	133.7	2.5	0.26
5	5	60	3.80	7.8	57.5	144.4	2.5	0.71
6	40	25	4.05	36.3	45.5	136.0	3.0	1.71
7	50	25	3.95	35.3	44.8	94.6	2.1	1.69

[a]Polymerization conditions: 30°C, ethylene pressure - 1 atm, toluene = 150 cm^3, $Cr(OBu^t)_4$ = 5.2×10^{-3} mmol, $AlEt_2Cl$ = 1.5 mmol.

[b]Calculated from Equation (2).

The polymerization of ethylene was carried out at 30°C by using the Cat-B/$MgCl_2$/$AlEt_2Cl$ system at different mole ratios of Mg/Cr. The results obtained are summarized in Table 2.[31] Both the molecular weight and polydispersity of polyethylene changed drastically by adding $MgCl_2$ as was observed in the Cat-A/$MgCl_2$/$AlEt_2Cl$ system. The number of the polyethylene molecules produced (N), which was roughly estimated by Equation (2), was also shown in Table 2,

$$N = Y/\bar{M}n \qquad (2)$$

where Y and Mn denote the polymer yield and the number average molecular weight. The N/Cr ratio (here, N is limited to the high molecular weight polyethylene produced with the newly formed highly active species) increased gradually with the polymerization time (Runs No. 2 - 5), indicating that some transfer reactions take place in this catalyst system. However, the N/Cr ratio in Run No. 4 (0.26) is much smaller than those in Runs No. 6 (1.71) and 7 (1.69). Moreover, the polydispersity of polyethylene obtained by using $MgCl_2$ is always very narrow, indicative of the homogeneity of active species. Therefore, it may be said that the increase in the activity with increasing Mg/Cr ratio is mainly attributed to the increase in the number of highly active species.

Table 3. Polymerization Results of Ethylene With the $Cr(OBu^t)_4$/$AlEt_2Cl$/MCl_x Catalysts[a]

MCl_x	Polym. Time (min)	Yield (g)	Activity (Kg/g-Cr·hr)	X[b]
$MgCl_2$	25	4.05	36.3	6.0
$MnCl_2$	25	1.94	17.9	7.5
$CoCl_2$	25	1.87	17.2	9.0
0	25	0.35	3.3	8.5[c]
$AlCl_3$	25	0.39	3.6	10.0
$NbCl_5$	25	0.28	2.6	17.6

[a]Polymerization conditions: 30°C, ethylene pressure = 1 atm., toluene = 150 cm^3, $Cr(OBu^t)_4$ = 5.2×10^{-3} mmol, $Al(C_2H_5)_2Cl$ = 1.5 mmol. M/Cr = 40

[b]Calculated from $X = (1 + 2x)X_0$ (X_0: Pauling's electronegativity of metal atom, X: that of metal ion, x: oxidation number).

[c]Electronegativity of Cr^{+2}

Ethylene was then polymerized at 30°C by using various MCl_x in place of $MgCl_2$. The results shown in Table 3 seem to be in good agreement with those obtained with the Cat-A/MCl_x/$AlEt_2Cl$ system (Figure 2).

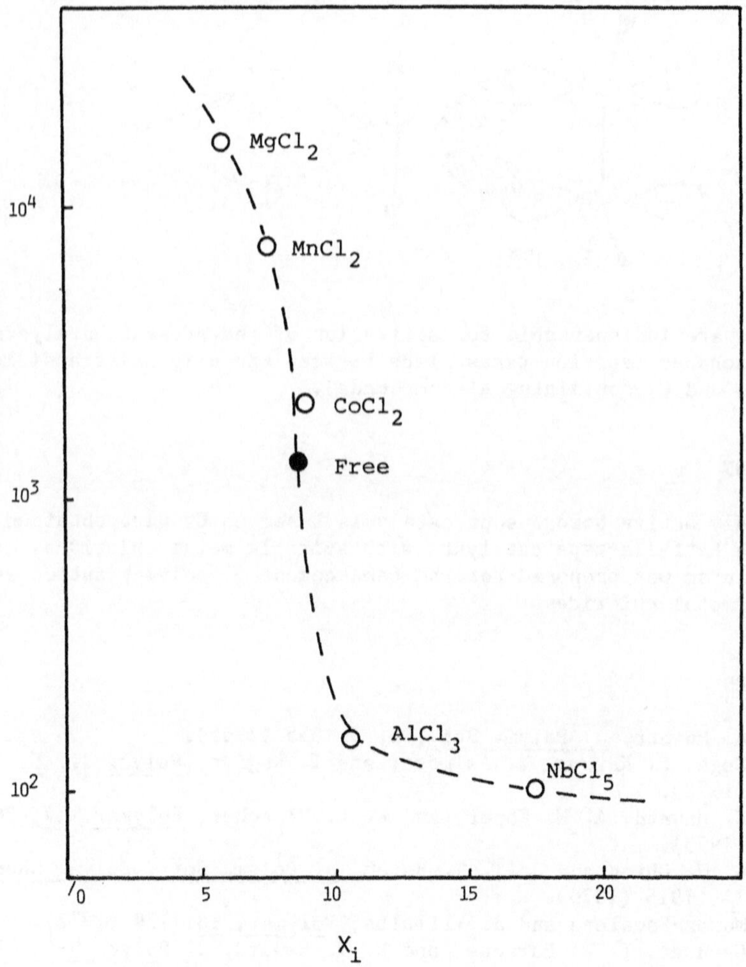

Fig. 2. Correlation Between Stationary Rate and Electronegativity of Metal Ion in MCl_x

Polymerization conditions are the same as in Figure 1 (M/Cr = 37.5)

EFFECT OF MCL_x ON THE POLYMERIZATION ACTIVITY

We have hitherto investigated the effect of carriers on the polymerization activity by using three different systems including the $Ti(OBu)_4/MCl_x/AlEt_2Cl$ catalyst. The results obtained seem to be in good agreement. Only the MCl_x which contain metal ions with the electronegativities below those of the active transition metal ions accelerate the polymerization. The results strongly suggest that the metal chlorides involving in the active species influence the electronic structure of the active transition metal ions by inductive effects. The electron-donating capability of the MCl_x to the active transition metal ions (T) stabilizes the coordination of an olefin monomer (M) by the back-donation of electron from T to M, resulting in an acceleration of a subsequent insertion of M into the T-P chain bond. In the following is shown a plausible model of such a highly active species[28] (T = active transition metal ion). To form such a highly active species, it may be essential for the transition metal compounds to possess Cl ligands. This should be the main reason why Cl-containing Al

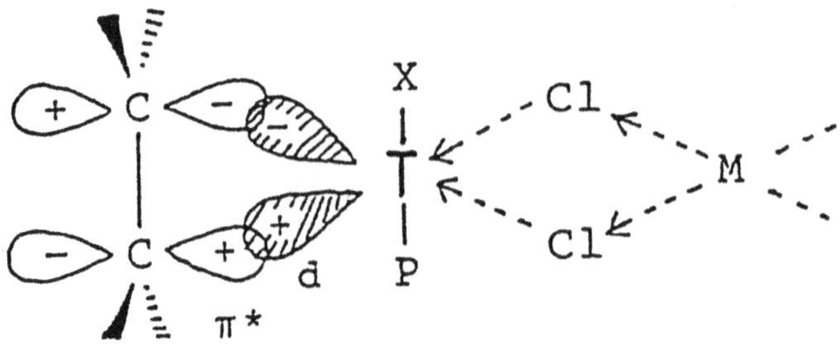

compounds are indispensable for activation of the present catalysts (The ligand-exchange reaction takes place between the original transition metal compounds and Cl-containing Al compounds).

CONCLUSION

Highly active homogeneous catalysts based on Cr were obtained by modifying the Battelle-type catalysts with suitable metal chlorides. A plausible mechanism was proposed for the enhancement of polymerization activity by using metal chlorides.

REFERENCES

1. W. A. Hewett, J. Polym. Sci., B, 5, 855 (1965).
2. K. Soga, S. Katano, Y. Akimoto, and T. Kagiya, Polym. J., 5, 128 (1973).
3. R. N. Haward, A. N. Koper, and K. L. Fletcher, Polymer, 14, 365 (1973).
4. J. C. W. Chien and J. T. T. Hsieh, J. Polym. Sci., Polym. Chem. Ed., 14, 1915 (1976).
5. A. Munoz-Escalona and J. Villalba, Polymer, 18, 179 (1977).
6. K. Gardner, I. W. Parsons, and R. N. Haward, J. Polym. Sci., Polym. Chem. Ed., 16, 1983 (1978).
7. E. Suzuki, M. Tamura, Y. Doi, and T. Keii, Makromol. Chem., 180, 2235 (1979).
8. E. W. Duck, D. Grant, and E. Kronfli, Eur. Polym. J., 15, 625 (1979).
9. K. Soga, T. Sano, and R. Ohnishi, Polym. Bull., 4, 157 (1981).
10. K. Soga, K. Izumi, M. Terano, and S. Ikeda, Makromol. Chem., 181, 657 (1980).
11. S. S. Ivanchev, A. A. Baulin, and A. G. Rodionov, J. Polym. Sci., Polym. Chem. Ed., 18, 2045 (1980).
12. N. Kashiwa, Polym. J., 12, 063 (1980).
13. H. L. Hsieh, Polym. J., 12, 596 (1980).
14. K. Soga and M. Terano, Polym. Bull., 4, 39 (1981).
15. P. Galli, L. Luciani, and G. Cecchin, Angew. Makromol. Chem., 94, 63 (1981).
16. Y. Giannini, Makromol. Chem. Suppl., 5, 216 (1981).
17. T. Keii, E. Suzuki, M. Tamura, M. Murata, and Y. Doi, Makromol. Chem., 183, 2285 (1982).
18. Y. Doi, M. Murata, K. Yano, and T. Keii, Ind. Eng. Chem. Prod. Res. Dev., 21, 580 (1982).
19. N. Kashiwa and J. Yoshitake, Makromol. Chem., Rapid Commun., 3, 211 (1982); Mokromol Chem., Rapid Commun., 4, 41 (1983).
20. G. D. Bukatov, S. H. Shepelev, V. A. Zakharov, S. A. Sergeev, and Yu. I. Yermakov, Makromol. Chem., 183, 2657 (1982).
21. J. C. W. Chien, J. C. Wu, and I. J. Kuo, J. Polym. Sci., Polym. Chem. Ed., 20, 2091 (1982).

22. A. D. Caunt, J. A. Licchelli, I. W. Parsons, R. N. Haward, and M.R.Y. AI-Hillo, Polymer, 24, 121 (1983).
23. P. C. Barbe, L. Noristi, G. Baruzzi, and E. Marchetti, Makromol. Chem., Rapid Commun., 4, 249 (1983).
24. Petrochemicals Ltd., Brit. Pat. 841, 822 (1955).
25. Shell, Brit. Pat. 904, 510 (1960).
26. Y. Doi, K. Soga, M. Murata, E. Suzuki, Y. Ono, and T. Keii, Polym. Commun., 24, 244 (1983).
27. K. Soga, Y. Doi, and T. Keii, Proc. 8th Intern. Congr. Catalysis, West Berlin, Vol. 5, p. 349 (1984).
28. K. Soga, the 189th ACS meeting, Miami Beach, Fl. USA (1985).
29. C. Alfred (Battelle Memorial Institute), Ger. Offen. 2606243.
30. K. Soga, S. I. Chen, T. Shiono, and Y. Doi, Polymer, in press.
31. K. Soga, S. I. Chen, T. Shiono, and Y. Doi, Polymer, in press.
32. Z. C. Alyea, J. S. Basi, D. C. Bradley, and M. H. Chrisholm, J. Chem. Soc., A, p. 772 (1971).
33. S. N. Gan, S. I. Chen, R. Ohnishi, and K. Soga, Makromol. Chem. Rapid Commun., 5, 535 (1984).
34. K. Tanaka, A. Ozaki, and K. Tamaru, Shokubai, 6, 262 (1964).

POLYOLEFIN CATALYSTS BASED ON TITANIUM AND VANADIUM CHLORIDES

H. L. Hsieh, M. P. McDaniel, J. L. Martin, P. D. Smith, and D. R. Fahey

Phillips Research Center
Bartlesville, OK

INTRODUCTION

The foundation of the Hogan-Banks' high density polyethylene (HDPE) process was (and still is) the chromium oxide/silica catalyst discovered in the early 1950's. Its extreme versatility has permitted its continuous modernization over the last 35 years to make a wide variety of competitive resins. That Phillips has remained the largest manufacturer of HDPE in the world to this day is a strong testament to the utility of this catalyst system.

Nevertheless, our research work has not focused entirely on Cr/silica systems. Other types of catalysts have also been explored and commercially developed. Our research has aimed simultaneously in three different directions, shown below: Each family of catalysts behaves differently, and each has its own application.

1. <u>Cr Oxide Catalysts</u>
   a. Modifications of Cr/Silica
   b. New Supports Replacing Silica
2. <u>Transition Metal Chloride Systems</u>
   a. Ti-Mg Chloride Combinations
   b. V Chlorides
   c. Ti-V Bimetallic Catalysts
3. <u>Organo-Transition Metal/Oxide Supports</u>
   a. Organo-Cr
   b. Organo-Ti, V, Zr

For years, a limitation in the Hogan-Banks HDPE technology was the inability to make high melt index resins in the slurry process, i.e. resins with a narrow molecular weight distribution which are needed in injection and rotational molding applications. Such resins were formerly made by a solution process, which is more energy intensive, because higher reactor temperatures were needed to enhance the melt index potential of ordinary Cr/silica catalysts. Therefore in the mid 1970's, while research on Cr/

silica continued, a separate group was set up at Phillips to explore catalyst systems based on transition metal chlorides, which are well adapted for this purpose.

The MI capability of Cr/silica was soon pushed into the desired range by the development of R/R activation, which has already been described.[1-3] But simultaneously, the second group discovered several metal chloride catalysts which offered other advantages, such as very high activity, and a glossy appearance on the surface of the molded part which customers find desirable. Therefore, Phillips has now commercialized some of these catalysts and continues to explore this area. Such catalysts are quite compatible with Cr/silica in the Hogan-Banks technology.

Our metal chloride-based catalysts often exhibit about ten times the activity of Cr oxide catalysts, based on the actual weight of the catalyst, but have about the same activity based on the active metal itself. They generally employ titanium and/or vanadium chlorides as the active ingredient, together with an alkylaluminum cocatalyst, and sometimes a special high surface area base is also used. One can modify them for producing polypropylene, other poly-1-olefins, or for linear low density polyethylene.

The third and newest class of catalysts, investigated at Phillips, consists of organotransition metal compounds deposited on acidic oxide supports. These compounds, although not active themselves, are believed to react with acidic surface hydroxyls where they become isolated and fixed to the support where they do develop activity. Organochromium compounds have been studied most, and are probably the most versatile, although they bear little resemblance to the Cr oxide systems described earlier. Often olefins, such as butene, hexene, or octene are produced along with the polyethylene and become incorporated into it. Branched polymers are obtained instead of the usual linear ones. Organo Ti, V, and Zr compounds have also been investigated. These differ from Ziegler catalysts in that they do not contain halide and do not require an aluminum alkyl cocatalyst. This class of catalysts may be preferred for applications which require the production of extremely high molecular weight polyethylene.

This report summarizes some of our work with transition metal chloride catalysts. Three families of catalysts are discussed, providing an example of a catalyst (1) based on titanium chloride, (2) based on vanadium chlorides, and (3) based on a combination of titanium and vanadium chlorides. These examples are not necessarily used commercially but they are similar to those used commercially, and were chosen to illustrate some method of polymerization control, or catalyst synthesis.

EXPERIMENTAL

The silica used in these experiments was Davison grade 952, having a surface area of about 280 m^2/g, and a total porosity of about 1.6 ml/g. The vanadium catalysts were also supported on Davison HPV alumina, having a surface area of 520 m^2/g and pore volume of about 2.8 ml/g. Aluminum phosphate supports were made by coprecipitation[4], and were only slightly less porous, having a surface area of 300-400 m^2/g and pore volume of 2.5-2.8 ml/g.

Polymerization runs were made mainly in an Autoclave Engineers one-gallon reactor containing 2 l of isobutane diluent and 1 ml of 15% trialkylaluminum. In the Ti-Mg/silica experiments, triisobutyl aluminum was used as a cocatalyst, whereas triethylaluminum was used with the vanadium and titanium-vanadium catalysts. The ethylene partial pressure was varied from 100 to 250 psi, depending on the particular series of experiments, and hydrogen partial pressure was varied from 0 to 100 psi.

The heterogeneity index (HI) values reported in Figure 6 are not actual GPC measurements of Mw/Mn. Instead they are a visco-elastic measure of shear response, which is also an indication of the broadness of the polymer molecular weight distribution. This measurement was then calibrated with the HI's of known standards. This procedure provides a very close estimate of Mw/Mn for polymers of this type.

TITANIUM-MAGNESIUM CHLORIDE COMBINATIONS

Many catalysts used commercially for making polyethylene or polypropylene contain titanium III or IV chlorides as the active component, often dispersed on a high surface area magnesium chloride matrix. The support can also be a close relative of $MgCl_2$, such as the hydroxychloride, alkoxide, hydroxide, or other magnesium compounds easily convertible to the chloride. We too, have explored a number of variations of this basic recipe.

Generally, they all produce narrow MWD polyethylene, and all are sensitive to hydrogen for regulation of molecular weight. Termination is by hydrogenation, producing almost completely saturated polymer end groups. The activity of these catalysts tends to parallel the surface area of the magnesium support, so that numerous ways of achieving a high surface area $MgCl_2$ have been developed. These include the following:

(1) ball milling
(2) swelling of $MgCl_2$ with alcohols or other chelates, followed by their removal
(3) dissolving the $MgCl_2$ as a complex and reprecipitating it by breaking the complex, changing solvents, etc.
(4) depositing the $MgCl_2$ on a high surface area oxide carrier.

Often other ingredients, such as aluminum chloride or organic compounds, are also incorporated to maintain the high surface area, or in the case of polypropylene catalysts, to regulate stereoregularity.

As an example of a Ti-Mg chloride catalyst, one of the latter systems[4] has been chosen in which the carrier is silica. Unlike most Ziegler catalysts this system is a highly versatile one, because many of the commercial polymer properties can be controlled by varying the silica pretreatment. Another advantage is good control of polymer particle size, due to the proper choice of silica. This depends on the simple reaction of dibutylmagnesium with $TiCl_4$.

[A] $$MgBu_2 + TiCl_4 \longrightarrow MgCl_2{:}TiCl_3$$

Butene and butane are also given off as by-products. The exact composition of the catalyst can vary somewhat, particularly if excess $TiCl_4$ is present, since it may also be adsorbed.

The above ingredients can be dispersed on silica by first impregnating the silica with dibutylmagnesium, then treating the combination with $TiCl_4$. Presumably the magnesium can react with surface hydroxyl groups.

[B]

Then, when $TiCl_4$ is added, a brown color usually forms, indicating reduction to $TiCl_3$. Several rinses with heptane removes excess $TiCl_4$.

We found that the amount of titanium adsorbed by this catalyst varies with the initial calcining temperature of the silica. This is shown in Figure 1. When the silica has been calcined at high temperatures, like 900C, only a small amount of titanium adsorbs, about 1.6% by weight. But silica samples dried at only 200-300C adsorb almost 5% Ti. This is undoubtedly because both the dibutylmagnesium and also the $TiCl_4$ react with hydroxyls on the silica surface, which is governed by the initial calcining temperature.

Figure 1 also shows how the hydroxyl population on the silica varies with temperature. These measurements were made by reacting the silica with methylmagnesium iodide, and volumetrically determining the amount of methane liberated.[5,6] Note that the hydroxyl population drops from about 4.5 OH/nm^2 for a fully hydrated sample, to about .8 OH/nm^2 after calcina- of 900C. Therefore, the hydroxyl population parallels the amount of titanium adsorbed by the catalyst.

Figure 2 tells something about the spacing of these hydroxyls. Here they have been classified as "paired" hydroxyls or as isolated "singles", according to how they react with $SiCl_4$ or other reagents.[7-11] Isolated hydroxyls react in a one-to-one ratio with $SiCl_4$, whereas "paired" hydroxyls are those which react in a two-to-one ratio. Note that a fully hydrated sample contains about 30% singles and 70% pairs. When the silica is heated, only the paired hydroxyls initially condense, while the singles population remains constant. At about 600-700C, however, the paired hydroxyls have all condensed and only singles remain. Upon further heating to 900C these also condense, although with more difficulty.

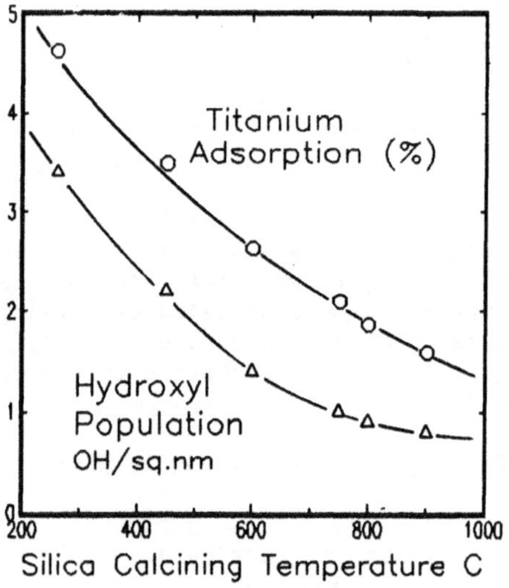

Fig. 1. Silica, calcined at the temperature shown was impregnated with dibutylmagnesium (.5% Mg), treated with a large excess of $TiCl_4$, and rinsed several times with heptane to make an active polymerization catalyst. The initial surface hydroxyl population after calcining, and also the amount of titanium adsorbed, is plotted against the calcining temperature.

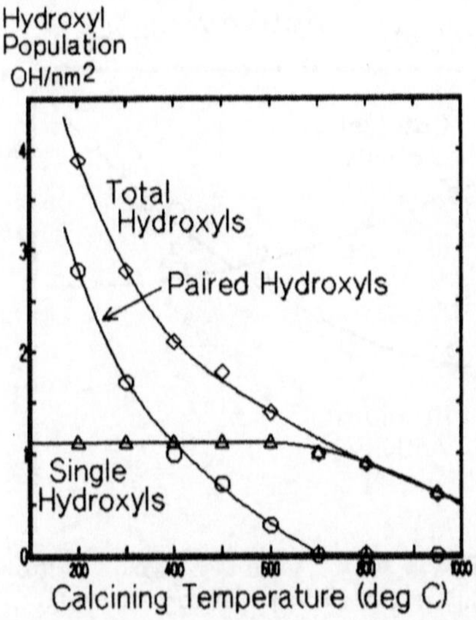

Fig. 2. The hydroxyl population on silica as a function of calcining temperature. "Paired" hydroxyls are those which react with $SiCl_4$ and other reagents in a 2:1 ratio. "Singles" react in a 1:1 ratio.

At 900C, only isolated hydroxyls remain, and one can imagine them reacting with $MgBu_2$ as in [B] above, where each hydroxyl is replaced by one titanium. However, for the experiments in Figure 1 only, 5% magnesium was impregnated, which is far less than the total OH population. Even at 900C only about half the hydroxyls react with magnesium. The remaining hydroxyls probably react directly with the excess $TiCl_4$ itself, as in [C] below. Both reactions [B] and [C] deposit one Ti per OH reacting, and this is actually observed in Figure 1. At the high calcining temperatures, approximately one titanium atom is adsorbed per hydroxyl group.

[C]

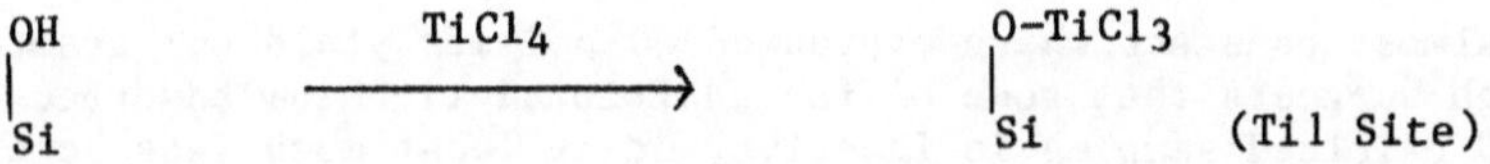

When the silica is calcined at low temperatures, like 200-300C, much more titanium is adsorbed, as shown in Figure 1, because there are more hydroxyls to react. However, the adsorption is now considerably less than one Ti per OH. This is reasonable because at 200-300C many of these hydroxyls are not isolated singles as at 900C, but exist in pairs. So a double reaction with the $MgBu_2$ or the $TiCl_4$ can occur, as in [D] and [E] below. Note that the titanium adsorption in Figure 1 is not proportional to the OH population. Instead it flattens out more at the low temperature end, which is evidence of the double reaction.

[D]

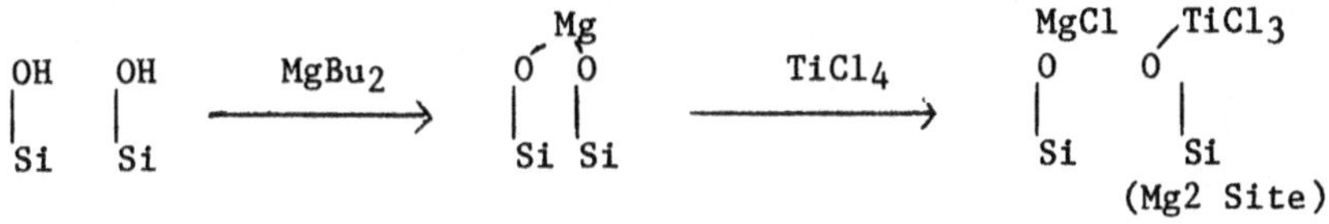

Although titanium is certainly the active metal for ethylene polymerization, the activity of the catalysts shown in Figure 1 does not parallel the amount of titanium adsorbed. This is shown in Figure 3, where activity is plotted against the silica calcining temperature. In fact the activity

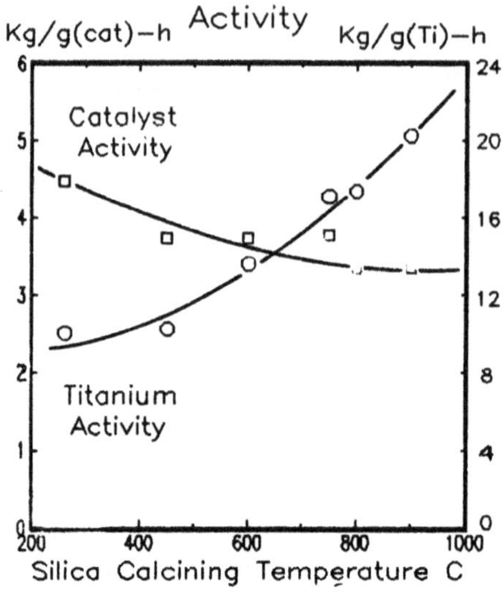

Fig. 3. The activity of Ti-Mg/Silica catalysts, each containing 5% Mg, is plotted against the calcining temperature of the silica. Overall catalyst activity is fairly independent of calcining temperature, but the more dehydrated samples adsorbed less $TiCl_4$, indicating a more efficient use of it.

[E] 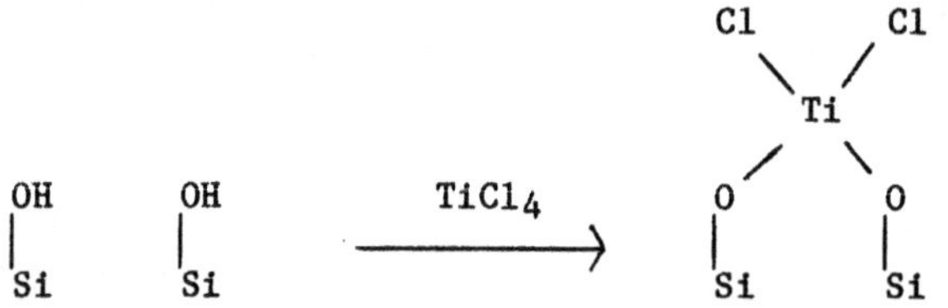

(Ti2 Site)

is almost constant where expressed as polymer yield per gram of catalyst, which suggests that some of the additional titanium adsorbed on the 200-300C calcined samples is inactive, or at least much less active. Since the attachment of titanium depends on a reaction with the surface, this cannot indicate clumping. Even at the highest loadings the titanium was monodispersed (a monolayer = 5% Ti). Instead, we believe that this behavior is an indication of the diversity of sites on the silica. In contrast, the unsupported counterpart ($MgCl_2$:$2TiCl_3$), probably contains a fairly uniform distribution of sites.

The series of catalysts shown in Figure 1 was made to contain 0.5% Mg, which is far less than saturation of the silica surface. Even on silica calcined at 900C, only about half the hydroxyls could react with $MgBu_2$ according to equation [B]. Those remaining hydroxyls, no doubt, react with $TiCl_4$ according to [C] to form a second type of site, let us call them Mg1 and Ti1 sites respectively. The Mg1 sites are probably far more active than Ti1 sites and therefore are responsible for most of the observed polymerization activity.

This can be confirmed by treating the silica with varying amounts of $MgBu_2$ which, of course, changes the relative amounts of the two types of sites. Table 1 summarizes the results of three experiments in which the magnesium level was varied. The silica had been calcined at 750C so that it would have only single hydroxyls. Therefore at 0% Mg loading Ti1 sites

Table 1. 750C Silica

Mg Content	Type of Sites	Activity (g/g-h)	Melt Index	Shear Ratio	Heterogeneity Index (Shear)
0%	Ti1	30	0.2		
0.5%	Mg1+Ti1	320	4.9	37	7.2
2.0%	Mg1	933	4.1	36	7.9

are formed exclusively, whereas at 2% Mg loading which saturates the surface, only Mg1 sites are obtained. As can be seen from Table 1, the Ti1 sites exhibit very little activity. Instead the catalyst activity is almost proportional to Mg1 concentration, which site-for-site is possibly 30 times more active.

Thus in Figure 3, above about 600C, where only isolated single hydroxyls are found, the activity per gram of catalyst is fairly constant because the Mg1 sites concentration is also constant. As the calcining temperature is decreased from 1000C, the extra hydroxyls react to form Ti1 sites which contribute little to the overall activity. However, the activity per gram of titanium, which is also shown in Figure 3, is greatly affected. As expected, it increases with calcining temperature because the site population becomes richer in Mg1 sites.

The molecular weight of the polymer produced also reflects the two different types of sites. The runs in Table 1 were made with a fixed amount of hydrogen added to the reactor as a chain transfer agent. Notice that the Ti1 sites yield much lower melt index polymer (higher MW) than the Mg1 sites, indicating a lower sensitivity to hydrogen. The Ti1 sites alone yielded .2 melt index polymer, but in the presence of Mg1 sites, which dominate the polymerization, due to their higher activity, melt index values in the 4-5 range are obtained. The molecular weight distribution of the polymers, shown in Table 1 as shear ratio (HLMI/MI) and heterogeneity index, are slightly broader than usually obtained from the similar catalysts made without the silica support, again suggesting a more diverse population of sites on silica.

On silicas calcined below 600C, pairs of hydroxyls also exist and can react. Therefore Table 2 presents the results of another series of experiments in which the magnesium loading was varied, but here the silica was calcined at only 260C. At 260C the OH population is so large that only about 10-15% of them can react with .5% Mg, mostly bifunctionally as in reaction [D]. The product of reaction [D], which we call Mg2 sites, is harder to define than previous sites. It is possible that in Mg2 sites the titanium is not associated with $MgCl_2$, and therefore Mg_2 sites are equivalent to Ti1 sites.

The presence of so many excess paired hydroxyls on the 260C silica in Table 2 also creates numerous Ti2 sites according to equation [E]. Again we suspect that the Ti2 sites are not very active, but in this case there are so many excess hydroxyls to react with $TiCl_4$ that the contribution from Ti1 and Ti2 sites does seem to become significant on silica calcined at 260C, because in Table 2 even the sample containing no magnesium has a respectable activity.

Note also in Table 2 that the amount of titanium adsorbed remains fairly constant, regardless of the magnesium loading. This suggests that

Table 2. 260C Silica

Magnesium	Titanium	Type of Sites	Activity (g/g-h)	Melt Index	Shear Ratio	Heterogeneity Index (Shear)
0%	4.5%	Ti2,Ti1	378	0.2	93	17.0
0.46%	4.6%	ALL	500	0.4	86	13.0
0.69%	3.9%	ALL	516	0.5	78	12.1
1.43%	4.3%	ALL	500	0.8	64	13.6
3.22%	4.9%	Mg1,Mg2	514	8.1	47	9.6

whether the hydroxyls react directly with $TiCl_4$ as in [E], or with $MgBu_2$ and then $TiCl_4$ as in [D], the adsorption ratio is about the same. If all the paired hydroxyls react bifunctionally (one Ti adsorbed per OH pair), and all the singles react too (one Ti per OH), then saturation should be about 4.5% Ti. As can be seen in Table 2, this is very close to the observed values, and it supports the postulated reaction [D].

Note in Table 2 that the activity of the 260C silica was little improved by increasing the amount of $MgBu_2$. Since most of the hydroxyls were paired, one would expect mainly Mg2 sites. This suggests that the Mg2 sites were no more active than Ti1 or Ti2, and in fact that they may be identical as postulated in [D].

The melt index values of the polymers, also shown in Table 2, tell a similar story. Except for the last sample, where the surface was saturated with magnesium, the melt index values are all very low, as would be expected from Ti1 or Ti2 sites. This suggests that the product of reaction [D] may not be associated with magnesium. Apparently Mg1 sites do not become significant until the surface is overwhelmed with $MgBu_2$.

Note also in Table 2 that the molecular weight distributions are considerably broader than those in Table 1. This suggests again that the 260C silica contains an even broader diversity of sites than seen on the 750C silica. This was even true when both silicas were saturated with magnesium and both yielded high melt index polymer.

A plot of activity versus magnesium loading is shown in Figure 4 for the two silicas. The 750C silica uses the magnesium more efficiently, probably entirely as Mg1 sites, because activity increases almost linearly with magnesium concentration. On the 260C silica, however, most appears to be wasted at Mg2 (=Ti1?) sites.

Thus in Figure 3, decreasing the calcining temperature also decreases the amount of Mg1 sites, which is compensated by the vast increase in the other less active sites. The result is a relatively flat activity per gram of catalyst, but a sharp decrease in activity per gram of titanium.

The polymer properties also reflect this diversity of sites. Figure 5 plots the melt index (MI) obtained from these same catalysts whose activity is shown in Figure 3. A high MI indicates a low molecular weight, that is, a high sensitivity to hydrogen as terminating agent. Note in Figure 5, that the MI increases with calcining temperature. This suggests that the Mg1 sites are more sensitive to hydrogen, and produce a higher MI, than the others. This is also evident in Tables 1 and 2, where catalysts with a high concentration of Mg1 sites also yielded highest MI polymer.

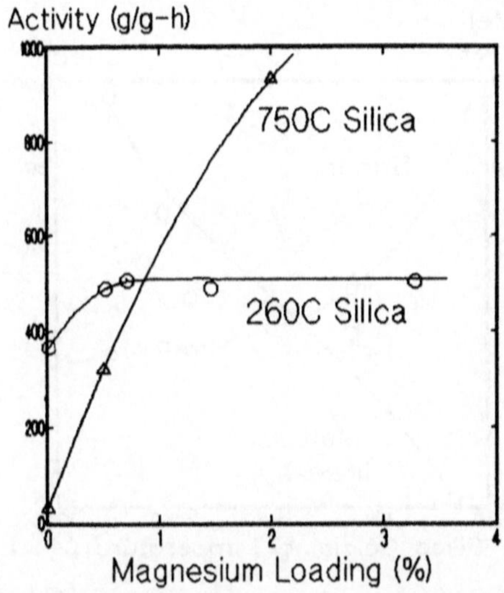

Fig. 4. Silica samples, calcined at either 260C or 750C, were impregnated with various levels of dibutylmagnesium and then treated with excess $TiCl_4$. The overall activity of each catalyst is plotted against Mg loading. The two quite different profiles indicate different types of bonding (paired vs single) which lead to several types of active sites. The more dehydrated silica makes more efficient use of the magnesium.

Although such MI variability is common among Cr/silica catalysts, it is relatively rare among Ziegler systems, where hydrogen is usually the only way of regulating MW. The unsupported version of this catalyst, $MgCl_2{:}2TiCl_3$ from equation [A], probably does not have such a wide variety of sites. Thus it produces a narrower molecular weight distribution than silica supported catalysts. This is shown in Figure 6 where the heterogeneity index (HI) obtained from melt rheology is plotted for both types of catalysts. The HI is a measure of the shear response of the polymer. A high HI generally indicates a broad WMD. In Figure 6, HI is plotted against melt index because for any family of polymers the flow is dependent on MW as well as MWD; the short chains tend to lubricate the long ones during flow. Notice that for any melt index value the polymers made from silica supported catalysts have a broader MWD (higher HI) than those not containing silica. This indicates a broader diversity of sites. Like Mgl sites, the unsupported catalyst also produced relatively high melt index.

Another indication of site diversity is seen by comparing Tables 1 and 2. Notice that both the HI's and shear ratios increase in Table 2, where the activity is less dominated by one type of site, Mgl. Finally, this change in site population is also evident in Figure 5, which plots the shear response values obtained from the silica supported catalysts, as a function of calcining temperature. The lower the calcining temperature, the more varied the site populations, and the higher the shear response, indicating broader MWD's.

We can further demonstrate the general validity of the above interpretation by chemically modifying the silica, that is, by artificially changing the OH population. Several examples of this are shown below.

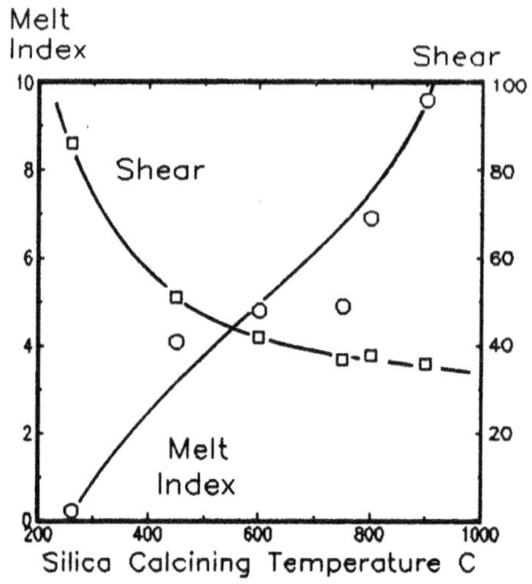

Fig. 5. Ti-Mg/Silica catalysts containing .5% Mg were allowed to polymerize ethylene in the presence of an equal amount of hydrogen to decrease the molecular weight. Evidence for multiple types of active sites is seen in the melt index of the polymers obtained. The higher the silica calcining temperature, the higher the melt index, indicating greater sensitivity to H_2. The shear response (HLMI/MI) reflects the broadness of the MW distribution, which in turn suggests diversity in the active site population.

One way of removing most or all of the surface hydroxyl groups at relatively low temperatures is to silate them, replacing them with a nonreactive hydrophobic group. In one experiment a virgin silica was treated with trimethylchlorosilane at 260C. Then it was impregnated with 0.5% $MgBu_2$ and treated with $TiCl_4$ as usual. Table 3 shows that this catalyst behaved more like a catalyst precalcined at 900C than 260C. It yielded a high melt index, and a narrow MWD. Since the hydroxyl population was greatly diminished, the catalyst adsorbed little titanium and produced almost entirely MgI sites. Hence the activity was greatly improved.

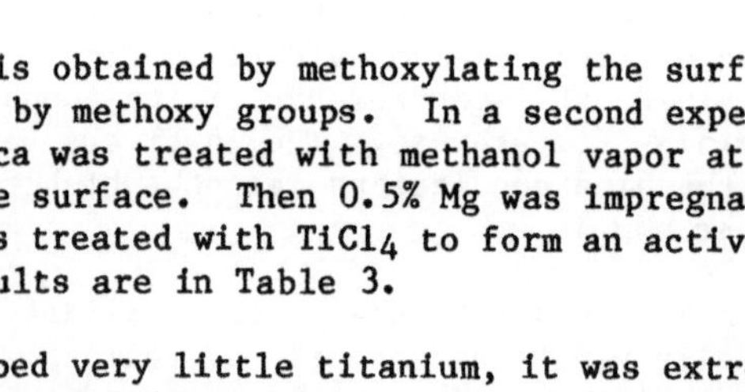

A similar effect is obtained by methoxylating the surface, which replaces the hydroxyls by methoxy groups. In a second experiment, also shown in Table 3, silica was treated with methanol vapor at 350C, causing esterification with the surface. Then 0.5% Mg was impregnated as before and the combination was treated with $TiCl_4$ to form an active catalyst. The polymerization results are in Table 3.

Although it adsorbed very little titanium, it was extremely active. Again the melt index was much higher than expected from the 350C calcining

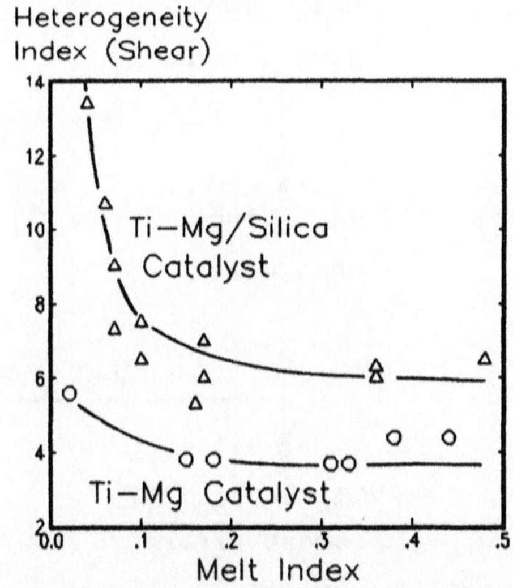

Fig. 6. That the silica supported catalyst has a more diverse population of sites than the unsupported version of the catalyst is clearly visible in the MW distribution of the polymers. Here a visco-elastic estimate of the heterogeneity index (Mw/Mn) is plotted against melt index.

temperature. Since this sample probably had no hydroxyl population at all, it is likely that only Mgl sites, or even the unsupported type A sites, were formed. This would explain the unusually low shear response. The amount of titanium adsorbed, however, is not consistent with $MgCl_2:2TiCl_3$, which should result if the dibutyl magnesium merely adsorbed onto the surface. One possible reaction which might explain this is shown below.

$$\underset{\text{Si}}{\overset{\text{O–Me}}{|}} \xrightarrow{MgBu_2} \underset{\text{Si}}{\overset{\text{Bu}}{|}} + BuMgOMe \xrightarrow{TiCl_4} ClMgOMe:TiCl_3 \text{ or } MgCl_2:TiCl_3$$

Thus we see that this catalyst recipe can be used to produce wide variations in polymer properties, depending on how the silica is pretreated. The range of catalysts can be further broadened by using mixtures of two types of silicas.

Table 3. Modified Silicas

Calcining Treatment	Ti Adsorbed	Activity (g/gTi-h)	Melt Index	Shear Response
260C	4.6%	12,100	0.2	80
750C	2.0%	15,000	4.9	37
900C	1.6%	21,100	6.9	36
Silated 260C	1.1%	35,300	5.9	36
Methoxylated 350C	0.8%	42,500.	7.9	32

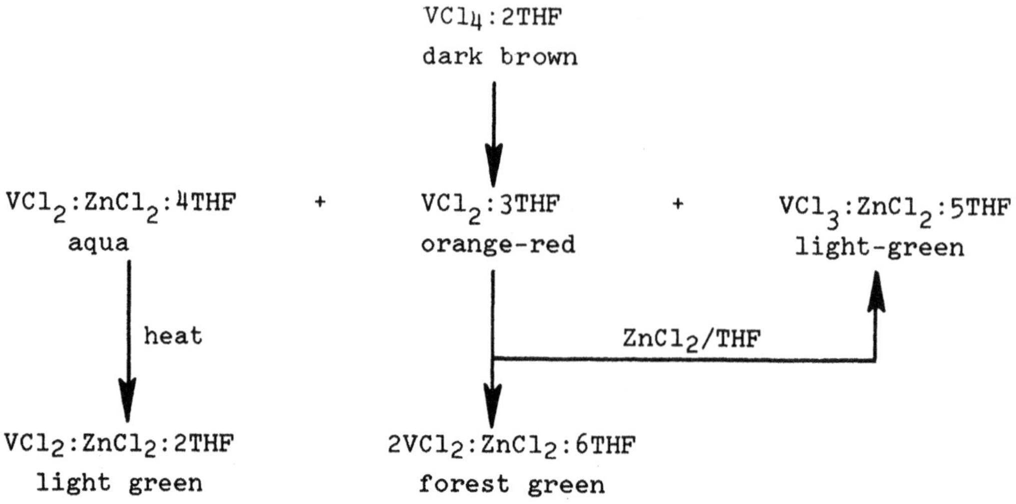

Fig. 7

VANADIUM CHLORIDE SYSTEMS

Paul Smith and Joel Martin showed that the zinc metal reduction of VCl_4 did not yield the simple VCl_3 or VCl_2 as had been expected from an examination of the literature. Instead, a whole spectrum of colored products was obtained, depending on reaction conditions and the amount of zinc used. These materials were later identified as complex chlorides of vanadium and zinc.[12] With some analytical effort, the scheme in Figure 7 was deduced.

Table 4. Activity of Vanadium-Zinc Complexes

Compound	Activity (g/gV-h)	Hydrogen (psi)	Melt Index	Shear Response
VCl_3	18,000	0	0	(IV=33)
	15,000	200	0	
VCl_3:3THF	24,000	0	0	-
	176,000	80	1.3	39
VCl_3:$ZnCl_2$:5THF	35,000	0	0	-
	59,000	80	1.6	38
VCl_2:$ZnCl_2$:4THF	279,000	0	0	(IV=25)
	258,000	80	1.0	119
VCl_2:$ZnCl_2$:2THF	247,000	80	1.1	-
2VCl_2:$ZnCl_2$:6THF	191,000	.0	0	(IV=19-30)
	279,000	80	2.1	45

*All runs made with TEA and $(CFCl_2)_2$ activator

Table 5. Catalyst = $VCl_2:ZnCl_2:4THF$

$CFCl_2-CFCl_2$	Hydrogen	Activity	Melt Index
NO	NO	34,000	0
NO	YES	7,500	14
YES	NO	279,000	0
YES	YES	152,000	28

*Activity in g/gV-h, all runs made with TEA cocatalyst

All of these materials were found to be highly active for the polymerization of ethylene. The participation of $ZnCl_2$ or THF may serve to promote activity, because the simple vanadium chlorides are not known for their high productivity. A summary of the activity of these compounds is shown in Table 4. Note that the divalent vanadium-zinc chloride compounds yielded considerably better activity than the straight VCl_3. Polymer properties also varied somewhat between compounds.

These vanadium chloride-based Ziegler catalysts behave somewhat differently from their titanium chloride cousins. All of these polymerization runs in Table 4 were again made with triethylaluminum cocatalyst, but all also contained a chlorocarbon activator, $(CFCl_2)_2$. The specific role of this activator in polymerization is not clear, but it usually improves the activity of vanadium based catalysts by at least several times.

Some of the runs in Table 4 also used hydrogen to decrease the molecular weight of the polymer. In the absence of the chlorocarbon activator, hydrogen nearly killed these catalysts, but in the presence of the activator, hydrogen was much less damaging to the activity. In fact, examples in Table 4 actually show an improvement by hydrogen, although this was not always true. This suggests that the vanadium active site is deactivated during termination, possible to a hydride, and the chlorocarbon activator somehow reactivates it. Table 5 summarizes this interaction between hydrogen and the activator.

$$V\text{-}CH_2\text{-}(CH_2)_n\text{-}CH_3 \longrightarrow \underset{\text{dead}}{V\text{-}H} + CH_3\text{-}(CH_2)_n\text{-}CH_3 \xrightarrow[\text{Activator}]{\text{TEA}} V\text{-}Et \ ?$$

These vanadium chloride catalysts were often more sensitive to hydrogen than titanium chloride systems, giving higher MI values. But in the absence of hydrogen, another difference becomes evident. While titanium chloride catalysts do produce high molecular weight polymer, with an inherent viscosity (IV) of say 13, those produced by the vanadium chlorides are even higher, often above IV=30. This combination of higher MW in the absence of hydrogen, and higher sensitivity to hydrogen, gives the vanadium based catalysts a wider effective melt index range.

These compounds were also dissolved in dichloromethane and impregnated onto high surface area oxide carriers. This usually further improved the activity. In Table 6 the activity of these compounds, supported on alumina, is shown. This time the activity is listed as yield of polymer per gram of total catalyst, including the support. Since these samples normally contained about 3% vanadium, one can see that the activity, based

Table 6. Alumina Supported Catalysts

Vanadium Compound	Hydrogen (psi)	Activity g/g cat-h	Melt Index g/10m	Shear
VCl_3:3THF	0	9,500	0	-
	80	2,600	.02	105
VCl_3:$ZnCl_2$:5THF	0	2,500	0	-
	80	900	.07	83
2VCl_2:$ZnCl_2$:6THF	0	40,500	0	-
	80	3,000	.92	45
VCl_2:$ZnCl_2$:4THF	0	10,000	0	-
	80	3,000	.11	55

*All runs made with TEA and chlorocarbon activator

on vanadium, is extremely high, in some cases approaching 1.5 million. Notice also in Table 6 that the support tends to lower the melt index and broaden the molecular weight distribution, in comparison to Table 4.

The compound 2VCl_2:$ZnCl_2$:6THF was found to be the most active compound on alumina, as it also was without a support. This compound was then tested on a number of other supports and the results are shown in Table 7. Silica was the least active, but there was only minor variations in activity between these catalysts. The most striking effect was on phosphated alumina, which greatly enhanced the melt index response to hydrogen. This is further evidence that the support is not simply an inert carrier. There is, apparently, some type of vanadium-support interaction.

BIMETALLIC CATALYSTS

It is interesting that neither the titanium nor the vanadium chlorides, alone, constitute an effective catalyst, but combined in a certain way they do form a powerful catalytic system.

Table 7. 2VCl_2:$ZnCl_2$:6THF on Various Supports

Support	Activity	Melt Index	Shear
Alumina	4400	52	73
Silica	2050	2.00	57
$AlPO_4$	3600	.23	69
H_3PO_4/Alumina	3700	15.1	--

*All runs made with TEA, chlorocarbon activator, and 20 psi hydrogen

When $Ti(OEt)_4$ is contacted with $VOCl_3$, an exothermic reaction occurs and a viscous red liquid is formed. The liquid is only poorly soluble in hexane, whereas both Ti and V starting materials are very soluble, so the product probably has polar character. Undoubtedly a ligand redistribution occurs and possibly a dimerization (to account for the color). If the reaction is carried out in hexane to moderate the heat evolved, the ligand redistribution process occurs more slowly. Our best guess of the structure of the red bimetallic compound is shown below:

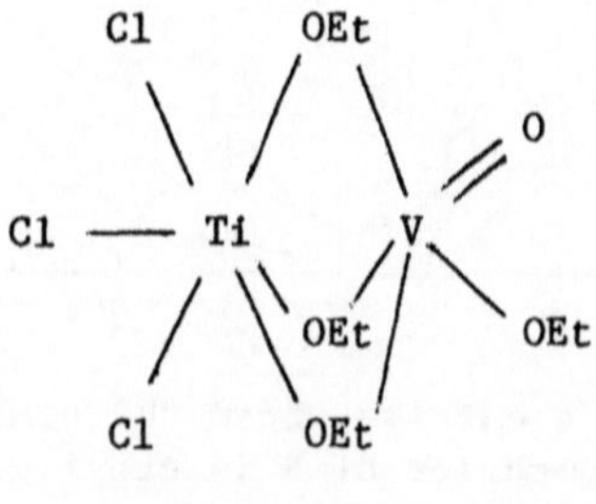

The catalyst is prepared from this red Ti-V complex by a two step process. In the first step the complex is reduced by an organoaluminum halide, giving a brown precipitate. In the final step this precipitate is treated with $TiCl_4$ to yield the active catalyst.

In Table 8, a series of comparative catalyst preparations and their polymerization activities are listed. The most active catalysts were made from $Ti(OEt)_4$:$VOCl_3$ combinations (Run 7). The essential elements can be seen clearly in the Table. Runs 1-2 show that neither the titanium nor the vanadium starting materials can function alone as an active catalyst. Run 3 demonstrates that alkoxy-groups are necessary, but Runs 4-6 show that a simple balance of halide and alkoxy-groups on titanium is not sufficient.

Table 8. Essentials of Ti-V Catalyst

Run	Reagents	Activity g/g cat-h
1	$TiCl_4$	5,400
2	$VOCl_3$/ether	GUM
3	$TiCl_4$:$VOCl_3$	7,000
4	$TiCl_3(OEt)$	400
5	$TiCl_2(OEt)_2$	320
6	$TiCl(OEt)_3$	160
7	$Ti(OEt)_4$:$VOCl_3$/ether	31,000
8	$2Ti(OEt)_4$:$VOCl_3$/ether	15,900
9	$Ti(OEt)_4$:$2VOCl_3$/ether	20,100

*All runs used EASC as reducing agent, and TEA as cocatalyst

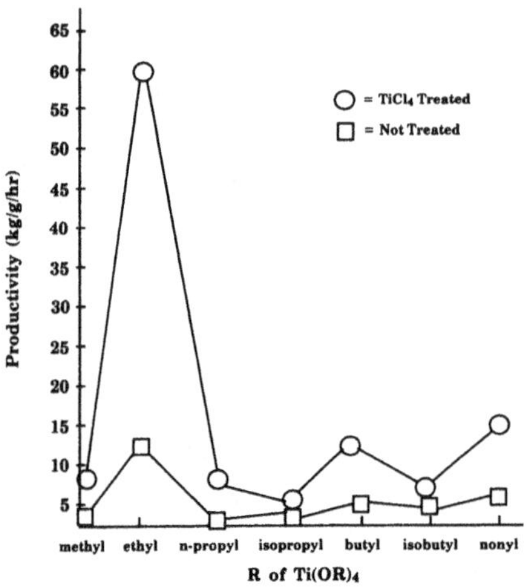

Fig. 8. In preparing a catalyst from the complex $Ti(OR)_4:VOCl_3$, the preferred choice of R is ethyl.

A combination of titanium, vanadium, and alkoxy-groups seem to provide the highest activity. The remaining Runs 7-9 show that the Ti:V molar ratio of 1:1 is best.

Not only are the alkoxy-groups essential to high productivity, but the choice of groups seems to be extremely limited. Only ethyl will do. This is plain in Figure 6, where the optimum Ti-V preparation was carried out with various $Ti(OR)_4$ selections. When R was ethyl the activity was 60,000 g/g-h. But when R was methyl or propyl the activity decreased to about 10,000 g/g-h. Other alkyls were similarly ineffective. The reason for this peculiar specificity is unclear.

Four different reducing agents have been examined in this catalyst preparation, and their effect is shown in Table 9. The reduction process is highly exothermic, and was therefore most effectively accomplished in an ice bath. Both ethylaluminum dichloride (EADC) and ethylaluminum sesquichloride (EASC) afforded catalysts, after a final $TiCl_4$ treatment, with high productivity. Reduction by EADC was less vigorous than by EASC. When the addition order was reversed, that is when the Ti-V complex was added to EASC, even better activity resulted. This suggests that, like any precipitated catalyst, the surface area and porosity strongly depend on the particular method of contacting the two reactants. The reduction step was most effective when the EASC was added in a one to one ratio with the complex.

Table 9. Reducing Agent for Ti-V Complex

Reducing Agent	Activity (g/g cat-h)
Dibutyl Magnesium	5,200
EADC	33,000
EASC	35,000
DEAC	19,000
EASC (Reversed)	56,500

*All runs made with TEA as cocatalyst

Numerous other variables were also optimized for this catalyst system, giving activities as high as 85,000 g/g-h. This corresponds to almost one million based on vanadium or titanium.

We are not sure which is the active metal of the system, perhaps both. In some ways this bimetallic catalyst behaves like a vanadium chloride catalyst. The melt index is highly responsive to hydrogen, and the productivity drops sharply with hydrogen in the absence of chlorocarbon activator. But nevertheless, the activator is not needed to achieve high productivity with this sytem, even with hydrogen, suggesting that titanium is really the active metal. If so, it is an unusual system because to be so active, titanium chloride systems ordinarily require the presence of magnesium chloride, which is absent here.

REFERENCES

1. M. P. McDaniel and M. B. Welch, J. Catal., 82, 98 (1983).
2. M. B. Welch and M. P. McDaniel, J. Catal., 82, 110 (1983).
3. M. P. McDaniel, M. B. Welch, and M. J. Dreiling, J. Catal., 82, 118 (1983).
4. M. P. McDaniel and M. M. Johnson, U.S. Patents 4,364,842; 4,364,854; and 4,364,855 (Dec. 1982).
5. J. J. Fripiat and J. Uytterhoeven, J. Phys. Chem., 66, 800 (1962).
6. M. P. McDaniel, J. Catal., 67, 71 (1981).
7. M. P. McDaniel, J. Phys. Chem., 85, 532 (1981).
8. M. P. McDaniel, J. Catal., 76, 17 (1982).
9. C. G. Armistead, A. J. Tyler, F. H. Hambleton, S. A. Mitchell, and J. A. Hockey, J. Phys. Chem., 73(11), 3947 (1969).
10. R. J. Peglar, F. H. Hambleton, and J. A. Hockey, J. Catal, 20, 309 (1971).
11. J. Kunawicz, P. Jones, and J. A. Hockey, Trans. Faraday Soc., 67(3), 848 (1971).
12. P. D. Smith, J. L. Martin, R. L. Bansemer, J. C. Huffman, and K. G. Caulton, Inorg. Chem., in press.

STEREOCHEMICAL CONTROL IN

ZIEGLER-NATTA CATALYSTS

S. E. Wilson, W. L. Callender,
and R. C. Job

Shell Development Company
Houston, Texas 77001

INTRODUCTION

The remarkably high level of stereoselectivity associated with magnesium chloride-based catalysts is realized only in the presence of a selectivity control agent (SCA), frequently an aromatic ester.[1] The usual effect of the SCA is to reduce activity while significantly enhancing stereoselectivity.[2] Fortunately, the reduction in productivity (65%) is not as great as the reduction in atactiticy (92%) (Figure 1).

A form of SCA is present both in the preparation of the procatalyst (magnesium) chloride - titanium tetrachloride - SCA) as well as in the cocatalyst (triethylaluminum - SCA). Whereas one ester may function best in the metathesis reaction used to generate the procatalyst, a structurally distinct ester is often better suited as the SCA in the cocatalyst (Figure 2).

Since the function of the SCA is a critical one in obtaining polypropylene of high isotacticity it is important to consider how the aromatic ester interacts with the other catalytic components. During the preparation of the procatalyst the ester plays a critical role in dispersing the titanium tetrachloride on small hexagonal crystallities of magnesium chloride. In the absence of an ester a catalyst of higher titanium loading and much lower activity is obtained. Solid state C-13 NMR reveals a carbonyl carbon chemical shift the same as what is obtained from ball-milling the ester with magnesium chloride. The chemical shift of the carbonyl carbon on the crystalline ester - titanium tetrachloride complex is shifted by about 4 ppm. These data are consistent with ester being bonded through the carbonyl oxygen to magnesium rather than to titanium in the procatalyst. The precise nature of any bonding interaction between the SCA and titanium remains obscure. Precedence for an ester preferentially bonding to magnesium rather than to titanium can be found in the well-characterized structure shown in Figure 3.[3]

The decay curve of the active catalyst delineates a rapid loss in activity with time (Figure 4).[4] Loss in activity appears to be a function of the catalytic components only and independent of the polymerization process as evidenced by invariance of the decay behavior when the system was starved of propylene for a short period during the polymerization (Figure 5).[5] It was of interest to determine during the decay process if

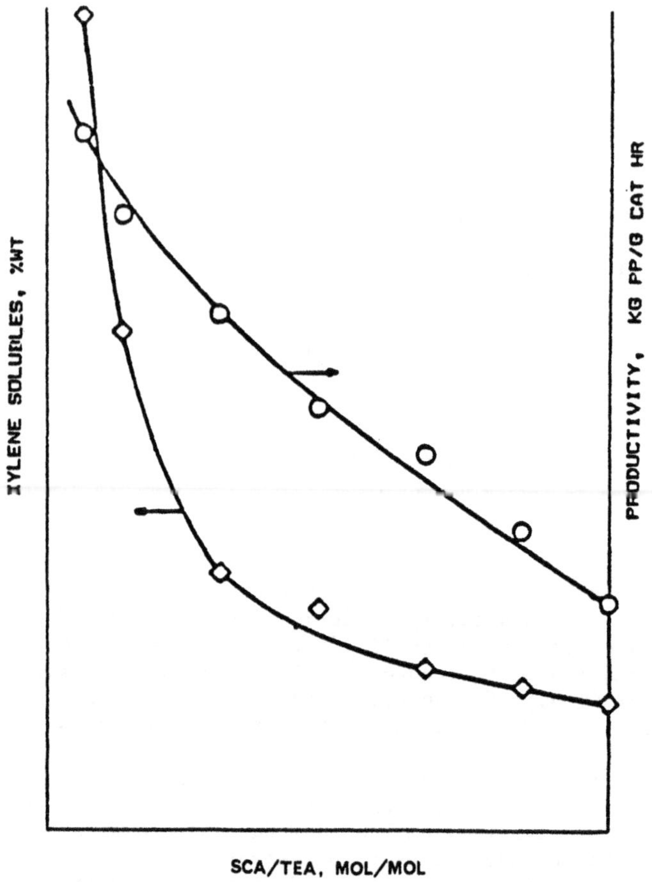

Fig. 1. Polypropylene Productivity and Xylene Solubles as a Function of SCA/TEA

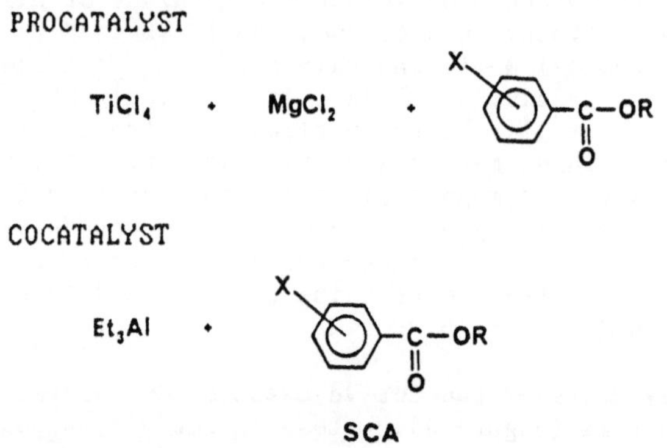

Fig. 2. Chemical Components of PP Catalyst

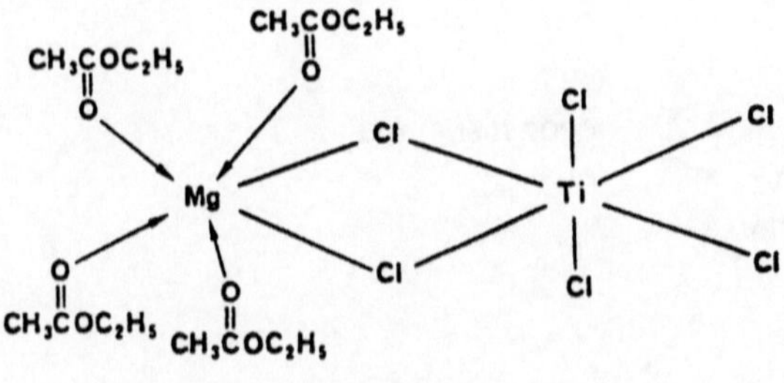

Fig. 3. Characterized Complex

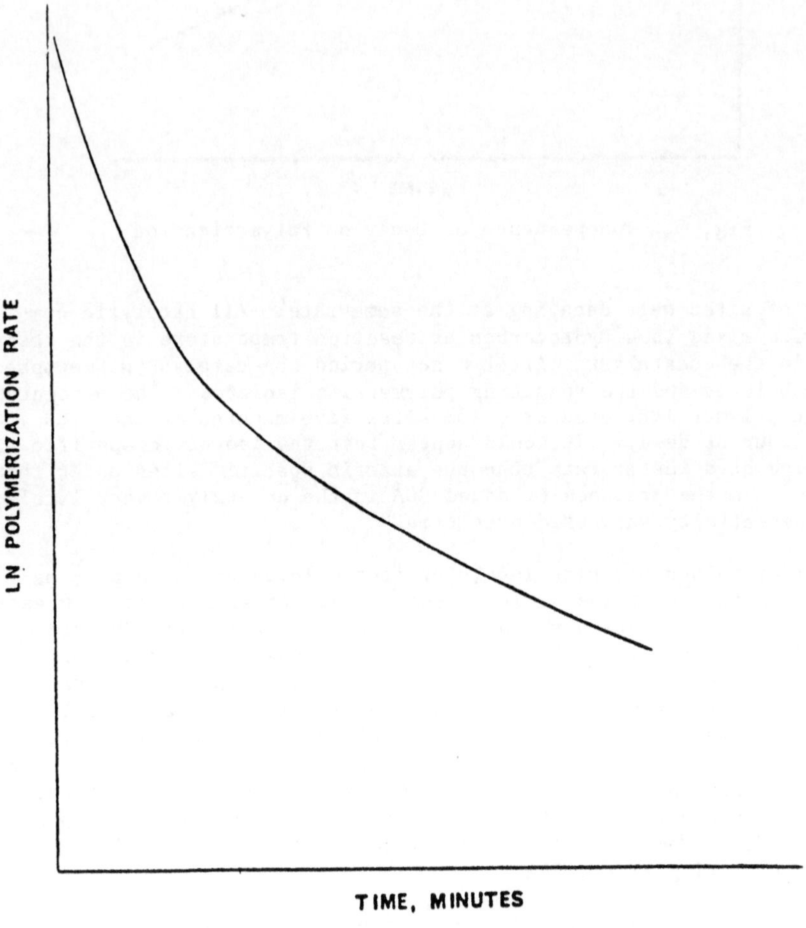

Fig. 4. Decay Behavior of PP Catalyst

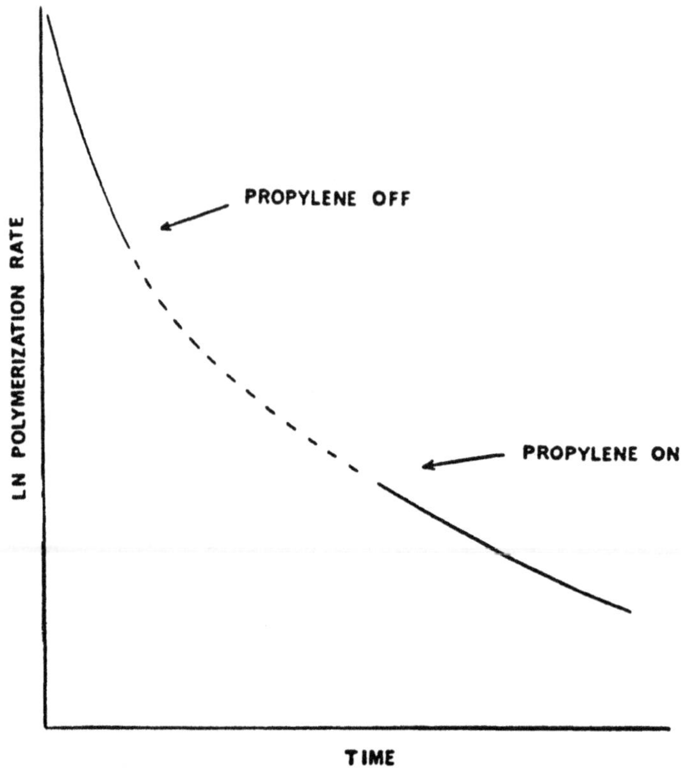

Fig. 5. Independence of Decay on Polymerization

all types of sites were decaying at the same rate. All catalytic components were mixed in a hydrocarbon at reaction temperature in the absence of a SCA in the cocatalyst. After a set period the catalyst was exposed to propylene briefly and the resulting polymer was isolated. The percentage of atactic polymer increased from 13% after five minutes of decay to 25% after one hour of decay. It would appear that the isotactic specific sites are decaying at a faster rate than the atactic specific sites under these conditions. In the presence of added SCA in the cocatalyst very little change in atacticity was noted over time.[6]

In order to achieve high levels of isotacticity an SCA has to be included with the cocatalyst. It is known that TEA will react with esters to provide bulky alkylation products as shown in Figure 6.[7] Therefore, the question emerges as to whether the ester or one of its reaction products with TEA serves as SCA. When the alkylation products were prepared independently and used in conjunction with TEA as the cocatalyst little effect was observed on stereochemical control. Consequently, the known alkylation products derived from reaction of the cocatalytic components do not serve as SCAs.[8] It has also been observed that when the procatalyst was pretreated with ester no polymerization occurred upon exposure to TEA. The evidence to date is consistent with the TEA-aromatic ester complex acting as the SCA.

The mechanistic interpretation of the effect of additional SCA in the cocatalyst has placed emphasis on the existence of two distinct types of sites. The published work of Pino involving the use of an optically active SCA is most educational in this regard.[9] The effect of the ester is to substantially reduce the number and/or the activity of nonstereoselective chiral centers which appear to be stronger Lewis acids than the highly

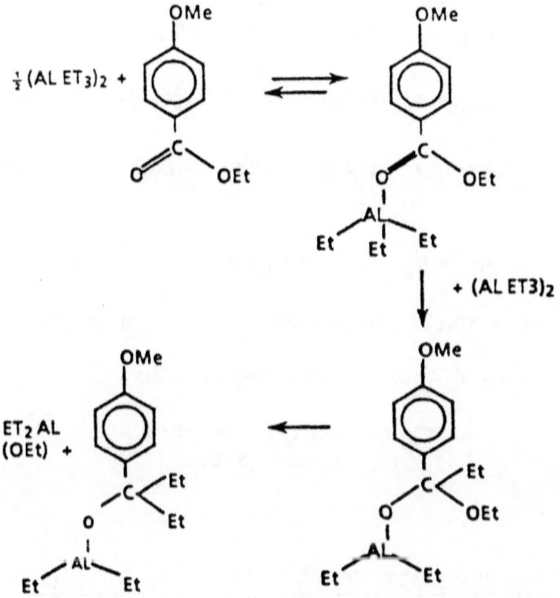

Fig. 6. Alkylation of an Ester with TEA

stereoselective chiral centers. The mechanism as shown delineates an active site which has no SCA bonded directly to it. An abundance of such sites is available as judged from the high level of isotactic polymer produced in the complete absence of an SCA in the cocatalyst.

Evidence in support of this contention can be found by comparison of isotactic polymer structure derived from several different types of catalysts, some of which contain no SCA. NMR analysis has shown these polymers to be essentially identical in terms of stereoregularity which is consistent with the sites being very similar.[10] Further evidence in favor of a dynamic equilibrium as depicted comes from decay studies of active catalyst as a function of TEA/ester. As stated earlier the presence of SCA has the effect of reducing activity while increasing isotacticity. In the context of the above equilibria the concentration of AlR_3 is increased relative to AlR_3B^* which ultimately drives each reaction to the left.

The question remains as to why the atactic specific site behaves as a much stronger Lewis acid than the isotactic specific site. One possibility is that the less sterically congested atactic specific site bonds to the ester in a stronger fashion while such bonding at the isotactic specific site would remain somewhat strained. However, the atactic polymer is still reasonably stereoregular which is indicative of more subtle steric differences between the two types of sites. The sites may differ in oxidation state or charge distribution. Correlation of decay with reduction of Ti(IV) to Ti(III) during the early stages of reaction lends credence to the notion that the highly active but fragile stereoselective sites at the outset of the reaction are Ti(IV) sites.[11] These would be expected to be strong Lewis acids not weak ones. A reasonable alternative to the above presents itself when one assumes that the isotactic polymerization can be halted by complexation of the SCA to magnesium rather than to titanium.

The carbonyl functional group donates more strongly to titanium than to magnesium as can readily be ascertained by comparison of the changes in the carbonyl stretching frequencies of ethyl benzoate upon complexation with appropriate metal halides (Figure 8). The bond strength of the com-

$$C_{solid} + AlR_3B^{\bullet} \underset{}{\overset{K_1}{\rightleftharpoons}} CB^{\bullet}_{solid} + AlR_3$$

$$C^{\bullet}_{solid(d)} + AlR_3B^{\bullet} \overset{K_2}{\rightleftharpoons} C^{\bullet}_{(d)}B^{\bullet}_{solid} + AlR_3$$

$$C^{\bullet}_{solid(l)} + AlR_3B^{\bullet} \overset{K_3}{\rightleftharpoons} C^{\bullet}_{(l)}B^{\bullet}_{solid} + AlR_3$$

$K_1 > K_2$ $\quad K_1 > K_3$ $\quad K_2 \neq K_3$

C = atactic specific site $\quad$ $B^{\bullet}$ = SCA

$C^{\bullet}_{(d)}$ and $C^{\bullet}_{(l)}$ = isotactic specific sites

Fig. 7. Equilibria of SCA and Active Sites

ETHYL BENZOATE COMPLEX	$\Delta\nu_{C=O}$
$MgCl_2$	-20 40 cm^{-1}
$TiCl_4$	-150 cm^{-1}
$AlMe_3$	-60 cm^{-1}

Fig. 8. Change in Carbonyl IR Stretching Frequencies Upon Complex Formation

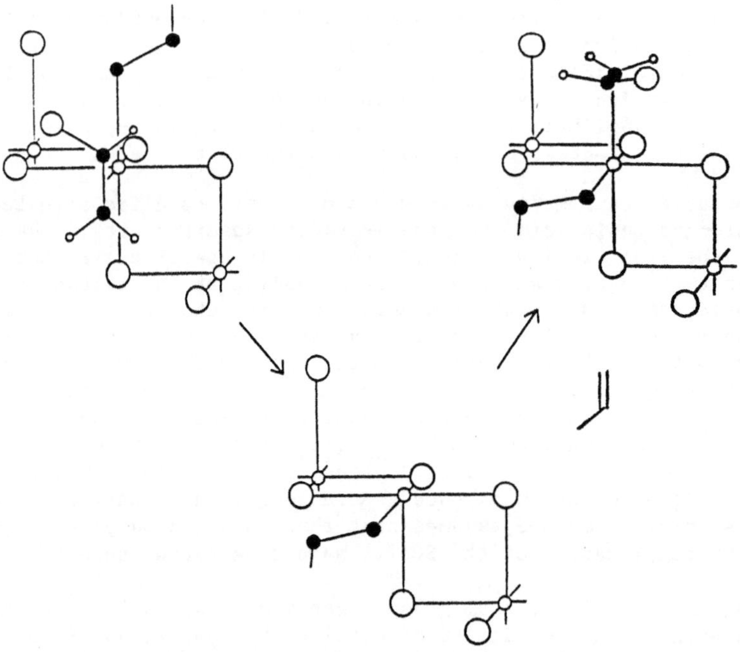

Fig. 9. Isotactic Specific Site

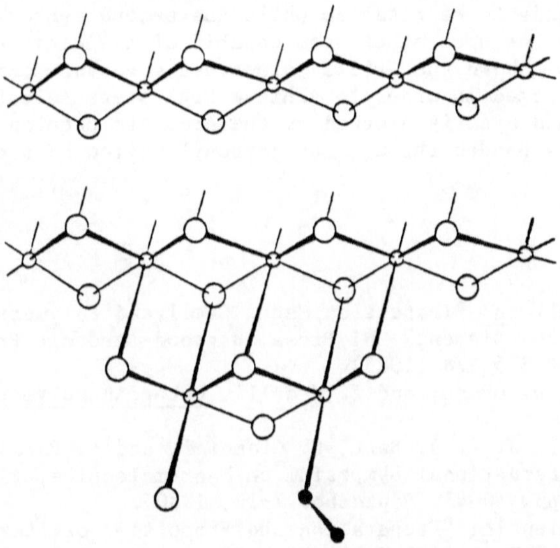

Fig. 10. $TiCl_3$ Dimer on $MgCl_2$ Monolayer Edge

plex between ethyl benzoate and TEA falls somewhere between these two.[12] It would be expected that ethyl benzoate could more easily transfer to an available titanium site than to a magnesium site. An equilibrium involving a competition of titanium and magnesium for the aluminum-bonded carbonyl should result in greater exposure of magnesium which is proposed to be somehow associated with the isotactic specific site. When the procatalyst was subjected to reaction with TEA reduction of Ti(IV) to Ti(III) was observed by ESR. The actual catalyst appears to be under increasing total reduction levels during the polymerization process. By analogy to homogeneous systems, it is suspected that bridged titanium-aluminum species are generated on the surface.

One of the recurring themes in stereochemical control is the importance of the polymer chain itself in influencing the facial specific insertion of propylene. The conformational preferences assumed by the chain are dictated by other members of the chiral site through steric induction. These notions are delineated in Figure 9 and have been discussed recently in a series of papers by Corradini, et al.[13] The "least nuclear motion" arguments offered by Cossee require that the second carbon atom of the polymeric chain should be as near as possible to the olefin just prior to insertion without introducing adverse steric repulsions. This requirement can be met as shown in Figure 9 when the first two carbon atoms of the polymeric chain lie in the same plane with titanium and three of its attendant ligands. The propylene bonds to titanium parallel to this plane with the double bond parallel to the titanium-chain bond. The most interesting feature of this depiction is that propylene uses the same enantiometic face independent of the site location of the chain. The chain is not required to migrate back to the original site as suggested in the early Cossee mechanism. This single appealing characteristic of the site leaves little room for error in the stereoselective generation of isotactic polymer and avoids the necessity of having the polymer chain migrating into a more sterically congested area.

Sites possessing the appropriate symmetry can easily be imagined by expanding the edge of a magnesium chloride crystallite. By way of illustration a site has been generated by projecting a Ti(III) dimer from the edge of the crystallite (Figure 10). Actually, only the polymerization center itself needs to be titanium while the second member of the dimer could be aluminum or some other atom capable of assuming an octahedral coordination. As shown the Ti(III) dimer has a vacant coordination site which must be blocked in order to achieve high stereoselectivity. It is possible that this site is blocked by the aromatic portion of the ethyl benzoate which is bonded through its carbonyl oxygen to a neighboring magnesium atom.

REFERENCES

1. B. L. Goodall in: "Transition Metal Catalyzed Polymerizations: Alkenes and Dienes," MMI Press, Harwood Academic Press, New York, N.Y., pp. 355-378 (1983).
2. E. W. Duck, D. Grant, and E. Kronfli, European Polymer J., 15, 625 (1979).
3. E. Albizzati, J. C. J. Bart, U. Giannini, and S. Parodi, Preprints of IUPAC International Symposium on Macromolecules, Florence, Italy, Vol. 2, pp. 40-43 (September 7-12, 1980).
4. J. C. W. Chien in: "Preparation and Properties of Stereoregular Polymers," D. Reidel Publishing Company, p. 113 (1979).
5. Y. Doi, M. Murata, and K. Yano, Ind. Eng. Chem., Prod. Res. Dev., 21, 580 (1982).
6. P. Galli, P. C. Barbe, and L. Noristi, Die Angewandte Makromoleckulare Chemie, 120, 73 (1984).
7. J. C. W. Chien and J. C. Wu, J. of Polym. Sci.: Polym. Chem. Ed., 20, 2445 (1982).
8. A. W. Langer, T. J. Burkhardt, and J. J. Steger, Polym. Sci. Technol., 19, 225 (1983).
9. P. Pino, G. Fochi, O. Piccolo, and U. Giannini, J. Am. Chem. Soc., 104, 7381 (1982).
10. Y. Doi, E. Suzuki, and T. Keii, "Transition Metal Catalyzed Polymerizations: Alkenes and Dienes," MMI Press, Harwood Academic Press, New York, N.Y., pp. 737-749 (1983).
11. J. C. W. Chien and J. C. Wu, J. of Polym. Sci.: Polym. Chem. Ed., 20, 2461 (1982).
12. T. Keii, E. Suzuki, M. Tamura, and Y. Doi, "Transition Metal Catalyzed Polymerizations: Alkenes and Dienes," MMI Press, Harwood Academic Press, New York, N.Y., pp. 97-113, (1983).
13. a) P. Corradini, V. Barone, R. Fusco, and G. Guerra, European Polymer J., 15, 1133 (1979).
    b) P. Corradini, G. Guerra, R. Fusco, and V. Barone, European Polymer Journal, 16, 835 (1980).
    c) P. Corradini, V. Barone, R. Fusco, and G. Guerra, J. of Catalysis, 77, 32 (1982).
    d) P. Corradini, V. Barone, G. Guerra, Macromolecules, 15, 1242 (1982).
    e) P. Corradini, G. Guerra, and V. Barone, European Polymer J., 20, 1177 (1984).

PARTICLE CONTROL OF SUPPORTED

ZIEGLER-NATTA CATALYSTS

A. Muñoz-Escalona, J. G. Hernández,
J. A. Gallardo, and A. Sustic

Laboratorios de Polímeros, Centro de Química, I.V.I.C.
Apartado, Nº 1827, Caracas 1010-A, Venezuela

INTRODUCTION

A good catalyst for olefin polymerization should exhibit, among other characteristics, good morphological control of the resulting polymers, as well as high activity, control of the molecular weight and molecular weight distribution. Thus, in regard to the polymer morphology, particles with spherical shape, having an average diameter of about 500μ, no fine content, narrow particle size distribution and high bulk density, these particles are very desirable either for fluidized gas-phase and slurry processes, because they can eliminate many problems in plant operation. To achieve good polymer particle control, catalysts with good mechanical properties and morphology have to be prepared, due to the fact that the catalyst imposes its shape on the polymer (replication phenomena).[1,2]

High active catalysts for olefin polymerization can be prepared by supporting $TiCl_4$ on $MgCl_2$ in the presence of an electron donor.[3] However, due to the low surface area of the $MgCl_2$ and also to the poor interaction with the $TiCl_4$, both components have to be intensively ball-milled, in order to activate them and introduce sufficient amounts of titanium into the crystal lattice of the $MgCl_2$. The result is a highly active catalyst exhibiting poor morphology.

Thus, very irregular particles having an appreciable amount of fine particles have been obtained by using the described catalyst procedure. This can be very critical in the slurry process because the columns used for solvent purification can become plugged during operation. In gas-phase processes, the use of catalysts with poor particle size distribution is also undesirable, especially in a fluid bed operation.

Catalysts with better ability in controlling polymer morphology can be prepared by a number of methods and routes. One of these methods uses the well defined morphology of an appropriate silica, silica-alumina or alumina, which is loaded with $TiCl_4$[4,5] or other active transition metal compound.[6,7] In the case of SiO_2 supported $TiCl_4$ catalysts, we have recently reported activities as high as 18 Kg.PE x $g.Ti^{-1}$ x $hr.^{-1}$ x $atm.^{-1}$ By simple impregnation of SiO_2 with $TiCl_4$, catalysts were synthesized with activities of 22.8 Kg.PE x $g.Ti^{-1}$ after 2 hr. polymerization time, at 50°C and 5 atm. ethylene pressure. This activity could be improved up to 34.8 Kg.PE x $g.Ti^{-1}$ when Davison 951 SiO_2 was impregnated twice with mixtures of $TiCl_4$ and $AlR_{3-x}Cl_x$. It is known that the best silicas used as supports are those with high surface area, good porosity and enough mechanical strength, so that they do not break up at the early steps of the polymerization. These characteristics, together with good particle size distribution and shape, are very important for the synthesis of catalysts having good control of the polymer morphology.

In this work, attempts were made to prepare high activity catalysts for ethylene polymerization with good control of the polymer morphology. To achieve these goals, Davison 951 silica was used for simultaneous impregnation with $TiCl_4$ and Grignard compounds, following procedures used for mixtures of $TiCl_4$ and $AlR_{3-x}Cl_x$.[8]

EXPERIMENTAL

Materials

Polymerization grade ethylene supplied by Matheson Gas Prod. was purified by passing through a series of columns containing activated R3-11 catalyst from BASF, Linde molecular sieve 4 Å dried at 350°C for several hours and P_2O_5. Nitrogen was also purified in the same way. Aluminumalkyl compounds ($AlEt_3$, $AlEt_2Cl$, $Al(iso\text{-}Bu)_3$ and $ZnEt_2$) supplied by Ethyl Corporation and $TiCl_4$ from Merck were used as received. C_4H_9MgCl was prepared in ether using C_4H_9Cl analytical grade from Merck and following usual procedures. The n-heptane from Merck used as solvent for catalyst preparation and polymerization was distilled over AlLiH in a N_2 stream and dried over sodium wire prior to use. Grace Davison commercial silica-gel grades 951 and 952, containing as a main impurity Al_2O_3 (0.15 wt.%), were used as supports after being dried under vacuum at 150°C during 4 hours.

Catalyst Preparation and Characterization

Catalysts were prepared by the reaction of the silica simultaneously with $TiCl_4$ and aluminumalkyl compounds in n-heptane as specified in Figure 1. Numbers given in brackets correspond to the catalyst prepared using silica 952 as carrier.[8]

Other catalysts were synthesized by similar procedures but substituting the aluminumalkyl compounds by C_4H_9MgCl. Thus, the silica 951 was coimpregnated by simultaneous reaction with $TiCl_4$ and C_4H_9MgCl in n-heptane as solvent, for 3 hr. at 50°C which stirred to obtain catalyst 569 (see Figure 2). Catalyst 570 was prepared by heating catalyst 569 for 4 hours at 450°C under vacuum. Catalyst 571 was obtained putting the catalyst 570 in n-heptane and adding $TiCl_4$ and C_4H_9MgCl again under vigorous stirring for 3 hr. at 800°C. The precipitate was separated by filtration, washed and dried under vacuum.

Catalysts were subsequently characterized by determining their surface area by BET method, surface acidity (Lewis and Brönsted Acid Centers) by the Webb method, based on the amount of chemisorbed NH_3, and metal content by elemental analysis and atomic absorption techniques.[4,9]

Polymerization Procedures

Polymerizations of ethylene were carried out in a propeller-stirred glass 1 l. autoclave (Büchi, Zürich, Schweiz) at 50°C under constant monomer pressure of 5 atm. 1 n-heptane. The polymerization rates were determined from the rates of monomer consumption, using a two-buret system similar to that described by Schnecko et al.[10] The apparatus is shown diagrammatically in Figure 3. Both catalyst components (supported Ti and

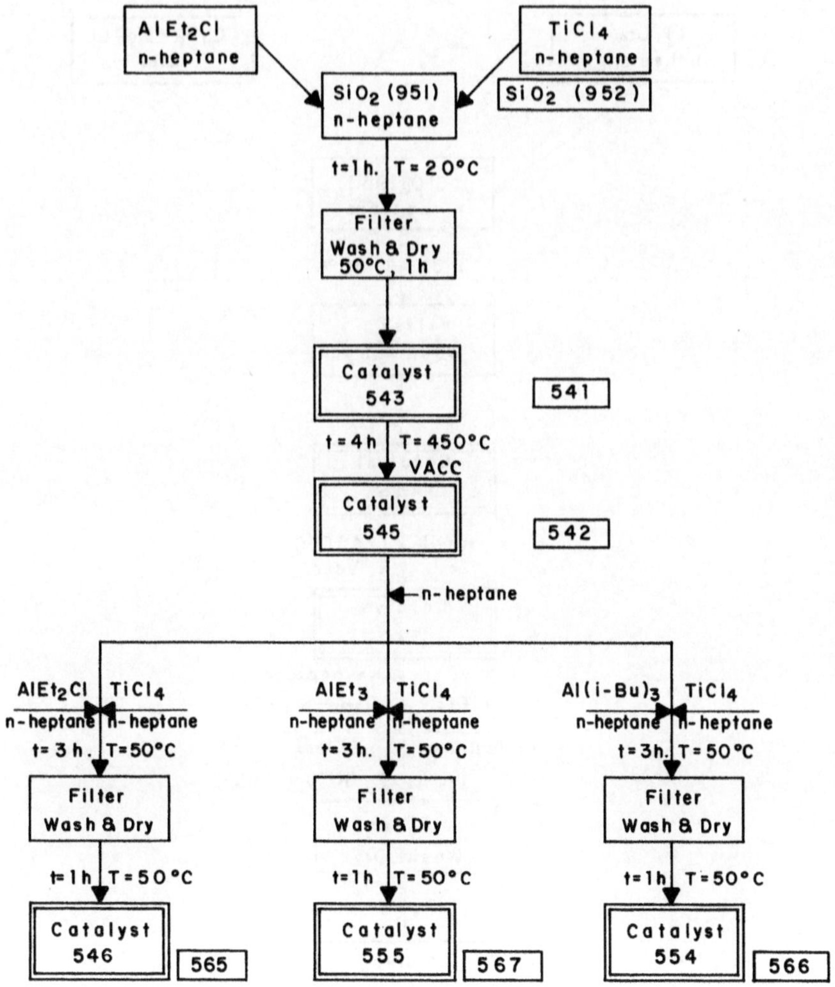

Fig. 1: Catalyst preparation by supporting $TiCl_4$ and aluminumalkyl compounds over SiO_2 951 and 952 (catalysts numbers are given in brackets).

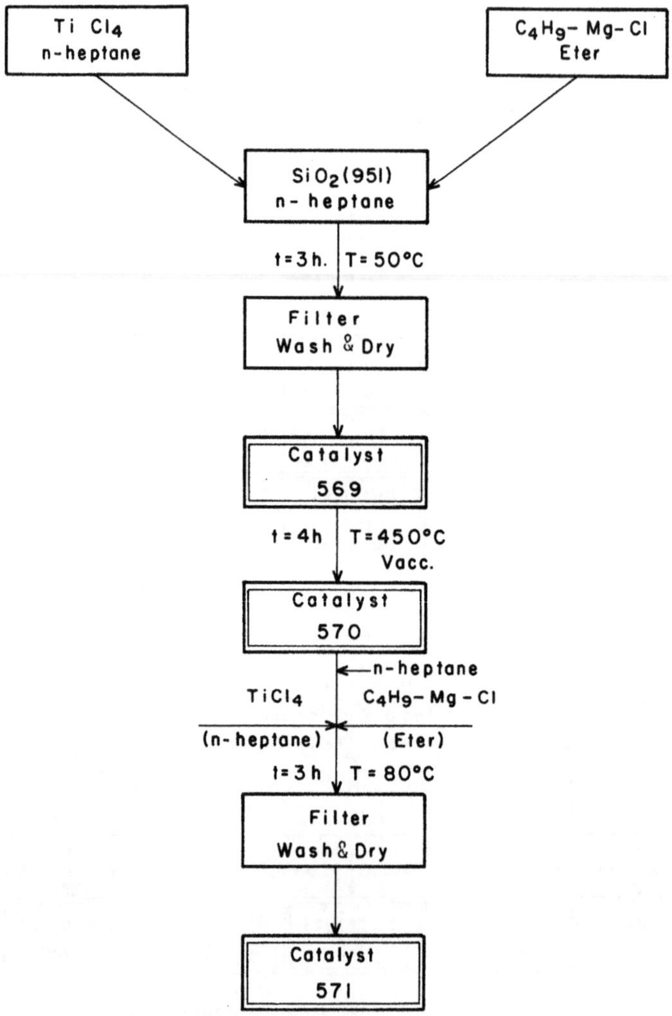

Fig. 2. Catalyst preparation by supporting $TiCl_4$ and C_4H_9MgCl Grignard over SiO_2 951.

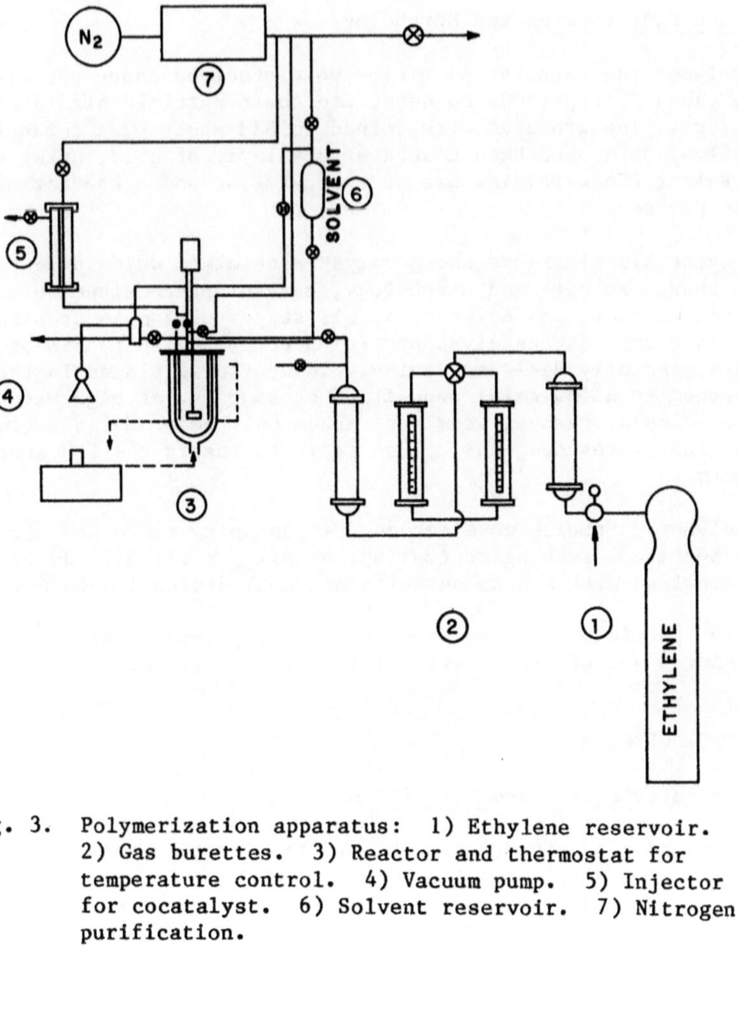

Fig. 3. Polymerization apparatus: 1) Ethylene reservoir. 2) Gas burettes. 3) Reactor and thermostat for temperature control. 4) Vacuum pump. 5) Injector for cocatalyst. 6) Solvent reservoir. 7) Nitrogen purification.

cocatalyst) were introduced separately in the reactor, putting first the solvent contained in reservoir 6, than the Ti containing solid in a glass ampoule, followed by pressurization with ethylene at 5 atm. The aluminum-alkyl compound was introduced with N_2 pressure using the injector 5.

The polymerization was timed just after breaking the ampoule. The Al/Ti ratio was kept constant in all polymerization at 10. A stirring speed of 1200 rpm was used in order to minimize mass transfer control during polymerization. After 3 hours, the reaction was quenched by introducing a solution of ethanol containing hydrochloric acid. The polymers were washed several times with water diluted ethanol and dried under vacuum at 50°C.

Particle Size Distribution and Morphology

The polymer and catalyst particles were observed under the electron microscope (SEM) Philips 500, to determine their particle size distribution and morphology. The granules were joined to SEM stubs by a conductive-adhesive silver paint and then coated with a layer of gold, using a vacuum unit from Balzer (Schweiz) in case of the catalyst and a sputtering apparatus for the polymer.

To prevent hydrolysis of the catalyst particles, which produces dramatical change in size and morphology, careful precautions were taken to avoid contact with air and moisture at all stages of sample preparation and handling. In a dry lab, catalyst particles were adhered to SEM stubs and mounted in a specially designed device. Then, it was placed in the vacuum unit and opened to allow metal deposition of samples, at high vacuum. Then, it was closed, removed from the vacuum unit under N_2 in a glove bag attached to the microscope, and opened again to insert the SEM stubs in the electron beam.

The polymer particles were embedded in an epoxy resin and sectioned, without disturbing, crushing or tearing, by using a Sorvell (du Pont) microtome provided with a diamond knife and cooled with liquid N_2.

Finally, particle size distribution of the polymers was also determined by using a set of sieves with different mesh sizes.

RESULTS AND DISCUSSION

Catalyst Characterization and Activities

According to a previous publication,[8] the surface areas of the silicas decrease by simple reaction of $TiCl_4$ with their hydroxyl groups.[4,9] Thus, in case of silica 951, we have reported that its surface area changes from 610 $m^2 \times g^{-1}$ to 350 $m^2 \times g^{-1}$, after supporting 6.7% wt. of Ti (catalyst 571). The surface acidity, however, increases from 0.35 to 8.7 mmol. $NH_3 \times g.cat.^{-1}$ Furthermore, when the silicas were coimpregnated simultaneously with $TiCl_4$ and $AlEt_2Cl$ the surface area decreased to about 280 $m^2 \times g^{-1}$ (catalyst 543), see scheme in Figure 1; while the surface acidity increased to 12.3 mmol.$NH_3 \times g.cat.^{-1}$ Finally, when the coimpregnated silica 951 was reimpregnated with $TiCl_4$-$AlEt_2Cl$ (catalyst 546), $TiCl_4$-$AlEt_3$ (catalyst 555) and $TiCl_4$ Al(iso-Bu)$_3$ (catalyst 554) respectively, the surface areas were about 150 $m^2 \times g^{-1}$, while surface acidities were between 7.7-15.3 mmol.$NH_3 \times g.cat^{-1}$, depending on the aluminumalkyl used.

The most active series of catalyst were obtained with silica 951 used as a support for the reimpregnation procedure. Al(iso-Bu)$_3$ was the most suitable aluminumalkyl compound, either for catalyst synthesis (catalyst

Table 1. Physio-Chemical Characterization Catalyst Prepared by Reaction of SiO_2 (951) with $TiCl_4$ -C_4H_9MgCl Grignard

Catalyst Nº	Surface Area (m^2/g)	Surface Acidity (mmol NH_3/g.cat.)	Ti (%w)	Mg (%w)
SiO_2 (951)	610	0.35	-	-
569	300	6.80	2.9	12.0
570	260	4.70	3.3	4.2
571	95	8.10	4.2	6.6

554) and cocatalyst. Activities as high as 18 Kg.PE x $g.Ti^{-1}$ x $hr.^{-1}$ x $atm.^{-1}$ were obtained using this catalyst activated with Al(iso-Bu).$_3$

In order to obtain catalysts with higher activities, Grignard compounds were used instead of aluminumalkyl compounds for catalyst synthesis and silica 951 was also used as a carrier. The synthesis route is shown in Figure 2.

The physico-chemical characterization of the catalysts are shown in Table 1. Coimpregnation and reimpregnation of SiO_2 951 with $TiCl_4$ and C_4H_9MgCl (catalysts 569 and 571, respectively), decreased the surface areas while the acidities increased. Thus, for the catalyst 571 containing 4.2% wt. Mg., the surface area decreased from 610 m^2 x g^{-1} to 95 m^2

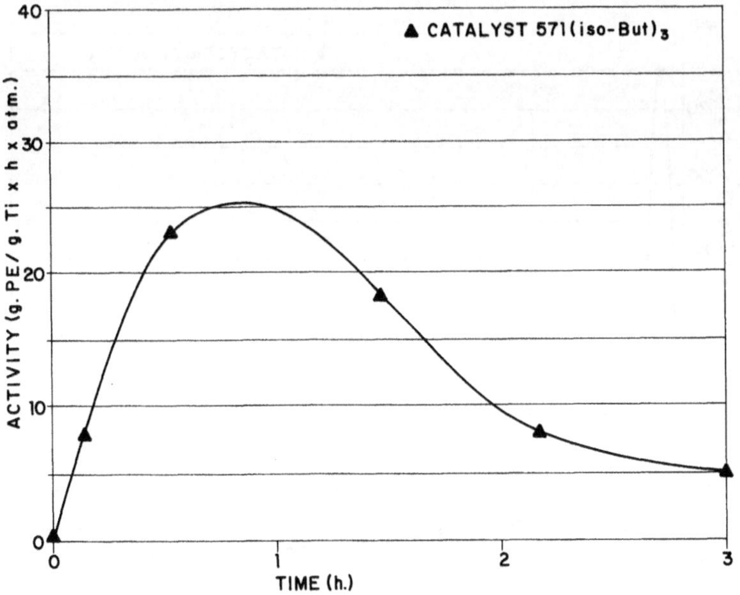

Fig. 4. Catalytic activity vs. polymerization time for catalyst 571: SiO_2 (951) $TiCl_4$ + C_4H_9MgCl/$TiCl_4$ + C_4H_9MgCl, activated with cocatalyst Al(iso-Bu)$_3$.

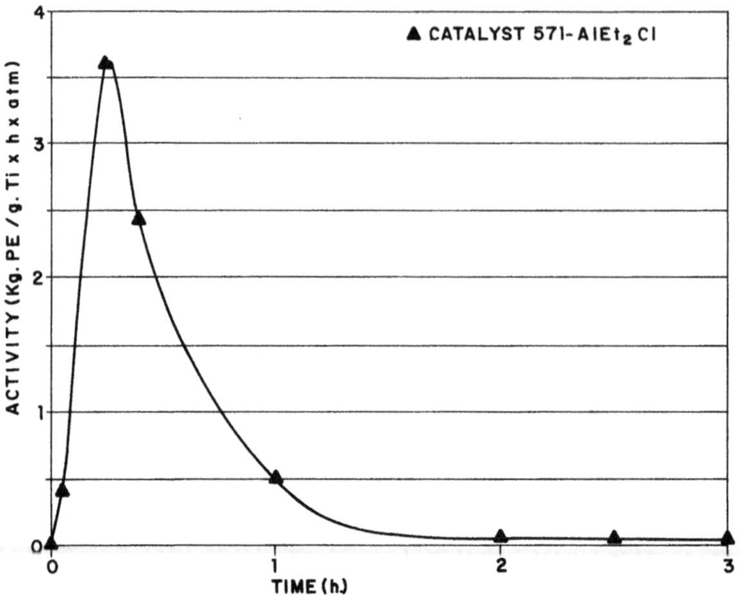

Fig. 5. Catalytic activity vs. polymerization time for catalyst 571, activated with cocatalyst $AlEt_2Cl$.

x g^{-1}. The surface acidity reached an increase from 0.35 mmol.NH_3 x $g.cat.^{-1}$ to 8.10 mmol.NH_3 x $g.cat.^{-1}$.

Catalysts 569 and 570 show very low activities for ethylene polymerization. Duck et al.[11] used a different method of synthesis consisting

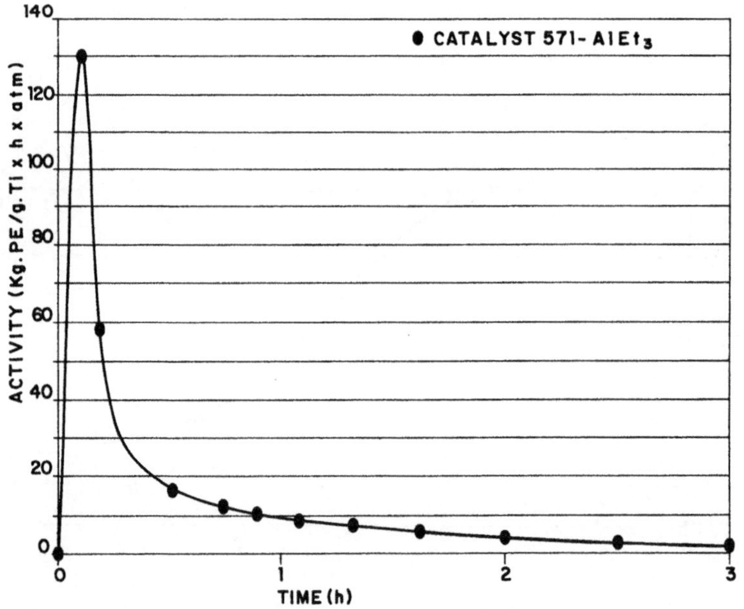

Fig. 6. Catalytic activity vs. polymerization time for catalyst 571, activated with cocatalyst $AlEt_3$.

in supporting PhMgBr on SiO_2 and obtained activities of about 4.0 Kg.PE x $g.Ti^{-1}$ x $hr.^{-1}$ x $atm.^{-1}$ Maximal activities were obtained near the stoichimetry $PhMgBr(Et_2O)_{0-40}$ $TiCl_4$. Catalysts prepared Et_2O were much less active. However, these activities are even less to those obtained using aluminumalkyl compounds for catalysts synthesis.[8] Higher activity catalysts were prepared by reimpregnating catalyst 570 with the $TiCl_4$ -C_4H_9MgCl mixture to give catalyst 571. This catalyst is very sensitive to the cocatalyst system used for activation. Thus, low activities were obtained using $Al(iso\text{-}Bu)_3$ and $AlEt_2Cl$ as cocatalyst (see Figures 4 and 5). In case of $Al(iso\text{-}Bu)_3$, activities in the order of several grams of PE x $g.Ti^{-1}$ x $hr.^{-1}$ x $atm.^{-1}$ were obtained. This behavior is very different from that of catalysts prepared with aluminumalkyl compounds, where the $Al(iso\text{-}Bu)_3$ was the best cocatalyst.[8] The activity increases to about 100 times by replacing $Al(iso\text{-}Bu)_3$ for $AlEt_2Cl$ as cocatalyst (Figure 5).

Additional increase can be obtained by using $AlEt_3$ as cocatalyst. Thus, an activity as high as 130 Kg.PE x $g.Ti^{-1}$ x $hr.^{-1}$ x $atm.^{-1}$ can be obtained after a 15 minute polymerization time, as shown in Figure 6. The kinetic curve exhibits a very strong decay, to a value lower than 10 Kg.PE x $g.Ti^{-1}$ x $hr.^{-1}$ x $atm.^{-1}$, after 1 hr. polymerization time.

It may be concluded that a very strong alkylating and reducing cocatalyst system is needed to obtain high activities. Catalytic activities increase in order of: $AlEt_3 > AlEt_2Cl > Al(iso\text{-}Bu)_3$.

Variation of Polymerization Activities with Cocatalytic Systems

The catalyst 571 is very sensitive to the cocatalyst system used for activation. Experiments were carried out in order to increase activities by changing cocatalyst compositions. Thus, $AlEt_3$ and $AlEt_2Cl$ mixtures were used as cocatalyst systems in the molar ratios of 1:1 to 100:1. Results are shown in Figures 7 and 8. The following conclusions may be

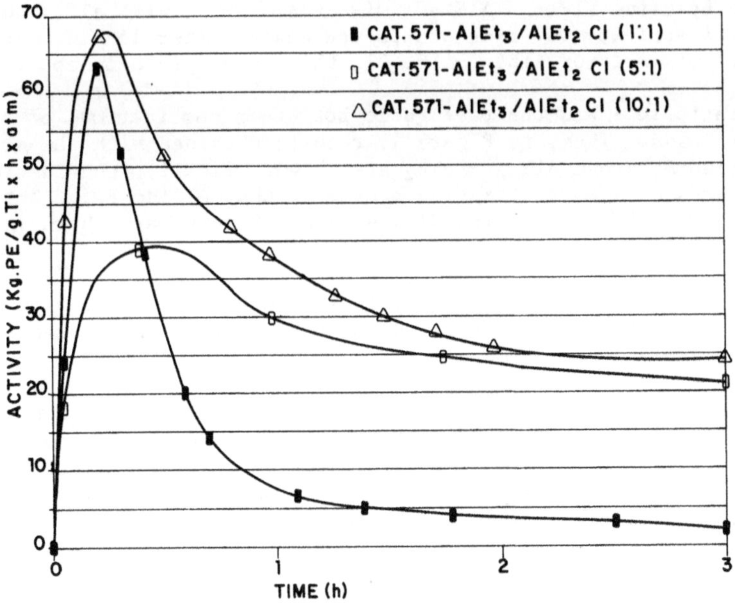

Fig. 7. Catalytic activity vs. polymerization time for catalyst 571, activated with $AlEt_3/AlEt_2Cl$ at different ratios: (■) 1:1; (□) 5:1; (△) 10:1.

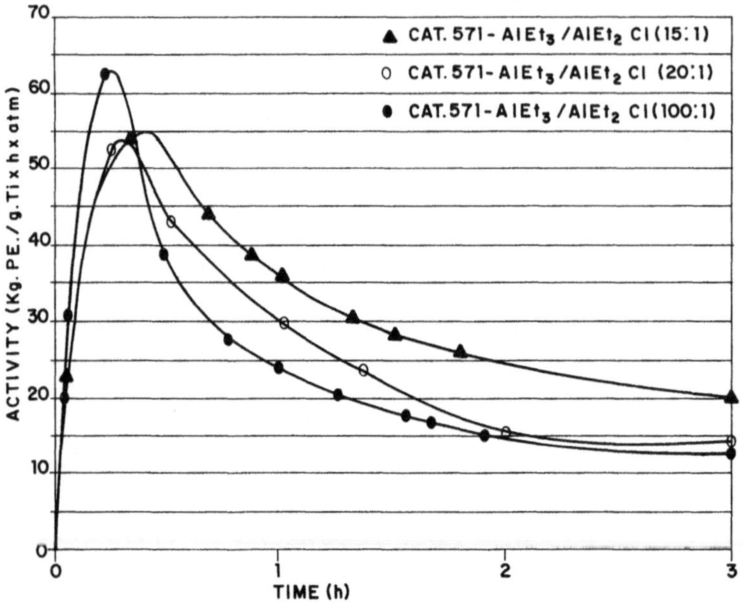

Fig. 8. Catalytic activity vs. polymerization time for catalyst 571, activated with $AlEt_3/AlEt_2Cl$ at different ratios: (▲) 15:1; (○) 20:1; (●) 100:1.

drawn: (a) By increasing the ratio $AlEt_3/AlEt_2Cl$ from 1:1 to 10:1, the stationary rate also increases, reaching a maximum of about 25 Kg.PE x $g.Ti^{-1}$ x $hr.^{-1}$ x $atm.^{-1}$ and decreasing at higher ratios (> 10). (b) The maximum activities of the decay curves obtained by using $AlEt_3$ $-AlEt_2Cl$ mixtures as cocatalyst systems, decrease in comparison to the maximum reached with $AlEt_3$ only. (c) The decay of the kinetic curves can be delayed by using $AlEt_3$ - $AlEt_2Cl$ mixtures. Thus, with $AlEt_3$ only, the curve of the activity vs. time starts decreasing after 15 min., while in the case of mixtures of $AlEt_3$ $-AlEt_2Cl$, this takes place in about 30 min.

The ratio 10:1 was the best ratio but other combinations of alkyl compounds were used. Thus, in Figure 9, results obtained with the mixtures $AlEt_3$ $-Al(iso\text{-}Bu)_3$ and $AlEt_3$ $-ZnEt_2$ are shown. Finally, considering that the $ZnEt_2$ acts as a more effective scavenger than $Al(iso\text{-}Bu)_3$, $ZnEt_2$ in combination with $AlEt_3$ and $AlEt_2Cl$ was used. Results are shown in Figure 10.

All of the above results may be explained by assuming the presence of different types of active centers on the catalysts, each of them is activated under various conditions.

Figure 11 summarizes the % of yield after 2 hrs. of polymerization time at 50°C polymerization temperature and 5 atm constant monomer pressure using different catalyst systems. By employing the cocatalyst formed by the molar ratio of $AlEt_3/AlEt_2Cl$, a yield of 400 Kg.PE x $g.Ti^{-1}$ was obtained. This activity lies in the range of commercial catalysts based on $MgCl_2$ as carrier.

The Nature of Active Centers

The chemistry occurring during catalyst synthesis and activation steps, is not understood because a wide range of reactions can take place

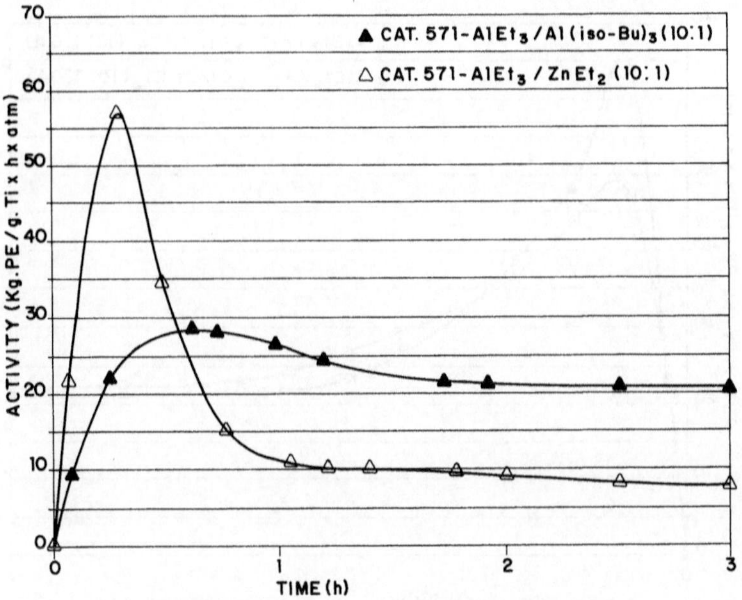

Fig. 9. Catalytic activity vs. polymerization time catalyst 571, activated with different cocatalytic systems: (▲) $AlEt_3/Al(iso\text{-}Bu)_3$, 10:1; (△) $AlEt_3/ZnEt_2$, 10:1.

between all types of chemical compounds present. Titanium species having different oxidation states, single and double bonded to the surface of the silica, as well as more or less strongly chemically and physically absorbed, can be formed. Each has different ligand environments, different chemical reactivities and different accessibilities to the reagents.[4]

Soga[5] suggested a plausible mechanism for the formation of active species in catalysts synthesized by supporting $TiCl_4$ on SiO_2 followed by reaction with $MgCl_2$, dissolved in THF. Chien et al.[3] used different analytical techniques and carried out systematic studies on $MgCl_2$, ball-milled with $TiCl_4$ and ethyl benzoate and produced titanium with oxidation states of +2, +3, and +4. The ester apparently moderates the alkylating power of $AlEt_3$ used as cocatalyst and avoids excessive formation of divalent titanium species. The following possible intermediary bimetallic complex may be formed in the preparation of SiO_2 supported catalysts by reimpregnation with $TiCl_4$ $-AlR_{3-x}Cl_x$ (e.g. $TiCl_4$ $-AlEt_3$):

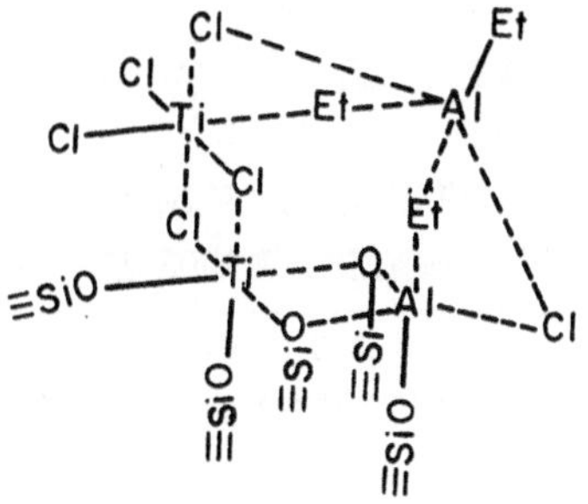

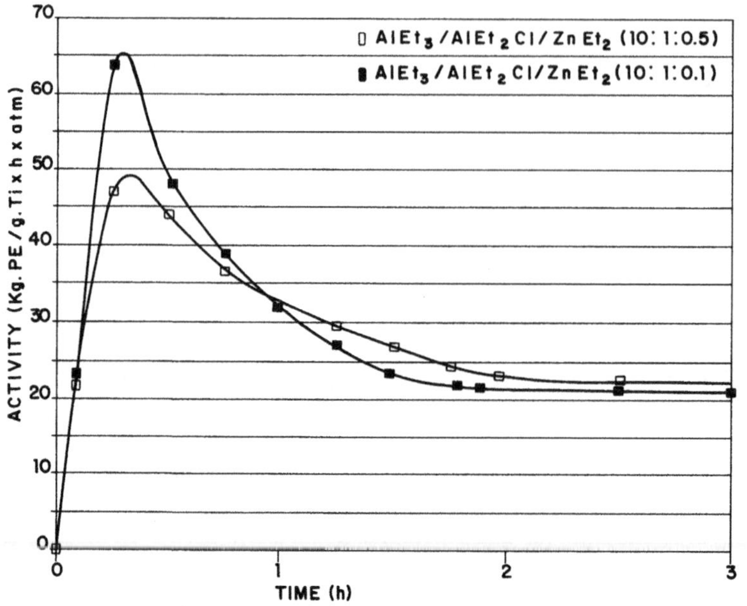

Fig. 10. Catalytic activity vs. polymerization time for catalyst 571, activated with the three component cocatalytic system $AlEt_3/AlEt_2Cl/ZnEt_3$ at different ratios: (□) 10:1:0.5; (■) 10:1:0.1.

In the co-impregnation step, the first layer of Al and Ti, single and double bonded to the SiO_2 is formed. By calcination at 450°C under vacuum, no reduction of titanium species was observed, due to a possible splitting of chlorine atoms. However, additional double bonded titanium

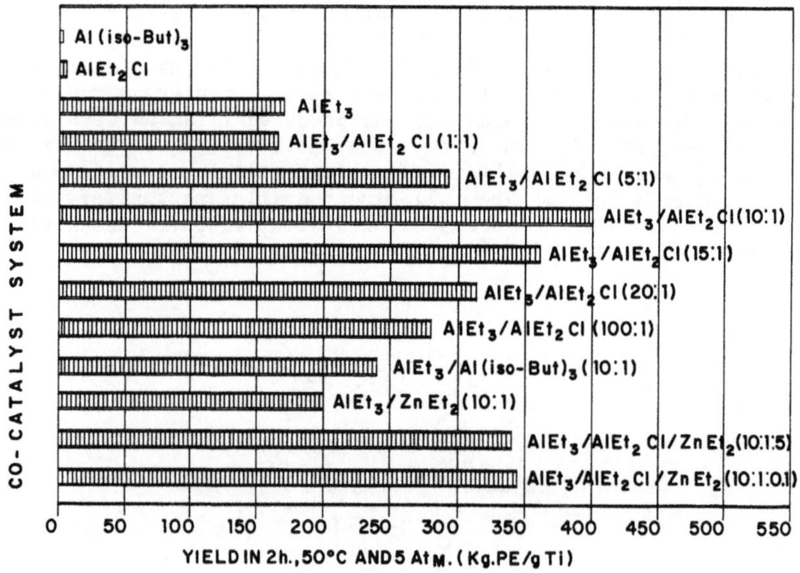

Fig. 11. Yields obtained after 2 hr. polymerization time, 50°C and 5 atm. with different cocatalytic systems.

moities were produced by the reaction of chlorine atoms of single bonded titanium with the remaining hydroxyl groups of the silica and removal of HCl.[12] Hence, the polymerization activity decreases. Furthermore, the above mentioned bimetallic complex formed in the reimpregnation step, undergoes rapid reduction giving rise to partially alkylated tiny crystals of Ti and Al, as revealed by electron-microscope.

In case of catalysts prepared by coimpregnation of $TiCl_4$ -C_4H_9MgCl mixtures on SiO_2 951, the following intermediate complex may be formed after heating at 450°C under vacuum (see Table 1):

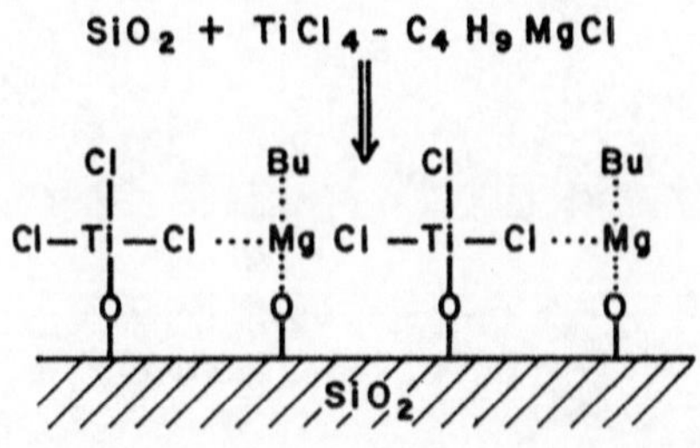

Therefore, a first chemically bonded layer of equivalent amount of Mg and Ti was formed. The ether remains coordinated to the complex. Similarly, by reimpregnation with the same $TiCl_4$ -C_4H_9MgCl mixture, a formation of a second layer of microcrystals of Ti containing Mg atoms inside their crystal lattice, may be produced on the surface of the first layer.

Morphology of Catalyst and Polymer Particles

Studies of the catalytic control of the polymer morphology are receiving considerable attentions in the Ziegler-Natta polymerization. Thus, average diameter, size distribution, shape, surface roughness, porosity and bulk density of polymer particles, can affect the polymerization rate, the polymer concentration in the reactor and polymer production. Furthermore, the particle size distribution and porosities of the polymer affect dispersion and mixing of additives and stabilizers, preventing degradations of the polymer during finishing and processing steps. Therefore, the ability of different catalysts in controlling these parameters was tested.

According to previous results[4,9,13] supported Ziegler-Natta catalysts, using SiO_2 as a carrier, control the polymer morphology more easily than $TiCl_3$-AA. SiO_2 needs to have good morphology and also enough mechanical strength in order to produce catalysts capable of controlling the polymer morphology. Thus, when SiO_2 has low mechanical strength, it undergoes dispersion into fine aggregates at the early stages of polymerization; giving rise to smaller polymer particles, broadening the particle size distribution and increasing the amount of fines.

Figure 12 shows the morphology of the catalyst 545, obtained by coimpregnation of SiO_2 951 with $TiCl_2$ -$AlEt_2Cl$ mixtures, followed by calcination at 450°C. The catalyst granules are rather compact, having nearly spherical shape and narrow particle size distribution in the range of 10-80 μ (Figure 12-a). As a result of the coimpregnation, the particle size of the SiO_2 951 increases slightly.[9] At higher magnification (Figures 12-b and 12-c) fine details of the catalyst surface can be observed. The cata-

(A)

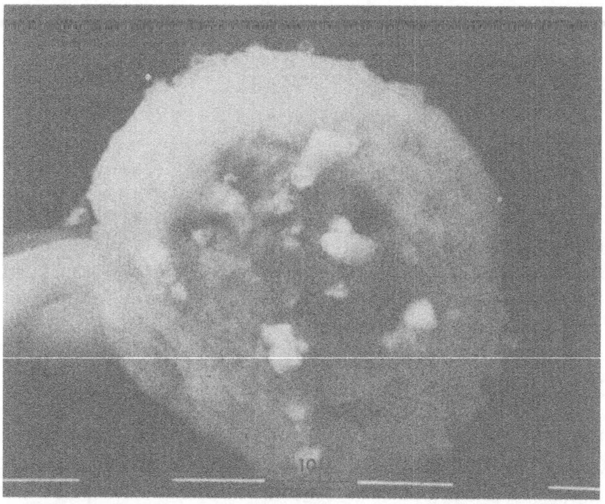

(B)

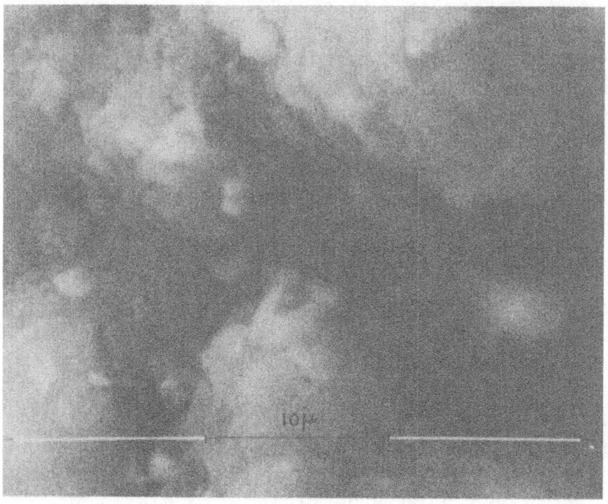

(C)

Fig. 12. (a), (b), and (c) SEM micrographs at different magnifications of catalyst 545.

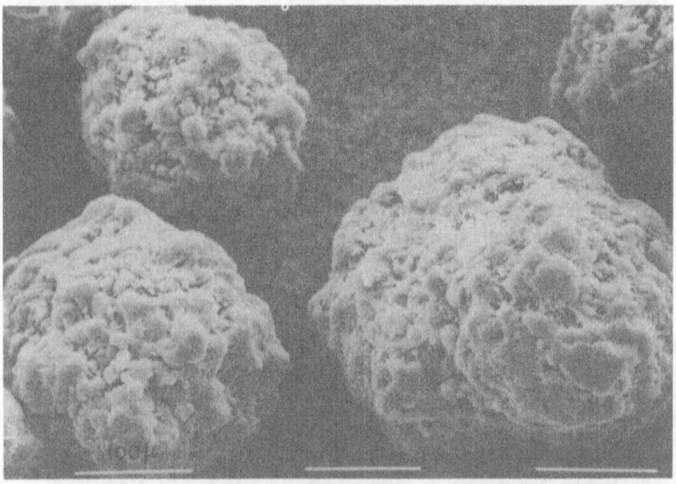

(A)

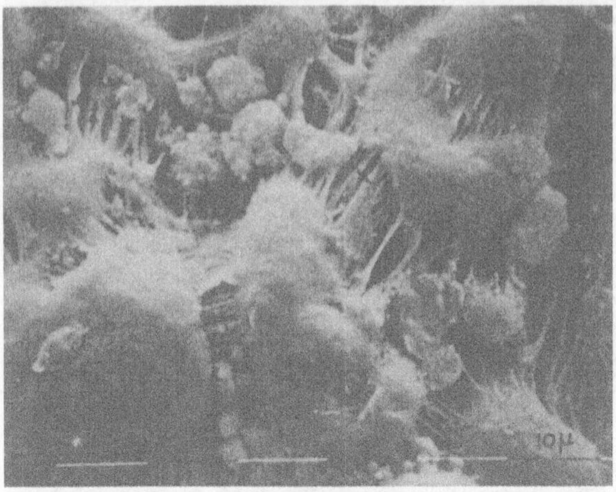

(B)

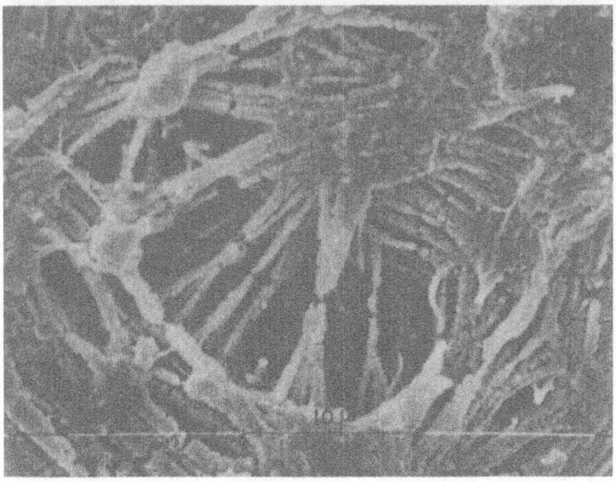

(C)

Fig. 13. (a), (b), and (c) SEM micrographs at different magnifications of PE obtained with catalyst 515 based on SiO_2 (951) $-TiCl_4$.

lyst granules have a rather smooth surface and are formed by spherical subparticles, in the order of magnitude of 0.2-0.5 μ, (Figure 12-c). These subparticles are imperfect microcrystals, having Ti and Al as chemical composition, as revealed by energy dispersive X-ray analysis. This was done by stopping the electron beam on their surface and using an EDAX unit attached to the microscope.

The microcrystals are strongly bonded to the catalyst support and, although they break up into smaller particles during polymerization, all of them stay together giving rise to a so called cobweb morphology first

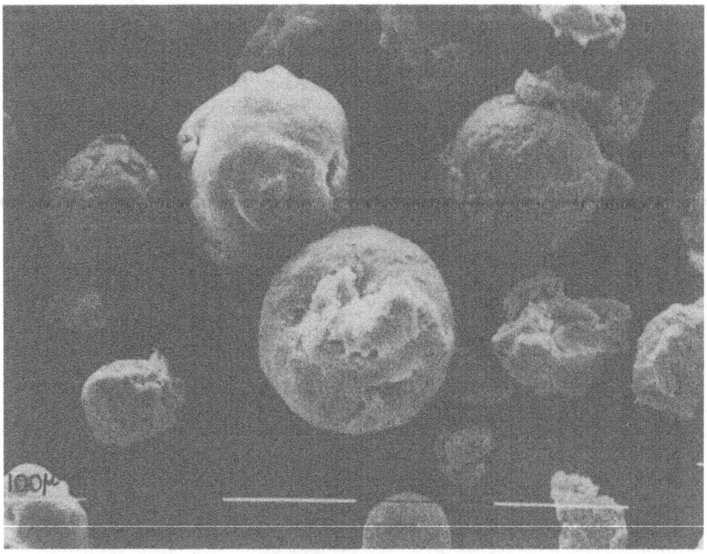

(A)

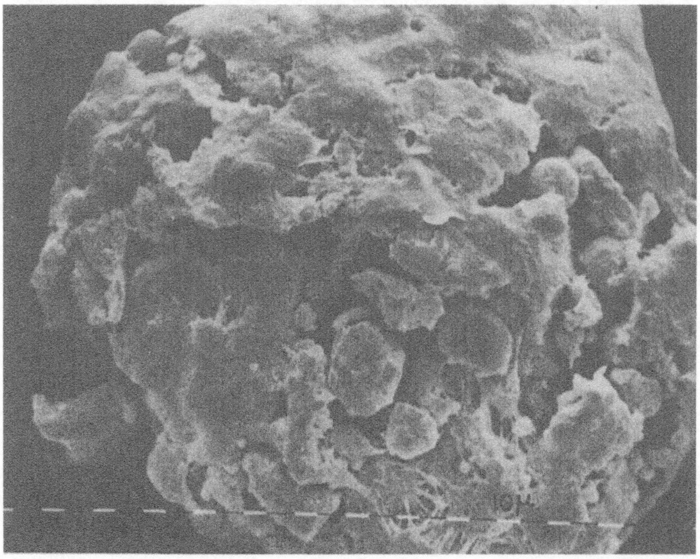

(B)

Fig. 14. (a) and (b) SEM micrographs of PE obtained with catalyst 548 based on SiO_2 (952) $-TiCl_4$.

described by Graff, Kortleve and Vonk.[14] According to these authors, the polymerization rate at the surface of the catalyst is initially much higher than in the interior, while later the opposite is true. As a result, a crust of polymer is first formed covering the particles. However, as the polymerization rate increases towards the center of the catalyst particles, they expand breaking the crust and forming cold-drawn threads of polymer which are interconnecting parts of the crusts. The resulting cobweb structure is illustrated in Figures 13 and 14 for the case of catalysts 515 and 548, both prepared by single impregnation of silicas 951 and 952 with $TiCl_4$, respectively. Similar polymer morphology was obtained with catalysts prepared by the coimpregnation of the same silicas with $TiCl_4$ -$AlEt_2Cl$ mixtures (catalysts 545 and 542). All these catalysts control polymer morphology. Thus, nearly spherical polymer granules are obtained with a narrow polymer size distribution (Figures 13-a and 14-a).

In case of catalyst 548, however, due to the fact that SiO_2 952 is weaker, some fine particles were obtained as seen in Figure 14-a. The particle sizes lies between 75-500μ in case of polyethylene obtained with catalyst 515 and 38-500μ for catalyst 548 (Figures 15 and 16). Taking into account the particle size of the catalyst, a replication factor (i.e. polymer size to catalyst size relationship) of about 10 times can be estimated.

In case of catalysts prepared by reimpregnation, the morphology changes drastically. The surface is very rough, resembling a macroporous sponge, as shown in Figure 17. The particle shape of the silica 951 does not change noticeably, but due to an excess of $TiCl_4$ -C_4H_9MgCl mixture, a certain amount of fine catalyst particles may be formed. Microcrystals with very broad size distribution, were also produced on the surface (Figures 16-b and 16-c). Energy dispersive X-ray analysis reveals that they are formed by a mixture of Ti and Mg. This morphology of catalyst surface was observed by reimpregnation of catalyst by transition metal compounds in attempting to control polymer molecular weights.[15]

SEM micrographs of polyethylene obtained with these types of catalysts, reveal very porous particles, resembling a sponge, as can be seen in Figures 18 and 19. The polymer particles are composed by agglomeration of globules with sizes ranging between 0.5-4μ in case of catalyst 554 (Figures 17-b and 17-c) and bigger (about 10μ) for the most active catalyst 571 (Figures 18-b and 18-c). The globules are themselves formed by agglomeration of subparticles of about 0.05-0.2 μ diameter (see Figure 18-c).

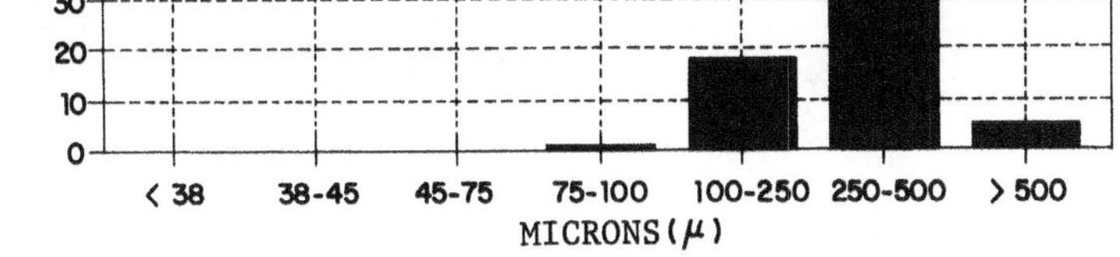

Fig. 15. (a) particle size distribution of PE obtained with catalyst 515

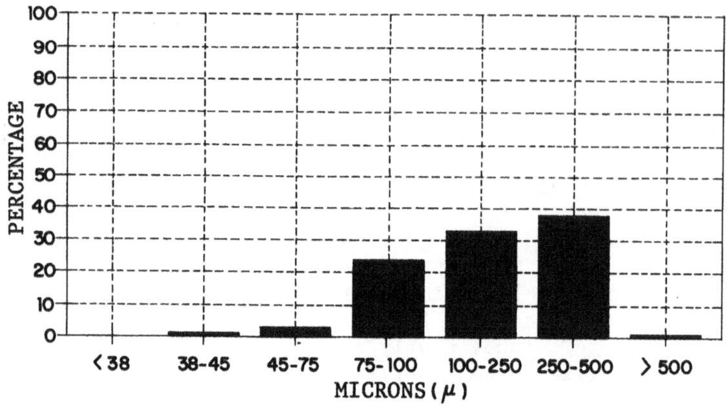

Fig. 15. (b) with catalyst 548.

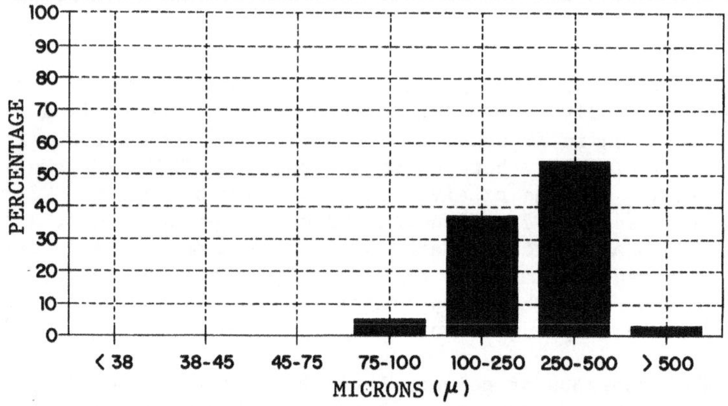

Fig. 16. (a) particle size distribution of PE obtained with catalyst 543 based on SiO_2 (951) $TiCl_4$ $-AlEt_2Cl$ by coimpregnation.

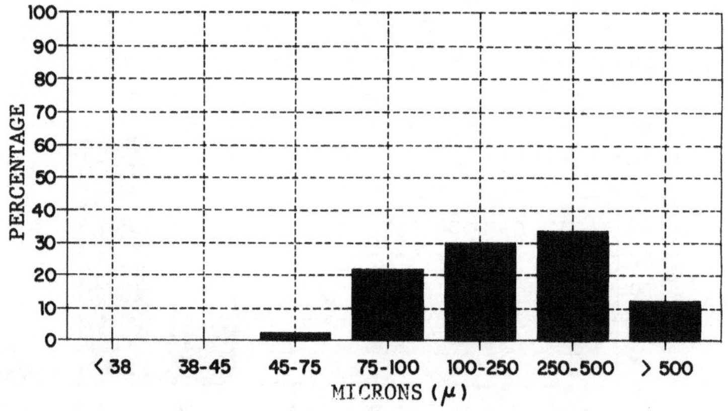

Fig. 16. (b) and PE obtained with catalyst 541 based on SiO_2 (952) $TiCl_4$ $-AlEt_2Cl$.

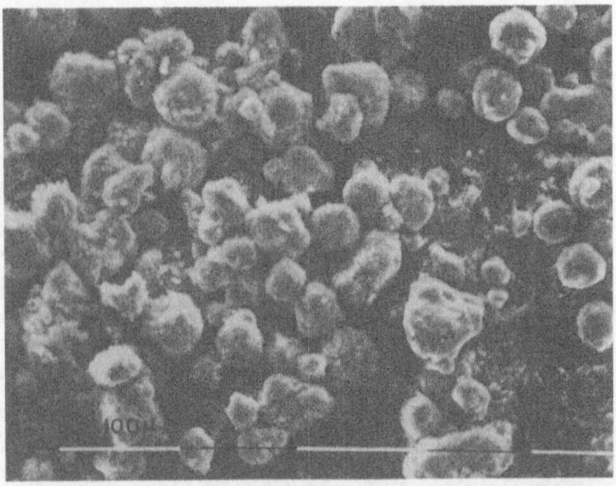

(A)

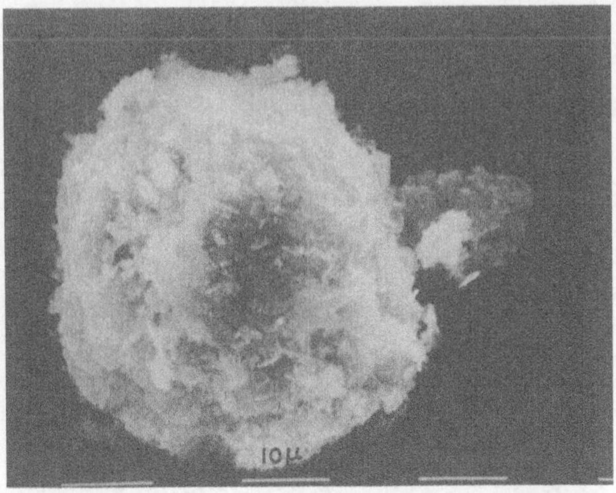

(B)

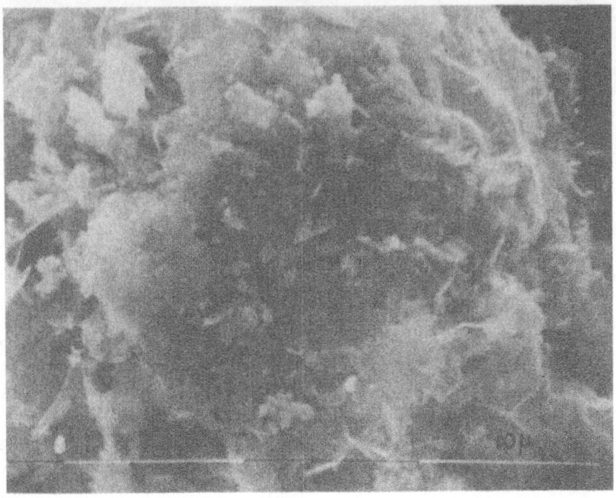

(C)

Fig. 17. (a), (b), and (c) SEM micrographs of catalyst 571 at different magnifications.

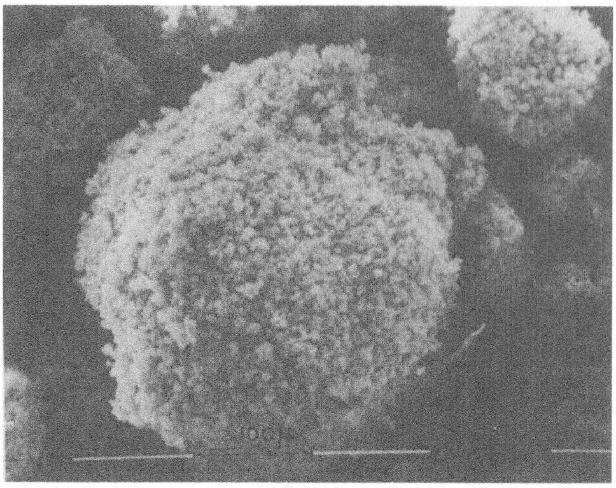

(A)

(B)

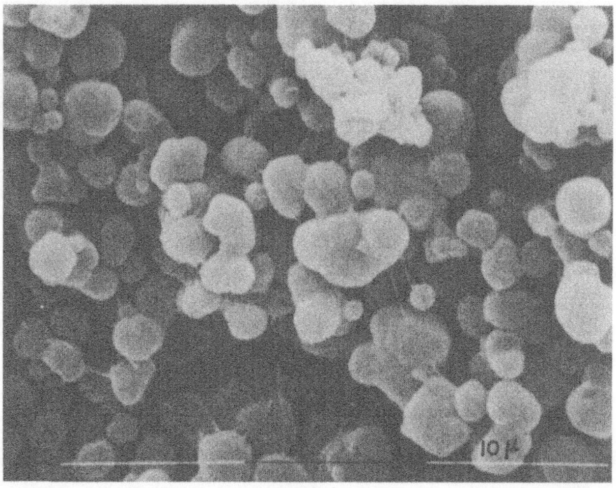

(C)

Fig. 18. (a), (b), and (c) SEM micrographs of PE obtained with catalyst 554 based on SiO_2 (951) $TiCl_4$ -$AlEt_2Cl$/ $TiCl_4$ - Al(iso-Bu)$_3$ and cocatalyst Al(iso-Bu)$_3$.

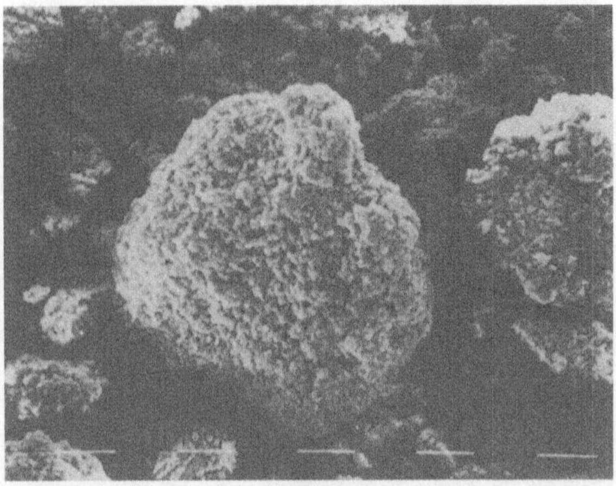

(A)

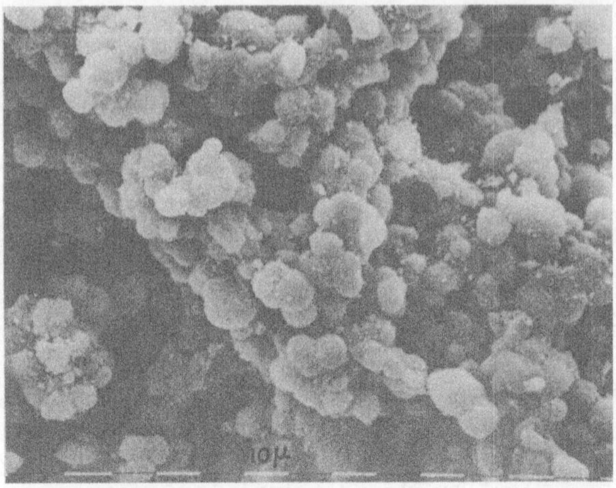

(B)

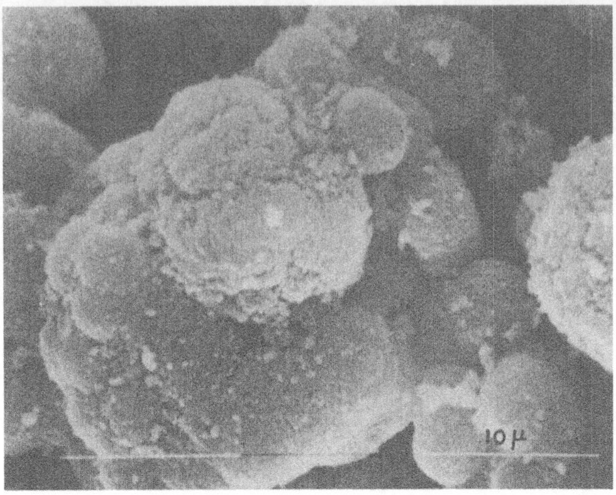

(C)

Fig. 19. (a), (b), and (c) SEM micrographs of PE obtained with catalysts 571 based on SiO_2 (951) $TiCl_4$ $-C_4H_9MgCl$/ $TiCl_4$ $-C_4H_9MgCl$ and cocatalyst $AlEt_3$ $-AlEt_2Cl$, 10:1.

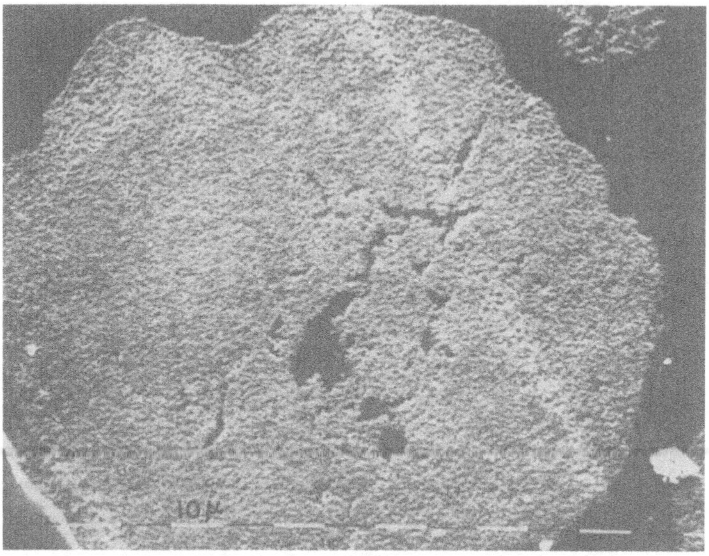

A

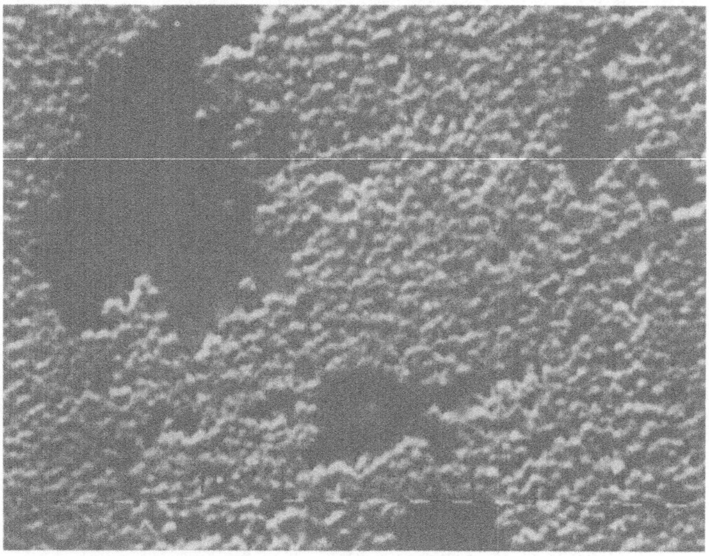

B

Fig. 20. (a) and (b) SEM micrographs at different magnifications of sectioned PE particle obtained with catalyst 548.

Details of the morphology inside the particles was obtained by cutting them with a microtome. All polyethylene particles are composed of globules with pores throughout the monomer which can diffuse during polymerization (see Figure 20).

CONCLUSIONS

Very active catalysts can be prepared by the reimpregnation of silica with mixtures of $TiCl_4$ -Aluminumalkyl compounds or $TiCl_4$-Grignards. The supported Ziegler-Natta catalysts control the nascent polymer morphology (size and shape). The silicas used as support should have good mechanical strength so that they do not break up during polymerization and form fine particles.

ACKNOWLEDGEMENT

The authors wish to thank Investigacion y Desarrollo, C.A. (Maracaibo, Venezuela) for secretarial assistance in the preparation of the manuscript.

REFERENCES

1. J. Boor, Jr., "Ziegler-Natta Catalysts and Polymerization," Academic Press, New York (1979).
2. P. Galli, L. Lucciani, and G. Cecchini, Angew. Makromol. Chem., 94, 63 (1981).
3. J. W. C. Chien, J. C. Wu, and C. I. Kuo, J. Polym. Sci. Polym. Chem. Ed., 20, 2019 (1982).
4. A. Muñoz-Escalona in: "Transition Metal Catalyzed Polymerizations - Alkene and Dienes," R. P. Quirk Ed., Harwood Academic Publishers, London, p. 323 (1983).
5. K. Soga in ACS meeting. Miami, Florida, U.S.A. (April 28-May 3, 1985).
6. D. G. Ballard, Adv. Catal., 23, 263 (1973).
7. Yu. I. Yermakov, B. N. Kuznetsov, and V. A. Zakharov, "Catalysis by Supported Complexes," Elsevier, Amsterdam (1981).
8. A. Muñoz-Escalona, J. G. Hernandez, and J. A. Gallardo in: "Catalytic Polymerization of Olefins," T. Keii, K. Soga, and J Doi Eds., Kodansha-Elsevier, Tokyo-Amsterdam (in press).
9. A. Muñoz-Escalona, J. G. Hernandez, and J. A. Gallardo, J. Appl. Polym. Sci., 29, 1187 (1984).
10. H. Schnecko, M. Reinmuller, K. Weirauch, W. Lintz, and W. Kern, Macromol. Chem., 69, 105 (1963).
11. E. W. Duck, D. Grant, A. V. Butchner, and D. G. Timms, Europ. Polym. J., 10, 77 (1973).
12. A. Muñoz-Escalona, J. G. Hernandez, and J. A. Gallardo, unpublished results.
13. A. Muñoz-Escalona, C. Villamizar, and P. Frias in: "Structure Properties Relationship of Polymeric Solids," A. Hiltner Ed., Plenum Press, New York, p. 95 (1983).
14. R. J. L. Graff, G. Kortleve, and C. G. Vonk, J. Polym. Sci. Part B 6, 735 (1970).
15. H. P. Boeme in Future Aspect of Olefin Polymerization meeting, Tokyo, Japan (July 4-6, 1985).

CATALYSTS OF ORGANIC POLYMER-BOUND Pd AND Rh COMPLEXES

Kiyotomi Kaneda, Toshinobu Imanaka, and Shiichiro Teranishi

Department of Chemical Engineering
Faculty of Engineering Science
Osaka University, Toyonaka, Osaka 560, Japan

INTRODUCTION

The use of homogeneous catalysts on an industrial scale leads to a number of practical problems, e.g., difficulties associated with corrosion, plating out on the reactor wall, and catalyst recovery from the reaction mixture. Chemically bonding metal complexes to an insoluble solid is one of the sufficient methods for conquering these difficulties.[1-3] This method is achieved by the treatment of metal ions with organic or inorganic polymers containing ligand atoms having a coordinating ability to metals. The catalysts, prepared in such manner, are situated between those usually classified as "heterogeneous" and as "homogeneous", and can be regarded as a new class of catalysts: "hybrid catalysts".

Organometallic complexes of palladium and rhodium catalyze a large number of organic reactions under mild conditions and the chemistry is well explored.[4-6] Therefore, we chose palladium and rhodium among transition metals, and tried to heterogenize their soluble catalysts by using polystyrene and cellulose as polymer supports. We have studied the preparation of polymer-bound metal complexes and their catalysis in various organic reactions; especially polymer-bound palladium complex prepared by the reaction of $PdCl_2$ with poly-4-diphenylphosphinomethylstyrene is a versatile catalyst, e.g., 1) hydrogenation of olefins and acetylenes,[7,8] 2) oligomerization of olefins and/or acetylenes,[9,10] 3) carbon-carbon bond formation,[11] and 4) telomerization of butadiene with nucleophiles.[12]

In this paper, our recent study on the catalysis of the polymer-bound palladium and rhodium complexes is summarized.[7-16]

EXPERIMENTAL

Preparation of Polymer-Bound Pd and Rh Complexes

Phosphinated Polystyrene-Bound Pd(II) Complexes. All of the phosphinated polystyrene-bound palladium(II) complexes were prepared in approximately the same manner. The synthesis of PC-2 is typical. Under a nitrogen atmosphere, phosphinated polymer (2.5 g, 8.5 mg-atom of P) was dissolved for 26 h in a refluxing 1:2 benzene-ethanol (100 ml) containing

Table 1. Analytical Data of the Polystyrene-Bound Palladium Complexes

Complex	Resin used[a)]	Percentage C	H	Cl	P	Empirical atomic ratio[b)] P/Pd
PC-1	P-1	61.17	4.57	9.20	6.68	1.23
PC-2	P-1	53.88	4.54	12.48	7.62	1.17[d)]
PC-2[c)]	P-1	53.97	4.01	12.11	7.29	1.10
PC-3	P-1	53.88	4.36	11.76	6.55	0.97
PC-4	P-2	56.05	4.33	11.83	6.39	1.04
PC-5	P-2	55.68	4.34	10.40	6.53	0.97
PC-6	P-2	55.90	4.25	11.00	7.09	1.10
PC-7	P-2	69.55	5.11	1.81	8.41	1.95

a) Anal. Found for P-1: C, 82.15; H, 6.06; P, 10.48; Cl, 0.0 and for P-2: C, 82.42; H, 6.25; P, 8.71; Cl, 0.54%. b) Pd content was calculated by difference. C) Recovered after olefin hydrogenation. d) Pd analysis was obtained on this complex: observed, 22%.

$PdCl_2$ (1.5 g, 8.5 mmol). A yellow product was precipitated. The product was recovered by filtration and washed with ethanol until the filtrate was colorless. The solid was dried for 50 h under vacuum. Analytical data for the various preparations of the polymeric complexes are summarized in Table 1.

Aminated Polystyrene-Bound Pd(II) Complexes. Both poly-4-dimethyl-aminomethylstyrene (2% divinylbenzene, 100-200 mesh, 1.40 mg-atom of nitrogen/g of polymer) and poly-4-pyrrolidinomethylstyrene (2% divinyl-benzene, 100-200 mesh, 1.35 mg-atom of nitrogen/g of polymer) were used as a polymer ligand. The preparation of poly-4-dimethylaminomethylstyrene-bound palladium complex with a metal loading of pd/N = 1.0 is typical. The aminated polystyrene (3.0 g, 4.2 mg-atom of nitrogen), bis(benzonitrile) dichloropalladium (1.61 g, 4.2 mmol), and acetone (20 ml) were placed in a flask under a nitrogen atmosphere. The mixture was stirred for 3 h at 50° C. The resulting polymer was filtered, and washed with acetone and ether, and dried under vacuum.

Phosphinated Polystyrene-Bound Pd(0) Complex. The polymer-bound palladium(II) complex (PC-6, 7.0 g, 14.8 mg-atom of Pd) was added to ethanol (50 ml) solution containing triphenylphosphine (9.1 g, 34.7 mmol). The mixture was stirred at room temperature for 3 h. Then hydrazine hydrate (6.0 g, 0.12 mol) was added to the mixture and stirring was continued for 2 h. The resulting polymer was filtered, washed with ethanol and ether, and dried under vacuum to give a dark green polymeric complex (PC-7). All the above procedures were carried out under an argon atmosphere. The result of analysis was shown in Table 1.

Phosphinated Polystyrene-Bound Rh(II) Complexes. Under a nitrogen atmosphere, the phosphinated polystyrene (0.836 g, 2.85 mg-atom of P) was dissolved at 60° C for 14 h in an ethanol solution containing $RhCl_3 \cdot 3H_2O$

Table 2. Analytical Data of Cellulose Phosphinite

	Tritylcellulose (g)	Pyridine (ml)	PPh_2Cl	Reaction Temp. (°C)	Reaction Time (days)	Analysis C	Analysis H	Analysis P
S-1	3	50	10	50	8	74.41	5.58	3.81
S-2	3	50	9	25	8	74.79	5.62	4.01
S-3	7	100	20	25	8	74.21	5.58	4.42
S-4[a)]	3	-	6	25	1	73.11	5.53	3.52

a) Pretreatment with n-BuLi (10 ml) in THF (50 ml).

(0.752 g, 2.85 mmol). After cooling, a dark green product was recovered by filtration and washed with ethanol until the filtrate was colorless. The solid was dried under vacuum.

Cellulose-Bound Pd Complexes. The phosphinite ligand containing the O-P bond was synthesized according to the following sequence. Hydrolysis of cellulose acetate and tritylation afforded 6-O-triphenylmethylcellulose. Analysis calculated for ($C_{25}H_{24}O_5$): C, 74.24; H, 5.98%. Found: C, 74.32; H, 5.82%. Treatment of the 6-O-triphenylmethylcellulose with diphenyl-

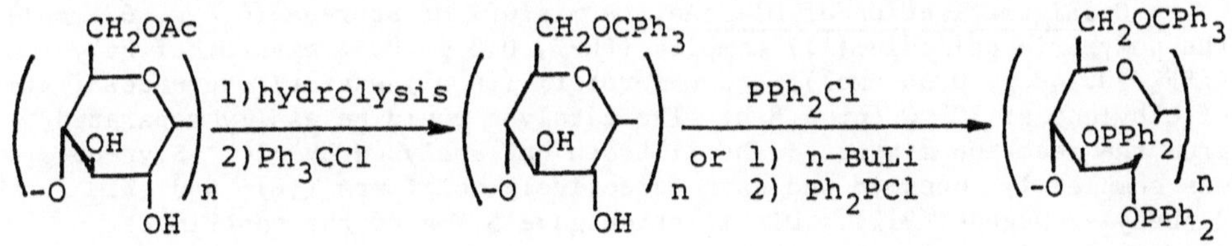

phosphinous chloride gave the cellulose phosphinite. Typical analytical results are shown in Table 2. The phosphinite formation is incomplete. However, pretreatment of triphenylmethylcellulose with n-BuLi followed by diphenylphosphinous chloride did not afford improved results (see sample S-4). This may be due to strong hydrogen bonding between hydroxyls of glucose units. Cellulose-bound palladium complexes were prepared by a bridge-splitting method which is similar to that in the polystyrene-bound palladium complexes. Under a nitrogen atmosphere, 6-O-triphenylmethylcellulose phosphinite (sample S-1, 1 g, 1.22 mg-atom of phosphorus) was dissolved for 24 h in 30 ml of refluxing benzene-ethanol (1:1) solution containing $PdCl_2$ (0.218 g, 1.23 mmol). The gelled mixture was poured into water-free methanol, whereupon a yellow product precipitated. The product was recovered by filtration and washed with ethanol until the filtrate was colorless. The solid was dried under vacuum. Anal: C, 43.90; H, 4.02; Cl, 7.91; P, 3.69; Pd, 14%.

The polymeric palladium(0) complex was prepared by the reduction of the polymeric palladium(II) complex with hydrazine in the presence of 2 mol excess of triphenylphosphine (Anal: C, 60.62; H, 5.01; P, 4.84; Pd, 5.44%).

Diphenylphosphinated cellulose was prepared as follows: Elemental analysis of the phosphinated cellulose showed C, 76.80; H, 5.47; P, 3.92%.

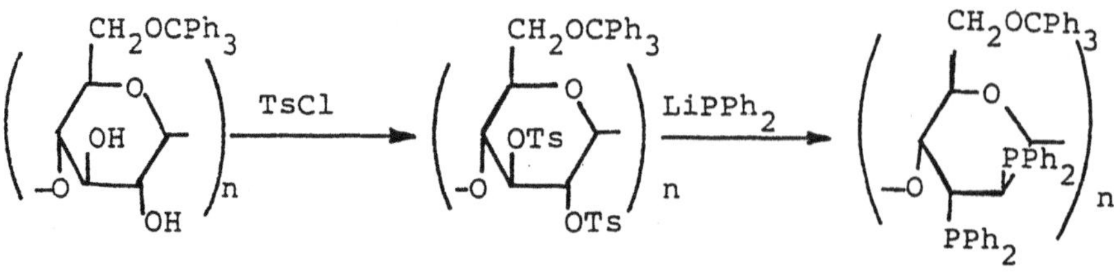

Procedure of Various Organic Reactions Catalyzed by Polymer-Bound Pd and Rh Complexes

Hydrogenation. The polymeric metal complex was added to a reaction flask and the system was alternately evacuated and flushed with hydrogen five times. A solvent and substrate were added through a sidearm entry port and stirring was begun. The volume of hydrogen uptake at constant pressure or the change in the hydrogen pressure at constant volume was measured. Samples were withdrawn from the reaction mixture at regular time intervals and were analyzed by glc.

Codimerization of Acetylenes and Allyl Halides. The reaction of phenylacetylene and allyl chloride is typical. To a suspension of the polymeric palladium(II) complex (PC-1, 0.10 g, 0.18 mg-atom of Pd) in allyl chloride (13.8 g, 180 mmol) was added phenylacetylene (1.8 g, 18 mmol) and stirring was continued for 48 h. Then the reaction mixture was filtered to remove the polymeric catalyst. After a usual work up, the codimer products were analyzed and isolated by GLC.

Cooligomerization of Olefins. A mixture of styrene (6.4 g, 62 mmol), the polymeric palladium(II) complex (PC-2, 0.3 g, 0.64 mg-atom of Pd), $AgBF_4$ (0.142 g, 0.64 mmol), and chloroform (16 ml) was stirred under 1 atm of ethylene at 50° C for 4.5 h. The catalyst could be easily separated from the reaction mixture. The filtrate was analyzed by glc. Styrene was completely consumed and gave selectively cotrimers ((E)- and (Z)-diphenyl-2-hexene; 91%). Distillation gave 5.7 g of the contrimers. Oligomerizations of other olefins were carried out in the same manner.

Vinylic and Acetylenic Hydrogen Substitution Reaction. The reaction of styrene with iodobenzene is typical. Styrene (2.60 g, 25 mmol), iodobenzene (5.70 g, 28 mmol) and tri-n-butylamine (4.7 ml) were placed in a reaction flask. The polymeric palladium(0) complex (PC-7, 0.10 g, 0.14 mg-atom of Pd) was added and the flask was flushed with argon. Then the mixture was stirred at 100° C for 2 h. The catalyst was separated from the reaction mixture by centrifugation, washed several times with ether in air, and reused for the same reaction. Glc analysis of the separated solution showed that (E)- and (Z)-stilbene had been produced in 78 and 8%, respectively, and that unreacted styrene remained. The solution was concentrated under reduced pressure to give a white solid, which was recrystallized from benzene-ethanol to give 3.15 g of (E)-stilbene (70%). Other reactions were carried out in the same manner.

Vinylic Halogen Substitution Reaction. (Z)-β-Bromostyrene (0.38 g, 2.1 mmol), the polymeric palladium(0) complex (PC-7, 0.025 g, 0.036 mg-atom of Pd), and benzene (20 ml) were placed in a reaction flask. The flask was flushed with argon. Then a titrated ethereal solution of methylmagnesium iodide (4.5 ml, 5.4 mmol) was added and the mixture was stirred at room temperature for 4 h. The solution was directly analyzed by glc. The starting bromostyrene was consumed completely to give (Z)- and (E)-β-methylstyrene at 98 and 2% yield, respectively. Similarly, the reaction of (E)-β-bromostyrene was carried out to give (E)- and (Z)-β-methylstyrene in 72 and 3% yield, respectively.

Telomerization of Butadiene with Nucleophiles. The polymeric palladium(0) complex was weighed in a 100 ml glass autoclave. A nucleophilic compound and solvent were added. The reaction system was flushed with nitrogen three times, and then butadiene was introduced into the reaction vessel. The reaction mixture was heated to an appropriate temperature with stirring. After the reaction, the catalyst was separated from the mixture by centrifugation. The homogeneous solution was treated with a usual workup. The yields of products were determined by glc analysis.

The Water Gas Shift Reaction. The reaction vessel was a 25 ml flask with a sidearm sealed with silicone rubber. It was attached to a closed reaction system with a condenser, a gas circulator, and a vent. The rhodium compound and aminated polymer were placed in the vessel, the reaction system (250 ml) was evacuated and flushed three times with CO and then charged to an initial CO pressure of 700 mmHg. To the vessel were succesively added 2-ethoxyethanol and H_2O. The mixture was heated with stirring to 80° C. Gas phase was analyzed periodically by glc.

RESULTS AND DISCUSSION

Characterization of Polymer-Bound Pd and Rh Complexes

Phosphinated Polymer-Bound Pd(II) Complexes.[7] The polymeric palladium(II) complexes were prepared by the bridge splitting reaction. An equimolar mixture of $PdCl_2$ and the phosphinated resin as phosphorus in benzene-ethanol was refluxed to give a yellow complex. In the solution of pure ethanol, palladium black deposited slightly during the reaction. The phosphinated resin dissolved in solvents such as benzene, tetrahydrofuran, and dioxane since the resin was not cross-linked. The polymeric palladium complex obtained, however, did not dissolve in all the solvents used in this study. This indicates that the polymer chains may be cross-linked with just a small portion of palladium atoms. From Table 1, the atomic ratio of phosphorus to palladium was near one in all the complexes prepared under various conditions. This result shows that each palladium atom can be attached to only phosphorous atom on the polystyrene.

The infrared spectrum of the polymeric complex in KBr disk contained one band not found in the spectrum of the polymer ligand. This band appeared at 351 cm^{-1}. The terminal trans Pd-Cl stretching band in several compounds of the general form $PdCl_2R_2$ (R is a tertiary phosphine or arsine) appears at 350 to 360 cm^{-1}.[17] The polymeric complex appears to exist exclusively in the trans configuration.

According to elemental analysis and infrared absorption spectroscopy, the oxidation state of the polymeric complex seems to be divalent. Furthermore, the electronic state was examined by use of X-ray photoelectron spectroscopy (XPS). The palladium 3d and chlorine 2p levels were measured for the polymeric compounds and other compounds (Table 3). The Pd(3d) binding energies of the polymeric complex are slightly higher than those of other divalent compounds and the Cl(2p) binding energy is higher than that of $PdCl_2(PPh_3)_2$ with a terminal Pd-Cl bond. The electronic state of Pd-Cl bond moieties in the polymeric complex as a whole is of lower electron density, and the electrons on palladium and chlorine atom may flow into the polymer ligand. The XPS spectra of the recovered complex did not vary appreciably after hydrogenation. In addition, the polymeric complexes did not show an electron spin resonance signal due to monovalent palladium species.

From the results of elemental analysis, IR, and XPS, the most likely structure of the polymeric palladium complex seems to be a three-coordinated Ⓟ-Pd(II)-Cl_2 (Ⓟ is the polymer ligand).

Table 3. Palladium 3d and Chlorine 2p Binding Energies (eV)[a]

Compound	Pd $3d_{3/2}$	Pd $3d_{5/2}$	Cl 2p
P -$PdCl_2$ (PC-2)	344.0	338.6	199.1
P -$PdCl_2$ (PC-2)[b]	343.8	338.5	199.2
P -Pd-PPh_3 (PC-7)	342.6	337.6	
$PdCl_2(PPh_3)_2$	343.6	338.3	198.6
$PdCl_2$	343.6	338.1	199.3
$Pd(PPh_3)_4$	341.4	336.2	
$Pd(CO)(PPh_3)_3$	343.2	337.9	
$Pd(dba)_2$	342.4	337.0	

a) The binding energy values are referenced to carbon 1s (285.0 eV), and the energy differences were determined with an accuracy of $\pm$ 0.2 eV.

b) Recovered complex after hydrogenation.

Phosphinated Polystyrene-Bound Pd(0) Complexes.[11] In ethanol solution containing two mole excess of triphenylphosphine, the polymeric palladium(II) complex was easily reduced with hydrazine hydrate to give a dark green polymeric complex (PC-7). From elemental analysis (PC-7 in Table 1), the polymeric complex is composed of one atom of palladium, one mole of the polymeric phosphine, and one mole of triphenylphosphine. Reduction with hydrazine hydrate in the absence of triphenylphosphine gave a very air-sensitive complex. Phenyllithium also could be used as a reductant. Leaching of the metal was not observed during the reduction. From Table 3, the palladium $3d_{3/2}$ and $3d_{5/2}$ binding energies for the reduced complex, Eb = -1.4 and -1.0, are lower than those of the starting polymeric palladium(II) complex, respectively. From the results for the three soluble palladium(0) complexes, it is clear that the palladium 3d binding energies depend strongly on the nature of ligands. Consequently, it is impossible to identify the reduced complex as a zerovalent one on the basis of its Pd 3d binding energies only. It, however, is evident that the shifts show that the reduction of the starting palladium(II) complex to the lower valent state has taken place. The Pd 3d spectrum of the reduced complex shows sharp lines with 2.7 and 2.5 eV Pd $3d_{3/2}$ and $3d_{5/2}$ energy widths at half-maximum intensity, respectively (Fig. 1). This shows the reduced complex to be composed of one species. Also the palladium is not metallic, because if it were, the energy widths would be broader. Furthermore, ESR signal assignable to Pd(I) species was not observed. Therefore, it is probable that the reduced complex is an unsaturated, two-coordinated palladium(0) complex with two vacant sites as shown in Scheme 1. However, mixtures of one-, two-, and three-coordinated palladium(0) complexes may exist. These results suggest that polymer chains sterically hinder the coordination of a sufficient number of phosphine ligands to form a coordinatively saturated complex and that they stabilize the unsaturated complex. The facile formation of unsaturated complexes using polymeric ligands is one of the advantages over homogeneous systems.

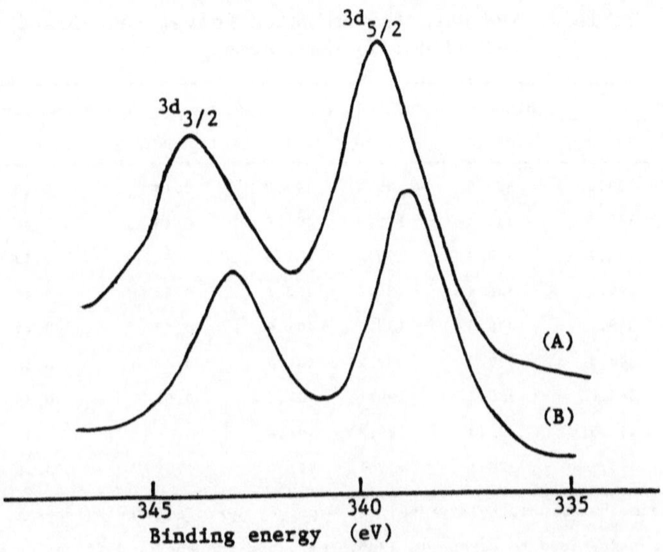

Fig. 1. XPS Spectra of the Phosphinated Polystyrene-Bound Pd(II) Complex (A) and its Reduced Complex (B)

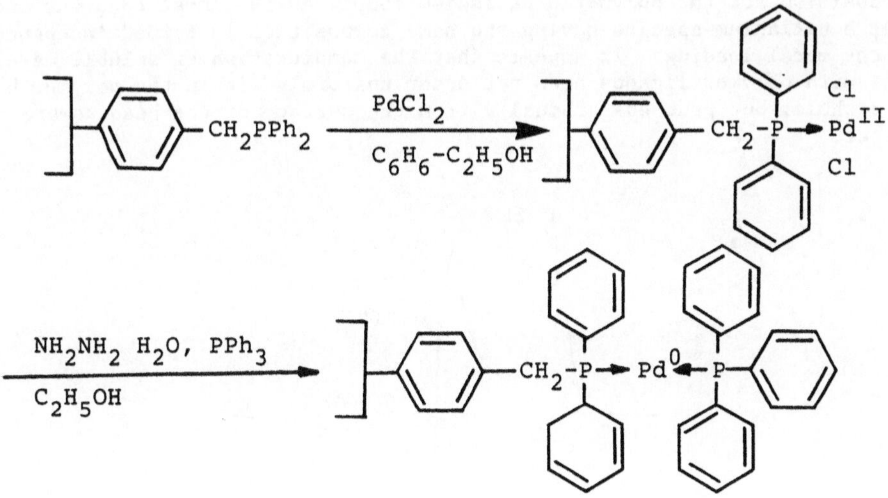

Scheme 1

Aminated Polystyrene-Bound Pd(II) Complexes.[14] Poly-4-dimethylaminomethylstyrene and poly-4-pyrrolidinomethylstyrene were used as polymer ligands. $PdCl_2(PhCN)_2$ reacted with the polymers to give brown polymer beads, neither of which showed C≡N IR stretching bands. The XPS spectral data of the polymeric complexes are listed in Table 4. The Pd 3d binding energies for all the polymeric complexes were almost constant and independent of the metal loading, and similar to those for divalent palladium complexes. Also, no appreciable difference in the Pd 3d binding energies was observed with different polymer ligands. Although all the chlorine atom in the polymeric complexes are present in the form of the terminal Pd-Cl bonds (334 cm^{-1}), the chlorine 2p binding energies are similar to that for $PdCl_2$. The peak intensity ratios Pd 3d/C 1s, Cl 2p/Pd 3d, and N 1s/Pd 3d are almost constant for both aminated polystyrene systems, and do not depend on the metal loading. This shows that the surface

Table 4. XPS Data for Aminated Polystyrene-Bound Palladium(II) Complexes

Ligand[a)]	Pd/N	Binding energy[b)] (eV)				Peak intensity ratio		
		Pd $3d_{3/2}$	Pd $3d_{5/2}$	Cl 2p	N 1s	Pd 3d/C 1s	Cl 2p/Pd 3d	N 1s/Pd 3d
(1)	0.3	343.8	338.5	199.4	402.0	0.44	0.48	0.081
(1)	0.4	343.6	338.4	199.3	401.8	0.45	0.45	0.098
(1)	0.5	343.7	338.4	199.2	401.9	0.44	0.43	0.097
(1)	0.7	343.6	338.4	199.4	402.1	0.43	0.46	0.094
(1)	1.0	343.7	338.3	199.2	401.9	0.43	0.44	0.099
(2)	0.3	343.5	338.2	199.2	401.7	0.48	0.38	0.079
(2)	0.5	343.6	338.2	199.1	401.4	0.53	0.38	0.078
(2)	0.7	343.6	338.3	199.3	401.8	0.55	0.39	0.075
(2)	1.0	343.5	338.1	199.1	401.8	0.51	0.36	0.079

a) (1) = poly-4-dimethylaminomethylstyrene; (2) = poly-4-pyrrolidinomethylstyrene. b) The binding energy values are referenced to carbon 1s (285.0 eV), and the energy differences were determined with an accuracy of ± 0.2 eV.

composition for the polymeric palladium complexes is almost constant and that a palladium species having the same composition is formed independent of the metal loading. It appears that the complexation of soluble metal ions with polymer ligands does not occur uniformly within the polymer beads as a whole, but proceeds gradually from the surface of the bead toward the center.

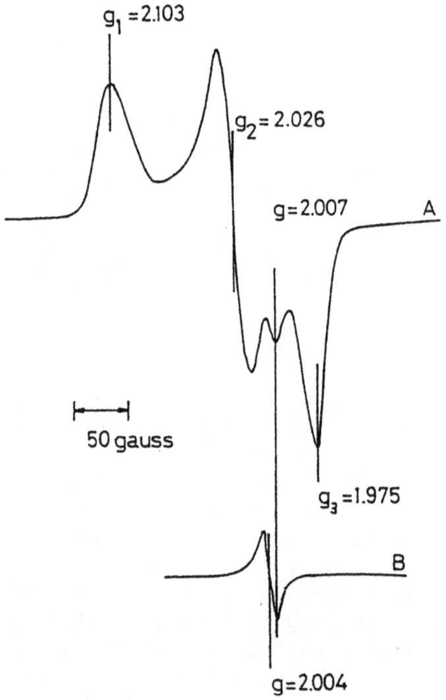

Fig. 2. ESR Spectra of the Phosphinated Polystyrene-Bound Rh Complex (A) and the Polymer Ligand (B)

Phosphinated Polystyrene-Bound Rh(II) Complexes.[13] All the elemental analysis of the polymeric rhodium complexes prepared under the various conditions shows that the complexes are composed of about one atom of rhodium, two atoms of chlorine, and one mole of the phosphine. The infrared spectra of the polymeric rhodium complexes in KBr disk contained one band not found in the spectrum of the polymer ligand. This band appeared at 334 cm^{-1} and is due to the terminal Rh-Cl bond.

The ESR spectra of the polymeric rhodium complex and the polymer ligand are shown in Fig. 2. It seems that the spectrum of the polymeric rhodium complex is derived from two species, because one species with g = 2.001 is also observed in the spectrum of the polymer ligand. The signals at g_1 = 2.103, g_2 = 2.026, and g_3 = 1.975 are similar to those previously reported for divalent rhodium compounds and especially, its average g value (g_{rms} = 2.094) is in good agreement with that of trans-$RhCl_2[P(o\text{-}toly)_3]_2$.[18] Accordingly, the signal of g_{rms} = 2.094 was assigned to divalent rhodium. In addition, the signal intensity of the divalent rhodium was unchanged in the course of the hydrogenation of olefins discussed in a later section.

The rhodium electronic state was examined by use of the XPS technique. The Rh $3d_{5/2}$ peak full width at half-maximum is ca. 3.0 eV and no shoulders. The rhodium complex has 3d binding energies similar to those of divalent rhodium compound, trans-$RhCl_2[P(o\text{-}toly)_3]_2$. In addition, the recovered rhodium complex after the hydrogenation of olefins displayed no appreciable change in its XPS spectra. The rhodium valence state is conclusively shown to be paramagnetic divalent.

From the above results, the most likely structure of the polymeric rhodium complex seems to be a three-coordinated Ⓟ-$RhCl_2$ (Ⓟ: polymer ligand) with three vacant sites around the central rhodium ion.

Cellulose Phosphinite-Bound Pd Complexes.[16] From elemental analysis of cellulose phosphinite-bound palladium complexes, the atomic ratio of palladium to chlorine and to phosphorus is close to 1:2:1. The electronic state of the palladium was examined by using XPS techniques. The palladium 3d binding energies (($3d_{3/2}$) = 343.8 eV, ($3d_{5/2}$ = 338.4 eV) were similar to those of the polystyrene-bound palladium(II) complex. The most likely structure of the polymeric palladium complexes seems to be a three-coordinated Ⓒ-$OPPh_2PdCl_2$(II) where Ⓒ is the cellulose support. The corresponding Pd(0) complex was prepared by a method similar to that described in the preparation of the polystyrene-bound Pd(0) complex.

Catalysis of Polymer-Bound Pd and Rh Complexes

Phosphinated Polystyrene-Bound Pd(II) Complexes. Hydrogenation.[7,8] The phosphinated polystyrene-bound palladium(II) complexes are extremely efficient catalysts for heterogeneous hydrogenation of olefins at ambient temperature and pressure of 1 atm or below. Functional groups such as keto-, hydroxy-, cyano-, nitro-, ether, ester, or carboxylic acid are not hydrogenated under the reaction conditions. The polymeric complex displayed a high activity for the hydrogenation of styrene, isoprene, 1,3-cyclooctadiene, and allyl compounds. The activity is very sensitive to the steric hindrance about the carbon-carbon double bond and to the molecular size of olefins, particularly in the case of cyclic olefins. Isoprene was selectively hydrogenated to monoenes and the monoenes were not hydrogenated to isopentane in the presence of the unreacted isoprene. The ratio of the 1,2- and the 1,4-addition was about 1:1 as judged by comparison of the rate of formation of 2-methyl-1-butene and 3-methyl-1-butene produced by the former with that of 2-methyl-2-butene produced by the latter.

This suggests that the half-hydrogenated state contains a π-allylic intermediate.

The polymeric catalyst displayed a high activity for hydrogenation in oxygen-containing solvents such as dimethylformamide, ethanol, tetrahydrofuran, and acetone. All of them solvate the catalyst species to cause an irreversible yellow-to-green color change in the catalyst. Presumably, the coordination of oxygen atom to palladium plays an important role in accelerating the catalytic activity. Dimethylsulfoxide and nitromethane, however, decreased the rate of hydrogenation. They coordinate much more strongly to palladium and disturb the coordination of hydrogen or an olefin to the metal. Ethanol being a poor swelling solvent for polystyrene seems to display a lower activity for hydrogenation than the better solvents such as benzene, acetone, and ethylacetate, since contraction of the polymer decreases the diffusion rate of a substrate into polymer channels. In our hydrogenation, however, ethanol displayed a higher activity than the better swelling solvents. This fact suggests that the contraction does not have much effect on the rate.

The overall rates of hydrogenation of cyclohexene, styrene, and 1, 3-cyclooctadiene were measured and the dependence of rates on hydrogen [P], catalyst [A], and olefin [S] concentrations, and on temperature was investigated. In the case of cyclohexene, the rate (R) can be correlated by the following equations.

$$1/R = a(1/[S]) + b, \quad R = c[P], \quad R = d[A],$$

a, b, c, d: constant

In the case of styrene and 1, 3-cyclooctadiene, the rate can be correlated as follows.

$$R = e[P], \quad R = F[A], \quad e, f: \text{constant}$$

The polymer-bound palladium complex was divalent before and after use for hydrogenation. When the catalyst was pretreated with hydrogen in the presence of a solvent, the color of the catalyst changed from yellow to dark green and the induction period in hydrogenation disappeared. Then a hydride complex seems to be formed by the heterolytic fission of hydrogen by replacing one of the chlorine on the palladium.

$$\text{(P)}\text{-PdCl}_2 + H_2 \longrightarrow \text{(P)}\text{-PdHCl} + HCl$$

We can present a formal scheme containing Pd-H species for the hydrogenation of cyclohexene.

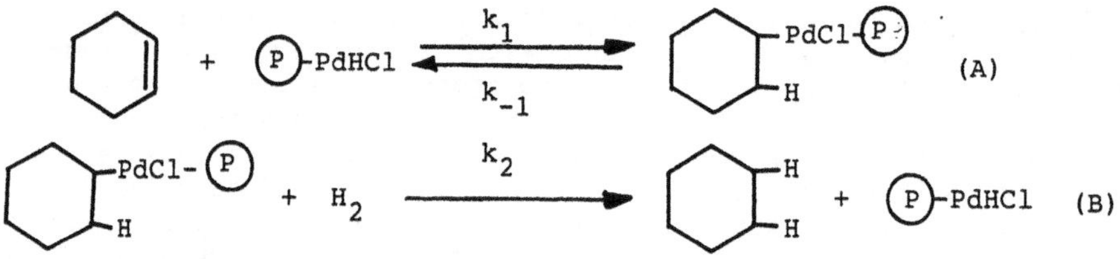

Scheme 2

Using the principle of steady state, the rate is

$$R = -\frac{d[S]}{dt} = \frac{k_1 k_2 [S]\,[P]\,[A]}{k_{-1} + k_1[S] + k_2[P]} \qquad (1)$$

Table 5. Kinetic Data for the Hydrogenation of Cyclohexene, Styrene, and 1, 3-Cyclooctadiene

Substrate	ΔH (kcal/mol)	ΔS (e.u.)	k_2^H/k_2^D	Frequency factor
Cyclohexene	3.9	-40.3	1.2	2.63×10^4
Styrene	7.9	-24.5	1.5	6.81×10^7
1,3-Cyclooctadiene	7.5	-26.2	1.7	3.11×10^7

where k_1, k_{-1}, and k_2 are the rate constants for the reaction equations (A) and (B). The above rate expression is found to reasonably accommodate most of the experimental data for hydrogenation of cyclohexene. If $k_1[S] \gg k_{-1} + k_2[P]$, Eq. (1) reduces to the simple form

$$R = -\frac{d[S]}{dt} = k_2[P]\,[A] \qquad (2)$$

The rate expression reasonably accommodates the experimental data for styrene and 1, 3-cyclooctadiene. The above assumption is reasonable because styrene and 1, 3-cyclooctadiene add to the metal hydride to form a more stable π-benzyl and π-allyl intermediate, respectively. Only the conjugated diene was selectively hydrogenated to corresponding monoenes in the presence of both a conjugated diene and a monoene, and the monoene was neither hydrogenated nor isomerized. This fact shows that the conjugated dienes coordinate more strongly to the metal hydride than monoenes. It seems that the fast hydrogenation rate of conjugated dienes may be attributed to the formation of π-allyl intermediate and to the activation by the interaction of the metal with the second double bond.

The kinetic data at 25°C for hydrogenation of cyclohexene, styrene, and 1, 3-cyclooctadiene obtained from these studies are given in Table 5. The activation entropies $\Delta S^{\neq}$ are negative. The decrease in entropy corresponds to a considerable loss of freedom due to fixation of the catalyst molecule on the polymer matrix. The kinetic characteristics of catalysis with this polymer-bound complex resemble those of heterogeneous catalysis rather than those of homogeneous catalysis. The polymeric complex is much more active for hydrogenation of olefins than an analogous homogeneous catalyst, $PdCl_2(PPh_3)_2$-$SnCl_2$.[19] The homogeneous catalyst hydrogenates nonconjugated dienes at a faster rate than conjugated dienes. On the other hand, the polymeric catalyst displays its reverse activity.

Bruner and Bailar have prepared a polymer-bound palladium complex by the reaction of dichlorobisbenzonitrile palladium(II) with phosphinated polystyrene with a 2% cross-linking.[20] The differences of the reactivity for olefin hydrogenation between our complex and Bailar's one are discussed in reference 7.

The phosphinated polystyrene-bound palladium(II) complex showed an effective catalytic activity for the hydrogenation of acetylenes. Table 6 summarizes the results of hydrogenation of 15 acetylenes. In most cases the rate of hydrogen uptake was nearly constant until near the complete consumption of acetylene. In all the cases, only the hydrogenation products and their double bond isomers were obtained, and hydrogenolysis and oligomerization products were not observed.

Table 6. Hydrogenation of Acetylenes with Polystyrene-Bound Pd(II) Complex[a)]

Run	Acetylene	Reaction time (min)	Conversion (%)	Products (%, yield) Olefins	Paraffin	Selectivity (%)[b]
1	1-Heptyne	57	98	1-, cis-2-, trans-2- (90) (2) (1)	5	95
2	2-Heptyne	22	99	cis-2-, trans-2-, cis-3- (90) (4) (3)	2	98
3	Phenylacetylene	36	100	(93)	7	93
4	1-Phenyl-1-propyne	106	89	cis-(59), trans-(4)	26	71
5	1-phenyl-1-propyne[c)]	100	89	cis-(64), trans-(3)	22	75
6	1-Phenyl-1-butyne	81	93	cis-(62), trans-(5)	26	67
7	1-Cyclohexyl-1-butyne	32	100	cis-(94), trans-(5)	1	99
8	Cyclohexenylacetylene	27	100	Vinyl-cyclohexene (81), Ethyl-cyclohexene (13), Vinyl-cyclohexane (6)	0	81
9	2-Methyl-1-buten-3-yne	251	100	Isoprene (85), Isopentene (15)	0	85
10	1,5-Hexadiyne	205	44	1-ene-5-yne, 1,5-diene, 1-ene (18) (13) (13)	0	41
11	1-Hexen-5-yne	400	90	1,5-diene (87), 1-ene (3)	0	97
12	Propargyl chloride	600	47	(47)	0	100
13	Propargyl alcohol	240	29	(29)	0	100
14	Propargyl alcohol[d)]	247	93	(92)	1	99
15	2-Methyl-3-butyn-2-ol	35	97	(92)	5	95
16	3-Methyl-1-pentyn-3-ol	33	94	(91)	3	97
17	Acetylenedicarboxylic acid dimethy ester	81	100	cis- (90), trans- (2)	8	92

a) Using catalyst, 7.0 mg-atom of Pd/liter; actylene, 0.35 mol/liter; 1 atm of hydrogen in benzene-ethanol (1:1), [13-(volume of substrate)] ml at 25 °C. b) Selectivity = (semihydrogenated products of the triple bond/products) x 100. c) Benzene was used as a solvent. d) Dimethylformamide was used as a solvent.

From Table 6, it is clear that the activity of the palladium complex for acetylene hydrogenation depends more strongly on the electron density of carbon-carbon triple bonds than on the steric hindrance about the bonds. The more the electron density in the triple bond decreases, the more the triple bond coordinates strongly to the metal. The triple bonds coordinate too tightly to the palladium for hydrogenation and consequently inhibit the catalysis to some degree.

Semihydrogenations of isolated triple bonds were accomplished in a high selectivity (Runs 1, 2, 7, and 12-17). The selective coordination of the triple bond to the complex is thought to lead to such success, because the strength of coordination of triple bonds is much greater than that of double bonds.

On the other hand, hydrogenation of triple bonds in a conjugated system did not gain successful results (Runs 4-6, 8, and 9). A similar result has often been observed with the Lindlar catalyst, which also shows a high selectivity for isolated triple bonds. However, reasons for this phenomenon are not clear. We think that the regiospecific addition of the

Table 7. Codimerization of Acetylenes with Allyl Halides

Acetylene	Codimerization Products (%, yield) Heterogenous reaction[a)]	Homogeneous reaction[b)]
$Ph\text{-}C\equiv CH$	$CH_2=CHCH_2CH=C(Ph)Cl$ (61) $CH_2=CHCH_2C(Ph)=CHCl$ (23)	$CH_2=CHCH_2CH=C(Ph)Cl$ (95)
$Ph\text{-}C\equiv CH$[c)]	$CH_2=CHCH_2CH=C(Ph)Br$ (79) $CH_2=CHCH_2C(Ph)=CHBr$ (13)	$CH_2=CHCH_2CH=C(Ph)Br$ (95)
$Ph\text{-}C\equiv CCH_3$	$CH_2=CHCH_2C(CH_3)=C(Ph)Cl$ (29)[d] $CH_2=CHCH_2C(Ph)=C(CH_3)Cl$ (13)[d]	$CH_2=CHCH_2C(CH_3)=C(Ph)Cl$ (97)
$HC\equiv CH$[e)]	$CH_2=CHCH_2CH=CHCl$ (75)[f)]	$CH_2=CHCH_2CH=CHCl$ (2330)[f)]
$n\text{-}Pr\text{-}C\equiv CH$	$CH_2=CHCH_2CH=C(n\text{-}Pr)Cl$ (78)	$CH_2=CHCH_2CH=C(n\text{-}Pr)Cl$ (90)
$n\text{-}Bu\text{-}C\equiv CH$	$CH_2=CHCH_2CH=C(n\text{-}Bu)Cl$ (82)	$CH_2=CHCH_2CH=C(n\text{-}Bu)Cl$ (90)
$Et\text{-}C\equiv C\text{-}Et$	$CH_2=CHCH_2C(Et)=C(Et)Cl$ (54)	$CH_2=CHCH_2C(Et)=C(Et)Cl$ (95)
$Ph\text{-}C\equiv C\text{-}Ph$	no reaction	$CH_2=CHCH_2C(Ph)=C(Ph)Cl$ (95)

a) Unless otherwise noted, acetylene : allyl chloride : catalyst = $10^2 : 10^3 : 1$, for 48 h, 20 °C. b) For 2 h at room temperature. c) Allyl bromide was used. d) For 700 h. e) In an acetylene atmosphere. f) Based on Pd and others are based on reactant acetylene.

palladium hydride to the triple bond determines the selectivity in a conjugated system.

Internal acetylenes were selectively hydrogenated to cis-olefins (Runs 2, 4-7, and 17). The double bond configuration of produced olefins is determined by the manner of addition of the palladium hydride to a triple bond to give a σ-vinyl complex; cis-addition of the palladium hydride to the triple bond leads to a cis-olefin and trans-addition to a trans-olefin, because the cis-trans isomerization of the σ-vinyl complex does not occur under reaction conditions and the following reaction proceeds with retention of the geometry. Therefore, the addition of the palladium hydride to the triple bonds seems to be exclusively in a cis-manner.

The selectivities observed here are comparable to those of the best catalyst systems for the selective hydrogenation of acetylenes to olefins reported earlier, homogeneous cationic rhodium and the Lindlar catalyst.[21,22] For example, the former reduces 2-heptyne to 2-heptene with 92 to 99% selectivities and the latter rarely shows selectivity above 90%. Furthermore, this polymeric catalyst is superior to the homogeneous rhodium catalyst in handling, because of its heterogeneity.

Codimerization of Acetylenes and Allyl Halides.[9] We found that the selective codimerization of various acetylenes and allyl halides proceeds smoothly with palladium complex catalysts under mild conditions.[23] The reaction provides a facile and useful 1, 4-diene synthesis without isomerization to 1, 3-dienes.

We tried to heterogenize the codimerization reaction by using the phosphinated polystyrene-bound palladium(II) complex. The polymeric catalyst has the activity for the codimerization of the various acetylene compounds except of diphenylacetylene, though the rate of dimer formation is slower than in the case of the homogeneous palladium catalyst (Table 7). Remarkable differences between the heterogeneous polymeric catalyst and the homogeneous catalyst are as follows: 1) In the case of the polymeric catalyst, the rate of codimerization is more strongly affected by the molecular size of acetylene compounds. The steric bulk of the substituent on acetylenic carbons controls the rate; $Ph\text{-}C\equiv CH > Ph\text{-}C\equiv C\text{-}CH_3 \gg Ph\text{-}C\equiv C\text{-}Ph$. A similar phenomenon has also been observed in the case of olefin hydrogenation. 2) In the case of phenyl-substituted acetylenes, e.g., phenylacetylene and methyl phenyl acetylene, internal phenyl-substituted codimers can be obtained by using only the polymeric catalyst. The product ratio of external phenyl-substituted codimer (I) vs. internal phenyl-substituted codimer (II) was found to be affected by the kind of solvents and conditions of preparation of polymeric complexes.

$$CH_2{=}CHCH_2X + Ph\text{-}C\equiv C\text{-}Y \xrightarrow[20°C]{Pd(III)\text{-polymer}} CH_2{=}CHCH_2CY{=}C\begin{smallmatrix}X\\Ph\end{smallmatrix} \quad (I)$$

$X = Cl, Br$

$Y = H, CH_3$

$$+ \; Ch_2{=}CHCH_2\underset{\displaystyle Ph}{C}{=}CXY \quad (II)$$

In ethyl ether, methanol, and cyclohexane, the formation of codimer(I) predominates, while the portion of codimer(II) increases in nitromethane, ethyl acetate, and dichloromethane. Using the polymer ligand with a lower phosphine-loading, the proportion of (II) increases. The above phenomena are under investigation at present, but it can be said that the origin of both (I) and (II) is derived by direction of Pd-X addition to the acetylenic triple bond.

Cooligomerization of Olefins.[10] Codimerization of olefins with transition metal complexes has been extensively studied. However, little is known concerning cotrimerization of olefins. We found that the phosphinated polystyrene-bound palladium(II) complex selectively catalyzes cotrimerization of styrene and ethylene or propylene and that the complex displays a higher reactivity for styrene, compared with homogeneous palladium complexes.

The reaction of styrene and ethylene (1 atm) catalyzed by the polymeric palladium(II) complex-$AgBF_4$ system selectively gave cotrimers (III) and (IV), where styrene was consumed completely. On the other hand, under 5 atm of ethylene 1:1 adducts of styrene (V) and ethylene (VI) were obtained in almost quantitative yield. All of the above oligomers contain internal phenyl groups. It is interesting that homooligomers of ethylene could not be detected under the reaction conditions. The polymeric complex separated after the reaction also showed the catalytic activity without adding $AgBF_4$ anew. In the absence of additive $AgBF_4$, the palladium complex showed no catalytic activity. Other compounds were tried as additives. $AgPF_6$ and $AgClO_4$ were active for the reaction, but AgOAc, $AgNO_3$, AgCl, $BF_3{\cdot}OEt_2$, and $AlCl_3$ showed no activity. The facts indicate that the function of the additive may be removal of a chlorine ligand to form a cationic palladium(II) complex.

In the reaction of styrene with propylene (1 atm), a similarly selective cotrimerization could be observed. It is noteworthy that the cotrimers containing a branched methyl group could not be detected.

This catalyst system also induced the homodimerization of styrene, ethylene, and propylene. Styrene gave (IX) (90% after 1 h). The dimerization rate of ethylene or propylene, however, was very slow under an atmospheric pressure.

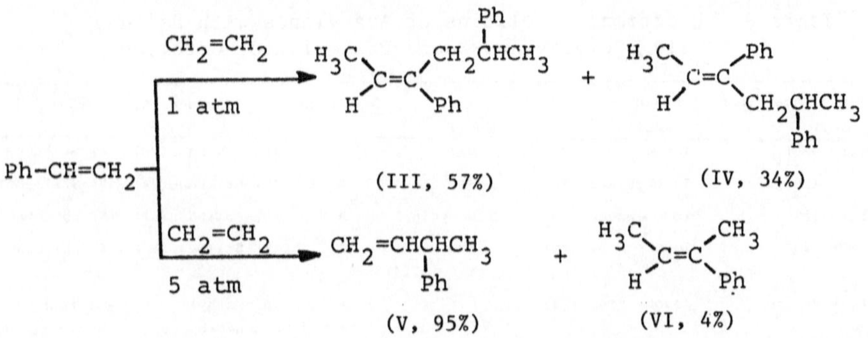

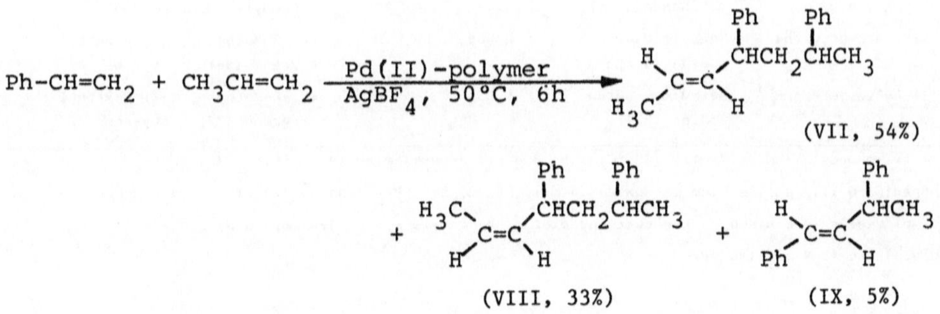

It is noted that the corresponding homogeneous system, $PdCl_2(PPh_3)_2$-$AgBF_4$, catalyzed no oligomerization.

The codimers containing an internal phenyl group are also obtained by using σ-aryl-$PdX(PPh_3)_2$-BF_3 or (π-allyl-PdCl)$_2$-Lewis acid-tertiary phosphine.[24,25] On the other hand, $PdCl_2$ and ($PdCl_2$·Ph-CH=CH_2)$_2$ give mainly external phenyl-substituted codimers.[26,27] However, these catalysts do not generate the cotrimers of styrene and ethylene or propylene.

Carbon-Carbon Bond Formation Reaction.[11] In order to test the catalytic activity of the polymer-bound palladium(0) complex,[28] three representative types of Pd(0)-induced reactions, all of which involve the oxidative addition of halides to the metal, were studied: (i) vinylic hydrogen substitutions with aryl halides,[29] (ii) acetylenic hydrogen substitutions with aryl halides,[30] (iii) vinylic halogen substitutions with Grignard reagents.[31]

The results are listed in Table 8. Most reactions were carried out under conditions similar to those in the corresponding homogeneous reactions. The polymeric complex displayed efficient catalytic activities in these reactions. In the reaction of styrene with iodobenzene (Run 1), stilbene was formed in 86% yield and (E)-stilbene could be easily isolated in 70% yield by recrystallization from a filtrate. Bromobenzene and chlorobenzene were used in place of iodobenzene in this study, but the corresponding stilbene was obtained in only trace yield. The polymeric complex was effective in catalyzing the acetylenic hydrogen substitutions (Runs 6 and 7).

The reactions of the isomeric β-bromostyrenes with methylmagnesium iodide proceeded quantitatively and showed a high stereospecificity (Runs 8 and 9).

The polymeric complex, which was separated from the reaction mixture

Table 8. Reactions of Olefins or Acetylenes with Halides Using Polystyrene-Bound Palladium(0) Catalyst

Run	Olefin or Acetylene (mmol)	Halide (mmol)	Catalyst (mmol)	Conditions[a)]	Product (%, yield)[b)]	
1	Styrene (25)	Iodobenzene (28)	0.14	A	(E)-Stilbene(78)	(Z)-Stilbene(8)
2	Styrene (25)	Iodobenzene (28)	Recycle-1	A	(E)-Stilbene(74)	(Z)-Stilbene(8)
3	Styrene (25)	Iodobenzene (28)	Recycle-2	A	(E)-Stilbene(61)	(Z)-Stilbene(7)
4	Styrene (25)	Iodobenzene (28)	Exposed to air (0.14)	A	(E)-Stilbene(63)	(Z)-Stilbene(9)
5	Methylacrylate (25)	Iodobenzene (28)	0.14	A	(E)-Methyl-cinnamate (68)	(Z)-Methyl-cinnamate (13)
6	Phenylacetylene (25)	Iodobenzene (28)	0.14	B	Diphenylacetylene (63)	
7	1-Pentyne (5)	Iodobenzene (5)	0.14	C	1-Phenyl-1-pentyne (59)	
8	(Z)-β-Bromostyrene (2.1)	Methylmagnesium iodide (5.4)	0.035	D	(Z)-β-Methyl-styrene (98)	(E)-β-Methyl-styrene (2)
9	(E)-β-Bromostyrene[c)] 2.1)	Methylmagnesium iodide (5.4)	0.035	D	(E)-β-Methyl-styrene (72)	(Z)-β-Methyl-styrene (3)

a) A = tri-n-butylamine (4.7 ml), 100 °C, 2 h; B = tri-n-butylamine (4.7 ml), 40 °C, 20 h; C = piperidine (1.5 ml), sodium methoxide (5 ml), 50 °C, 5 h; D = benzene (20 ml), room temperature, 24 h. b) Yields were based on the starting olefin or acetylene, and determined by glc. c) Containing 12 % (Z)-isomer.

by centrifugation in air, could be reused with some loss of activity (Runs 1-3). However, it seems that the loss of activity is not due to the lowering of the catalytic activity itself, but to the loss of polymer catalyst in the preparation process. Even after the catalyst had been exposed to air for 6 days, its catalytic activity did not remarkably decrease (Run 4).

Telomerization of Butadiene with Nucleophiles.[12] The selective telomerization of 1, 3-dienes and nucleophiles is characteristic of palladium(0) compounds as catalysts and can scarcely be achieved by other metal compounds. The application of the telomers to the synthesis of useful natural products had become of interest in recent years.[5] We have found that the phosphinated polystyrene-bound palladium(0) complexes catalyzed the telomerization of butadiene with various nucleophiles such as alcohols, amines, carboxylic acids, phenols, water, and silane. The main products were adduct telomers (2:1) of butadiene and nucleophiles except for reaction with formic acid. For example, the reaction of butadiene with secondary amines gave two regioisomers of 1-amino-2, 7-octadiene and 3-amino-1, 7-octadiene, and 1, 3, 7-octatriene was a minor product.

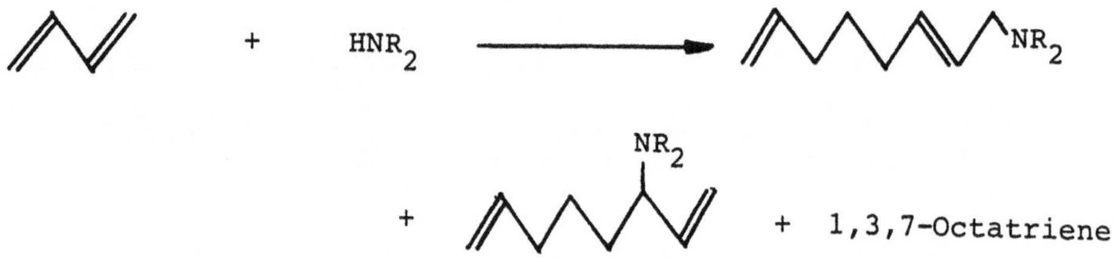

Table 9. Recycling of Polystyrene-Bound Pd(0) Catalyst

Catalyst	Overall yield[b)] (%)	Products(%) Ia	IIa	III
Fresh	97	92	6	2
Recycle-1	99	98	1	1
Recycle-2	72	98	1	1

a) Reaction conditions: polymeric catalyst (P/Pd ratio of 4), 0.1 mg-atom of Pd; morpholine, 80 mmol; 85 °C, 5 h.

b) Based on butadiene used. Ia = 1-morpholino-2,7-octadiene, IIa = 3-morpholino-1,7-octadiene, III = 1,3,7-octatriene.

The reactivity of amines decreases in the order morpholine > piperidine > diethylamine > diisopropylamine. Similar reactivity orders have been found when homogeneous palladium catalysts were used.[32,33] The catalyst recycling study with a polymeric palladium complex of a P/Pd ratio of 4 was carried out in the reaction with morpholine. Data are shown in Table 9. The polymeric palladium complex could be reused with some loss of its catalytic activity. The remarkable change in the product distribution could not be observed during each run.

In the case of formic acid, (formyloxy)octadienes of 2:1 adduct could not be detected, and 1, 7-octadiene was exclusively formed. In contrast, it has been reported that 1, 6-octadiene was exclusively obtained in the presence of the homogeneous $Pd(OAc)_2$ catalyst.[34] The highly selective formation of 1, 7-octadiene may be explained by a mechanism involving a (formato)palladium intermediate; evolution of CO_2 in a formato complex affords a π-allylpalladium hydride, followed by reductive elimination, and then a preferential attack of hydrogen on C-3 of an allylic moiety to give 1, 7-octadiene (Scheme 3).

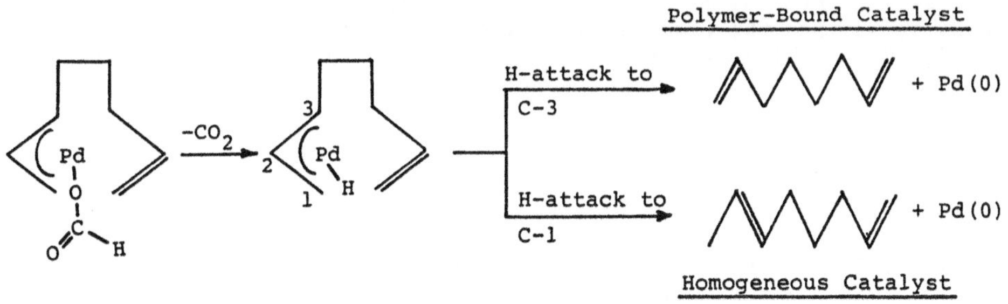

Scheme 3

Our polymer-bound palladium(0) complex is an active catalyst not only for carbon-carbon bond formation involving the oxidative addition of organic halides but also the telomerization of butadiene with various kinds of nucleophiles.

Phosphinated Polystyrene-Bound Rh(II) Complexes.[13] The polymeric rhodium complexes are efficient catalysts for hydrogenation of monoenes and dienes at ambient temperature and pressure of 1 atm, but display no activity for acetylene hydrogenation. The activity is very sensitive to the steric hindrance about the carbon-carbon double bond: 1-heptene > 2-heptene, and to the molecular size of substrate: 1-octene > 1-dodocene, cyclohexene > cyclooctene. Functional groups such as keto-, hydroxy-, cyano-, nitro-, ester, and carboxylic acid are not hydrogenated under the reaction conditions.

The polymeric rhodium complex was divalent before and after use for olefin hydrogenation. Furthermore, the intensity of the ESR signal of the rhodium complex scarcely changed during the reaction. The divalent rhodium complex acts as a catalyst, being kept intact; that is to say, the oxidation state does not change during the reaction. The induction period in the hydrogenation was observed, and never disappeared even when the catalyst was pretreated with hydrogen or olefin. This suggests the presence of at least two reaction equilibria involving hydrogen or the olefin. The migration of double bond was observed during the reaction, but did not occur in the absence of hydrogen. This result shows that a σ-alkyl complex exists as an intermediate, which is formed probably by the addition of the rhodium-hydride to the olefin.

We can propose a mechanism containing Rh(II)-H species for the hydrogenation of styrene and cyclohexene.

$$\textcircled{P}\text{-RhCl}_2 + H_2 \overset{K_1}{\rightleftharpoons} \textcircled{P}\text{-RhHCl} + \text{HCl}$$

$$\textcircled{P}\text{-RhHCl} + \text{Olefin} \overset{K_2}{\rightleftharpoons} \textcircled{P}\text{-Rh(alkyl)Cl}$$

$$\textcircled{P}\text{-Rh(alkyl)Cl} + H_2 \xrightarrow{k_3} \textcircled{P}\text{-RhHCl} + \text{Alkane}$$

Scheme 4

Using the principles of steady state, the rate R is

$$R = -\frac{d[S]}{dt} = \frac{k_3 K_1 K_2 [S][H_2][A]}{[HCl] + K_1[H_2] + K_1K_2[H_2]\ [S]}$$

where K_1 and K_2 are the equilibrium constants, k_3 is the rate constant, and $[H_2]$, $[S]$, and $[A]$ are hydrogen, the olefin, and catalyst concentration, respectively. If $[HCl] \ll K_1[H_2] + K_1K_2[H_2][S]$, the equation reduces to the form

$$R = -\frac{d[S]}{dt} = \frac{k_3 K_2 [S][H_2][A]}{1 + K_2[S]}$$

This rate expression reasonably accommodates kinetic results for styrene and cyclohexene. The kinetic data at 25°C for the hydrogenation of styrene are $\Delta H^{\neq} = 4.9$ kcal mol^{-1}, $\Delta S^{\neq} = -37.1$ e.u., and frequency factor = 1.29×10^5. These kinetic data are similar to those previously mentioned in the hydrogenation of olefins with the polymeric palladium complexes.

Aminated Polystyrene-Bound Rh Cluster Complexes.[15] The water gas shift reaction (WGSR) plays an important role in the adjustment of H_2:CO ratio in synthesis gas.

$$H_2O + CO \longrightarrow H_2 + CO_2$$

Commercial heterogeneous catalysts for the WGSR consist mainly of iron oxides or copper-zinc oxide mixtures, which have been operated at high temperatures (200-400°C).[35] Low temperature operation for the reaction is favorable from the standpoints of thermodynamic equilibrium and the lower thermal input requirement. We have found that $Rh_6(CO)_{16}$-diamine systems show high catalytic activity for the WGSR under the mild conditions.[36] Also, $Rh_6(CO)_{16}$-diamine and pyridine systems catalyze hydrogenation of nitrobenzene and unsaturated carbonyl compounds, and carbonylation of olefins.[37] Suitable choice of amines and pyridines as additives attains highly selective reactions which cannot be obtained in the case of molecular hydrogen.

We attempted to heterogenize the rhodium catalyst systems using aminated polymers which are synthesized from treatment of chloromethylated polystyrene with various amines. The WGSR using various kinds of aminated polystyrenes and poly-4-vinylpyridines in the presence of $Rh_6(CO)_{16}$ catalyst is shown in Table 10. Use of diethylenetriamino polymer gave the highest activity for the WGSR (Run 4). In a series of ethylenediamino polymers, the activity decreases in the following order: 2-aminopropylamino > 2-aminoethylamino > 2-(dimethylamino)ethylamino polymer. The similar effect of substituents can be observed in homogeneous systems. In the case of diethylenetriamino polymer, the 2-aminoethylamino moiety is situated apart from the polymer backbone. So, the diamino moiety might easily coordinate to rhodium metal to facilitate hydrogen generation, when compared to that of other ethylnediamino polymers. The low activity in the case of triethylenetetraamino polymer may be caused by cross-linking of the tetramine to other chloromethyl moiety. It seems that the heterogenization of ethylenediamine using polystyrene does not cause any great lowering of the WGSR activity in homogeneous systems (Runs 1 and 6). Use of poly-4-vinylpyridines gave moderate activity. In olefin hydrogenation using phosphinated polystyrene-bound titanium and rhodium complexes, cross-linking of polymers is preferable because of site isolation of metal active species.[38,39] But, in the present WGSR, cross-linking does not lead to a good result. Pittman et al. have adopted Amberlyst A-21 as a polymer support in the WGSR and nitrobenzene hydrogenation.[40] Our prepared diethylenetriamino polymer is much more efficient for the WGSR than the Amberlyst A-21 under the present conditions.

In effect of solvent, 2-ethoxyethanol and dioxane are the best. Sulfolane, dimethylformamide, and nitromethane are poor solvents. $Rh_2(CO)_4Cl_2$ and $Rh_6(CO)_{16}$ complexes have the highest catalytic activity; the activity of rhodium complexes decreases in the following order: $Rh_2(CO)_4Cl_2$ > $Rh_6(CO)_{16}$ > $Rh_4(CO)_{12}$ > $RhCl(PPh_3)_3$.

In the case of the $Rh_6(CO)_{16}$-diethylenetriamino polymer system, the polymeric rhodium complex of dark green powder, which was separated from the reaction mixture under a CO atmosphere, could be reused with some loss of activity. The IR spectrum of the recovered polymer gave two strong absorption bands as shown in Fig. 3. The polymeric rhodium catalyst decomposed slowly on exposure to the air to a pale yellow complex which lost the catalytic activity. The treatment of the exposed rhodium complex with CO at 10 atm renewed two strong bands at 1960 and 1805 cm^{-1}. These two bands are characteristic of $[Rh_{14}(CO)_{25}]^{4-}$.[41] It is notable that increasing H_2O amount in the WGSR led to another complex with 1990, 1960, 1830, and 1800 cm^{-1}. This polymeric rhodium complex may involve $[Rh_{14}H(CO)_{25}]^{3-}$,[42] which showed a higher activity for the WGSR than did $[Rh_{14}(CO)_{25}]^{4-}$. The relationship between the two rhodium complexes is as follows:

$$\text{(P)}-[Rh_{14}(CO)_{25}]^{4-} + H^+ \longrightarrow \text{(P)}-[Rh_{14}H(CO)_{25}]^{3-}$$

Table 10. The Water Gas Shift Reaction Using $Rh_6(CO)_{16}$-Aminated and Pyridinated Polymer Systems[a)]

Run	Aminated and pyridinated polymer (mg-atom of N)	Turnover number (T.N.)[b)]
1	$NH_2-CH_2-CH_2-$(P)[c)] (R_1, 5)	11
2	$NH_2-CH(CH_3)-CH_2-$(P) (R_2, 5)	15
3	$CH_3-N(CH_3)-CH_2-CH_2-$(P) (R_3, 5)	4
4	$NH_2-(CH_2)_2NH(CH_2)_2-$(P) (R_4, 5)	31
5	$NH_2-(CH_2)_2NH(CH_2)_2NH(CH_2)_2-$(P) ($R_5$, 5)	5
6[d)]	$NH_2-CH_2-CH_2-NH_2$ (5)	13
7	Poly-4-vinylpyridine (0% cross-linked, 5)	4
8	Poly-4-vinylpyridine (0% cross-linked, 10)	16
9	Poly-4-vinylpyridine (2% cross-linked, 10)	8
10	Poly-4-vinylpyridine (22% cross-linked, 10)	1
11	Poly-4-vinylpyridine (0% cross-linked, 0.1)	3
12	Amberlyst A-21 (0.1)	0.6

a) $Rh_6(CO)_{16}$, 0.05 mmol; H_2O, 0.72 ml, 2-ethoxyethanol, 7 ml; P_{CO}, 700 mmHg; 100 °C, 24 h. b) T.N. = mol of H_2/mol of catalyst, 24 h. c) (P):

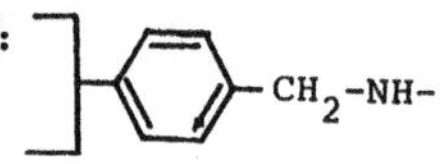

d) Diamine monomer was used.

The $[Rh_{14}(CO)_{25}]^{4-}$ complex could be observed also in the homogeneous system.

Filtrate separated from the polymeric rhodium complex in the reaction mixture showed no catalytic activity for the WGSR and addition of excess PPh_3 to the reaction mixture strongly depressed the activity. So, we think that the present WGSR is catalyzed by rhodium cluster anions bound to polymeric ammonium ions.

Functions of the aminated polymers in the WGSR are: 1) generation of OH^- (basicity), 2) stabilization of rhodium cluster anions with polymeric ammonium ion, and 3) coordinating activation of amines to rhodium intermediate species.

Cellulose-Bound Pd and Rh Complexes.[16] We have been interested in the use of cellulose as an insoluble support in place of polystyrene, not only because it is a cheap natural resin but also because it is a chiral ligand. As mentioned in the former section, the cellulose-bound palladium complex has unsaturated coordinating sites about palladium, which can be expected to be an active catalyst for various kinds of organic reactions. In order to test this theory, hydrogenation of olefins and acetylenes using the cellulose-bound palladium complexes was carried out at 25°C. The initial rates of hydrogenation uptake for representative olefins are compared with those using the polystyrene-bound palladium(II) complex in Table 11. The cellulose-bound catalyst shows a high activity for the hydrogenation of conjugated olefins and allyl compounds. The reactivity is sensitive to steric hindrance about the carbon-carbon double bond of the substrates.

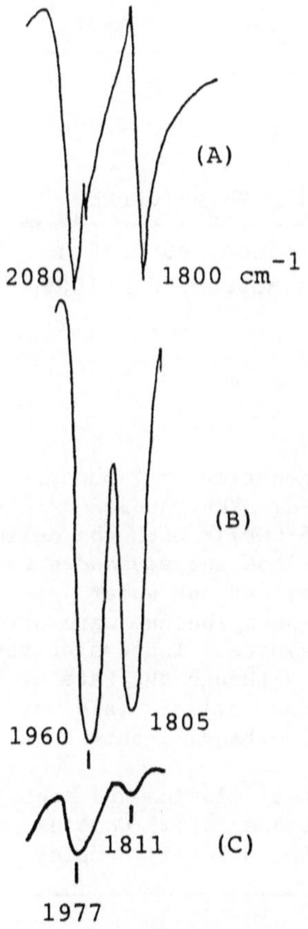

Fig. 3. IR Spectra of Rh Complexes (KBr disk).
(A) : starting $Rh_6(CO)_{16}$ complex
(B) : freshly recovered Rh complex
(C) : the recovered complex exposed to air for 10 days.

Phenylacetylene was hydrogenated to styrene in over 95% selectivity. The effect of solvent was examined in the hydrogenation of styrene. The polymeric palladium catalyst displays a high catalytic activity for this hydrogenation in solvents such as alcohols, THF, and acetone. Interestingly, in the case of the palladium catalyst derived from the reaction of cellulose containing no attached phosphine moieties with $PdCl_2(PhCN)_2$, such a remarkable solvent effect could be observed; even cyclohexane, a poorly coordinating solvent, gave a moderate rate for the hydrogenation. This solvent phenomenon is due to the catalysis of metallic palladium. During the hydrogenation, the color of the catalyst was grey, and it is probable that metallic palladium is dispersed in the cellulose.

The cellulose phosphinite-bound palladium(II) complex, which was separated from the reaction mixture in air, can be recycled twice and still maintains ca. 85% of hydrogenation rate for the fresh catalyst. The XPS spectra of the recovered complex did not vary appreciably.

The green polymeric palladium(0) complex showed catalytic activity for the Heck reaction and telomerization of butadiene with nucleophiles:

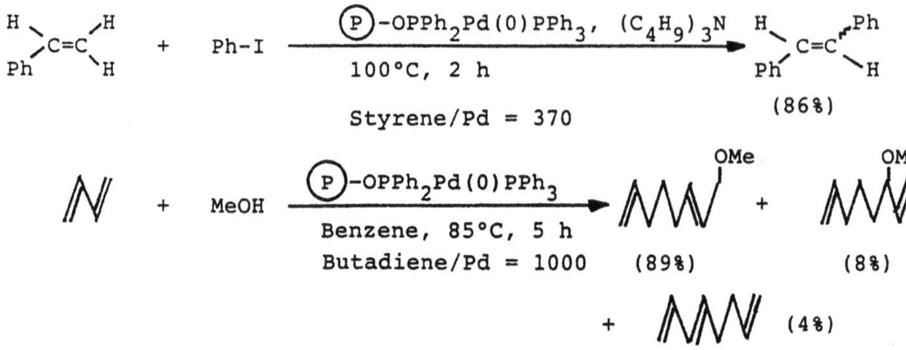

Next, asymmetric hydrogenation of 2-phenyl-1-butene with the polymeric rhodium complex was attempted. The rhodium catalyst was prepared in situ by the treatment of $[Rh(1, 5\text{-}COD)]_2$ with the cellulose phosphinite in benzene solution. 2-Phenyl-1-butene was added to the catalyst solution. These hydrogenations were carried out under 4-40 atm of H_2 and 80-100°C. The optical yields of [R]-2-phenylbutane were strongly dependent on the P/Rh ratio of the catalyst system. Increasing the P/Rh ratio tended to give higher optical yields, although the rate of hydrogenation slowed appreciably; at P/Rh = 27, the optical yield was 30% in 63% conversion. Asymmetric hydrogenation of 2-phenyl-1-butene using

Table 11. Hydrogenation of Olefins and Acetylene Using Cellulose Phosphinite-Bound Pd(II) Complex and Phosphinated Polystyrene-Bound Pd(II) Complex

Substrate	Initial rate (ml/min)	
	Cellulose-Pd(II)[a)]	Polystyrene-Pd(II)[b)]
Styrene	6.20	8.10
Mesityl oxide	0.62	0.31
Methyl vinyl ketone	7.09	5.80
1-Heptene	3.66	3.94
2-Heptene	1.00	0.59
1,3-COD	7.56	5.26
1,5-COD	0.65	0.93
Allyl alcohol	5.87	5.72
Phenylacetylene	3.82	3.53

a) Catalyst, 2.63×10^{-2} mg-atom of Pd; substrate, 4.36 mmol, methanol, [13 -(volume of substrate)]ml, 1 atm of H_2, 25 °C.

b) Catalyst, 4.16×10^{-2} mg-atom of Pd; substrate, 4.36 mmol: benzene-ethanol (1:1), [13-(volume of substrate)]ml, 1 atm of H_2, 25 °C.

the phosphinated cellulose ligand was carried out by a method similar to that used with the phosphinite. A 40% optical yield of [R]-2-phenylbutane was obtained at a P/Rh ratio of 22.5.

CONCLUDING REMARKS

We heterogenize soluble palladium and rhodium compounds using chemically modified polystyrene and cellulose as supports. This complexation method, using the polymeric ligand, is an effective tool for the preparation of coordinatively unsaturated metal complexes. The polymeric palladium and rhodium complexes are stable under a hydrogen or oxygen atmosphere, compared with the corresponding homogeneous metal complexes, and also use of polymeric ligands can stabilize large metal clusters; treatment of $Rh_6(CO)_{16}$ compound with aminated polystyrene under the WGSR conditions gave Rh_{14}-carbonyl cluster anions in ion-pairing form with polymeric ammonium ion. These polymeric metal complexes can catalyze various kinds of organic reactions. Especially, the phosphinated polystyrene-bound palladium complex is a versatile catalyst for 1) hydrogenation, 2) oligomerization, 3) telomerization, and 4) carbon-carbon bond formation reaction. The polymeric metal complexes can be reused with small loss of catalytic activities.

We have been much interested in the use of cellulose as an insoluble chiral ligand. Our results show that modified cellulose can be used as a polymer ligand to heterogenize metal complexes. Although high optical yields cannot be obtained in the asymmetric hydrogenation at present, phosphinated cellulose systems are promising chiral ligands for various metal-catalyzed reactions. In order to attain high optical yields, syntheses must be devised to increase the phosphorus content of cellulose.

There are many reports concerning the catalysis of polymer-bound metal complexes, but yet little success has been achieved in determining structure of polymeric metal complexes. We emphasize that the characterization of the polymeric metal complexes using various spectroscopic methods, e.g., FTIR, solid-state NMR, XPS, and EXAFS, is required in order to direct efforts at design of highly efficient catalysts, since the concept of the catalysis and the complexation for the polymeric metal complexes are not the same as those for the soluble metal complexes.

REFERENCES

1. N. K. Mathur, C. K. Narung, and R. E. Williams, "Polymers as Aids in Organic Chemistry," Academic Press, New York (1980).
2. C. U. Pittman, Jr., _in_ G. Wilkinson, ed., "Comprehensive Organometallic Chemistry," Vol. _8_, Pergamon Press, Oxford (1982), p. 553.
3. N. L. Holy, _in_ L. H. Pignolet, ed., "Homogeneous Catalysis with Metal Phosphine Complexes," Plenum, New York (1983), p. 443.
4. J. P. Collman and L. S. Hegedus, "Principles and Applications of Organotransition Metal Chemistry," University Science Books, California (1980).
5. J. Tuji, "Organic Synthesis with Palladium Compounds," Springer-Verlag, Berlin (1980).
6. L. H. Pignolet, "Homogeneous Catalysis with Metal Phosphine Complexes," Plenum, New York (1983).
7. M. Terasawa, K. Kaneda, T. Imanaka, and S. Teranishi, _J. Catal._, _51_, 406 (1978).
8. M. Terasawa, K. Kaneda, H. Yamamoto, T. Imanaka, and S. Teranishi, _J. Catal._, _57_, 315 (1979).

9. K. Kaneda, T. Uchiyama, M. Terasawa, T. Imanaka, and S. Teranishi, Chem. Lett., 449 (1976).
10. K. Kaneda, M. Terasawa, T. Imanaka, and S. Teranishi, Tetrahedron Lett., 2957 (1977).
11. M. Terasawa, K. Kaneda, T. Imanaka, and S. Teranishi, J. Organomet. Chem., 162, 403 (1978).
12. K. Kaneda, H. Kurosaki, M. Terasawa, T. Imanaka, and S. Teranishi, J. Org. Chem., 46, 2356 (1981).
13. K. Kaneda, M. Terasawa, T. Imanaka, and S. Teranishi, Chem. Lett., 995 (1976).
14. M. Terasawa, K. Sano, K. Kaneda, T. Imanaka, and S. Teranishi, J. Chem. Soc., Chem. Commun., 650 (1978).
15. K. Kaneda, M. Kobayashi, T. Imanaka, and S. Teranishi, Chem. Lett., 1483 (1984).
16. K. Kaneda, H. Yamamoto, T. Imanaka, and S. Teranishi, J. Molecular Catal., 29, 99 (1985).
17. D. Adams, "Metal-Ligand and Related Vibrations: A Critical Survey of the Infrared and Raman Spectra of Metallic and Organometallic Compounds," 1st Ed., Edward Arnold, London (1967).
18. M. A. Bennet and P. A. Longstaff, J. Am. Chem. Soc., 91, 6266 (1969).
19. H. Itatani and J. C. Bailar, Jr., J. Am. Oil Chem. Soc., 44, 147 (1967).
20. H. Bruner and J. C. Bailar, Inorg. Chem., 12, 1465 (1973).
21. R. R. Schrock and J. A. Osborn, J. Am. Chem. Soc., 98, 2143 (1976).
22. For a review, see E. N. Marvell and T. Li, Synthesis, 457 (1973).
23. K. Kaneda, H. Kobayashi, Y. Fujiwara, T. Imanaka, and S. Teranishi, J. Org. Chem., 44, 55 (1979).
24. H. Nozima, N. Kawata, Y. Nakamura, K. Maruya, T. Mizoroki, and A. Ozaki, Chem. Lett., 1163 (1973).
25. T. Ito, K. Takahashi, and Y. Takami, Nippon Kagakukaishi, 1097 (1974).
26. M. G. Barlow, M. J. Bryant, R. M. Haszeldine, and A. G. Mackie, J. Organomet. Chem., 21, 215 (1970).
27. K. Kawamoto, A. Tatani, T. Imanaka, and S. Teranishi, Bull. Chem. Soc. Jpn., 44, 1239 (1971).
28. Catalysis of polymeric palladium(0) complexes has been also studied by other groups: a) B. M. Trost and E. J. Keinan, J. Am. Chem. Soc., 100, 7779 (1978); b) C. U. Pittman and Ng. Q., J. Organomet. Chem., 153, 85 (1978); c) R. C. Card and D. C. Neckers, Inorg. Chem., 17, 2345 (1978).
29. R. F. Heck and J. P. Nolley, Jr., J. Org. Chem., 37, 2320 (1972).
30. H. A. Dieck and R. H. Heck, J. Organomet. Chem., 93, 259 (1975).
31. M. Yamamura, I. Moritani, and S. I. Murahashi, J. Organomet. Chem., 91, 259 (1975).
32. S. Takahashi, T. Shibano, and N. Hagiwara, Bull. Chem. Soc. Jpn., 41, 454 (1968).
33. M. Green, G. Scholes, and F. G. A. Stone, J. Chem. Soc., Dalton Trans., 309 (1978).
34. S. Gardnev and D. Wright, Tetrahedron Lett., 163 (1972).
35. C. L. Thomas, "Catalytic Processes and Proven Catalysts," Academic Press, New York (1970), p. 104.
36. K. Kaneda, H. Hiraki, T. Imanaka, and S. Teranishi, J. Molecular Catal., 9, 227 (1980).
37. a) K. Kaneda, M. Hiraki, T. Imanaka, and S. Teranishi, J. Molecular Catal., 12, 385 (1981); b) K. Kaneda, M. Yasumura, M. Hiraki, T. Imanaka, and S. Teranishi, Chem. Lett., 1763 (1981); c) K. Kaneda, T. Imanaka, and S. Teranishi, Chem. Lett., 1465 (1983); d) K. Kaneda, M. Yasumura, T. Imanaka, and S. Teranishi, J. Chem. Soc., Chem. Commun., 935 (1982).
38. W. D. Bonds, Jr., C. H. Brubaker, Jr., E. S. Chandrasekaran, C. Gibbons, R. H. Grubbs, and L. C. Kroll, J. Am. Chem. Soc., 97, 2128 (1975).

39. J. P. Collman, L. S. Hegedus, M. P. Cooke, J. R. Norton, G. Dolcetti, and D. N. Marquardt, J. Am. Chem. Soc., 94, 1789 (1972).
40. R. C. Ryan, G. M. Wilemon, M. P. Dalsanto, and C. U. Pittman, Jr., J. Molecular Catal., 5, 319 (1979).
41. S. Martinengo, G. Cianti, A. Sironi, and P. Chini, J. Am. Chem. Soc., 100, 7096 (1978).
42. G. Ciani and A. Sironi, J. Organomet. Chem., and S. Martinengo, J. Organomet. Chem., 192, C-42 (1980).

THE MECHANISM AND STABILITY OF SUPPORTED ZIEGLER CATALYSTS

R. Spitz, L. Duranel*, and A. Guyot

CNRS - Laboratoire des Materiaux Organiques
BP 24 - 69390 Lyon Vernaison (France)

INTRODUCTION

A family of supported Ziegler catalysts, suited for propene polymerization were studied for more than ten years, beginning in the early 70's. The solid precatalyst used for all the studies can be described by a common formula: $MgCl_2$-E.D.-$TiCl_4$, in which the electron donor (E.D.) is almost always an aromatic ester (A.E.), generally ethylbenzoate (E.B.). The preparation chiefly differs in the way the three components are reacted together by milling or impregnation, the temperatures, the durations and the ratios used.[1]

Some other additives are described in the patent literature, but do not seem to really change the trends of the system. In all cases, in spite of the progress achieved during recent years, this type of precatalyst used in propene polymerization shows a fast decrease in the polymerization rate with all the common alkylaluminum derivatives used as cocatalysts (mostly: aluminumtriethyl or triisobutyl, mixed in some cases with dialkylaluminum chloride). The decrease is observed at low or high pressure in the usual polymerization temperature range (60-80°C). Some increase in the decay rate is achieved when the catalytic system is made more isospecific by adding an E.D. to the cocatalyst.[1,2]

Figure 1 shows typical kinetic curves at 4 bar propene pressure emphasizing the effect of Al/E.D. ratio and Figure 2 shows that the kinetics in a propene liquid pool have the same shape. The high performance precatalyst used here is obtained by comilling $MgCl_2$ and E.B. followed by excess $TiCl_4$ impregnation and drying in an improved way.[3] Figure 3 shows the kinetic curves of this system using trioctylaluminum and diethylaluminum chloride as cocatalysts.

On the contrary, ethylene polymerization gives rather constant and even sometimes increasing kinetic curves depending on the polymerization temperature. An example is given in Figure 4 with the same catalytic system.

---

*Present address: ATOChem - SNEA(P)
BP 34 LACQ - 64170 ARTIX (France)

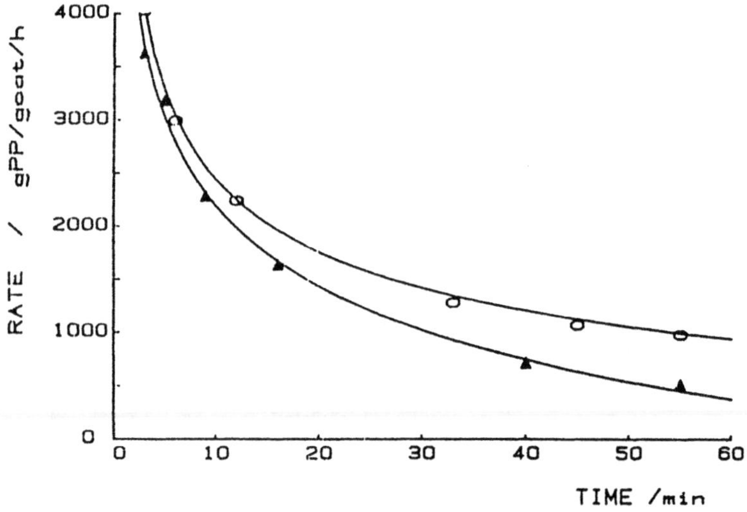

Fig. 1. Propylene Polymerization Kinetic. 4 bars, 63°C,
TEA 3 mmol/l Electron
Electron donor: ethylparatoluate (EPT)
0 Al/EPT = 4 (heptane insoluble fraction HI = 95.2%)
AL/EPT = 3 (HI = 96%)

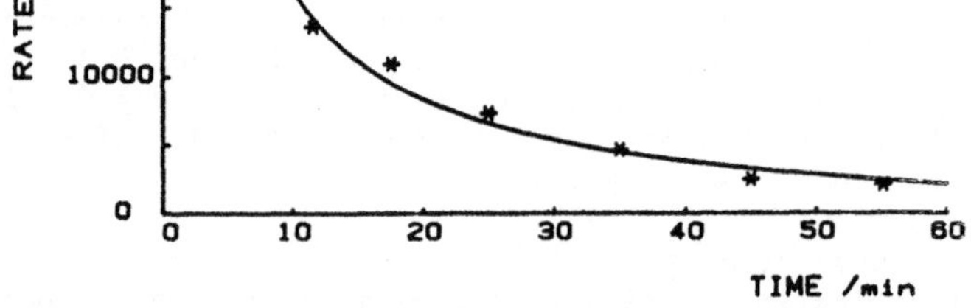

Fig. 2. Propylene Polymerization Kinetic. Liquid pool 66°C, 26 bars
TEA = 5 mmol/l; Al/EPT = 3 (HI = 96.2%)

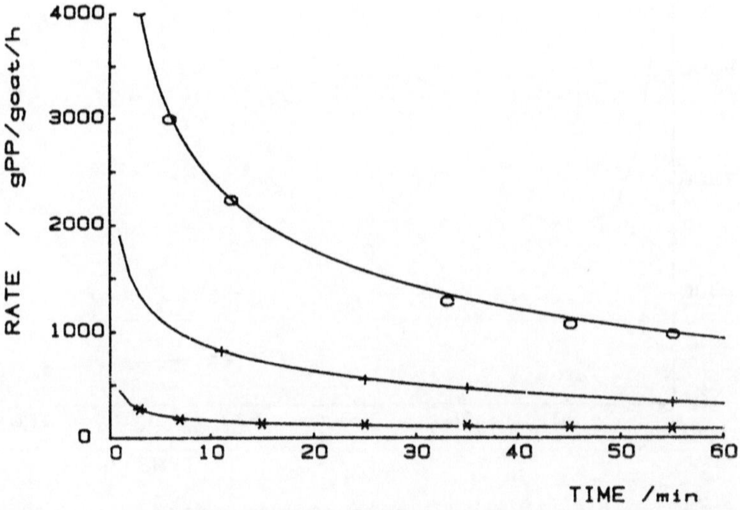

Fig. 3. Propylene Polymerization Kinetic. 4 bars, 63°C
0 TEA = 3 mmol/l Al/EPT = 4 (HI = 95.2%)
* Diethylaluminum chloride = 10 mmol/l
Al/ethylbenzoate = 4 (HI = 94%)
+ tri-n-octylaluminum = 10 mmol/l
Al/ethylanisate = 4 (HI = 90.5%)

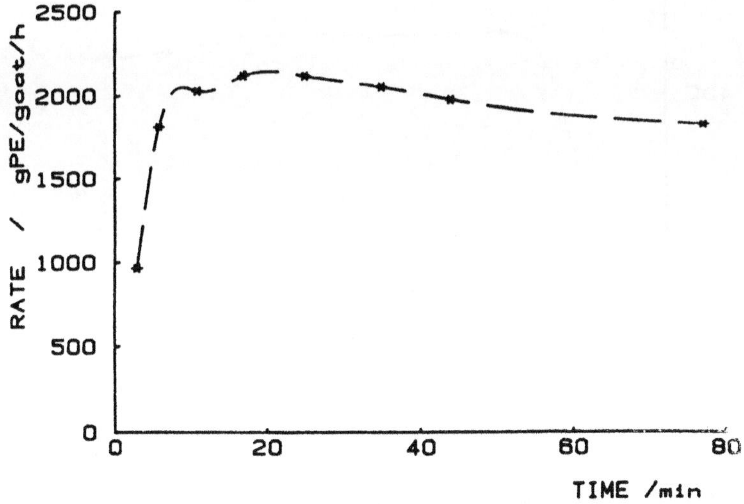

Fig. 4. Ethylene Polymerization Kinetic - 80°C, 3 bar pressure
TEA = 3 mmol/l

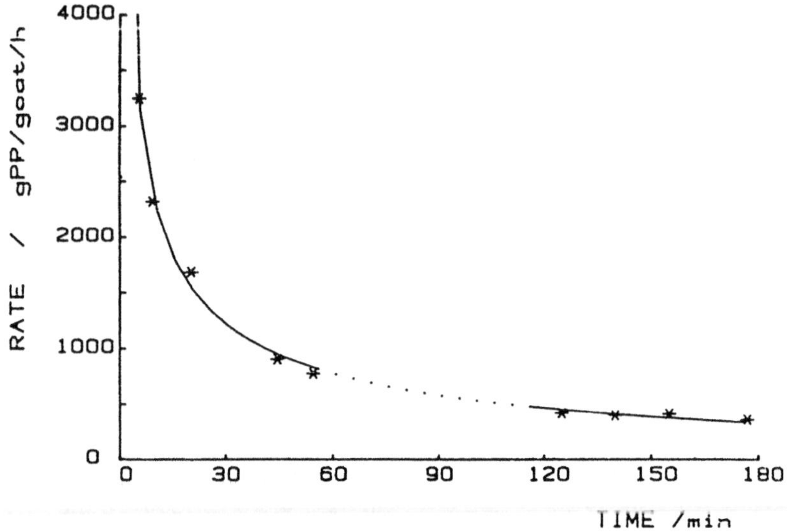

Fig. 5. Propylene Polymerization Kinetic - 63°C, 4 bars TEA = 5 mmol/1 - The polymerization is interrupted for 1 hour between 60 and 120 minutes (No monomer pressure . . .)

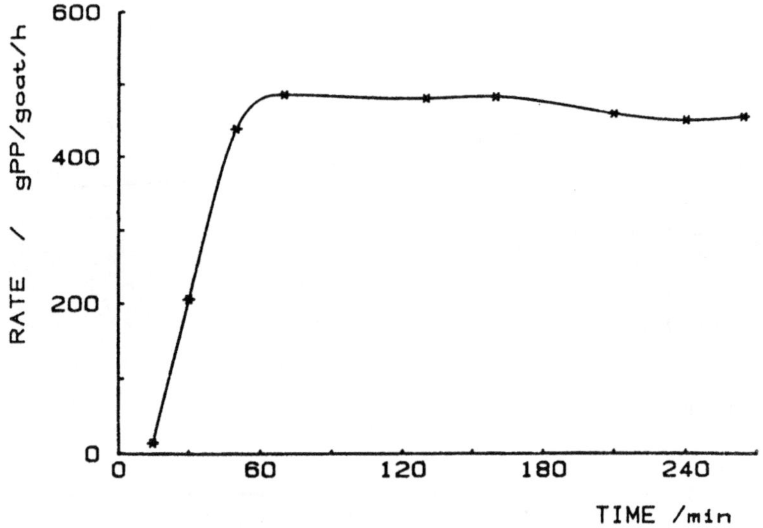

Fig. 6. Propylene Polymerization Kinetic with Isoprenylaluminum: 4 bars, 60°C, (Al) = 10 mmol/1 - (Al)/(Ethylanisate) = 4 (HI at 90 minutes = 92.2%)

The reason for the decrease in propene polymerization rate was discussed by many authors. Two kinds of explanations are generally presented: a chemical change in the active site, for instance a reduction of Ti(III) to Ti(II)[4] or a diffusion control due to the growth of the polymer particle.[5] Although a diffusion process would help to explain the broad molecular weight distribution observed, it is rather easy to show that most of the activity decrease is not due to the polymerization process itself, as it can be shown by aging experiments of the polymerization medium with no or almost no monomer present.[1,6] An example is given in Figure 5: the polymerization medium is "frozen," without any effect on the rate decay.

So, a chemical explanation of the deactivation of the catalyst is to be found. The nature of the process involved remains unknown and the classical mechanism of Ziegler polymerization gives no information on the rate decrease.

A cocatalyst, aluminumalkyl (A.A.) was widely used for ethylene polymerization and is very different from the usual A.A.: isopropenylaluminum (IPRA). IPRA has some bulky groups result from the cyclization side reactions of the synthesis.[7] But the most important structural effort is that each IPRA molecule comprises two or more bridged Al, differing from alumoxanes by longer bridges than in their pure hydrocarbon composition. IPRA is thought to be rather difficult to use because of its poorly-defined structure. Some change is observed when it is bought from different producers.

Figure 6 shows a typical kinetic curve obtained with IPRA. Compared to aluminumtriethyl (TEA), two important differences are observed: i) the activity does not start at the time of the mixing of pre- and cocatalyst but 5 to 25 minutes later; ii) after it has reached a maximum, the activity remains constant for hours.

The polypropylene obtained with IPRA is not different from that usually obtained: the low isotactic index obtained without E.D. (63% compared to 50-55% obtained with TEA) can be increased up to 93-94%, ethyl anisate gives the best results (Table 1). Other properties of the polymer (molecular weight and distribution, powder density and melt-index) are in the usual range. The productivity can be the same as with TEA for reaction times of 2 to 4 hours, depending on the conditions and the desired tacticity.

Table 1. Polypropylene Obtained with IPRA (4 Bar Total Pressure, 60°C, Heptane Suspension, 90 Min., Cocatalyst Complexed with Ethyl Anisate)

Complexation Ratio Al/Ethyl Anisate	Productivity g PP/g Catalyst	Isotactic Index (% Insoluble in Heptane)
6	710	86.7
4	550	92.2
3	300	93.5

Much information can be directly or indirectly deduced from the comparison of the two A.A. examples given in this paper.

The first result is a confirmation that diffusion control does not limit the polymerization rate: even at high polymerization activity. No decrease in polymerization rate is observed when the size of the polymer particle becomes very large and under "steady" conditions, the molecular weight distribution seems to be larger than in the "unsteady" conditions of polymerization cocatalyzed by TEA.

The three following experiments will be used to assess some aspects of the polymerization mechanism:

Dilution of Reaction Cocatalyzed by TEA

At very low A.A. concentration, no or almost no activity is observed. For instance, in the case at 0.5 mmoles TEA/liter at 60-65°C and 4 bar pressure. Figure 7 illustrates the effect of the dilution of the reaction medium by pure heptane, starting from 5 mmoles/l and reaching 0.5 mmoles/l. The activity is destroyed after a short time and one can predict that the delay in activity loss is the time needed to extract TEA from the polymer particle. This experiment suggests that the active centers, not only need the A.A. to be activated, but also react with a structure comprising TEA in a reversible complexation equilibrium with the active titanium species. The active centers are bimetallic. Bimetallic models for Ziegler catalyst active sites were proposed previously, but seldom with really sound arguments in the case of heterogeneous systems, unlike the homogeneous one.[8] Rodriguez and Van Looy's model applies to first generation catalysts[9] and Keii et al showed that the active centers work with the last cocatalyst with which the catalyst is contacted when the preceding cocatalysts are eliminated by washing.[10]

TEA Exchange

Adding IPRA to a TEA-cocatalyzed polymerization converts the kinetics into the IPRA shape within a rather short time (Figure 8). The activity

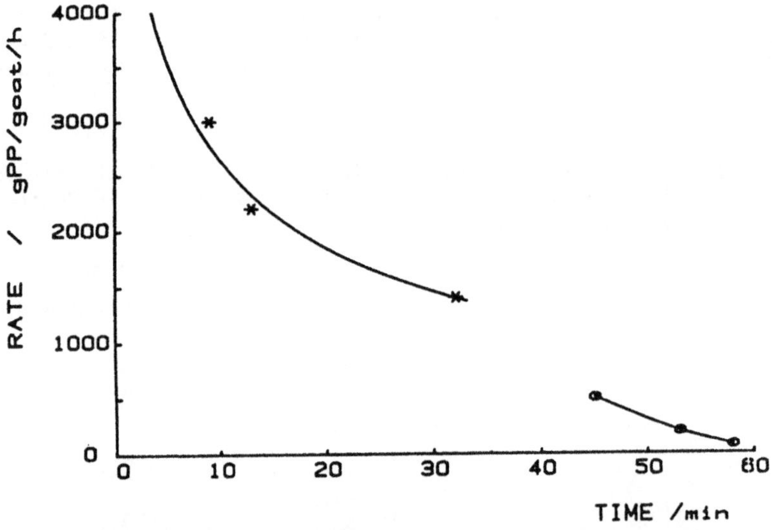

Fig. 7. Propylene Polymerization Kinetic - 63° C, 4 bars
Dilution at 40 minutes
* before dilution - TEA = 5 mmol/l Al/EPT = 4
0 after dilution - TEA = 0.5 mmol/l Al/EPT = 4

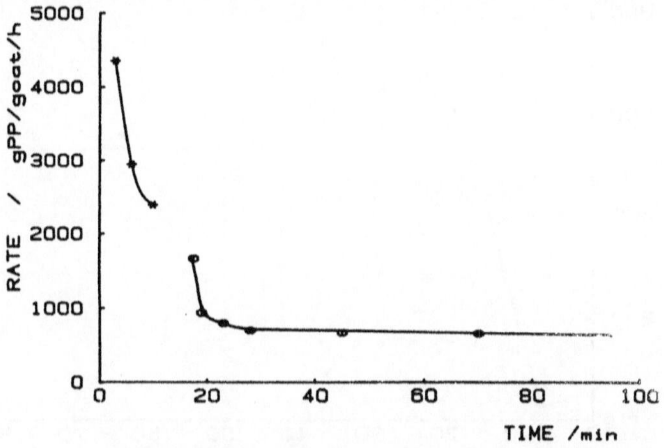

Fig. 8. Propylene Polymerization Kinetic - 60°C, 4 bars
Competition between two cocatalysts
* TEA = 5 mmol/l Al/EPT = 4
0 IPRA = 10 mmol Al/l added at 15 min. Al/EPT = 4

remains constant and the TEA left in the medium seems to be ignored by the catalyst.

Dilution Experiment with IPRA

IPRA cocatalyzed polymerization activity is much more sensitive to aluminum concentration than TEA (Table 2).

The dilution experiment shown in Figure 9 is very different from the last A.A.: after the heptane dilution, a constant activity is recovered, but a detailed study of this last reaction is outside the scope of this paper.[11]

An interpretation of the three experiments: dilution effect with TEA, no dilution effect with IPRA, domination of IPRA together with TEA can be summarized in the following way:

i) the active center is a bimetallic titanium-aluminum complex in reversible equilibrium with free A.A. in the solution

ii) with TEA, the exchange between free and complexed aluminum is very fast

iii) in the case of IPRA, the exchange is rather slow. This is probably due to its bridged structure: IPRA has a low mobility and is desorbed only when all links with the catalyst surface are broken at the same time (no dilution effect)

iv) in a competition between a slow and a fast desorbing species, the slow desorbing one fills all the adsorption sites after a time: this explains why IRPA hides TEA when present by displacement of the Ti-TEA complex.

More information can be deduced from the former conclusions concerning some aspects of polymerization.

i) retardation of the beginning of polymerization in the case of IPRA: it is generally admitted that the first step of the active site for-

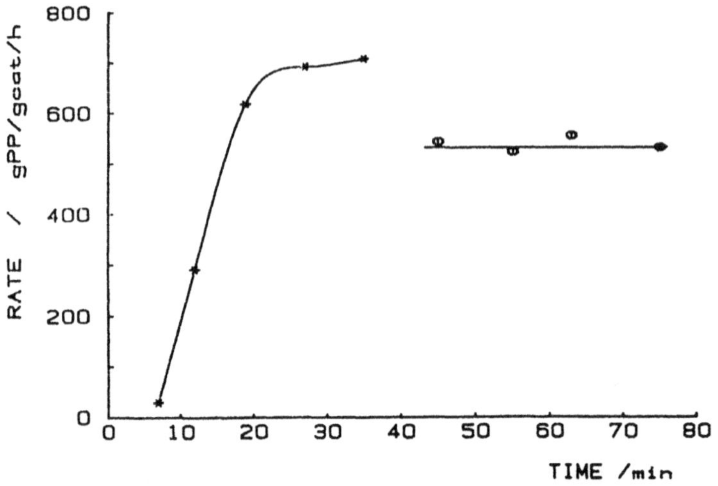

Fig. 9. Propylene Polymerization Kinetic - 60°C, 4 bars - Dilution at 40 minutes; cocatalyst IPRA complexed by ethylbenzoate
* 10 mmol Al/l (Al/EB = 4)
0 1 mmol Al/l

mation implies a reaction between the titanium chloride and A.A., probably leading to a reduction of Ti(IV) to Ti(III). In the Cossee-type models, a chlorinated A.A. must be desorbed from the active center before it is able to work with high activity. We used a chlorinated derivative of IPRA as a model and observed a very low activity in propene polymerization. Due to

Table 2. Effect of A.A. Concentration

TEA[a] mmole/l	Productivity g/PP/g catalyst	Isotactic Index (HI)
1	1730	94.3
3	1850	95.2
5	1760	93.7
10	1570	92.4
IPRA[b]		
3	160	87.4
10	450	80.7
20	490	79.2

[a]polymerization conditions: 4 bars, 63°C, 60 min., cocatalyst complexed with ethyl paratoluate (EPT), Al/EPT = 4.

[b]4 bars, 60°C, 90 min., cocatalyst complexed with ethyl anisate Al/EA = 4.

Table 3. Effect of TEA Concentration on Productivity and Rate

[Al] mmole/l	Productivity g PP/g cat/h	Rate(g PP/g cat/h) at 2 min.	at 60 min.
10	1040	4200	310
5	1500	6450	650
3	1900	6900	780

the stability of the titanium-IPRA complexes the desorption of the chlorinated by-product of the active center formation requires much more time than with TEA and retards the beginning of polymerization.

ii) stability and decrease in the polymerization rate: the active polymerization centers are destroyed during polymerization by an ill-defined mechanism (independently from the polymerization itself) but depending on the nature and also on the A.A. concentration in the case of TEA, as described on Table 3). This irreversible process, involving A.A., should not be confused with the reversible poisoning observed with an aromatic ester for instance.[16] The irreversible deactivation is probably due to reduction or ligand exchange on titanium. This kind of side reaction can be expected to occur during the time of building or breaking of the titanium-aluminum complex, and not when a stable complex is built. So, the higher the frequency of the complex formation, the lower the stability of the polymerization rate. The difference between TEA and IPRA cocatalyzed reaction is easily explained in this way, and clearly related to the bimetallic character of the active centers. It is to be noted that the activity observed with IPRA is really constant only in the usual temperature range (60-65°C) at 70°C or higher, a slightly decreased activity is observed (Figure 10).

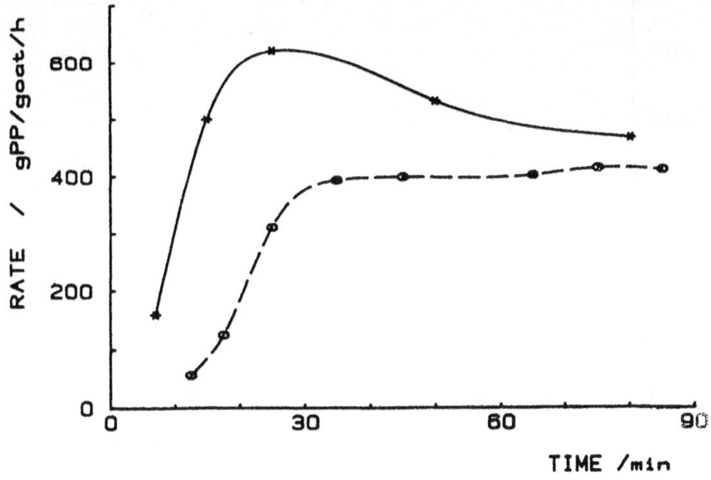

Fig. 10. Propylene Polymerization Kinetic - 4 bars - Temperature effect - Cocatalyst IPRA
* 70°C (Al) = 10 mmol/l Al/EPT = 3.5
0 60°C same conditions

iii) stereospecificity of the catalytic system: in the last part of this paper, we will consider results related to the control of isospecific polymerization. Isotactic polypropylene (up to 94 isotactic index or higher) can be obtained with IPRA as it is the case with TEA (Table 1) but the activity is much more sensitive to the Al/E.D. ratio with IPRA. This sensitivity can be related to the aromatic ester-A.A. complex formation equilibrium: with TEA, the equilibrium (followed by the two carbonyl bands of free and complexed ester on the IR spectrum) is displaced towards complex formation, and no free A.E. can be detected in the usual (2.5-4) Al/E.D. ratio.[12] With IPRA, free E.D. is observed even at a 4:1 ratio, so that the catalyst is poisoned at ratios under 4:1. The poor complexation is probably related to the IPRA structure: steric effect of bulky groups or reduced molar concentration for the same Al concentration. The fact that the isotactic index is not really higher than with TEA implies that free E.D. does not directly contribute to the isotactic control.

During polymerization, when E.D. is added to the system in the steady part of the reaction (Figure 11), activity is reversibly decreased and can be recovered by restoring the initial Al/E.D. ratio. The reversible poisoning by free E.D. was observed with the unsteady polymerization cocatalyzed by TEA.[1,11] It is to be noted that the decrease of the activity when E.D. is added is very fast (adsorption of free E.D. on the active titanium) and recovering of the activity is slower (desorption of E.D.).

During polymerization with IPRA, as it is the case with TEA,[11] the isotactic index continuously decreases (Table 4). The contribution of each step of polymerization to the isotactic index of the polymer can be recomputed and in the case described on Table 4, the stereospecificity of the system is certainly very low at the end of the reaction. The activity being constant, the loss in stereoselectivity cannot be associated with a different stability of iso- and non-isospecific active centers. It must be associated with a change in the cocatalytic solution. It is related to the destruction of the A.A.-A.E. complex through an intermolecular reduction mechanism[12,13] which is favored in the case of IPRA by the presence of a second aluminum atom.

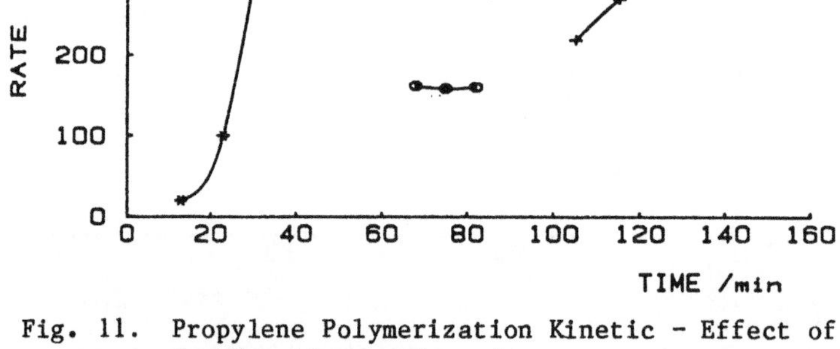

Fig. 11. Propylene Polymerization Kinetic - Effect of Complexation ratio:
* Al/ethylanisate = 4
0 ethylanisate added at 60 min. Al/EA = 3
\+ IPRA added at 95 min. Al/EA = 4

Table 4. Variation of Isotacticity During Polymerization

(cocatalyst: IPRA 60°C, 4 bars)

Duration (min.)	P gPP/gcat.	Isotactic Index	Intermediate Isotactic Index
90	550	92.3	92.2 (0 ⟶ 90)
270	2000	77.3	71.6 (90 ⟶ 270)

CONCLUSION

The use of IPRA as a cocatalytic component in propene polymerization with the classical supported Ziegler catalyst gives considerable information on these systems: first, it establishes the possibility of stable polymerization activity without drastic loss in the isotactic control, and no diffusion control. In fact, some counterparts of the IPRA structure reduce the practical interest of the result. It can also be a tool for studying the active center structure and helps understand how the cocatalyst works in the system: it builds an active titanium aluminum complex in equilibrium with the A.A. solution. The complex is itself in equilibrium with the A.A.-A.E. complex formation. The origin of the isotactic control is discussed elsewhere[1,13] but the use of IPRA helps exclude number of proposed mechanisms including selective poisoning of aspecific titanium centers, or any difference in stability of the active centers and points out the effect of the cocatalytic solution on the isotactic control.

It is rather easy to obtain stable ethylene polymerization activity with the same catalysts and almost any A.A. The question of the existence of different active centers for ethylene and propylene polymerizations and the ability of these catalysts to make copolymers remains an important field of research.

ACKNOWLEDGEMENT

This work was supported by ATOCHEM and ELF.

REFERENCES

1. R. Spitz, J. L. Lacombe and A. Guyot, J. Polym. Sci., Polym. Chem. Ed., 22, 2641 (1984).
2. T. Keii, E. Suzuki, M. Tamura, and Y. Doi, "Transition metal catalyzed polymerization," MMI Symposium Ser., 4A, 97 (1983).
3. To be published.
4. N. Kashiwa and J. Yoshitake, Makromol. Chem., 185, 1133 (1984).
5. T. W. Taylor, K. Y. Choi, H. Yuan, and W. H. Ray, "Transition metal catalyzed polymerization," MMI Symposium Ser., 4A, 191 (1983).
6. R. Spitz, J. L. Lacombe, M. Primet, and A. Guyot, "Transition metal catalyzed polymerization," MMI Symposium Ser., 4A, 421 (1983); R. Spitz, J. L. Lacombe, and A. Guyot, J. Polym. Sci., Polym. Chem. Ed., 22, 2625 (1984).
7. H. Hoberg, H. Martin, R. Rienacker, K. Zosel et K. Ziegler, Brennstoff Chemie, 50, 217 (1969).
8. G. Henrici Olive and S. Olive, Makromol. Chem., 121, 70 (1969).

9. L. A. M. Rodriguez and M. M. Van Looy, J. Polym. Sci., Al(4), 19711 (1966).
10. T. Kohara, N. Shinoyama, Y. Doi, and T. Keii, Makromol. Chem., 180, 2235 (1979).
11. A. Guyot, R. Spitz, L. Duranel, and J. L. Lacombe, "Future Aspects of Olefin Polymerization," Symposium - Tokyo (July 1985).
12. R. Spitz, J. L. Lacombe, M. Primet, and A. Guyot, J. Polym. Sci., Polym. Chem. Ed., 22, 2611 (1984).
13. L. Duranel, Thesis Lyon (1984).

ALCOHOL-MODIFIED MAGNESIUM TITANIUM CATALYSTS

R. E. Hoff, T. J. Pullukat, and R. A. Dombro

Norchem, Inc.
Rolling Meadows, Illinois 60008

INTRODUCTION

Reaction products of silica with solutions of dibutylmagnesium compounds can react with titanium tetrachloride to form free-flowing solids. When prepared in proper ratios, these solids are highly active as ethylene polymerization catalysts.[1] Active sites are bonded to the silica, and in the slurry synthesis of polyethylene, polymer grows associated with silica particles giving control over replication.

There is a combination of high activity and particle size control which makes such catalysts appear attractive for commercial use, especially in the energy-efficient Phillips particle form process. High density polyethylene, with a narrow molecular weight distribution, can be synthesized in this particle form process.[2]

The silica supported magnesium-titanium catalysts are fundamentally different from the thermally activated chromium catalysts. With chromium catalysts, the polymer molecular weight and molecular weight distribution can be significantly changed by variation in the pore size of the silica. In contrast, with the silica-magnesium-titanium catalysts, there is little effect with changes in silica micro-geometry.[3]

Another distinction is in the effect of polymerization temperature on molecular weight. Typical differences are shown in Figure 1 where melt index has been used as an indication of molecular weight. It is clear that changing synthesis temperature is a more effective MW control means with chromium catalysts.

Hydrogen is a more efficient chain transfer agent with silica-magnesium-titanium catalysts. Furthermore, it has been discovered that the incorporation of alcohols alters the chemical structure of such catalysts, so that at the same hydrogen concentration, lower molecular weight polyethylene is formed.[4,5] The synthesis of high density polyethylene with a melt index of 60 or higher can be achieved within the temperature limits of the process. At the same time, the desirable narrow molecular weight distribution is retained.

It is known that various alcohols are effective modifiers. Apparently, the most important interaction is between the active sites and hydroxyl groups with the alcohol carbon structure being of secondary importance.

This report is a study of the effect of hydrogen on ethylene polymerization by the silica-magnesium-titanium catalysts. A comparison is made between alcohol-modified and alcohol-free catalysts. Valence state analysis and EPR spectra of the catalysts are considered in connection with reactivities and molecular weight.

EXPERIMENTAL

Materials

Nitrogen was vaporized from a liquid reservoir, and passed through successive columns of molecular sieves, hydrogen-activated copper catalyst, and molecular sieves.

The silica grade which was used for catalyst synthesis had a surface area of 300 m^2/g, a pore volume of 1.65 cm^3/g and an average particle size of 1.40×10^{-2} cm. It was heated under nitrogen fluidization at 400°C for 5 hours and then kept under nitrogen.

Heptane was Phillips pure grade. It was sparged with N_2 for about 16 hours then passed through a column of 13X molecular sieves into a storage container. A layer of additional 13X molecular sieve pellets was kept in the storage container and the container was blanketed with N_2.

A solution of dibutylmagnesium-triethylaluminum complex in heptane was obtained from Texas Alkyls. The Mg/Al ratio was 6.5.

ACS reagent grade 1-butanol, obtained from a laboratory supply house, was transferred to an oven-dried bottle, purged for two hours by N_2

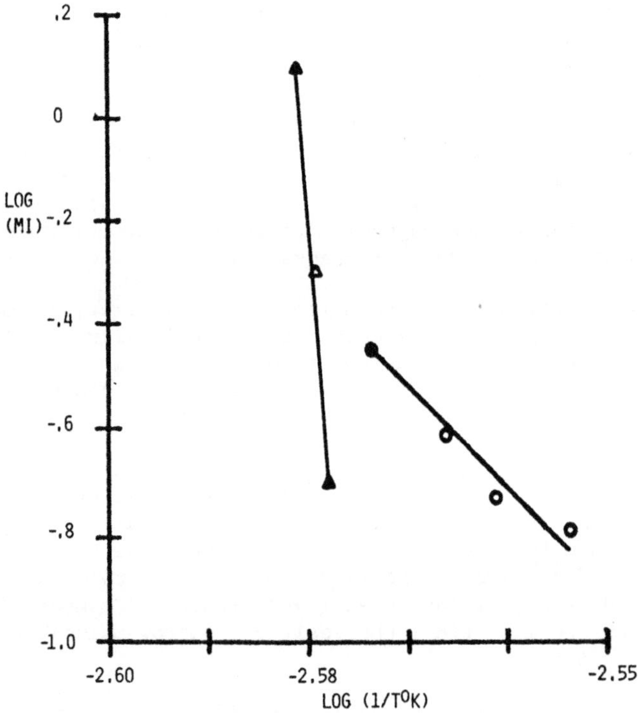

Fig. 1. A Comparison of Polymerization Temperature Effect on MI

bubbling, and stored under N_2 over a layer of freshly activated molecular sieves.

Titanium tetrachloride, 99.9%, was transferred to an oven-dried bottle inside an inert atmosphere box. The cover of the bottle was fitted with a tee and ball valve so that long needles could be inserted without admitting room air. The titanium tetrachloride in this bottle was bubbled with N_2 for two hours before use in catalyst synthesis.

Trisobutyl aluminum (TIBAL) from Texas Alkyls was used as the cocatalyst in all polymerization experiments.

Ethylene and isobutane for the polymerization experiments were acquired from Norchem's high density polyethylene plant. These were passed through various columns for removal of trace amounts of impurities.[6]

Linde ultra high purity grade hydrogen was passed through columns of molecular sieves, an Englehard Deoxo catalyst, and molecular sieves again.

The reagents for the trivalent titanium analysis were purchased from laboratory supply companies. Standard ceric sulfate solution was supplied by G. Frederick Smith Chemical Company. 1,10-phenanthroline ferrous sulfate complex as a 0.025 solution was the indicator.

Catalyst Synthesis

The catalysts were synthesized in standard 3-neck round bottom flasks. The left and right joints were fitted, respectively, with a N_2 source regulated to 5 psig and a vent to a mineral oil bubbler. The vent was also used to add the ingredients and to remove the finished catalyst. A Teflon paddle, attached to a glass shaft, was used as a mixer. All of the glassware was dried in an oven at 120°C at least overnight before use. It was assembled while hot and purged with N_2 while cooling to room temperature.

The desired amount of dried silica was transferred to the flask. A 10 ml volume of heptane was added by syringe for each gram of silica. This amount of solvent allows for easy slurring of the silica. A volume of the dibutylmagnesium complex solution was then added. The mixture was then stirred for 30 minutes at room temperature. Normally, if an alcohol was added it was introduced at this stage. Then after an additional 30 minutes the titanium tetrachloride was added.

If an alcohol was not added, the titanium tetrachloride was syringed into the reaction mixture after the first 30 minutes. In one catalyst synthesis 1-butanol was added after the titanium tetrachloride. After the final ingredient the catalyst preparation was stirred for another 30 minutes at room temperature. The flask was then heated in an oil bath at 95 - 100°C for about an hour under flowing N_2. The solvent evaporated leaving the finished catalyst as a dry granular material.

The amount of dibutylmagnesium complex was such that there was 1.75 millimoles of Bu_2Mg per gram of silica and the Mg/Ti ratio was 1.0. The amount of 1-butanol was varied.

Polymerization Tests

The polymerization reactor system has been described.[7] A quantity of the catalyst was transferred to the stainless steel vessel which was under N_2 purge and heated at the desired temperature. The trisobutyl aluminum cocatalyst was added so that the TIBAL/Ti ratio was 5.0. The vessel was closed and 500 ml isobutane, hydrogen as required, and ethylene to maintain the operating pressure were admitted. Hydrogen was added only at the start of the reaction but ethylene was added as needed to keep the pressure constant during polymerization. Ethylene flow was measured by a differential pressure device. The polymerization temperature was maintained $\pm 1°C$ of the set temperature by adjusting the pressure above refluxing methanol in the reactor jacket.

Trivalent Titanium Analysis

A titration cell was prepared from a bottle with a special screw cap lid (Thomas Scientific catalog number 9388-R30 modified). The design was such that the cell could be kept free of air by a gentle flow of nitrogen.

Catalyst samples could be added without exposure. After addition of the catalyst sample 10 ml of oxygen-free isopropyl alcohol was added to deactivate the organometallic species. The combination of alcohol and catalyst sample was stirred with a magnet bar while first 25 ml of 2N hydrochloric acid solution and then 25 ml of 0.1 N ferric ammonium sulfate in 2N hydrochloric acid were added. About 0.5 ml of 0-phenanthroline indicator solution was added, then the amount of divalent iron present was determined by titration with 0.10 N ceric solution. The 0-phenanthroline indicator color change was sharp and from orange to green in the presence of excess ferric ions.

The catalyst sample size was from 0.5 to 1.0 g.

The method of titration was evaluated with samples of hydrogen-reduced $TiCl_3$ and found to give excellent accuracy. The first color change that lasts about 20 seconds is the correct end point. Unless the ceric solution is in substantial excess the green color reverts to orange in a short time.

This method is a modification of that used by Carrick and coworkers.[7] The first modification was the initial addition of isopropyl alcohol. This was done to avoid the violent reaction which occurs when water solutions contact this type of catalyst. Addition of alcohol for this purpose is described by Lehr.[8]

The second modification was the substitution of ceric sulfate for potassium permanganate. This substitution made an indicator necessary. Ceric oxidant was adopted because of its broader suitability than permanganate.[9] The 0-phenanthroline complex (Ferroin) color change occurs from 1.2 to 1.3 volts (vs the normal hydrogen electrode) which is at the equivalence point in Fe^{+2} - Ce^{+4} titrations.

Electron Paramagnetic Resonance

The catalyst samples were sealed into 4 mm diameter quartz tubes as dry solids. A Varian E-9 spectrometer was used. Strong pitch in the dual cavity provided a reference for field calibration and also for the estimation of signal intensity.

Melt Index and GPC

The melt index values were determined by ASTM D-1238-52T. Samples were stabilized with 0.5 weight percent 2,6-di-t-butyl-p-cresol (BHT).

RESULTS AND DISCUSSION

Table 1 shows the effect of alcohols on catalyzed ethylene polymerization. Table 2 shows the time average reactivities and melt indices for

Table 1. Ethylene Polymerization With Various Alcohol-Modified Silica Magnesium-Titanium-Catalysts[a]

Alcohol	ROH/Mg	Polymerization Temperature °C	H_2 Partial Pressure (psi)	Reactivity g/g Catalyst	Melt Index
None	----	102	0	11000	<0.00
None	----	102	50	5600	0.58
None	----	102	100	5300	2.92
Benzyl	0.50	102	50	2200	1.78
Benzyl	0.50	102	100	1555	10.6
Cyclohexyl	0.50	102	50	3600	1.87
Cyclohexyl	0.50	102	100	1100	10.9
t-Butyl	0.50	102	50	4196	0.95
t-Butyl	0.50	102	100	2490	8.13
n-Butyl	0.50	105	50	3200	1.44
n-Butyl	0.50	105	100	1290	24.7

[a]Data from Ref (4)

ethylene polymerization with a silica-supported magnesium catalyst. It also contains these values for a catalyst synthesized identically except for the addition of 1-butanol. In the catalyst preparation, the added 1-butanol was in equal molar ratio to the added dibutylmagnesium.

At the various partial pressures of hydrogen, the reactivities of the two catalysts were similar. The data are plotted in Figure 2. At each hydrogen pressure, the reactivities may either be considered essentially the same, or the butanol-modified catalyst may be slightly more reactive. Obviously, each catalyst type diminishes in reactivity with increasing hydrogen, although the ethylene partial pressure was kept constant.

The decrease in reactivity is consistent with the literature for early Ziegler catalysts and for high activity catalysts containing magnesium. It is true that there are reports of rate increases due to hydrogen. But these have been for propylene,[10,11] 1-butene,[12] and 4-methyl pentene-1.[13] In the case of ethylene polymerization or copolymerization, Schindler,[14] for $TiCl_4$ plus $isoBu_2AlH$, and Bohm,[15] for $TiCl_4/MgCl_2/Et_3Al$, observed decreases in rates. Guastalla and Gianini[16] reports rate increases for propylene but decreases for ethylene.

Figure 3 displays a typical rate curve of ethylene flow for the experiments in Table 2. At the different hydrogen levels, the curve shifts up or down but the shape does not significantly change. In the polymerization test procedure, the catalyst is in contact with ethylene in the 2 to 3 minute period of pressure build up. During this short time flow rates cannot be measured. However, the polymer made during this short time is a negligible contribution to yield after about 30 minutes and in any case, there is no reason to suspect disproportionate flow rate during this blind period. Because of the similarity in flow curves it is believed that the time average reactivities of Table 2 are adequate for comparison rates.

Table 2. A Comparison of the Effect of Hydrogen on Ethylene Polymerization

1-BuOH/Mg	Hydrogen Partial Pressure psi	Reactivity[a]	Melt Index	High Load Melt Index
0	0	1685	0.00	0.00
0	25	1100	0.011	0.56
0	50	869	0.091	4.68
0	100	515	0.60	30.2
1.0	0	1863	0.00	0.13
1.0	25	1141	0.31	14.6
1.0	50	900	0.81	35.8
1.0	75	699	3.77	166
1.0	100	614	13.9	NA

[Conditions: 102°C, Ethylene Partial Pressure 165 psi, Titanium Concentration 4.2 to 12 mg/L, TIBAL cocatalyst at Al/Ti = 5.0, 500 ml isobutane, 40 minute reaction time]

[a]yield/(g-cat-hr) calculated from 40 minute reactions.

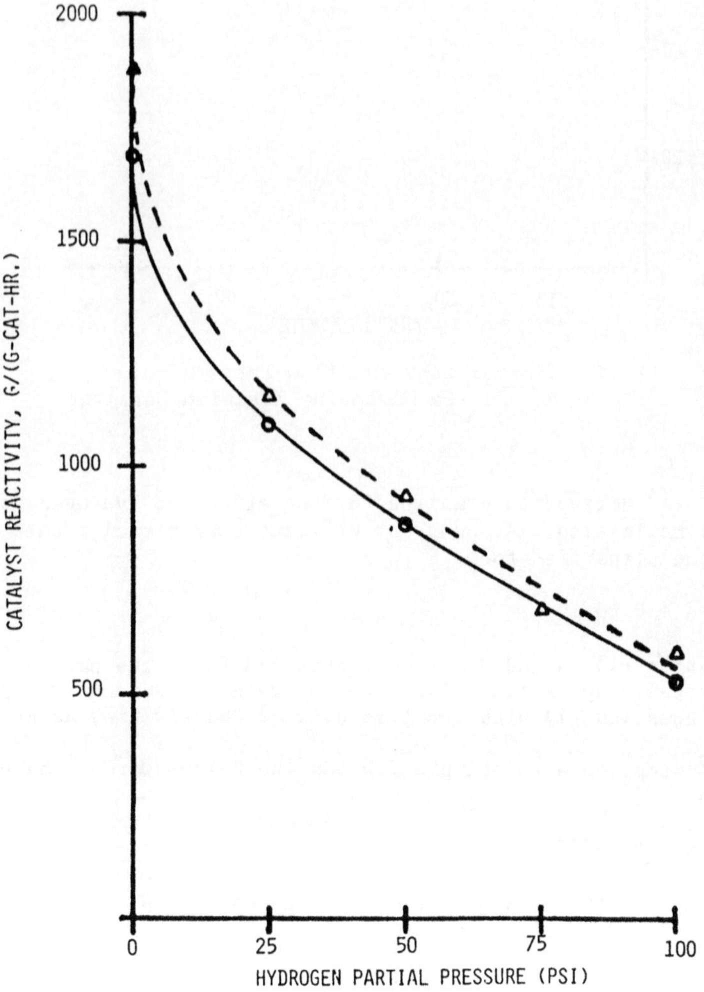

Fig. 2. Change in Catalyst Reactivity with Hydrogen Partial Pressure

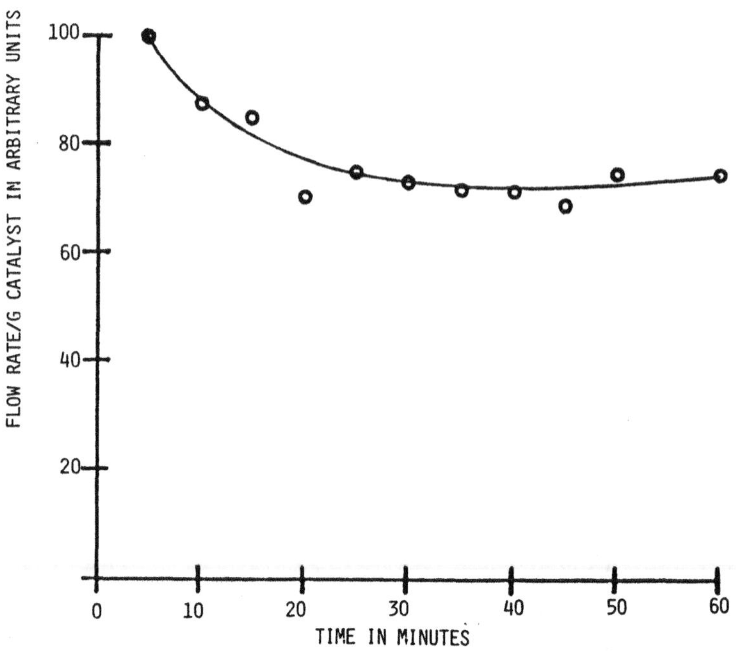

Fig. 3. Typical Ethylene Flow Pattern For Silica-Magnesium Titanium Catalyst

Schlinder[14] derived an equation for the effect of hydrogen in ethylene polymerization. At constant ethylene concentration this equation reduces to the simplified form:

$$1/Rp = A + B\ P_H \tag{1}$$

In Equation (1), A and B are constants and P_H is the partial pressure of hydrogen. Rp is the polymerization rate. Figure 4 is a plot in the form of Equation (1) with the time average reactivities as Rp values.

Natta[17] proposed a relation which has the half-power of the hydrogen pressure.

$$1/Rp = A + B\ P_H^{1/2} \tag{2}$$

If the average reactivities of Table 2 are taken for Rp, then linear regression provides the following:

Catalyst	Constants* A	Constants* B	Type	Correlation Coefficient
Alcohol-Free	5.62	0.134	P_H	0.995
Alcohol-Free	4.49	1.217	$P_H^{1/2}$	0.927
With Butanol	5.69	0.109	P_H	0.997
With Butanol	4.53	1.08	$P_H^{1/2}$	0.971

*Times 10^4

According to the correlation coefficients, the equation with P_H gives a better fit for each catalyst. Keii also favors his equation for ethylene polymerization.[18]

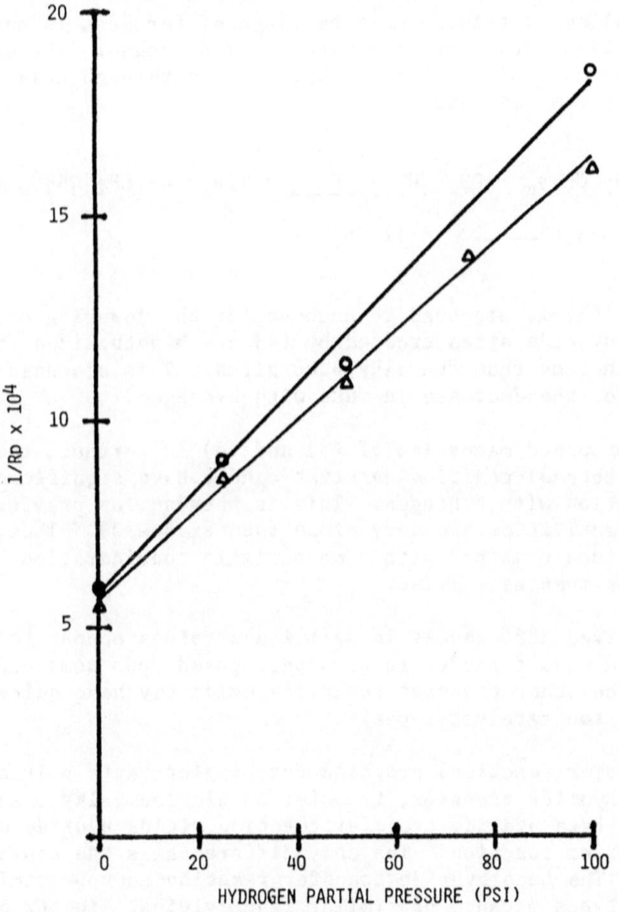

Fig. 4. Dependence of Reciprocal Rate on H_2 Partial Pressure

Whether the proper equation involves P_H, that is molecular hydrogen, or $P_H^{1/2}$, atomic hydrogen, the constants for each catalyst are similar. In particular, the B values are similar for the two catalyst types, with the alcohol modified catalyst having the lower values. This supports direct observation that the alcohol modified catalyst suffers a similar, or slightly less, loss in reactivity with hydrogen increase. This point is developed further in connection with molecular weight.

GPC of polymer samples made with the alcohol free catalyst indicated that the molecular weight distribution was only slightly affected by hydrogen pressure changes.[19] When the molecular weight distribution is the same, melt index changes are an approximately correct indication of molecular weight changes. In a simple approach the melt index is inversely proportional to the polymer viscosity, and the log of the viscosity is in linear relation to the log of some average molecular weight.[20]

In the absence of hydrogen, the polymers which were made had melt index values too low to be measured, except for the HLMI of the polymer made with the alcohol-modified catalyst. Also, the HLMI/MI ratios for each catalyst in Table 2 were essentially constant with changes in hydrogen pressure. The HLMI values are, therefore, as good an indication of molecular weight changes as the MI values. The butanol-modified produces lower molecular weight polymer than the other catalyst in the absence of hydrogen as well as at various hydrogen concentrations.

This implies that there must be large differences in one or more other transfer reactions which are independent of hydrogen. The mechanism for the control of molecular weight by hydrogen is through transfer reactions often cited in the literature.[17]

$$* \ Ti-(CH_2 \ CH_2)_m - CH_2 \ CH_3 \xrightarrow{H2} * \ TiH + H-(CH_2 \ CH_2)_m \ CH_2 \ CH_3 \quad (3)$$

$$* \ TiH + C_2H_4 \longrightarrow * \ Ti \ CH_2CH_3 \quad (4)$$

Reaction (3) is proposed to account for the lowering of molecular weight. The hydride sites created by (4) are hypothesized to be slower to react with ethylene than the alkylated sites. This slowness then is the explanation for the decrease in rate with hydrogen.

If the proposed mechanism of (3) and (4) is correct, then the alcohol-free and the butanol-modified catalyst cannot have significantly different rates of reaction with hydrogen. This is because, as previously pointed out, their reactivities are very close (see Figure 3). Thus, from a reactivity comparison combined with a mechanistic consideration, the hydrogen transfer rates must be similar.

The observed differences in melt index values appear to be the result of similar hydrogen transfer rates superimposed upon some other transfer reactions. The other transfer reactions evidently have quite different rates for the two catalyst types.

The transfer reactions proposed for Ziegler-Natta polymerization are usually beta-hydride transfer, transfer to aluminum alkyl, and transfer to monomer. The beta hydride transfer reaction yields hydride sites like the hydrogen transfer reaction. The only difference is the source of the hydride ion. The beta hydride transfer reaction must be similar for the two catalyst types because of similar reactivities. In our experiments with supported magnesium-titanium catalysts no consistent trend on MI values has been found with changes in TIBAL concentration or TIBAL/Ti ratios. It is unlikely that transfer to aluminum alkyl is important for these catalysts. Only transfer to monomer remains as explanation for the difference in molecular weights in the absence of hydrogen.

Additional experiments were done with catalysts of different 1-butanol contents and, for one catalyst, a different order of addition. Some data from this set appear in Table 3.

The polymerization tests in Table 3 were done at an ethylene partial pressure of 215 psi instead of the 165 psi of Table 2. Higher reactivities in Table 3 are due to increased ethylene concentration.

The catalysts were prepared with the same silica and the same 1.75 millimoles of Bu_2 Mg per gram. The Mg/Ti ratio in each was 1.0. Only the 1-butanol was varied. The melt index increased up to BuOH/Mg ratio of 1.0 then decreased at the 2.0 ratio. The reactivity also changed in a similar pattern.

The trivalent titanium contents percent of total titanium are listed in Table 3. The remainder of the titanium is tetravalent. Divalent titanium has not been detected in catalysts of this type.

The maximum reactivity and maximum melt index occur with 43% Ti^{+3} and 57% Ti^{+4}. Neither Ti^{+3} nor Ti^{+4} correlate with melt index or reactivity.

Table 3. Ethylene Polymerization with 1-Butanol Modified Silica-Magnesium-Titanium Catalysts

BuOH/Mg	Reactivity[a]	Melt Index	HLMI/MI[b]	Percent Ti^{+3}[c]
0	920	0.08	45.5	94
.5	902	0.22	39.4	76
1.0	1475	0.67	34.0	43
2.0	1219	0.42	31.8	3.2
1.0[d]	1147	.43	32.5	86

Conditions: 102°C, Ethylene partial pressure 215 psi, hydrogen partial pressure 50 psi, titanium concentration 4.2 to 12 mg/L, TIBAL cocatalyst at Al/Ti = 5.0, 500 ml isobutane, 60 minute reaction time

[a]Yield/(g cat.-hr)
[b]High load melt index/melt index
[c]Weight percent Ti^{+3} relative to total titanium
[d]In this preparation, the 1-butanol was added after the titanium tetrachloride

The last catalyst in Table 3 is the same as would be obtained by adding 1-butanol to the first catalyst at the ratio to Bu_2Mg of 1.0. Obviously, this is not the equivalent to adding butanol before the titanium chloride. Melt index, reactivity, and content of Ti^{+3} are all different.

Table 4. EPR g-Value for Alcohol-Modified Silica-Magnesium-Titanium Catalysts

1-BuOH/Mg	Temperature	g-Values
0	RT	1.948
0	77°K	1.943
0.5	RT	1.96, 1.945, 1.904
0.5	77°K	1.94, 1.907
1.0	RT	1.97, 1.945, 1.907
1.0	77°K	1.97, 1.95, 1.89
2.0	RT	2.001, 1.967, 1.948, 1.87
1.0*	RT	1.944
1.09*	77°K	1.945

*Reversed addition

EPR spectra were obtained for the purpose of studying the effect of the increasing amounts of 1-butanol on the titanium ligand field. Table 4 lists g-values which were found.

Chien and Wu[21] have studied the EPR of a titanium chloride-magnesium chloride composition which also contained ethyl benzoate, p-cresol, and triethyl aluminum reaction products. They found two g values for this catalyst from its 25°C spectrum, 1.945 and 1.913. There were no changes at 77°K except for an intensity increase. Most of the intensity increase could be accounted for by the Curie law.

Room temperature EPR spectra of the alcohol-free catalyst of Table 3, and the catalyst with 1-BuOH/Mg of 1.0 appear in Figure 5. The alcohol-free catalyst has only a single absorption with g = 1.948. Its spectrum does not change at 77°K except for the expected increase in intensity. The 1-butanol catalyst has three g values: 1.97, 1.945, and 1.907. At 77 K the g values are 1.97, 1.95, and 1.89. The spectrum shape is also different at low temperature. If the 1.907 corresponds to 1.89, then this absorption is narrower at 77°K.

The ~1.95 g absorption in the catalysts of this work probably correlates with the 1.945 g of Chien. This absorption does not change in spectra at room temperature versus liquid nitrogen temperature as indicated by Chien. However, the alcohol-modified catalysts have spectra with an absorption of lower g and which narrows at 77°K. This peak is also shifted to lower g at BuOH/Mg of 2.0 relative to the ratio of 1.0.

The 1.96 value is detectable only as an assymetry with BuOH/Mg of 0.50, but it is a definite shoulder on the side of the 1.945 peak at the ratio of 1.0. At the ratio of 2.0, the 1.967 g appears to be the most intense trivalent titanium signal. The g value of 2.001 corresponds to an extremely narrow peak which is apparently due to a hydrocarbon radical.

The EPR spectral intensity has not been determined for the catalysts of this work. However, Chien[21] found that only about 20 percent of the trivalent titanium was EPR observable. It is reasonable to assume that in the silica-supported catalysts of this work there is also a large fraction of Ti^{+3} which does not give an EPR signal. However, at the same time, it is also reasonable to assume that alcohol affects all of the trivalent titanium in at least an analogous way.

It appears that in the alcohol-free catalyst there is only a single type of trivalent titanium which has an EPR signal. There is no difference in the EPR shape between room temperature and 77°K. Evidently, this trivalent titanium has restricted rotation due to bonding to the silica-dibutylmagnesium reaction product.

In the alcohol-modified catalysts, there is a change in the EPR spectra shape with temperature. The species having the g value of 1.945, which does not change with temperature, is still present. However, other portions of the spectra change. One interpretation is that there is at least one other form of trivalent titanium which has more freedom of rotation.

CONCLUSION

The reactivities of the alcohol-free silica-magnesium catalyst and the butanol-modified catalyst decrease with hydrogen partial pressure as shown by Schindler,[14] and Keii.[18] The butanol-modified catalyst, however, which gives the lower molecular weight at each hydrogen level, has reactivities

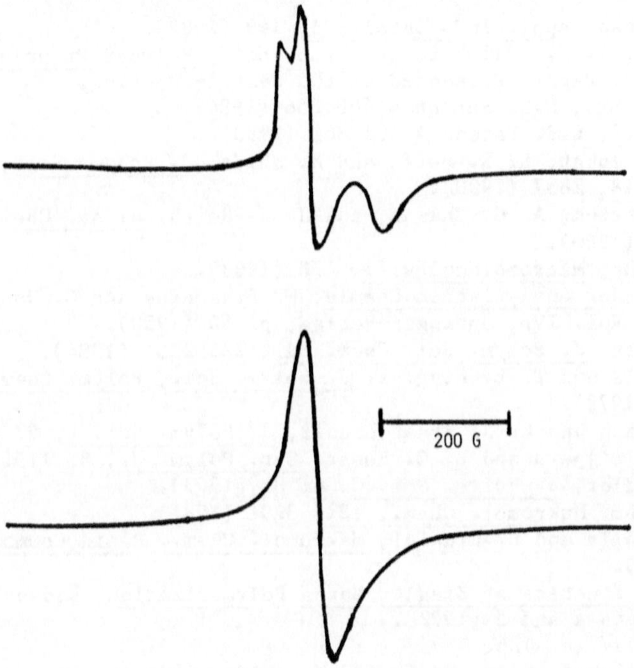

Fig. 5. EPR Spectra of Catalysts at 25°C

slightly greater than the alcohol-free catalyst. According to the proposed mechanism for molecular weight control and reactivity decrease, there can be similar reactivities only if the steady state concentration of hydride sites is about the same. Hence, without compensating effects for which there is no evidence, the rates of the reaction of growing sites with hydrogen must be similar for the two catalyst sites. The lower molecular weights from the alcohol-modified catalysts must be due to differences in other transfer reactions which are independent of hydrogen or hydride formation. The differences in high load melt index in the absence of hydrogen should be noted in this connection.

Since the other transfer reactions must not involve hydride formation, the most likely process is transfer to monomer. The EPR spectra have not been quantitatively analyzed. However, they indicate that the alcohol-modified catalysts have some trivalent titanium which has more rotational freedom. Since 1-butanol was added in molar ratios to dibutylmagnesium of 0.5, 1.0, and 2.0, it is probable that alcohol affected all of the titanium whether EPR active or not.

Thus, it must also affect the titanium which is in the active sites. The trivalent titanium analyses showed that valence state, that is Ti^{+3} or Ti^{+4}, did account for changes in molecular weight or reactivity. We suggest that transfer to monomer is facilitated by the greater rotational freedom which results from the alcohol modifications. β -hydride transfer and hydrogen transfer reactions are not significantly changed.

REFERENCES

1. M. Shida, T. J. Pullukat, and R. E. Hoff, U.S. Patent 4,263,102 (1981).

2. J. P. Hogan, Appl. Ind. Catal., 1, 149 (1983).
3. M. Shida, T. J. Pullukat, and R. E. Hoff, Polymer Preprints, 24, 110 (1983), Papers Presented at the Seattle Meeting.
4. R. A. Dombro, U.S. Patent 4,458,056 (1984).
5. R. E. Hoff, U.S. Patent 4,402,861 (1983).
6. T. J. Pullukat, R. E. Hoff, and M. Shida, J. Polym. Sci., Polym. Chem. Ed., 18, 2857 (1980).
7. W. L. Carrick, A. G. Chasar, and J. J. Smith, J. Am. Chem. Soc., 82, 5319 (1960).
8. M. H. Lehr, Macromolecules, 1, 178 (1968).
9. Handbuch der Analytischen Chemie, W. Fresenius and G. Jandar, Third Part, Vol. IVb, Springer-Verlag, p. 50 (1950).
10. J. F. Ross, J. Polym. Sci. Chem. Ed., 22, 2255 (1984).
11. V. W. Buls and T. L. Higgins, J. Polym. Sci., Polym. Chem. Ed., 11, 925 (1973).
12. C. D. Mason and R. J. Schaffhauser, J. Polym. Sci. B, 9, 661 (1971).
13. E. M. J. Pijpers and B. C. Roest, Eur. Polym. J., 8, 1151 (1972).
14. A. Schindler, J. Polym. Sci. C, 4, 81 (1963).
15. L. L. Bohm, Makromol. Chem., 182, 3291 (1981).
16. G. Guastalla and U. Gianini, Makromol. Chem., Rapid Commun., 4, 519 (1983).
17. T. Keii, Kinetics of Ziegler-Natta Polymerization, Kodansha, Tokyo, Chapters 4 and 5 (1972).
18. See Ref. 17, p. 126.
19. T. J. Pullukat, R. E. Hoff, and M. Shida, 1985 ACS Central Regional Meeting in Akron, Ohio (June 1985).
20. R. A. V. Raff and K. W. Doak, Crystalline Olefin Polymers Part I, Interscience Publishers, p. 488 (1965).
21. J. C. W. Chien and J.-C. Wu, J. Polym. Sci., Polym. Chem. Ed., 20, 2461 (1982).

POLYMERIZATION OF OLEFINS WITH MAGNESIUM CHLORIDE-SUPPORTED CATALYSTS

James C. W. Chien

Department of Polymer Science and Engineering
Department of Chemistry
University of Massachusetts, Amherst, MA 01003

INTRODUCTION

There is a resurgence of interest in Ziegler-Natta coordination catalysis of olefin polymerization owing to the development of $MgCl_2$-supported high mileage catalysts. The research activities have intensified, as witnessed by several symposia dealing, either in part or entirely, with the subject. This volume is the result of one such international symposium. Review articles have also appeared. For instance, a historical perspective of stereospecific polymerization of propylene by Pino and Mulhaupt,[50] discussions on the advances in the manufacturing of polypropylene by Galli et al.,[28,29] and surveys on patent and scientific literature by Dumas and Hsu[26] and Choi and Ray.[18]

The $MgCl_2$ supported catalyst family has many members. This is because $MgCl_2$ can be synthesized by different reactions and there is a diversity of modifiers and promoters. Also the order of administration of them can significantly alter the performance of the catalyst. Sometimes, the polymerization behavior of $MgCl_2$-supported catalysts can differ substantially. Unfortunately, comparison of published results is not made easier by brevity of the reports which often contain insufficient detail description of how a catalyst is prepared.

There are two reasonable approaches to study catalysts of great complexity. The first would be to begin with the minimal system, for instance $MgCl_2/TiCl_4$-$AlEt_3$, and study the effect of each additive in various combinations or permutations. This approach is probably the who which led scientists at Montedison and Mitsui Petrochemicals to the discovery of the optimum catalyst for the manufacturing of polypropylene. However, to take this route and to merely observe the cause effect relationship between catalyst modification and performance would not contribute toward basic understanding of the catalysts.

In contrast, if one is to fully characterize each catalyst modification, the effort required would be far beyond the limited resources available to an academic research group.

The second approach is to concentrate on an optimized system. This entails a more modest undertaking and we have taken this course of research. We have studied the patent literature and distilled from them a particular protocol[13] for the preparation of a catalyst $MgCl_2/EB/PC/AlEt_3/TiCl_4-AlEt_3/MPT$ (CW-catalyst) where EB, PC and MPT are abbreviations for ethylbenzoate, p-cresol, and methyl-p-toluate, respectively. This catalyst is believed to include most of the claimed innovations. We have omitted other claimed modifiers either because they are redundant or believed to have marginal significance.

Our first objective is to characterize the catalyst with all the techniques available.[13,16] In so doing, we have developed or demonstrated techniques for catalyst characterizations. The kinetics of propylene polymerization were investigated in detail and the polymers characterized.[5,10,15]

Then the polymerization of propylene was compared with that of ethylene (Chien and Bres, 1986), the latter acting by either $AlEt_3$ alone or $3AlEt_3/MPT$. The results permit us to postulate structures of different active sites and propose mechanisms for polymerizations. Supplemented by results of other important studies, we are able to present a rather detailed description of the CW-catalyst and its mode of actions.

We wish to interject two cautionary notes. There are still gaps in our understanding of the system which are filled with spectulative hypothesis. Also the discussion of this paper pertains strictly to the CW-catalysts, any application of the postulated catalyst structures, measured kinetic parameters, and hypothesized reaction mechanisms to other $MgCl_2$-supported system should be done only after all the differences between it and the CW catalyst have been taken into careful consideration.[21,24]

CW-CATALYST

Magnesium Chloride

$MgCl_2$ has been found to be the ideal support, its host lattice can support chiral titanium-active sites. This has been widely attributed to near isomorphism of $MgCl^2$ and $TiCl_3$ crystals. There are two crystal forms of $MgCl_2$: the cubic close-packed form, which is very similar to the structure of γ-$TiCl_3$ and the hexagonal close-packed form, which is very similar to the structure of α-$TiCl_3$.[47,1,27]

Crystals of both $TiCl_3$ and $MgCl_2$ consist of two planes of chlorine atoms sandwiching a plane of metal atoms; the titanium atoms occupy two-thirds of the octahedral sites in $TiCl^3$ while the magnesium atoms occupy all the sites. The dimensions of the layers are similar in the two compounds: $3a_{MgCl_2} \sim a_{TiCl_3}$ and $c_{MgCl_2} \sim c_{TiCl_3}$. Furthermore, through mechical treatment, $MgCl_2$ can be partially distorted by a $\pm$ 60° rotation of the Cl-Mg-Cl triple layers.[30] A similar distortion of $TiCl_3$ can be effected by ball-milling.

There are three faces of $MgCl_2$ crystalline of interest. The (100) plane has four coordinate exposed Mg atoms (Figure 1b). The (110) face of $MgCl_2$ and the (100) face of $TiCl_3$ are very similar both having five-coordinate metal atoms at the surface (Figure 1c). Both basal planes (001) of $TiCl_3$ and $MgCl_2$ are occupied entirely by Cl atoms; there are no exposed metal atoms. It will be shown below that the dimeric Ti_2Cl_6 epitactically placed on the (110) face of $MgCl_2$ is the most likely candidate for stereospecific active sites.

$MgCl_2$ also possesses desirable morphology as a support. Mercury porosimetry[17] showed it to be a highly porous polycrystalline material probably due to a large number of uniformly distributed micro-cracks. It is sufficiently resistant to fracture by mechanical manipulation and can be

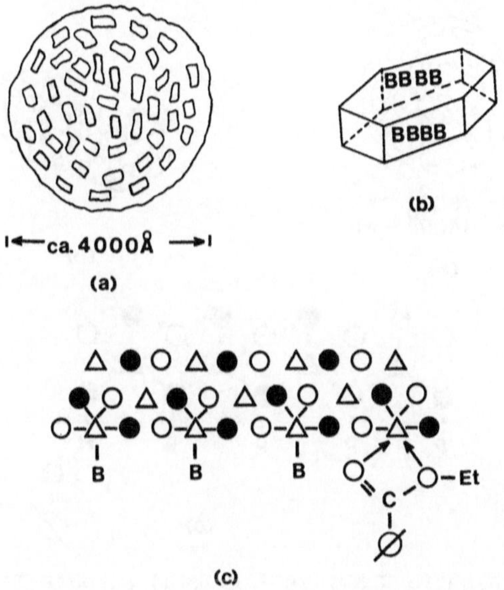

Fig. 1. Schematic Drawings for: (a) $MgCl_2$ crystallite; (b) the (110) cut; (c) the (100) cut. Symbols (●) Cl atoms up; (O) Cl atoms down; (Δ) Mg atoms; (◯) Ti atoms; (◇) Al atoms; (▭) VCP; (β) internal Lewis base; (P) phenol; (AP) aluminum phenoxide; (R) alkyl group

used for instance in gas phase polymerization, but it is weak enough so that it is broken down during polymerization. The porous morphology permits the diffusion of monomer into the interior of the catalyst particles and serves to determine the morphology of the polymer as well. This replication phenomenon, which has long been known for the $TiCl_3$ catalyst[41] has been observed also for the $MgCl_2$-supported catalyst.[32,44]

A third desirable characteristic of $MgCl_2$ is its low electronegativity among metal halides. Supports having electronegativities less than that of $TiCl_3$ are thought to promote the coordination of monomer molecule to Ti via π-back donation. Soga et al.[53] have shown that $MgCl_2$, $MnCl_2$ and $CoCl_2$ which have lower electronegativity than $TiCl_3$, according to ESCA measurement, increase productivity of olefin polymerization. In contradistinction, highly electronegative metal halides act to impede polymerization.

Finally, $MgCl_2$ is a rather chemically inert material so it can be left in the polymer without the usual requirement of deashing. On the other hand, though $MnCl_2$ and $CoCl_2$ also increase productivity, they are expected to catalyze autoxidation of polypropylene and must be removed if used as supports.[36,37]

Internal Lewis Base (B_i)

That ball-milling of $MgCl_2$ produces disordered crystal structure has been recognized.[31,33] Reduction of crystallite size also occurs during milling, but reaches a rather large limiting size. As crystallite size decreases, its surface energy increases to cause recrystallization. Therefore, co-milling $MgCl_2$ with $TiCl_4$ increases surface areas initially but decreases after long milling time.[33,30] The recrystallization can be inhibited by an additive which complexes with the substrate to reduce its surface energy. This is one of the main functions of B_i.

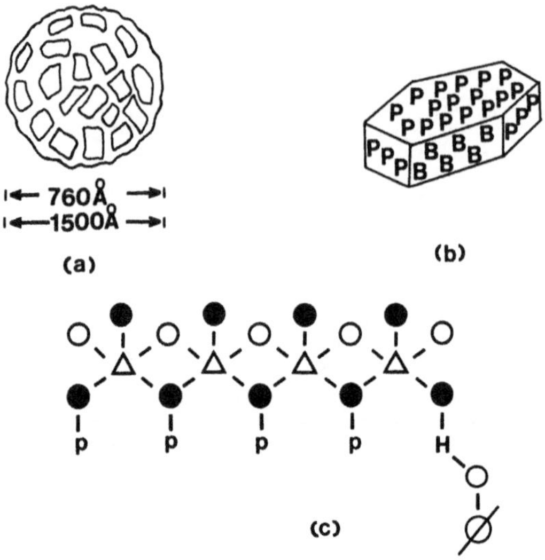

Fig. 2. Schematic Drawings for: (a) support I $(MgCl_2)_1$ $(EB)_{0.15}$; (b) crystallite; (c) (110) surface

Co-milling of $MgCl_2$ with EB can produce minute crystallites of $MgCl_2$ (20 to 40 Å, designated as support I) the dimensions, of which do not decrease further throughout the subsequent steps of catalyst preparation.[17] FTIR[17] showed that the ν(C=0) band at 1719 cm^{-1} and ν(C-0) bands at 1279 and 1109 cm^{-1} of free ethyl benzoate are shifted from the former to lower frequency of 1683 cm^{-1} and from the latter to high frequencies at 1329 and 1306 cm^{-1} in support I. Coordination of the oxygen atoms to Mg has been proposed for the complexation.[13]

EB cannot complex with the Cl atoms in the (001) planes. The Mg atoms in the (110) face of $MgCl_2$ are more acidic than those in the (110) faces. It is postulated that EB preferentially coordinates with the former as depicted in Figures 2b and c. Complexation of EB with the (100) surface Mg atoms cannot be excluded but any interactions would be very weak, and such EB molecules would be displaced by the subsequent addition of *p*-cresol.[43]

The crystallites of support I adhere to form large aggregates (Figure 2a). This is evidence that the (001) surfaces or the (100) surfaces have no complexed EB. The approximate dimensions of the (110) surfaces can accommodate a monolayer of EB in support I with a $(MgCl_2)_1(EB)_{0.15}$ composition.[13] The function of B_i is by no means to act as antiagglomerant only. Its other function will be discussed below.

Internal Alcohol

In the preparation of the CW-catalyst, support I is reacted with *p*-cresol to form support II with a composition of $(MgCl_2)_1(EB)_{0.15}$ $(PC)_{0.5}$. There are 5-fold increases in BET surface area and of internal pore surface and two to three fold increases in volme. The PC, in support II, is resistant to washing and may be complexed to Cl atoms by hydrogen bonding (Figure 3c). The fact, that there is no loss of EB, implies that PC and EB do not compete for the same surface sites. It is reasonable to assume that PC molecules are adsorbed on the (001) and (100) planes (Figure 3b). This reduces the surface energies of these planes resulting in the breakdown of aggregates of $MgCl_2$ crystallites as depicted in Figure 3a. The large amount of PC in support II is consistent with the large combined area of the (001) and (100) planes.

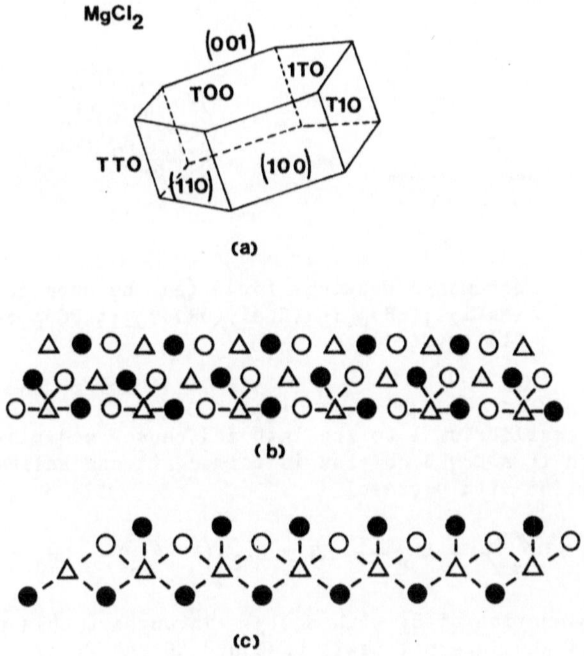

Fig. 3. Schematic Drawings for: (a) support II $(MgCl_2)_1(EB)_{0.15}(PC)_{0.5}$; (b) crystallite; (c) (100) surface

Most $MgCl_2$-supported catalyst preparations do not involve an internal alcohol. Very large surface area catalyst can be prepared without it. Therefore, the role of internal alcohol is a subtle one which may be related to the incorporation of small amounts of $AlCl_3$ in the catalyst or it may assist the initial partial reduction of epitactically-placed $TiCl_4$. However, if internal alcohol is used, then a subsequent treatment of aluminum alkyl is necessary, otherwise it will react with $TiCl_4$ to form titanates (vide infra).

Internal $AlEt_3$

Most $MgCl_2$-supported catalysts are activated with AIR_3 and AIR_3 and are not involved in the preparation of the catalyst. However, the CW-catalyst is prepared by reacting support II with $AlEt_3$ (half a mole of $AlEt_3$ to a mole of p-cresol). There was quantitative evolution of ethane.[13] The process increases BET surface area by two to three fold (Figure 4a). The average structure for the reaction product is EtAl $(OC_6H_4CH_3)_2$, which should still act to prevent regrowth of $MgCl_2$ by virtue of complexation of the electron deficient Al with the Cl atoms of the (001) and (100) surfaces (Figure 4b).

There is no loss of B_i in the reaction with $AlEt_3$ which might be expected to occur by Lewis acid base complexation:

$$AlEt_3 + EB \rightleftharpoons AlEt_3 \cdot EB \quad (1)$$

This reversible equilibrium cannot compete with the rapid and irreversible reactions such as:

$$AlEt_3 + HO-C_6H_4-CH_3 \longrightarrow Et_2Al-O-C_6H_4-CH_3 + C_2H_6 \quad (2)$$

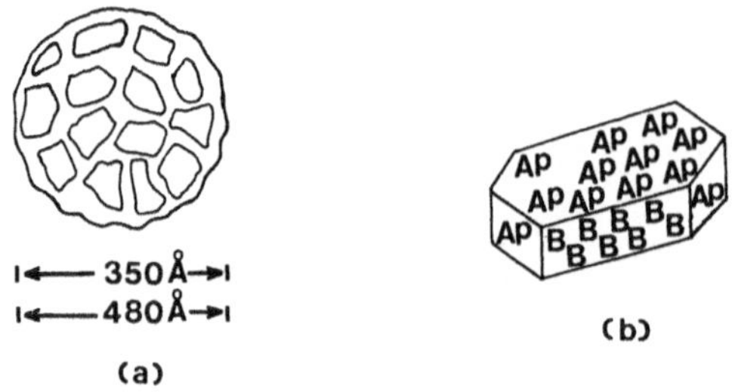

Fig. 4. Schematic Drawings for: (a) the support III $(MgCl_2)_1(EB)_{0.15}$-$(EtAl)(OR)_2)_{0.25}(PC)_{0.25}$; (b) crystallite

which can shift equilibrium 1 to the left and cause recomplexation of B_i with $MgCl_2$. Also if $AlEt_3EB$ complex is formed, it can reliberate ethyl benzoate by reaction with p-cresol

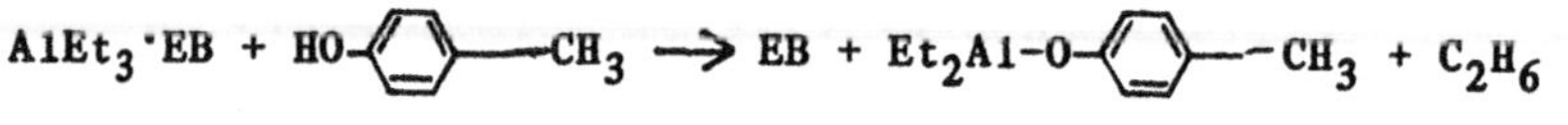

(3)

and cause recomplexation of B_i with $MgCl_2$. Throughout this paper, ethyl benzoate is the internal Lewis base and EB and B_i are used interchangeably.

The $AlEt_3$ to p-cresol ratio of 1:2 is recommended in the patent literature. This is probably the optimal ratio. Obviously, a ratio of 1:3 would consume all the ethyl groups and the resulting $Al(OR)_3$ is incapable of reducing Ti(IV) to Ti(III) in the final step of the catalyst preparation. For ratios smaller than 1:3, the excess p-cresol would react with $TiCl_4$ to produce $TiCl_n(OR)_{4-n}$ compounds which are unlikely to coordinate epitactically to the $MgCl_2$ support. Also equimolar amounts of $AlEt_3$ and p-cresol would form Et_2AlOR which would have higher reducing power than $EtAl(OR)_2$ to produce more Ti(II) species in the reaction with $TiCl_4$. Finally, amounts of $AlEt_3$ in excess of p-cresol are clearly detrimental. Any free $AlEt_3$ may extract the B_i by reaction 1 as observed by Busico et al.[3]

Optimum amounts of internal alcohol and aluminumalkyl combination serve to both break down the support aggregate and to produce a moderate alkylating/reducing agent in support III.[52,54]

Epitactical Coordination of Titanium Halides

Several reactions occur when support III is reacted with either neat $TiCl_4$ or a concentrated $TiCl_4$ solution in inert diluent. $TiCl_4$ reacts with the (110) surfaces by:

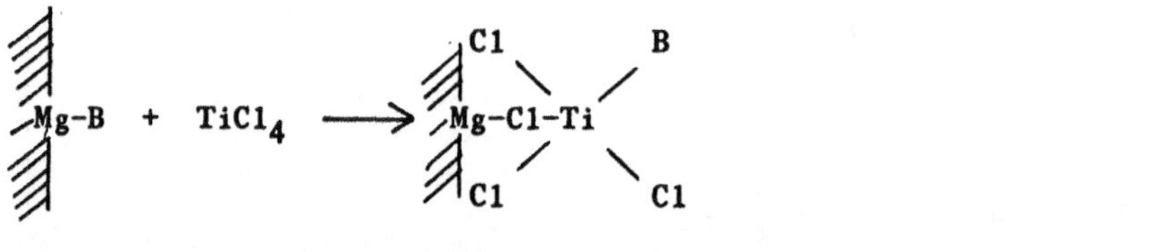

(4)

The reaction is based on the fact that $MgCl_2$ and $TiCl_4$ are known to form bimetallic chlorides. Since $TiCl_4$ is a strong Lewis acid, it will complex EB if the coordination state of Ti permits. The composition of the catalyst contains 0.10 to 0.15 mole of EB per mole of $MgCl_2$; Figure 5b depicts possible structures of the epitactically placed Ti species on the (110)

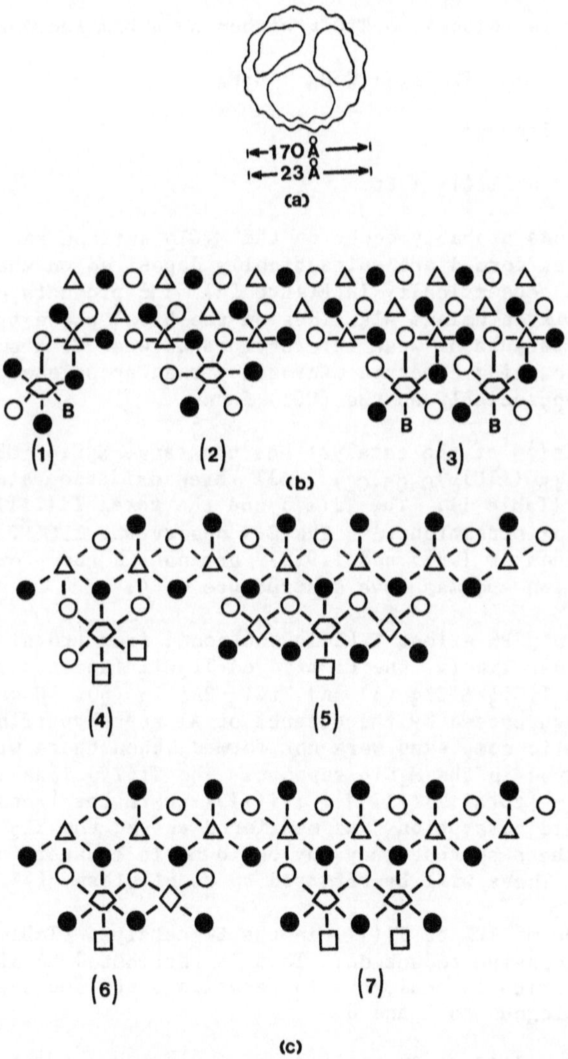

Fig. 5. Schematic Drawings for: (a) the CW-catalyst; (b) (110) surface; (c) (100) surface

surface. The main features are that they are tetravalent and coordinatively saturated. Some proposed species have coordinated B_1 as in 1 and 3 while 2 does not. There are both $TiCl_4$ and Ti_2Cl_8 which may or may not have adjacent titanium species. The latter are referred to as clustered sites.

The reaction of $TiCl_4$ with the basal plane and the other lateral surfaces is more complex because the lateral surfaces are occupied by alkylaluminum phenoxides. The following reactions occur, but not necessarily in the order given. These are the alkylation of $TiCl_4$:

$$TiCl_4 + ElAl(OR)_2 \longrightarrow EtTiCl_3 + ClAl(OR)_2 \qquad (5)$$

and the formation of $AlCl_3$ with the excess $TiCl_4$.

$$ClAl(OR)_2 + 2TiCl_4 \longrightarrow AlCl_3 + 2Cl_3Ti(OR) \qquad (6)$$

The Ti(IV) alkyl is reduced to $TiCl_3$ either by a bimolecular proces

$$2EtTiCl_3 \longrightarrow Ti_2Cl_6 + C_2H_6 + C_2H_4 \quad (7)$$

or by homolytic cleavage

$$EtTiCl_3 \qquad TiCl_3 + Et^{\cdot} \quad (8)$$

The above reactions probably occur on the $MgCl_2$ surface and the $TiCl_3$, Ti_2Cl_6, $AlCl_3$ etc. formed are epitactically deposited on the (100) surface as shown schematically in Figure 5c. The products of reactions of $TiCl_4$ with the alkylaluminum alkoxides on the basal plane are not deposited on it because of weak interactions between Cl atoms. Barbe et al.[2] showed by scanning electron microscopy that propylene polymerization does not occur appreciably on the (001) faces.

The composition of the catalyst has the range $MgCl_2$ $(EB)_{0.1-0.15}$ $(TiCl_{3.5})_{0.05-0.15}$ $(AlCl_3)_{0.05-0.1}$. All three oxidation states are present in the catalyst (Table 1). The Ti(II) and the total Ti(III) were determined by two redox titrations.[13] The EPR observable Ti(III) species, with axial g values of 1.945 and 1.913[15] are not in the proximity of another Ti(III) ion and may have a structure as **4**.

The amount of EPR silent Ti(III), obtained from [total Ti(III)] - [EPR Ti(III)], are, most likely, the Cl bridged Ti_2Cl_6 (species **7**). Also shown in Figure 5c are $TiCl_3 \cdot AlCl_3$ (**5**) and $TiCl_3 \cdot 2AlCl_3$ (**6**). However, their presence is not supported by the absence of Al superhyperfine splitting. If these bimetallic complexes were not formed, then there will be $AlCl_3$ or Al_2Cl_6 dissolved in the $MgCl_2$ support. The Ti(IV) ions were obtained from [total Ti] - [total Ti(III)] - [Ti(II)] where the [total Ti] was determined by atomic absorption. As mentioned above, the Ti_2Cl_6 may be surrounded by others so that they may be found in clusters of varying numbers of Ti_2Cl_6. These will be referred to as clusters of **7**.

The presence of 12% of Ti(II) in the CW-catalyst (Table 1) indicate that there is excessive reduction. This is attributed to alkylation of Ti(III) species which is analogous to reaction 5 and subsequent reduction of processes analogous to 7 and 8.

One important difference between sites on the two kinds of lateral surfaces is that the Ti(III) and Ti(II) species (Figure 5c) are coordinatively unsaturated whereas the Ti(IV) species (Figure 5b) are coordinatively saturated.

Finally, the x-ray diffraction pattern of the CW-catalyst has the same 20 reflections and sharpness as $MgCl_2$ before ball-milling with EB. The rotational disorder formation, which had been thought to cause the shift and broadening of reflections, may be annealed by the $TiCl_4$ treatment. The question is not settled regarding whether $TiCl_3$ is merely epitactically placed on the $MgCl_2$ lateral surface or whether it enters into the $MgCl_2$ crystallite to form a solid solution in the δ-$TiCl_3 \cdot 0.33AlCl_3$ catalyst.[28] The former is favored because ca. 93% of the titanium acts to initiate ethylene polymerization (vide infra).

Activation of the CW-Catalyst

The CW-catalyst is activated by $AlEt_3$ together with methyl-p-toluate which will be referred to as an external Lewis base (B_e). The Ti(IV) ions, in the catalyst, are reduced to lower oxidation states. This results in proportionate increases in Ti(II), EPR observable Ti(III), and EPR silent Ti(III) ions (Table 1). These ions are alkylated and the Lewis base

Table 1. Oxidation States of Titanium[a] in the CW-Catalyst

Catalyst	Ti(II)[b]	Ti(III) EPR Observable	Ti(III) EPR Silent	Ti(IV)
before activation	12%	9%	33%	45%
after activation[a]	18%	19%	55%	8%

[a]In percent of total titanium; [b]activated with $3AlEt_3$/MPT.

complexes with the more acidic ions, on the (110) plane of $MgCl^2$. Figure 6 depicts schematically the alkylated Ti(III) ions which are coordinatively saturated on the (110) plane and unsaturated on the (100) plane. In particular, species 7' are chiral[20] while species 4' are achiral.

If $AlEt_3$ is used alone as the activator without B_e, then the Lewis base coordinated to the acidic Ti ions may be dissociated. In that case 1', 2' and 3' can become coordinatively unsaturated.

NUMBER OF ACTIVE SITES

The principal objective in the R and D effort on Z-N catalysis is to improve the productivity, i.e., productivity of stereoregular polypropylene. Productivity can be expressed either in gm polymer/gm catalyst or gm polymer/gm Ti. For the classical $TiCl_3$ catalyst, the productivity is the same, in either unit. These two quantities become vastly different for supported catalyst containing very low amounts of Ti.

The advent of $MgCl_2$-supported high-mileage catalysts immediately raises the question of whether there is a very large number of active sites [Ti*] or if each site has a very high rate constant of propagation, R_p, or both as compared to the classical $TiCl_3$ catalysts. The two quantities are not independent of each other but are related by

$$R_p = k_p\,[Ti^*]\,[M] \tag{9}$$

Therefore, k_p can be calculated only when [Ti*] is known. However, there is no method which is entirely satisfactory for the determination of [Ti*]. The techniques of active site counting, their limitations and uncertainties have been reviewed.[55,56]

The active site concentration varies with polymerization time. There are two stages of polymerization, initiated by $MgCl_2$-supported catalysts, the durations, of which, vary with experimental conditions. In the initial 20 to 50 min., the polymerization either begins with a maximum rate, $R_{p,m}$ or there is a build up of rate reaching $R_{p,m}$ within 5 to 20 min. The former is favored by high aluminum activator to titanium ratio (A/T) and elevated temperatures; the latter prevails for low A/T and lower temperatures, particularly in the absence of B_e.[12] In both instances, T_p decreases rapidly soon after reaching the maximum rate. The second stage of polymerization is characterized by R_p which either decays very slowly or remains almost constant. At very low temperature, i.e. 0° for propylene polymerization,[8] R_p is slow and constant throughout the course of a batch polymerization.

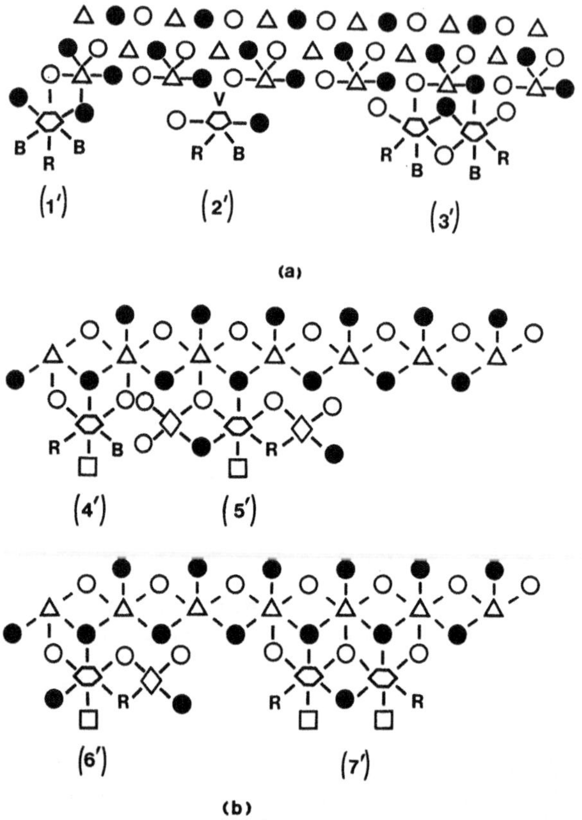

Fig. 6. Schematic Drawings for the CW-Catalyst After Activation: (a) (110) surface; (b) (100) surface

The rapid initial decay of R_p has been reported by many investigators. This has sometimes been attributed to limitation of monomer diffusion. Two definitive experiments ruled out the phenomenon due to limitation of diffusion of monomer to the active sites, through the barrier of crystalline polyolefins, which encapsulate the catalyst. The first[38,22] was the demonstration that the polymerization profile, i.e. R_p versus t, is not affected by intermittent removal and reintroduction of monomer. In other words, the decay of R_p continues, even when there is no polymerization.

The second evidence is the comparison of R_p versus t[5] for the polymerization of propylene versus that of decene-1. The stereoregular poly (decene-1) is soluble in heptane at 50°.

The two systems showed the same decrease of activity, i.e. the rate constant k_t' obtained by second order plot of R_p^{-1} versus t agreed within $\pm$ 10%. Other less direct evidence, such as the lack of effect of titanium concentration on catalyst efficiency,[25] and that the rate of catalyst deactivation is not related to productivity[31] also argue against diffusion limitation.

The initially formed active sites have been designated as the first kind, Ti_1^*, and those remained after long polymerization time as active site of the second kind, Ti_2^*. The former is characterized by rapid decay and the latter is relatively stable.

Among the various methods for active site counting, radioactive ^{14}CO has been said to react selectively and very rapidly with the propagating Ti*-P bond.[58,59] Mejzlik and coworkers[42,43] pointed out that the process is reversible and slow. Very careful radiotagging of propagating chains with ^{14}CO and $^{14}CO_2$ by Warzelhan et al.[57] showed the presence of different types of active centers, some of which react rapidly but reversibly with these reagents and others react much more slowly. This is supported by the fact that R_p does not decrease linearly with the quantity of Co added.[22] Most of the above studies were made on α-$TiCl_3$ or δ-$TiCl_3 \cdot 33AlCl_3$ catalysts.[61]

We found that if propylene[6] and ethylene polymerizations[11] were terminated by removal of monomer and then ^{14}CO injected, it took about an hour to reach about half of the ultimate ^{14}CO specific activity. The specific activity reached a constant value after > 20 hrs of contact time with ^{14}CO in ethylene polymerization mixture; the specific activity of polypropylene was still increasing after 45 hrs of reaction. Therefore, the reaction behavior of Ti*-P species, initiated by $TiCl_3$,[57] and by the CW-catalyst[6,11] are the same.

If the Ti_1^* sites will react with ^{14}CO with reasonable rapidity, then the initial increase of R_p and subsequent decay should be reflected by corresponding increase and decrease of ^{14}C specific activity. Experimentally, it was observed that the initial ^{14}C specific activity was very low even at $R_{p,m}$. This was true for both 1 hr and 45 min contact time of ^{14}CO with the polymerization mixture. The maximum ^{14}C specific activity was reached only after the decay of R_p is nearly over, i.e., radiotagging of Ti_1^*-P does not occur.

This should not be construed to mean that CO cannot inhibit monomer propagation, which it in fact does, albeit at least reversibly in part. Instead, the results may mean either the CO complex is too unstable

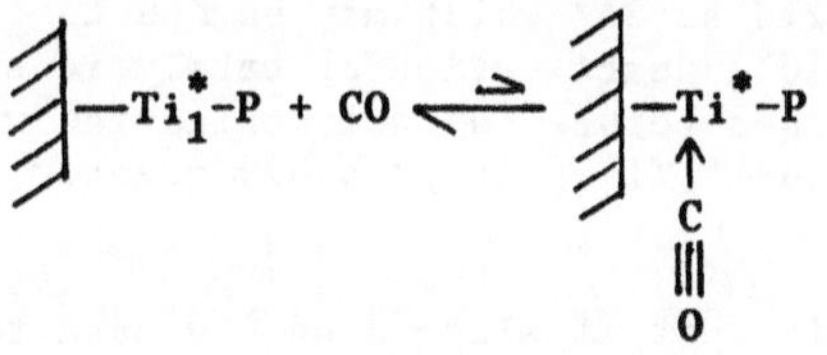

(10)

or it is too stable because of the formation of a bridged carbonyl

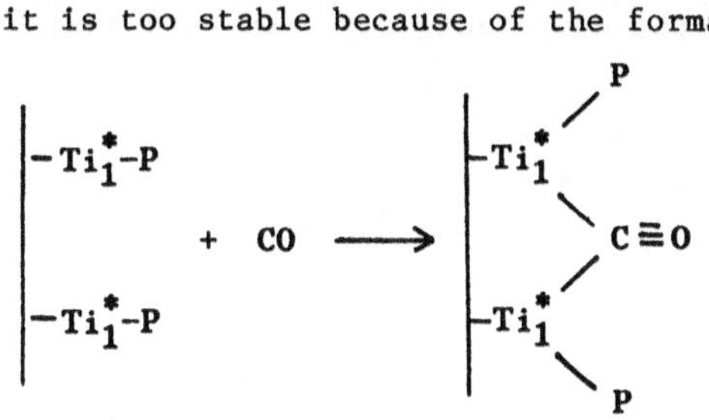

(11)

so that there is no insertion of CO into the Ti*-P bond. The second alternative is more probable because CO does inhibit polymerization. In any case, the technique is unsuited to determine $[Ti_1^*]$.

The reactions of CH_3O^3H with both metal polymer bonds, [MPB] = [Ti-P] + [Al-P] are rapid and quantitative:

$$[MPB]_t = [Ti^*]_o + \int_o^t k_{tr}^A [Ti^*][A]^* dt \qquad (12)$$

Table 2. Concentration of $[Ti_1^*]_o$ in Olefin Polymerizations Initiated by the CW-Catalyst[a]

Monomer	Propylene		Ethylene	
Temp. °C	$[Ti_1^*]_{o,i}$	$[Ti_1^*]_{o,a}$	$[Ti_1^*]_o$	$[Ti_1^*]_o^{-B_e}$
25	0.047	0.019	0.21	0.54
50[b]	0.067-0.13	0.063-0.14	0.50	0.93
70	0.11	0.17	0.42	0.61

[a]Active site concentration in mole/mole Ti, subscript i and a denote stereospecific and non-stereospecific, and superscript $-B_e$ is polymerization activated by $AlEt_3$ alone in the absence of external Lewis base; [b]range of values for polymerizations using different A/T ratios.

Consequently, it was possible to obtain $[Ti_1^*]$ by extrapolation of the $[MPB]_t$ versus t data to zero time; the values thus obtained were found to be in good agreement with those obtained by an alternative kinetic method.[7] The results are given in Table 2.

At optimum conditions such as temp. = 50° and A/T = 167, about half all the titanium initiates ethylene polymerization within a short time (< 5 min), after the introduction of the cocatalyst $3AlEt_3$/MPT. Since Ti, in all oxidation states, can catalyze ethylene polymerization, the active centers may be the coordinatively unsaturated species 4', 5', 6', and 7'. The coordinatively saturated species 1', 2', and 3 are probably inactive under these conditions. The $[Ti^*]_o$ is only 21% at 25° which may be due to the Arrhenius temperature dependence. At 70°, deactivation of catalytic sites began immediately, so minutes have elapsed before the polymerization was quenched for radioassay. Thus, the actual $[Ti^*]_o$ is probably greater than that given in Table 2.

An important result of this work is that if $AlEt_3$ alone is used to catalyze ethylene polymerization, then ca. 93% of the Ti are active. In the absence of B_e, the ligand B_i of the coordinatively saturated species 1', 2', and 3' is dissociated,

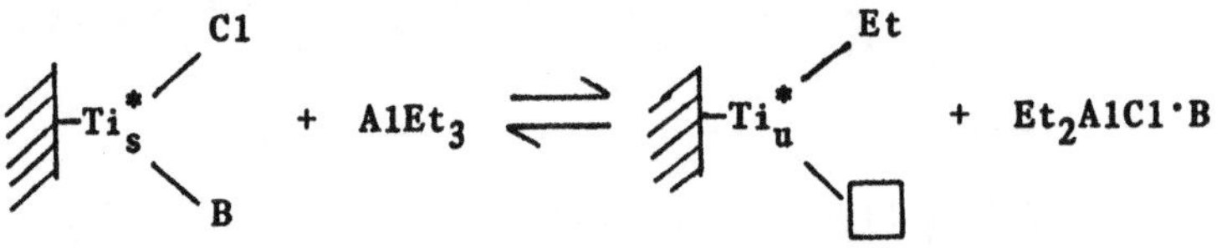

(13)

Subscripts s and u denote coordinatively saturated and unsaturated Ti*, respectively, B is a Lewis base, and □ represents a vacant coordination position, VCP. The occurrence of ligand dissociation prior to initiation is the reason why it takes > 20 min after the addition of $AlEt_3$ without B_e for a polymerization to reach $R_{p,m}$. From these results, we can postulate two functions for the Lewis base. The B_i complexes with those strongly acidic Ti sites, while the B_e assures that these sites remain coordinatively saturated. In the absence of B_e, the B_i ligands are dissociated. The resulting Lewis acid Ti sites can polymerize ethylene, but in the case of propylene, these sites can only produce inactive stereo-irregular polymers.

We can ignore the coordinatively saturated Ti ions in propylene polymerizations because B_e was always used as a coactivator. If it is assumed that the coordinatively unsaturated sites are those found on the (100) face of $MgCl_2$, then they are the Ti ions in the lower valence states in Table 1. However, divalent Ti ions are known to be inactive for propylene polymerization. This leaves a maximum of ca. 42% Ti(III) as potential active sites. Table 2 showed that there are as much as 13% isospecific sites and about the same amount of non-specific sites.

Therefore, about 15% of the Ti(III) sites are inactive under these conditions and these may be the ones coordinated with $AlCl_3$ (species 5' and 6). Caunt has shown that the Lewis acid $EtAlCl_2$ acts as an inhibitor in the $TiCl_3$-$AlEt_2Cl$-catalyzed propylene polymerization. Ti(III) ions having high Lewis acidity due to other structural origins are also candidates for the inactive sites. These sites are rendered inactive when B_e fills the VCP.

According to the above hypothesis, the chiral species 7' may be postulated to have stereospecific and stereoelective catalytic sites. The achiral species 4' is the non-stereospecific active sites. The rate of formation of Ti_1^* is first order with respect to [Ti] and [A] and has an activation energy of 12 kcal $mole^{-1}$.

The Ti_1^* sites are unstable and decay rapidly with a decrease of R_p. This process is complete in 30 to 50 min depending upon experimental conditions. However, the polymerization continues and we have designated the active sites here as the Ti_2^*. These Ti_2^* sites differ from the former mainly by their stability and of reactivity with ^{14}CO. Their other polymerization kinetic parameters are remarkably similar, whenever they can be

Table 3. Concentration of $[Ti^*]_o$ in Olefin Polymerizations Initiated by the CW-Catalyst[a]

Temp. °C	$[Ti_2^*]_{o,i}$	$[Ti_2^*]_{o,i}/[Ti_1^*]_{o,i}$	$[Ti_2^*]_{o,a}$	$[Ti_2^*]_{o,a}/[Ti_1^*]_{o,a}$
25				
50[b]	0.023-0.032	0.18-0.36	0.024-0.051	0.17-0.53
70	0.026	0.24	0.052	0.31

Temp. °C	$[Ti_2^*]_o$	$[Ti_2^*]_o/[Ti_1^*]_o$	$[Ti_2^*]_o^{-B_e}$	$[Ti_2^*]_o^{-B_e}/[Ti_1^*]_o^{-B_e}$
25	0.09	0.43	0.16	0.30
50[b]	0.10	0.20	0.24	0.26
70	0.08	0.19	0.15	0.25

[a,b]As in footnotes for Table 1.

compared (vide infra). The values of $[Ti_2^*]_o$ are summarized in Table 3 as determined from the asymptotic ^{14}C specific activity, following ^{14}CO labeling. Kinetic analysis gave results in agreement with radiolabelling.[7]

The ratio of $[Ti_2^*]_o/[Ti_1^*]_o$ lies mostly between 0.2 and 0.3. This implies that about one Ti_2^* site remained following the decay of three to five Ti_1^* sites. In two experiments, the ratios are greater than A/T is 334, especially for the non-specific sites.

The deactivation processes usually involve two adjacent active sites. For the simple case of cluster of Ti_1^* sites arranged in a linear array, the process will result in the transformation of a fraction of the Ti_1^* sites to Ti_2^* sites. For instance, in the case of six Ti_1^* sites, deactivation of two pairs of adjacent Ti(III) species by either reduction or oxidation leaves two Ti(III) species without the other Ti(III) ions adjacent to them.

A remarkable observation is that both $[Ti_1^*]_{o,i} \sim [Ti_1^*]_{o,a}$ and $[Ti_2^*]_{o,i} \simeq [Ti_2^*]_{o,a}$. Since the yields of atactic polypropylene are only a few percent of the total polymer, the non-specific sites must have $k_{p,a}$ value which is only about a tenth of $k_{p,i}$ for the isospecific sites. This difference is also reflected in $\bar{M}_n$ for the stereoregular polymers which are more than then ten times greater than that of the refluxing heptane soluble polymers.

There is an interesting effect of hydrogen which is monomer dependent. In propylene polymerization with 48 torr of H_2, active site concentrations as high as $[Ti_1^*]_{o,i}$ = 19% and $[Ti_1^*]_{o,a}$ = 15% have been observed with proportionately higher rates of polymerization. In other words, H_2 can create new active sites having the same k_p. It will also be shown that the isotacticity index (I.I. = percent of refluxing heptane insoluble polypropylene) is not changed by H_2. Exactly what the origin of the hydrogen activated sites is not known because there are many possibilities and the chemistry of adsorbed hydrogen can be complicated. For instance if adsorbed hydrogen dissociates to form H^+ then it can reduce some of the Ti(IV) ions present in the system. However, if the dissociation is to H^-, then it can convert some of the Ti(II) ions to Ti(III) hydrides. Finally, the activation may not be a redox process but rather a simple one of converting strong Lewis acidic Ti(III) ion, i.e., species **5'** or **6'**, to less acidic Ti(III) hydrides analogous to the reactions involved in chain transfer with hydrogen (vide infra). This has the effect of lowering the Lewis acidity of the Ti(III) ion and facilitating the initiation of olefin polymerization.

The effect of H_2 on $[Ti_2^*]$ in propylene polymerization is opposite to the above. The $[Ti_2^*]_o$ values obtained by ^{14}CO counting (48 hrs tagging time) at P_{H_2} = 48 torr are 0.01 and 0.015 mole/mole Ti for the isospecific and nonspecific sites, respectively. These values are only about one-half of the $[Ti_2^*]_o$ in the absence of hydrogen. This is attributable to H_2 assistance in catalyst deactivation.

The effect of H_2 on normal ethylene polymerization, i.e., activated by $AlEt_3$ with B_e causes 50% increase in $R_{p,m}$ at P_{H_2} = 48 torr. This suggests a corresponding increase of $[Ti_1^*]$. Yet, for long time of polymerization with H_2, R_p is smaller by about a factor of three as compared to polymerizations without H_2. Therefore, H_2 also reduces $[Ti_2^*]$ in ethylene polymerization.

In the absence of B_e, hydrogen has the effect of only decreasing R_p of ethylene polymerization. The formation of new catalytic sites by H_2 only takes place when $3AlEt_3$/MPT is used in the activation but not without the external Lewis base.

CHAIN PROPAGATION

There has been continual speculations about the mechanism of stereospecific polymerization of α-olefins. What is known for certain is that the titanium ion must be a chirality center which could maintain its absolute configuration and discriminate sharply between the two prochiral faces of the α-olefin during the insertion. What is of concern here is the activity of the catalytic centers, i.e., the intrinsic rate constants of propagation, in order to decide whether the catalytic centers in the $MgCl_2$-supported catalyst is much more active than those in the classical Ziegler-Natta catalyst or if there are many more of them in the former. We have obtained values of $k_{p,i}$, $k_{p,a}$, $k_{p2,a}$ for propylene polymerization and k_{p1}, k_{p2}, $k_{p1}^{-B_e}$ and $k_{p2}^{-B_e}$ for ethylene polymerization (Table 4) where subscripts i, a, 1, and 2 designate isospecific and nonspecific sites of first and second kind, respectively, and superscript $-B_e$ signifies the absence of B_e as coactivator.

The values of k_{p1i} are 94 to 265 M sec^{-1} between 25° and 70° for stereospecific polymerization of propylene, ΔE_p is 5.2 kcal mole^{-1}. The values are about two-fold larger than those reported for the classical Ziegler-Natta catalysts. The k_p values for δ-$TiCl_3$-$AlEt_2Cl$ at 75° is 71 M^{-1} sec^{-1}; it is 90 to 100 M^{-1} sec $^{-1}$ for δ-$TiCl_3 \cdot 0.33AlC_3$ activated by AlR_3.[19,60] The energy of activation is 5.5 kcal mole^{-1}. For a simpler supported catalyst system $MgCl_2$/EB/$TiCl_4$-$AlEt_3$, Doi et al.[23] estimated k_p values of ca. 340 M^{-1} sec^{-1} at 41° and E_p = 5.3 kcal mole^{-1}. Giannini[31] found k_p for the $MgCl_2/TiCl_4$-$AlEt_3$ for long polymerization times, the same as that of the δ-$TiCl_3 \cdot 0.33AlCl_3$-$Al(iBu)_3$ catalyst. However, the k_p values are thought to be greater for short polymerization times with the supported catalyst. Considering the doubtful accuracies for the estimation of k_p, one can only say that the catalytic centers in the $MgCl_2$-supported catalysts are only marginally more active than the $TiCl_3$ catalyst.

An important finding of this work is that the number of nonspecific active sites are about the same as the number of isospecific sites, and that the k_{p1a} and k_{p2a} values are 5 to 16 times smaller than the k_{p1i} and k_{p2a} values. Therefore, the CW-catalyst produces polypropylene with high stereoregularity. A possible reason for the slow propagation rate, for the achiral site, may be the result of the necessity for activation of the olefinic double bond by chemisorption (complexation) by the transition metal on the catalyst surface.[46,4,40] Yet, there is a dearth of direct evidence for olefin π complex formation. If the propagation processes are written as

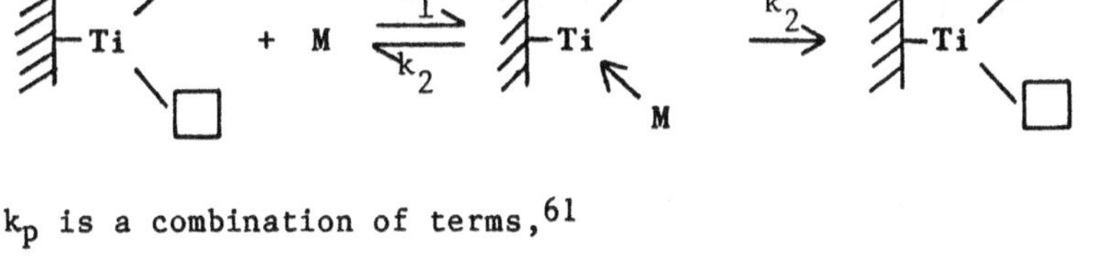

(14)

then k_p is a combination of terms,[61]

$$k_p = k_1 \, k_2 \, (k_2 + k_{-1} + k_1 \, [M]) \tag{15}$$

For π olefin complex of low stability toward dissociation, i.e. $k_{-1} \gg k_2 + k_1 [M]$, we have

$$k_p = k_2 k_1 / k_{-1} \tag{16}$$

and small k_p for the propagation, even if the insertion rate constant k_2 is not much different between sites. Therefore, unstable π olefin complex

Table 4. Rate Constants by Propagation[a,b,c,d]

Monomer	Propylene				Ethylene			
Temp. °C	$k_{p_1 i}$	$k_{p_1 a}$	$k_{p_2 i}$	$k_{p_2 a}$	k_{p_1}	$k_{p_1}^{-B_e}$	k_{p_2}	$k_{p_2}^{-B_e}$
25	94	8			530	270	780	596
50	160	11	86	6.7	310	303	720	743
70	265	48	113	20	430	280	1380	999
ΔE_p, kcal $mole^{-1}$	5.2	7.8				6.8		9.1

[a,b]As footnotes of Table 1; [c]R_p is in M^{-1} sec^{-1}; [d] subscript 1, 2 for active sites Ti_1^* and Ti_2^* a and i for non-specific and isospecific sites, respectively, superscript $-B_e$ indicates the absence of B_e.

with the achiral sites may be reason of the small values of $k_{p,a}$. If the achiral sites have the structure proposed for 4' (Figure 6) then the coordinated Lewis base may destabilize the olefin π complex by the competitive binding

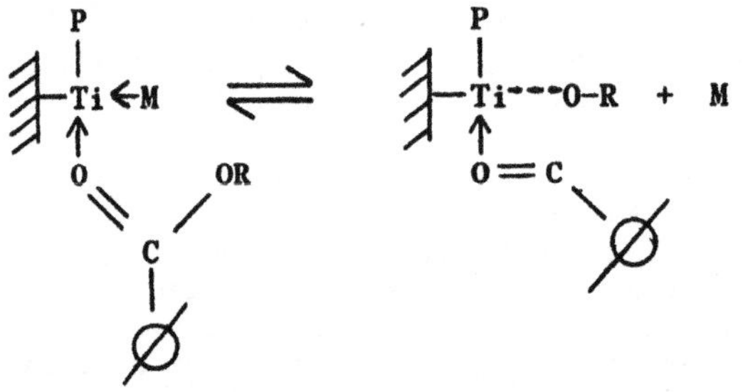

(17)

The k_p values for ethylene polymerization (Table 3) are two to ten times greater than those for isospecific propylene polymerization depending upon the conditions. The former is not significantly different for activation with $AlEt_3$ alone or with B_e. Therefore, B_e mainly affects the number of active sites but not the intrinsic reactivity of sites which already have coordinated B_i or others having no affinity for Lewis bases.

CHAIN TRANSFER WITH $AlEt_3$

Chain transfer with organoaluminum compounds were first shown by Natta et al.[48] to be an important process and that it is the monomeric species which is responsible, by virtue of the dependence of rate on $[A]^{1/2}$. Thus, one expects significant influence by Lewis bases, because FTIR showed complex formation of A•B.[17]

If no B is used either in the preparation of catalyst or in the activation, then the only equilibrium is

$$Al_2Et_6 \underset{}{\overset{K_D}{\rightleftharpoons}} 2AlEt_3 \tag{18}$$

The chain transfer may be written simply as

$$\text{-Ti*-P} + AlEt_3 \longrightarrow \text{-Ti*-Et} + Et_2AlP \qquad (19)$$

when a catalyst is synthesized with B_i, activation with $AlEt_3$ alone would result in the dissociation 13.

If an $AlEt_3$ and B_e are used as coactivators, there are equilibrium between A•B, 2A•B, A, and A_2. Ethyl benzoate and $AlEt_3$ has been reported to form both 1:1 and 1:2 complexes.[49] The additional equilibria are

$$A{\cdot}B \rightleftharpoons A + B \qquad (20)$$

$$2A{\cdot}B \rightleftharpoons A{\cdot}B + A \qquad (21)$$

which can have three consequences. A decrease in [A] in the polymerization media would decrease the rate of transfer 19. The presence of B would shift the equilibrium 20 to the left and inhibit the dissociation of complexed B_i. Finally, as the amount of [B] increases with respect to $AlEt_3$ the productivity decreases sharply.[51,39]

We have measured the rate constants of transfer k^A_{tr} which are given in Table 5. The processes are slowest for nonspecific sites in propylene polymerization. Perhaps the same hypothesis 18 applies here as well

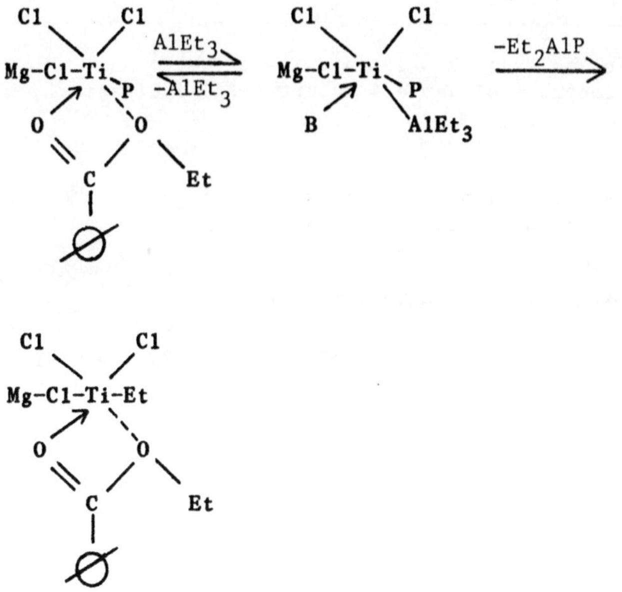

(22)

The values of k^A_{tr} for the isospecific sites are five to ten times greater by comparison. Chain transfer for the nonspecific sites increases ten-fold going from 50° to 70° as compared with a four-fold increase in k^A_{tr} values for isospecific sites.

Chain transfer is faster in ethylene polymerization than in propylene polymerization. This difference is greater for the Ti^*_z sites (Table 5). Omission of B_e greatly increases k^A_{tr}, as expected according to the above

Table 5. Rate Constants of Chain Transfer to $AlEt_3$[a,b,c,d]

Monomer Temp. °C	Propylene $k^A_{tr_1 i}$	$k^A_{tr_1 a}$	$k^A_{tr_2 i}$	$k^A_{tr_2 a}$
25	48	9.1		
50	21±2.6	2±0.2	11±1.4	1.2±0.1
70	91	24	38	10
ΔE_{tr}, kcal $mole^{-1}$				

Monomer Temp. °C	Ethylene $k^A_{tr_1}$	$k^A_{tr_1}$ $(-B_e)$	$k^A_{tr_2}$	$k^A_{tr_2}$ $(-B_e)$
25	80	300	120	670
50	21	500	48	1230
70	310	1800	1000	4700
ΔE_{tr}, kcal $mole^{-1}$	8.6		8.6	

[a,b]Same footnotes as in Table 1; [c]k_{tr} is in units of sec^{-1} x 10^5; [d]same as in footnote of Table 4 except $(-B_e)$ indicates the absence of B_e.

discussion. This process is dominant under this condition so that other chain transfer cannot be measured in the absence of B_e.

CHAIN TRANSFER WITH MONOMER AND BY β-HYDRIDE SHIFT

The rate constants of chain transfer with monomer, k^M_{tr}, and by β-hydride shift, k^β_{tr}, can be determined from the variation of $\bar{M}_n$ with monomer pressure. These measurements were made at 50°.[8] The monomer transfer rate constants are nearly the same for both monomers with propylene $k^M_{tr_1} = 1.34 \times 10^{-2}\ M^{-1}\ sec^{-1}$ and $k^M_{tr_2 i} = 7.2 \times 10^{-3}\ M^{-1}\ sec^{-1}$ and with ethylene $k^M_{tr_1} = 3.6 \times 10^{-3}\ M^{-1}\ sec^{-1}$ and $k^M_{tr_2} = 8.3 \times 10^{-3}\ M^{-1}\ sec^{-1}$. The MW of atactic polypropylene is independent of monomer pressure[8]. Apparently, the transfer process with monomer, for the nonspecific sites, is much slower than the other chain transfers.

Chain transfer via β-hydride shift occurs much more readily in propylene polymerization than in ethylene polymerization; the values of k^β_{tr} are $2.7 \times 10^{-1}\ sec^{-1}$ and $7.2 \times 10^{-4}\ sec^{-1}$, respectively. This may be attributable to the more labile β tertiary hydrogen in the propagating chain of polypropylene than the corresponding secondary hydrogen in ethylene polymerizing chain. Therefore, in the propylene polymerization, the rates of chain transfer are in the order of $R^A_{tr} < R^M_{tr} < R^\beta_{tr}$; the order in ethylene polymerization with B_e is $R^A_{tr} < R^M_{tr} \sim R^{\beta}_{tr}$. In the absence of B_e, R^A_{tr} is greater than the other transfer processes.

TRANSFER WITH HYDROGEN

The classical $TiCl_3$ catalyst produces polypropylene with $\bar{M}_w\bar{M}_n$ up to 20. The very high MW fraction makes the polymer difficult to process. Hydrogen is commonly used to control the MW. Natta[45] found for the α-$TiCl_3$-$AlEt_3$ catalyst that MW depends on $P_{H2}^{1/2}$. We have determined the rate of this transfer process to 50°C. The values of rate constants in units of $torr^{1/2}\ sec^{-1}$ are: $k^H_{tr1i} = 5.1 \times 10^{-3}$; $k^H_{tr2i} - 7.2 \times 10^{-3}$; $k^H_{tr1}3 = 2.7 \times 10^{-3}$ and $k^H_{tr2a} = 4.4 \times 10^{-3}$ in propylene polymerizations. Therefore, the differences are less than a factor of 2.5 for k^H_{tr} regardless of the sterospecificity of the active site or whether it is of the first or second kind.

The H_2 chain transfer rate constants in ethylene polymerizations are the same as in propylene polymerizations. At 50° $k^H_{tr1} = 4.0 \times 10^{-3}$ $torr^{1/2}\ sec^{-1}$ and $k^H_{tr2} = 5.1 \times 10^{-3}\ torr^{1/2}\ sec^{-1}$. Furthermore, these values are not dependent on whether an external Lewis base is present or not.

STEREOREGULARITY AND POLYDISPERSITY OF POLYMER

The CW catalyst has several remarkably characteristics. One is that the isotacticity index is very high; it is 96.2 at 50° and A/T = 167. Furthermore, I.I. remains constant during the entire course of polymerization. In column 4 of Table 6 are given the average I.I. values of samples taken at 5, 10, 20, 40, 60 and 90 min each in triplicates. The standard deviation given is that for 18 determinations. The same is true for the other polymerization conditions though the average values of I.I. are smaller at both high and low temperatures or A/T.

The CW catalyst is more stereospecific than the simpler $MgCl_2$-supported catalysts the I.I. values for some examples of the latter systems are given in Table 7. The $MgCl_2/TiCl_4$-$AlEt_3$ system produces polypropylene with I.I. of only 57%. Furthermore, the I.I. is as low as 23% during the first 60 sec. of polymerization. The stereospecificity is not changed if $TiCl_3$ replaces $TiCl_4$. The use of more Lewis acidic $AlEt_2Cl$, as activator, lowers I.I. to 45%. Inclusion of B_e as a coactivator raises I.I. to only 63-72%. Therefore, external Lewis base is not a cure all if many strongly Lewis acidic sites are present in the catalyst. With only B_i, the $MgCl_2$/EB/$TiCl_4$-$AlEt_3$ produces 72% stereoregular polymers. These comparisons showed the CW-catalyst to be definitely superior in stereoselectivity.

Another remarkable characteristic is the narrow MW distribution. The stereoregular polymers obtained at 0° to 50° have $\bar{M}_w/\bar{M}_n \simeq 2$ for the most probable distribution. They are broader at low and high A/T ratios and for 70° polymerizations (Table 6) but the largest $\bar{M}_w/\bar{M}_n$ value is 2.78. For those polymerizations, given the most probable distribution for the isotactic polymers, there is probably only one chemically distinct catalytic site. As the catalyst systems contain fewer modifiers (Table 7), the polydispersity increases. The $\bar{M}_w/\bar{M}_n$ values are 6.3 and 13.1 for the $MgCl_2$/$TiCl_3$-$AlEt_3$ and $MgCl_2/TiCl_3$-$AlEt_2Cl$ systems, respectively. It is reasonable to suppose that there are a multitude of active sites in these catalysts.

The other remarkable characteristics of the CW-catalyst are the invariance of both I.I. and $\bar{M}_w/\bar{M}_n$ with time of polymerization. The constancy

Table 6. Isotacticity and Polydispersity of Polypropylenes Obtained with the CW-Catalyst

Temp. °C	[M], M	A/T	Isotacticity[a] index	$\bar{M}_w/\bar{M}_n$[b]
0	1.46	167		2.04±0.14
25	1.3	167	88.6±1.9	1.84±0.17
50	0.65	334	93.8±0.3	2.35±0.15
50	0.65	84		2.12±0.07
50	0.65	42	90.5±0.6	2.55±0.06
50	0.75	167	95.1±1.4	2.06±0.19
50	0.65	167	96.2±0.9	2.08±0.17
50	0.55	167	95.5±0.5	2.08±0.20
50	0.45	167	95.4±0.6	2.29±0.14
50	0.35	167	96.1±0.6	1.98±0.09
70	0.315	167	84.0±2.4	2.78±0.17

[a]Fraction of polypropylene insoluble in refluxing heptane, the values are average of samples taken at 5, 10, 20, 40, 60 and 90 min polymerization each in triplicates with standard deviations; [b]polydispersity b GPC for the isotactic fraction of the above samples.

of $\bar{M}_w/\bar{M}_n$ suggests that R_p/R_{tr} is time independent. The time invariance of I.I. implies that the ratios $R_{p,i}/R_{p,a}$ and $R_{tr,a}$ between the isospecific and nonspecific sites remain unchanged during the course of polymerization of the CW-catalyst.

The effect of H_2 on the stereoselectivity and polydispersity of propylene polymerization is small in the case of the CW-catalyst. Table 8 shows that for the typical polymerization time of 60 min, the I.I. = 94.6 ± 0.8 independent of H_2 pressure. This suggests that the active site concentration ratios $[Ti_1^*]_i/[Ti_1^*]_a$ and $[Ti_2^*]_i/[Ti_2^*]_a$ are not appreciably altered by the presence of H_2. The values of $\bar{M}_w/\bar{M}_n$ = 2.02 ± 0.13 for the accumulated isotactic polymers is also not dependent on hydrogen concentration. This implies that there is only one chemically distinct isospecific active site in the presence or absence of hydrogen. For the first 6 min of polymerization, there appears to be a slight decrease of I.I. and a small increase of $\bar{M}_w/\bar{M}_n$ with p_{H2}.

CATALYST DEACTIVATION

Even after it has been established that the decay of R_p is due to catalyst deactivation, there are no definite ideas about the actual kinetics nor about the mechanisms of the processes involved. Most of the exper-

Table 7. Isotacticity and Polydispersity of Polypropylenes Obtained with Other $MgCl_2$ Supported Catalysts

Catalyst	Temp. °C	Isotacticity index	$\frac{\bar{M}_w}{\bar{M}_n}$ [a]	Reference
$MgCl_2/TiCl_4$-$AlEt_3$	60	57		c
$MgCl_2/TiCl_4$-$AlEt_2Cl$	60	45		c
$MgCl_2/TiCl_4$-$AlEt_3$/EB	60	64		c
$MgCl_2/TiCl_3$-$AlEt_2Cl$	65	61	13.1	d
$MgCl_2/TiCl_3$-$AlEt_3$	65	58	6.3	d
$MgCl_2/TiCl_3$-$10AlEt_3$/EB	65	87	7.1	d
$MgCl_2/TiCl_4$-$AlEt_3$ [b]	60	23±2.1		e
$MgCl_2/TiCl_4$-$4AlEt_3$/EB [b]	60	63-72	2.84±0.19	f
$MgCl_2$/EB/$TiCl_4$-$AlEt_3$	41	72±5.1	3.3 ±0.4	g,h,i
$MgCl_2$/EB/$TiCl_4$-$AlEt_2Cl$	41		4.1	i

[a]For the stereoregular polymer fraction; [b]short time of polymerization up to ca. 60 sec; [c]Vermel et al. (1980); [d]Soga et al. (1979); [e]Kashiwa and Yoshitake (1983); [f]Kashiwa and Yoshitake (1984); [g]Suzuki et al. (1979); [h]Keii et al. (1984).

imental results can be analyzed according to kinetics second order in [Ti*]. For very short time of propylene (< 10 min) polymerization, the initial activity decay has been said to be better described by third order kinetics.[38] However, Kashiwa and Yoshitake[34,35] found no loss of R_p during the first 60 sec of propylene polymerization for $MgCl_2$-supported catalysts. At high temperatures, the initial rapid second order decay is followed by a first order or another slower second order process.[7,38] For the polymerization at low A/T ratio, i.e., 42, Chien and Kuo[7] found first order decay of R_p. There is no loss of activity for polymerization at 0°.

One may inquire about what is the meaning of kinetic order of active site decay when they are fixed in the support surface. In fact the concentrations of active sites are not intentionally varied. Instead, one assumes that R_p is proportional to [Ti*] and plots $R_{p,t}^{-1} - R_{p,o}^{-1}$ versus time, where $R_{p,o}$ is the maximum rate. When this plot is linear, the catalyst deactivation is said to be kinetically second order. The slope of the plot divided by $k_p[M]$ is k_t. Analogous methods may be used to obtain rate constants of deactivation for the other kinetic orders. Consequently, the kinetic order here means the number of neighboring active sites involved in a deactivation event. When there are clusters containing several Ti_1^* sites there can be simultaneous deactivation of two sites or there may be three sites (vide supra). The deactivation process leave sites Ti_2^* which are

Table 8. Effect of Hydrogen on the Stereospecificity and Polydispersity of CW-catalyzed Propylene Polymerization[a]

p_{H_2}	6 min polymerization		60 min polymerization	
torr	I.I. %	$\bar{M}_w/\bar{M}_n$ [b]	I.I. %	$\bar{M}_w/\bar{M}_n$ [b]
0	96.0	2.10	96.0	2.03
12	95.6	1.98	95.6	1.98
24	95.6	2.07	95.6	1.92
36	95.2	2.36	94.7	1.97
48	93.6	2.37	94.0	1.82
60	95.2		94.0	
72	92.6	2.50	93.8	2.09
84	93.8		93.4	
96	94.8	2.51	94.4	2.18
108	92.4		94.7	
120	92.0	2.36	94.7	2.20
Av. ± S.D.		2.28 ± 0.2	94.6 ± 0.8	2.02 ± 0.13

[a]Polymerization conditions: [Ti] = 2.4×10^{-4} M, A/T - 167, $[C_3H_6]$ = 0.65 M, temp. = 50°; [b]for the stereoregular polymer fraction.

without neighboring active sites for second order deactivation. Consequently, they are much more stable and are responsible for the relatively constant polymerization, following the initial rapid decay of R_p. The Ti_2^* sites can decay by first order kinetics, such as homolysis or reduction of aluminumalkyls. When the catalyst is activated by small amounts of aluminumalkyls, at low A/T, or at low temperature, there are more isolated active sites than clustered sites. Then the catalyst may decay by first order kinetics or not at all.

For the soluble catalyst $(\pi\text{-}C_5H_5)_2TiCl_2\text{-}AlMe_2Cl_4$ the termination process was shown to be bimolecular reductive disproportionation

$$2(\pi\text{-}C_5H_5)_2\underset{\displaystyle P}{\overset{\displaystyle Cl}{Ti}}\cdot AlMe_2Cl \longrightarrow 2(\pi\text{-}C_5H_5)_2TiCl\cdot AlMe_2Cl + P_{+H} + P_{-H} \qquad (23)$$

Until recently it was assumed previously thhat analogous reactions occur in

$MgCl_2$-supported catalyst systems. However, several important observations showed this to be untenable.

The MPB concentration at time t is

$$[MPB]_t = [Ti^*]_t + \frac{k_{tr}^A Y_t}{k_p [M]} \left(\frac{K_D^{1/2} [A]^{1/2} K_{ab}}{1 + K_D^{1/2} [A]^{1/2} K_{ab}} \right) \quad (24)$$

where $[Ti^*]_t$ is the concentration of propagating species at time t, Y_t is the polymer yield, and K_{ab} is the absorption coefficient of monomeric $AlEt_3$. The time dependence of [MPB] during the course of polymerization may be due either to formation of new MPB by chain transfer with $AlEt_3$ or the production of chains not bonded to a metal atom through chain transfer with monomer and hydrogen and via β-hydride elimination contained in the second term on the right-hand-side of eq. 25.

However, during early stage of the polymerization, the yield is low and the contribution of $[Ti^*]_t$ to $[MPB]_t$ is large. This is also the period of rapid decrease of $[Ti_1^*]$. In all the measurements[6,8,9,11,12] there was no loss of [MPB] during this period but rather there is only either a slightly steady or no increase of $[MPB]_t$ with t. This means that the principal mechanisms of catalyst deactivation is not via a process like eq. 23, but are those which retains the Ti-P bond.

It was recognized only very recently[7,11,12] that the catalyst deactivation can involve monomer, i.e.

$$-\frac{d[Ti_1^*]}{dt} = \{k_t^M [M]^{1/2} + k_t^H [p_{H_2}]^{1/2}\} [Ti_1^*]^2 \quad (25)$$

The values of these rate constants have been obtained at 50° for isospecific polymerization of polypropylene
$(torr)^{-1/2}$ sec^{-1}.

For ethylene polymerization, the hydrogen-assisted terminations have rate constants of 1.7 and $01.4M^{-1/2}$ $torr^{-1/2}$ sec^{-1} with and without B_e, respectively. The participation of monomer is catalyst deactivation has not yet been demonstrated for ethylene polymerization in the absence of hydrogen as it has been shown in the case of propylene. Hence, we can only give the apparent rate constant k_t' without inclusion of the monomer dependence. The values at 25°, 50° and 70° are 4.2, 3.3, $6.6M^{-1}$ sec^1, respectively, for normal ethylene polymerization; they are 1.85, 0.86, and 1.54, respectively, in the absence of B_e.

The mechanisms of deactivation of catalytic sites are extremely difficult to elucidate for heterogeneous catalysts. For the classical Ziegler-Natta catalysts, the active sites are so few that any trace amount of impurities or inhibitors produced during the activation or polymerization may poison them. With the increase of number of active sites in the $MgCl_2$-supported catalysts, it is still doubtful that the mechanism of deactivation can be elucidated because of the nature of chemisorbed species are not known. We can only offer some speculations.

For the monomer-assisted catalytic site deactivation, the rate is proportional to $[M]^{1/2}$. Therefore, the monomer adsorbed on the surface (s) is probably dissociated

$$C_3H_6(s) \rightleftharpoons F(s) + G(s) \quad (26)$$

Several dissociative processes may be postulated. If F is an allyl radical (s) and G is a H·(s), then the deactivation may be represented by

$$P\text{-}Ti^{*}(s) + CH_2{=}CH\text{-}CH_2\cdot(s) \longrightarrow P\text{-}Ti\text{-}CH_2\text{-}CH{=}CH_2(s) \quad (27)$$

$$P\text{-}Ti^{*}(s) + H\cdot(s) \longrightarrow P\text{-}Ti\text{-}H(s) \quad (28)$$

Another possibility is that if F(s) is the vinyl radical (s) and G(s) is the methyl radical (s), the deactivation may be proposed to be

$$P\text{-}Ti^{*}(s) + CH_2{=}CH\cdot(s) \longrightarrow P\text{-}Ti\text{-}CH{=}CH_2(s) \quad (29)$$

$$P\text{-}Ti^{*}(s) + CH_3\cdot(s) \longrightarrow P\text{-}Ti\text{-}CH_3(s) \quad (30)$$

In the above processes, the inactive species are Ti(IV). Alternatively, the process could be reductive, i.e.,

$$P\text{-}Ti^{*}\text{-}Cl(s) + CH_2{=}CH\text{-}CH_2\cdot(s) \longrightarrow P\text{-}Ti(s) + CH_2{=}CH\text{-}CH_2Cl \quad (27')$$

$$P\text{-}Ti^{*}\text{-}Cl(s) + H\cdot(s) \longrightarrow P\text{-}Ti(s) + HCl \quad (28')$$

$$P\text{-}Ti^{*}\text{-}Cl(s) + CH_2{=}CH\cdot(s) \longrightarrow P\text{-}Ti(s) + CH_2{=}CHCl \quad (29')$$

$$P\text{-}Ti^{*}\text{-}Cl(s) + CH_3(s) \quad P\text{-}Ti(s) + CH_3Cl \quad (30')$$

If F is CH=CH and G is CH_4, then the deactivation may be via oxidative coupling.

$$2P\text{-}Ti^{*}(s) + CH{=}CH \longrightarrow P\text{-}Ti\text{-}CH{=}CH\text{-}Ti\text{-}P(s) \quad (31)$$

The mechanism of hydrogen-assisted deactivation depends upon the nature of the dissociated species on the catalyst. Either oxidation, reduction, or formation of bridge hydrides could cause deactivation.

It would be rather difficult to ascertain what is the change of oxidation state occurs during the deactivation Ti_1^* sites. Redox titration techniques, used to determine the Ti oxidation states in the catalyst,[13] cannot be applied to polymer encapsulated catalyst. Methods such as ESCA or EXAF would require hard vacuum which may alter the system. Indirect evidence obtained with model systems, i.e. internal olefins, may be an approach worth pursuing.

Based on the above discussion on catalyst deactivation, there should not be marked differences between the Ti_1^* and Ti_2^* sites. They have comparable kinetic parameters for chain propagation and all the chain transfer processes. The main difference is the absence of neighboring active sites for Ti_2^* and therefore its greater stability than the Ti_1^* sites. The other difference is their reactivity toward CO. If the failure of Co incorporation is the formation of stable carbonyl compounds, as suggested in eq. 11, then this reaction may be less likely if the active Ti(III) ion is Cl bridged to either inactive Ti(IV) or Ti(II) ions.

CONCLUSION

The high performance of the $MgCl_2$-supported CW-catalyst is achieved by complicated and sophisticated synthetic methods. Hydrolysis GC-MS, FTIR and elemental analysis has been used to follow the changes in the chemical composition of the support, BET and mercury porometry was used to monitor the increases in the gross and internal surface areas, X-ray diffraction was used to obtain estimates for the crystallite dimension, EPR and redox titration were used to determine the distribution of Ti oxidation states, ^{14}CO and CH_3O^3H to radiotag the propagating chains and metal polymer bonds, and GPC was used to give the MW distributions. The results of characterization led to an in-depth understanding of the physical, morphological and chemical structures of the catalysts and possible nature of the active sites.

This catalyst was used in propylene and ethylene polymerizations and rate constants of propagation and chain transfer processes determined. The effects of hydrogen and Lewis base were investigated. These results were interpreted in terms of the postulated active sites.

This paper summarizes those solid experimental facts, active site concentrations, polymer properties, and kinetic parameters. There are, however, much speculative discussions which we hope will stimulate further research. These interpretations are meant for this particular CW-catalyst and any application to other $MgCl_2$-supported catalysts should be done with caution and deliberation.

ACKNOWLEDGEMENT

Acknowledgement is made to the donors of the Petroleum Research Foundation, administered by the American Chemical Society for support of this research.

REFERENCES

1. G. Allegra, Nuovo Cimento, 23, 502 (1962).
2. P. C. Barbe, L. Noristi, and G. Baruzzi, Makromol. Chem. Rapid Commun., 4, 249 (1983).
3. V. Busico, P. Corradini, L. deMartino, A. Proto, V. Savino, and E. Albizzati, 186, 1279 (1985).
4. J. C. W. Chien, J. Am. Chem. Soc., 81, 86 (1959).
5. J. C. W. Chien, T. Ang., and C. I. Kuo, J. Polym. Sci. Polym. Chem. Ed., 23, 723 (paper no. VI of the series) (1985).
6. J. C. W. Chien and C. I. Kuo, J. Polym. Sci. Polym. Chem. Ed., 23, 731 (paper no. VII of the series) (1985a).
7. J. C. W. Chien and C. I. Kuo, J. Polym. Sci. Polym. Chem. Ed., 23, 761 (paper no. VIII of the series) (1985b).
8. J. C. W. Chien and C. I. Kuo, J. Polym. Sci. Polym. Chem. Ed., submitted (paper no. IX of the series) (1986a).
9. J. C. W Chien and C. I. Kuo, J. Polym. Sci. Polym. Chem. Ed., submitted (paper no. X of the series) (1986b).
10. J. C. W. Chien and T. Ang., J. Polym. Sci. Polym. Chem. Ed., submitted (paper no. XI of the series).
11. J. C. W. Chien and P. Bres, J. Polym. Sci. Polym. Chem. Ed., submitted (paper no. XII of the series) (1986a).
12. J. C. W. Chien and P. Bres, J. Polym. Sci. Polym. Chem. Ed., submitted (paper no. XIII of the series) (1986b).
13. J. C. W. Chien, J. C. Wu, and C. I. Kuo, J. Polym. Sci. Polym. Chem. Ed., 20, 2019 (1982).
14. J. C. W. Chien and J. C. Wu, J. Polym. Sci. Polym. Chem. Ed., 20, 2445 (1982a).
15. J. C. W. Chien and J. C. Wu, J. Polym. Sci. Polym. Chem. Ed., 20, 2461 (1982b).
16. J. C. W. Chien, J. C. Wu, and C. I. Kuo, J. Polym. Sci. Polym. Chem. Ed., 21, 725 (1983a).
17. J. C. W. Chien, J. C. Wu, and C. I. Kuo, J. Polym. Sci. Polym. Chem. Ed., 21, 737 (1983b).

18. K.-Y. Choi and W. H. Ray, Rev. Macromol. Chem. Phys., C-25, 1 (1985).
19. N. B. Chumaevskii, V. A. Zakharov, G. D. Bukatov; G. I. Kuznetsova, and Yu.I. Yermakov, Makromol. Chem., 177, 747 (1976).
20. P. Corradini, V. Barone, R. Fusco, and G. Guerra, Gaz. Chim. Ital., 113, 601 (1983).
21. P. Corradini, G. Guerra, R. Fusco, and V. Barone, Eur. Polym. J., 16, 835 (1980).
22. Y. Doi, M. Murata, K. Yano, and T. Keii, Ind. Eng. Chem. Prod. Res. Dev., 21, 580 (1982).
23. Y. Doi, M. Murata, and K. Yano, Ind. Eng. Chem. Prod. Res. Dev., 21, 580 (1982).
24. Y. Doi, E. Suzuki, and T. Keii, Makromol. Chem. Rapid Commun., 2, 293 (1981).
25. E. W. Duck, D. Grant, and E. Kronfli, Eur. Polym. J., 15, 625 (1979).
26. C. Dumas and C. C. Hsu, Rev. Macromol. Chem. Phys., C-24, 355 (1984).
27. A. Ferrari, A. Braibauti, and G. Bigliardi, Acta Cryst., 16, 846 (1963).
28. P. Galli, P. C. Barbe, and L. Nortisti, Angew. Makromol. Chem., 120, 73 (1984).
29. P. Galli, L. Luciani, and O. Cecchim, Angew. Makromol. Chem., 94, 63 (1981).
30. P. Galli, P. Barbe, G. Guidetti, R. Zannetti, A. Martorana, A. Marigo, M. Bergozza, and A. Fichera, Eur. Polym. J., 19, 19 (1983).
31. U. Giannini, Makromol. Chem. Suppl., 5, 216 (1981).
32. F. J. Karol, B. E. Wagner, L. J. Levine, G. L. Goeke, and A. Nushay, this volume (1986).
33. N. Kashiwa, Polym. J., 12, 603 (1980).
34. N. Kashiwa and J. Yoshitake, Makromol. Chem. Rapid Commun., 4, 41 (1983).
35. N. Kashiwa and J. Yoshitake, Polym. Bull., 12, 99 (1984).
36. T. Keii, Y. Doi, E. Suzuki, M. Tamura, M. Murata, and K. Soga, Makromol. Chem., 185, 1537 (1984).
37. T. Keii, E. Suzuki, M. Tamura, and Y. Doi, Proc. MMI Int. Sym. on Transition Metal Catalyzed Polymerization: Unsolved Problems, Midland, p. 97 (1981).
38. T. Keii, E. Suzuki, M. Tamura, M. Murata, and M. Doi, Makromol. Chem., 183, 2285 (1982).
39. T. Keii, Y. Doi, E. Suzuki, M. Tamura, M. Murata, and K. Soga, Makromol. Chem., 185, 1537 (1984).
40. D. B. Ludlum, A. W. Anderson, and C. E. Ashby, J. Am. Chem. Soc., 96, 1737 (1979).
41. P. Mackie, M. N. Berger, B. M. Grieveson, and D. Lawson, J. Polym. Sci., B5, 493 (1967).
42. J. Mejzlik and M. Lesna, Makromol. Chem., 178, 261 (1977).
43. M. Mejzlik, M. Lesna, and J. Majer, Makromol. Chem., 184, 1975 (1983).
44. Munoz-Escalona and J. G. Hernandez, this volume (1986).
45. G. Natta, Chim. Ind. (Milan), 41, 519 (1959).
46. G. Natta, Angew. Chem., 68, 393 (1956).
47. G. Natta, P. Corradini, and G. Allegra, J. Polym. Sci., 51, 399 (1961).
48. G. Natta, I. Pasquon, J. Svab, and A. Zambelli, Chim. Ind. (Milan), 44, 621 (1962).
49. S. Pasynkiewicz and K. Strowieyski, Rocz. Chem., 41, 1139 (1967).
50. P. Pino and R. Malhaput, Angew. Chem. Int. Ed. Engl., 19, 857 (1980).
51. P. Pino, G. Guastalla, B. Rotzinger, and R. Malhaupt, Proc. MMI Int. Sym. on Transition Metal Catalyzed Polymerization: Unsolved Problems, Midland, p. 435 (1981).
52. K. Soga, M. Terano, and S. Zkeda, Polym. Bull., 1, 849 (1979).
53. K. Soga, S. I. Cheng, and Y. Doi, this volume (1986).
54. E. Suzuki, M. Tamura, Y. Doi, and T. Keii, Makromol. Chem., 180, 2235 (1979).

55. P. J. T. Tait, Proc. MMI Int. Sym. on Transition Metal Catalyzed Polymerization: Unsolved Problems, Midland (1981).
56. Ye. Ye. Vermel, V. A. Zakharov, Z. K. Bukatova, O. P. Shkurina, L. G. Yechevskaya, E. M. Moroz, and S. V. Sudakova, Polym. Sci. USSR, 23, 23 (1980).
57. V. Warzelhau, T. F. Burger, and D. J. Stein, Makromol. Chem., 183, 489 (1982).
58. Yu. I. Yermakov, V. A. Zakharov, and G. D. Bukatov, Proc. 5th Int. Congr. on Cat., 1, 399 (1973).
59. Yu. I. Yermakov and V. A. Zakharov, Coordination Polymerization, Ed. J. C. W. Chien, Academic Press, N.Y., p. 91 (1975).
60. V. A. Zakharov, G. D. Bukatov, N. B. Chumaevskii, and Yu. I. Ermakov, Kin. I. Kat., 18, 848 (1977).
61. V. A. Zakharov, G. D. Bukatov, and Yu. I. Yermakov, Polym. Sci. Tech., 19, 267 (1983).

POLYMERIZATION OF OLEFINS WITH

MAGNESIUM CHLORIDE-SUPPORTED CATALYSTS

S. Sivaram and P. R. Srinivasan

Research Centre
Indian Petrochemicals Corporation Limited
Baroda INDIA 391 346

INTRODUCTION

Activated magnesium chloride has been established as an ideal support for titanium-based high efficiency olefin polymerization. One of the earliest disclosed methods was to comill anhydrous magnesium chloride and titanium tetrachloride for prolonged periods either alone[1] or in presence of an electron donor, such as ethyl benzoate (EB).[2,3] A later variant involves comilling anhydrous $MgCl_2$ with an electron donor (typically EB) followed by treatment with $TiCl_4$ in the liquid phase.[4] In some cases prior to reaction with $TiCl_4$, the milled $MgCl_2$.EB complex is also treated with a second electron donor and an organoaluminum compound.[5,6]

Unlike these processes which involve physical milling of $MgCl_2$ (physical activation), more recent literature has alluded to the possibility of support activation using only chemical reactions with $MgCl_2$ (chemical activation).[7] The ability to activate $MgCl_2$ without use of mechanical forces is believed to be responsible for the development of supported titanium catalysts capable of yielding polyolefins with spherical shape and controlled particle size distribution.[8]

In spite of the rapid advances in practical catalysis, a detailed understanding of the nature of morphological changes occurring in the support during the process of activation and its effect on catalyst performance has only recently started to emerge. The primary effect of milling $MgCl_2$ with $TiCl^4$ is to break the layered structure of $MgCl_2$ crystal and to create rotational disorders of Cl-Mg-Cl triple layers.[9-11] With prolonged milling, crystallite sizes progressively decrease, whereas surface area initially increases, up to a certain time, beyond which it either remains constant or decreases.[9,12] This has been attributed to crystal aggregation. $MgCl_2$ and $TiCl_4$ are believed to result in the formation of solid solutions similar to that formed by $TiCl_3$ with $AlCl_3$.[13]

Milling $MgCl_2$ with EB also results in the formation of a new crystalline phase. Conflicting reports have appeared on the nature of changes occurring in the support, as a result of this process. Whereas some authors observed a modest increase in surface area,[14,15] Sergeev et al. reported a decrease in the surface area.[16] Crystallite sizes were reported to decrease progressively with increased milling time.

The observation that the reduction in crystallite size was not accompanied by an increase in surface area led Chien to suggest that the small crystallites are held together as an aggregate by complexation with EB. Treatment of $MgCl_2$.EB complex with $TiCl_4$ resulted in the part by the removal of EB as $TiCl_4$.EB complex, penetration of $TiCl_4$ into the $MgCl_2$ lattice and a large increase in surface area.

It has been suggested that the resultant cleavage surfaces of $MgCl_2$ viz., the (110) and (101) plane contain magnesium atoms which are coordinatively unsaturated which can form strongly bonded surface complexes with $TiCl_4$.[13] This concentrates titanium on the lateral faces and crystal edges of $MgCl_2$, similar to that found in $TiCl_4$.[17] Ethyl benzoate thus assists in breaking down the $MgCl_2$ crystallites and stabilizing the surface complex between electropositive magnesium atoms bound to titanium through double halogen bridges.[18] Although ethyl benzoate has been most widely used, it appears that its role is not specific and many other Lewis bases are capable of similar effects.

In contrast, no report has appeared thus far on the nature of chemical, physical and structural changes occurring in $MgCl_2$ during the process of chemical activation. The present paper attempts to delineate these changes occurring at the various steps involved in catalyst preparation.

EXPERIMENTAL

Material

$MgCl_2.6H_2O$ (s. Merck, India), ethyl benzoate (S. Merck, India), $SiCl_4$ (Aldrich) and diethylaluminum chloride (Ethyl Corp.) were used as received. $TiCl_4$ (S. Merck, India) was distilled under N_2 prior to use. All solvents were dried by distilling over triethylaluminum and were stored under N_2 over sodium wire. All other reagents used in the study were purified by standard methods. All operations involving preparation of catalysts were carried out inside a dry box under a stream of high purity nitrogen.

Methods

Powder x-ray diffraction was obtained with a Phillips instrument operating at 40 K.V and 20 ma using a Ni filtered Cu-K_α radiation. The sample was covered with a thin Mylar film and loaded into the diffractometer. The diffraction pattern was recorded in the range $5° < 2\theta > 60°$. The crystallite sizes were measured in the (101), (001) and (110) direction using the Laue-Scherer equation. The BET surface area was determined using a Carlo Erba Sorptomat Model 1810. The porosity was calculated from surface area values. The estimation of Mg and Cl was carried out using EDTA and argentimetric titrations. Titanium estimation was carried out volumetrically by redox titration.[21]

Anhydrous Magnesium Chloride

$MgCl_2.6H_2O$, 150g was placed in a three necked round bottom flask, fitted with a mechanical stirrer, a nitrogen inlet and outlet. The flask was heated to 135°C with stirring, under a steady stream of nitrogen, for 48 hr. The resulting white powder was preserved under nitrogen. Analysis, Exptl: Mg=25.42%, Cl=73.9%; Calc; Mg=25.49%, Cl=74.5%.

Preparation of $MgCl_2$.6EtOH Adduct

To 5g (52 m moles) of anhydrous $MgCl_2$, 50 mL of absolute ethanol was added and the mixture was refluxed under N_2 for 6 hr. The solution was concentrated and allowed to cool to 0°C. The separated solid was filtered and dried in vacuo. Yield: 19.5g; Analysis, Exptl: Mg=6.50%, Cl=19.08%; Calc. Mg=6.54%; Cl=19.12%.

Preparation of $MgCl_2$.2EHA Adduct

To 5g (52 m moles) of anhydrous $MgCl_2$, 50 ml of absolute ethanol was added and the mixture was refluxed under N_2 for 6 hr. The solution was concentrated under vacuum, cooled to 0°C and the separated solid filtered and dried in vacuo. Yield: 21g; Analysis, Exptl: Mg=6.81%, Cl=19.3%; Calc. Mg=6.84%, Cl=19.9%.

Reaction of $MgCl_2$.6EtOH with EB

To 5g (13 m moles) of $MgCl_2$, 6EtOH in 50 ml heptane was added 0.5 ml (3 m moles) of EB. The mixture was heated to 70°C. The solid was then filtered, washed and dried in vacuo. Yield: 5.4g; Analysis for $MgCl_2.(EtOH)_{6.0}(EB)_{0.15}$, Exptl: Mg=6.08%, Cl=18.01%; Calc. Mg=6.17%, Cl 18.03%.

Reaction of $MgCl_2$.6EtOH with STC and EB

To 5g (13 m moles) of $MgCl_2$.6EtOH was added 20 ml of STC and the mixture was heated at 60°C for 10 hr. The HCl liberated during the reaction was quantitatively estimated. The excess STC was distilled off 50ml. Heptane and 0.5 ml EB was added and the reaction continued for another 4 hr. The solid was separated, washed with heptane and dried in vacuo. Yield: 7.1 g; Analysis for $MgCl_2.(EtOH)_{4.8}[Si(OEt)_4]_{1.1}(EB)_{0.2}$, Exptl: Mg=4.0%, Cl=12.1%; Calc: Mg=4.2%, Cl=12.3%.

Reaction of $MgCl_2$.6EtOH with DEAC and EB

To 5g (13 m moles) of $MgCl_2$.6EtOH was added 20 ml of a 15% by wt. solution of DEAC in heptane and the mixture was heated to 90° for 6 hr. 0.8 ml, EB was added and the reaction continued for another 4 hr. The solid was filtered, washed and dried in vacuo. Yield: 6.8g; Analysis for $MgCl_2\ (EtOH)_{3.5}\ (AlEt(OEt)_2)_{2:3}\ (EB)_{6.15}$ Exptl: Mg=7.6%, Al=15.0%, Cl=12.8%; Calc: Mg=4.0%, Al=10.1%, Cl=11.55%.

Similar procedures were followed for the reaction of $MgCl_2$.2 EHA adduct with STC and DEAC.

Reaction of Activated Supports with $TiCl_4$

To 2g of $MgCl_2.(EtOH)_{4.8}\ (Si(OEt)_4)_{1.1}\ (EB)_{0.2}$ complex in 100 ml of heptane was added 8 ml of $TiCl_4$ and the mixture was heated for 3 hr at 80°C. The resulting solid was filtered through a sintered glass crucible, washed with heptane until the filtrate was free of $TiCl_4$ and dried in vacuo. Yield: 3.1 g; Analysis: Ti=2.2%; Mg=8.2%; Cl=22.4%.

RESULTS AND DISCUSSION

Anydrous Magnesium Chloride

Anhydrous magnesium chloride prepared by drying $MgCl_2.6H_2O$ at 135°C showed a hexagonal close packed structure (hcp) similar to δ-$TiCl_3$. (Figure 1a). The XRD showed a strong (101) reflection at d=2.71 Å (32°), (110) reflection at d=1.82 Å (50°) and (001) reflection at d=5.85 Å (15°). The sample had an initial BET surface area of 20-30 m^2/g versus a value

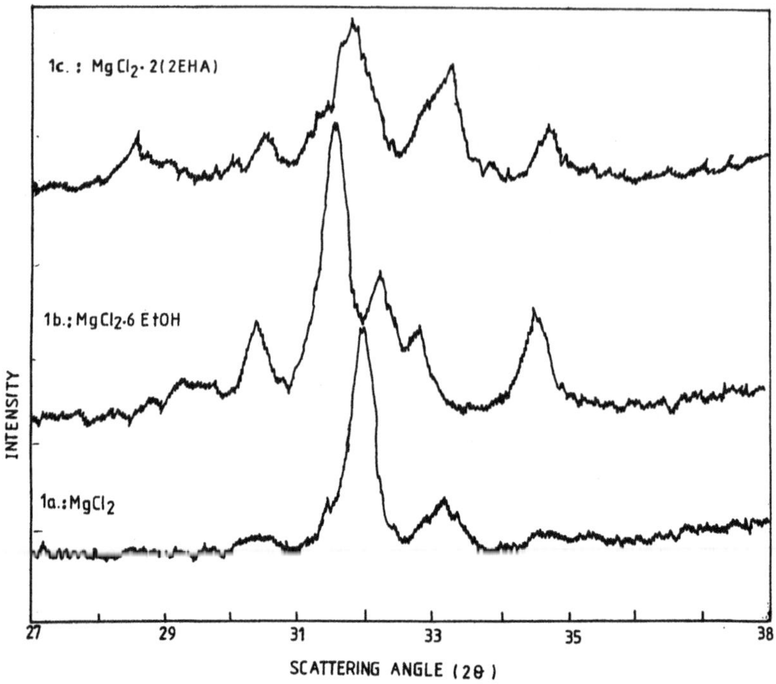

Fig. 1. X-ray Powder Diffraction of $MgCl_2$.Lewis Base Adduct

of 1-2 m^2/g reported for commercial $MgCl_2$.

A hcp structure for $MgCl_2$ has been rarely reported.[19] The most frequently observed structure is the cubic closed packing (ccp), which is thermodynamically more stable and is characterized by a strong (104) reflection at d=2.56 Å (35°) and less intense reflections at (003) plane (d= 5.85 Å, 15°) and (110) plane (d=1.82 Å, 50°).

Reaction of Anhydrous Magnesium Chloride with a Lewis Base

The first step in the process of chemical activation, usually involves the reaction of magnesium chloride with a Lewis base. In our study, we chose a series of aliphatic alcohols and phenols, as Lewis bases. It was found that alcohols solubilized anhydrous $MgCl_2$ near their boiling points resulting in the formation of well defined molecular complexes (Table 1). The BET surface area increased two to four fold and the crystallite sizes decreased. Ethylene glycol and phenol were generally less effective than monohydric aliphatic alcohols in destroying the crystalline nature of $MgCl_2$. Alcohols with higher carbon numbers as well as branching caused a more efficient disruption of $MgCl_2$ layer structure (Table 2).

The XRD spectra of $MgCl_2$.6EtOH and $MgCl_2$.2EHA are shown in Figures 1b anc 1c.

Reaction of $MgCl_2$.ROH Adducts with a Second Additive

Two adducts viz, $MgCl_2$.6EtOH and $MgCl_2$.2EHA were reacted with one the following additives: ethylbenzoate (EB), silicon tetrachloride (STC) and diethylaluminumchloride (DEAC).

Reaction of $MgCl_2$.6EtOH with EB led to the formation of a new compound, $MgCl_2.(EtOH)_6.(EB)_{0.15}$. Repeated washing with an inert

Table 1. Reaction of Anhyd.$MgCl_2$ With Lewis Bases

R-OH	TEMP., °C	COMPOSITION	S.AREA, m^2-g^{-1}	D_{101}, Å	D_{001}, Å	D_{110}, Å
-	-	$MgCl_2$	30	300	130	68.4
ETHANOL	70	$MgCl_2$.6EtOH	70	186	98	65.3
2-ETHYL HEXANOL(EHA)	120	$MgCl_2$.2EHA	120	158	79	64.6
HEXANOL	150	$MgCl_2$.2 HEXANOL	105	170	100	66.1
p-CRESOL	200	$MgCl_2$.*p*-CRESOL	85	195	113	66.8
2,6-*Di*-*t*-BUTYL-*p*-CRESOL(BHT)	120	$MgCl_2$.BHT	42	254	123	67.0
ETHYLENE GLYCOL	140	$MgCl_2$.EG	92	175	108	66.8

Table 2. Effect of Branching in Alcohol on $MgCl_2$.Alcohol Properties

R-OH	TEMP. °C	COMPOSITION	S.AREA, m^2g^{-1}	D_{101}, Å	D_{001}, Å	D_{110}, Å
n- BUTANOL	100	$MgCl_2$.2.5(BUTANOL)	95	183	107	66.9
s-BUTANOL	110	$MgCl_2$.2(*s*-BUTANOL)	90	180	105	66.4
t-BUTANOL	100	$MgCl_2$.2(*t*-BUTANOL)	110	165	84	64.4

Table 3. Reaction of $MgCl_2$.6EtOH With a Second Additive

ADDITIVE	COMPOSITION	S.AREA, m^2g^{-1}	D_{101}, Å	D_{001}, Å	D_{110}, Å
NIL	$MgCl_2$.6EtOH	70	186	98	65.3
EB	$MgCl_2 \cdot (EtOH)_6 \cdot (EB)_{0.15}$	148	132	79.3	60.1
STC	$MgCl_2 \cdot (EtOH)_{4.8}(Si(OEt)_4)_{1.1}$	172	110	54.8	59.6
STC+EB	$MgCl_2 \cdot (EtOH)_{4.8}(Si(OEt)_4)_{1.1}(EB)_{0.2}$	193	92	43.1	58.2
DEAC	$MgCl_2 \cdot (EtOH)_{3.5}(EtAl(OEt)_2)_{2.3}$	154	125	65.7	59.7
DEAC+EB	$MgCl_2 \cdot (EtOH)_{3.5}(EtAl(OEt)_2)_{2.3} \cdot (EB)_{0.15}$	165	100	50.3	58.5

EB: ETHYL BENZOATE STC: SILICON TETRACHLORIDE DEAC = DIETHYLALUMINUM CHLORIDE

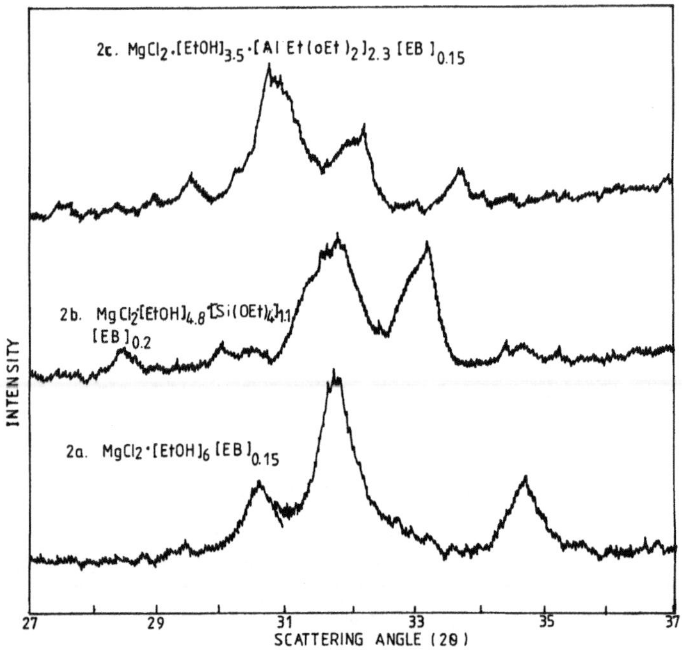

Fig. 2. X-ray Powder Diffraction of the Reaction Product of $MgCl_2$.Lewis Base With Second Additive

Table 4. Comparison of Properties of $MgCl_2$.EB Adduct Obtained by Physical Milling and Chemical Reaction

COMPOSITION	MILLING TIME, HR	D_{110}, Å	S.AREA, m^2g^{-1}	REFERENCE
$MgCl_2\cdot(EB)_{0.17}$	NIL	120	70	16
	2	80	6	
$MgCl_2\cdot(EB)_{0.15}$	4	85	6.5	15
$MgCl_2\cdot(EtOH)_6\cdot(EB)_{0.15}$	NIL	60	148	THIS WORK

Table 5. Reaction of $MgCl_2.2EHA$ With a Second Additive

ADDITIVE	COMPOSITION	S.AREA, m^2g^{-1}	D_{101}, Å	D_{001}, Å	D_{110}, Å
NIL	$MgCl_2 \cdot 2EHA$	120	158	79	64.6
EB	$MgCl_2 \cdot (EHA)_2 \cdot (EB)_{0.14}$	142	144	69.8	59.2
STC	$MgCl_2 \cdot (EHA)_{1.5} \cdot (Si(OR)_4)_{0.5}$	150	134	61.2	58.3
STC+EB	$MgCl_2 \cdot (EHA)_{1.5}(Si(OR)_4)_{0.5} \cdot (EB)_{0.15}$	155	130	56.8	58.0
DEAC	$MgCl_2 \cdot (EHA)_{1.1} \cdot (EtAl(OR)_2)_{0.9}$	148	149	68.3	59.4
DEAC+EB	$MgCl_2 \cdot (EHA)_{1.1}(EtAl(OR)_2)_{0.9} \cdot (EB)_{0.14}$	153	136	58.8	58.3

EHA: 2-ETHYL HEXANOL EB: ETHYL BENZOATE STC: SILICON TETRACHLORIDE DEAC: DIETHYLALUMINUM CHLORIDE - OR : 2-ETHYL HEXANOL

not change the composition indicating complexation. The composition of this adduct, with respect to EB, is similar to that which Chien and co-workers reported obtained after ball milling $MgCl_2$ and EB for 162 hr.[20] The reaction with EB did not displace any of the alcohol already complexed with $MgCl_2$, but caused an increase in surface area and further reduction in crystallite sizes (Table 3, Figure 2a).

Of the three important planes, namely (110), (101) and (001), the (001) plane represent the plane perpendicular to chlorine layer while (110) and (101) represent the planes along the chlorine layers. In the structure of $MgCl_2$, individual Cl-Mg-Cl layers are held by weak van der Waal's forces, whereas within layers, stronger ionic and even covalent bonds dominate.

As observed during milling, treatment of $MgCl_2$ with EtOH and EB appears to cause a breakdown of crystallites, both in the (001) and in (110) direction. However, the breakdown is more rapid in the energetically favored (001) direction than in the (110) direction. Earlier reports on physical milling of $MgCl_2$ with EB indicate that prolonged milling causes the small $MgCl_2$ crystallites to aggregate, resulting in a smaller surface area than anticipated[15,17] (Table 4). In the case of chemical activation, such crystallite aggregation does not occur and results in small crystallite dimensions having high BET surface area.

It is also observed that the value of particle size along the (110) direction is smaller with the $MgCl_2.(EtOH)_6\ (EB)_{0.15}$ compared to milled samples of $MgCl_2.(EB)_{0.15}$. Similar results were obtained with $MgCl_2.2EHA$ (Table 5). The reaction of $MgCl_2$. alcohol adducts with STC was conducted 60°C. HCl was evolved and part of the ethanol (or 2-ethylhexanol) complexed with $MgCl_2$ was converted to $Si(OR)_4$.

The composition and properties of the new compounds formed are shown in Table 4 and 5. Further reaction of this compound with EB led to almost quantitative 90% adsorption of EB (0.2 moles/mole $MgCl_2$), further increase in surface area and a decrease in crystallite size (Figure 2b).

Use of silicon halides in the preparation of high efficiency olefin polymerization catalysts has been recently reported.[7,21] It was suggested that the role of silicon halide is to react with the alcohol and at least partially remove it from the $MgCl_2$.alcohol adduct.[7] Our results show that only about 20 mole % of the alcohol is lost from the $MgCl_2$.alcohol adduct. Under similar conditions free ethanol reacted quantitatively with STC giving silicon tetraethoxide.

Silicon tetraethoxide by itself caused negligible change in the properties $MgCl_2$. However, STC and EB caused substantial changes in the crystal structure of $MgCl_2$. STC was far less effective in the case of $MgCl_2$. EHA adduct although it displaced approximately 25 mole % of the alcohol. (Table 5).

In the $MgCl_2$. alcohol adducts, the crystallites are probably held together by complexation with alcohols in the form of larger agglomerates. The role of STC is to remove these alcohols by reaction, thus causing the collapse of the agglomerates. The reason why STC is more effective with the ethanol adduct compared to 2-EHA adduct may be due to the relatively smaller size of primary crystallites being held together by 6 moles of complexed ethanol as against 2 moles of 2-ethylhexanol.

The reaction of $MgCl_2$. alcohol adduct with DEAC also resulted in the partial displacement of alcohol (Table 4 and 5). However, reaction with DEAC and subsequently with EB, resulted only in a marginal increase in surface area and a decrease in crystallite sizes (Figure 2c). It is apparent that whereas DEAC is superior to STC in displacing the alcohol from the adduct, STC is far more effective in activating the support. Further treatment of these compounds with EB causes absorption of EB and a further increase in surface area/decrease in crystallite sizes.

<u>Reaction of Activated $MgCl_2$ with $TiCl_4$</u>

The various adducts prepared previously were treated with excess $TiCl_4$ at 80°C in heptane as solvent. The composition and properties of the product after washing free of excess $TiCl_4$ are shown in Table 6. XRD pattern (Figure 3) shows broadening of the peaks around $2\theta=32.3°$, indicating the diffusion of $TiCl_4$ and further cleavage of $MgCl_2$ surfaces.

In the case of catalyst containing aluminum, a shift in the peaks around $2\theta=32°$ and the appearance of new peaks around $2\theta=25°$, indicate a process of annealing and cocrystallization of $MgCl_2$ with $TiCl_3$, similar to the observation of Chien.[15] Treatment with $TiCl_4$ also caused an increase in surface area and catalyst porosity. It was found that as much as 50-60% of added titanium is fixed on the support. The efficiency of fixing on a chemically activated support is superior to what has been reported for supports prepared by milling $MgCl_2$ with $TiCl_4$ or EB.[12,16] The ultimate surface area and crystallite size of the catalyst prepared during the present study were similar to those obtained by earlier workers using physical techniques for $MgCl_2$ activation.

CONCLUSIONS

1. Alcohols and phenols solubilize anhydrous magnesium chloride with the formation of well-defined molecular complexes. This is accompanied by increase in BET surface area and decrease in crystallite size, indicating that the layer structure of $MgCl_2$ is disrupted. The smaller crystallites of $MgCl_2$ are held together in the form of layer aggregates by complexation (hydrogen bonding) with alcohols.

2. Alcohols with higher carbon numbers and branching appear more efficient in this process. Phenols are less effective. However, ethanol may be preferred for the formation of stable crystallite agglomerates.

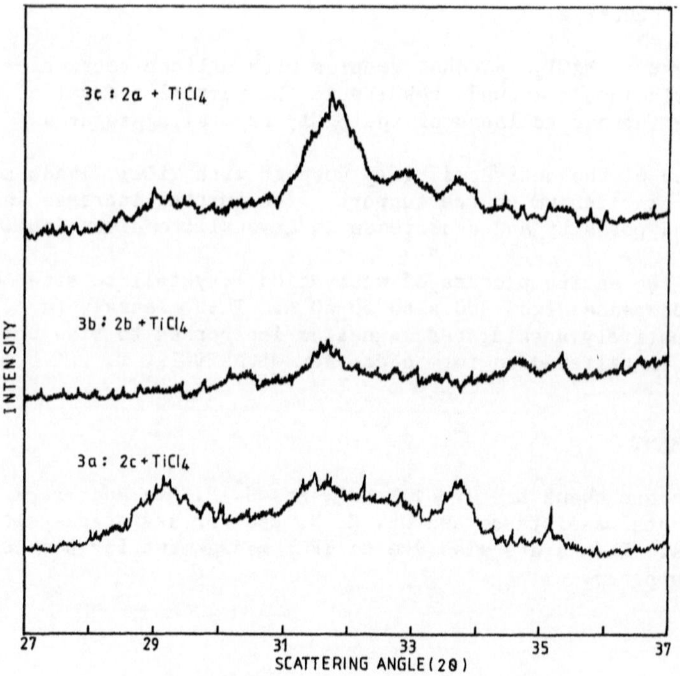

Fig. 3. X-ray Powder Diffraction of $TiCl_4$ Treated Final Product

Table 6. Reaction of Activated Support With Titanium Tetrachloride

PRECURSOR	Ti,%	S.AREA, m^2g^{-1}	POROSITY, Cm^3g^{-1}	D_{101}, Å	D_{001}, Å	D_{110}, Å
$MgCl_2\cdot(EtOH)_6\cdot(EB)^a_{0.15}$	2.0	194	1.03	65	38.7	61.5
$MgCl_2\cdot(EtOH)_{4.8}\cdot(Si(OEt)_4)_{1.1}$ $(EB)^b_{0.2}$	2.2	242	1.38	52	30.5	60.1
$MgCl_2\cdot(EtOH)_{3.5}(EtAl(OEt)_2)_{2.3}$ $(EB)^c_{0.15}$	1.9	186	0.94	58	40.3	62.1

[a] Mg=12%; Cl = 24.6% [b] Mg = 8.2%; Cl= 22.4% [c] Mg = 9.7%; Cl= 22.6%

3. Treatment of the $MgCl_2$. alcohol complex with ethylbenzoate causes further disruption of the $MgCl_2$ structure. Ethyl benzoate does not displace the complexed alcohol. Unlike physical milling, reduction in crystallite size during chemical activation is not accompanied by reduction in surface area, indicating that crystal reaggregation may not be occurring.

4. Treatment of $MgCl_2$. alcohol complex with silicon tetrachloride or organoaluminum compound, results in the partial removal of alcohol, causing further collapse of the $MgCl_2$ crystal aggregates.

5. Reaction of the activated $MgCl_2$ support with $TiCl_4$, leads to efficient fixing of titanium on the support, with further increase in surface area and porosity and a decrease in crystallite dimensions.

6. During the entire process of activation, crystallite size in the (101) plane decreases from 300 Å to 50-60 Å. This cleavage face, containing coordinatively unsaturated magnesium ion bonded to five chlorine atoms, is believed to form a complex with $TiCl_4$.

ACKNOWLEDGEMENT

The authors thank Dr. J. D. Patel, Mr. J. H. Oza and Mr. R. M. Shah for experimental assistance and Dr. K. V. Rao for assistance with x-ray measurements. Thanks are also due to IPCL management for permission to publish this paper.

REFERENCES

1. Ger. 2,153,520 to Montedison S.p.A, CA 77:62505s (1972).
2. Ger. 2,347,577 to Montedison S.p.A, CA 83:59870a (1972).
3. Ger. 2,638,429 to Montedison S.p.A, CA 86:140706x (1975).
4. Belg. 848, 427 to Montedison S.p.A, CA 87:68893v (1977); Japan 79,118,484 to Mitsubishi Industries, CA 92:59450z (1979).
5. Ger. 2,809,318 to Montedison S.p.A, CA 89:216056h (1978).
6. N. Kashiwa and J. Yoshitake, Makromol. Chem., 185, 1133 (1984).
7. U.S. 4,227,589 to Montedison S.p.A, CA 86:172169d (1981); Ger. 3,022,738 to Mitsui Petrochemical Ind., CA 94:122303n (1981); H. L. Hsieh, Catal. Rev. - Sci. Eng., 26, 631 (1984).
8. P. Galli, L. Luciani, and G. Cecchin, Die Angew. Makromol. Chem., 94, 63 (1981).
9. P. Galli, P. C. Barbe, G. Guidetti, R. Zanetti, A. Martorana, A. Marigo, M. Bergozza, and A. Fichera, Eur. Polym. J., 19, 19 (1983).
10. C. Dumas and C. C. Hsu, J. Macromol. Sci. - Rev. Macromol. Chem. Phys., C-24, 355 (1984).
11. U. Giannini, Makromol. Chem. Suppl., 5, 216 (1981).
12. N. Kashiwa, Polymer J., 12, 603 (1980).
13. P. Galli, P. C. Barbe, and L. Noristi, Die Angew. Makromol. Chem., 120, 73 (1984).
14. B. Keszler, G. Bodor and A. Simon, Polymer, 21, 1037 (1980).
15. J. C. W. Chien, J-C. Wu, and C-I. Kuo, J. Polym. Sci. Polym. Chem. Ed., 21, 737 (1983); J. C. W. Chien, Catal. Rev. - Sci. Eng., 26, 613 (1984).
16. S. A. Sergeev, G. D. Butakov, V. A. Zakharov, and E. M. Moroz, Makromol. Chem., 184, 2421 (1983).

17. J. Boor, J. Polym. Sci., Pt A-1, 9, 617 (1971).
18. F-J. Karol, Catal. Rev. - Sci. Eng., 26, 557 (1984).
19. S. Xiao, S. Cai, H. Lui, S. Zhang, and L. Zhou, Tsui Hua Hsuch Pao, Polymer, 1(4), 291 (1980).
20. J. C. W. Chien, J-C. Wu, C-I. Kuo, J. Polym. Sci., Polym. Chem. Ed., 20, 2019 (1982).
21. Japan 80-1510 10 to Mitsui Petrochemical Industries, CA 94:122292h (1980).

POLYMERIZATION OF PROPYLENE WITH MAGNESIUM CHLORIDE-SUPPORTED CATALYSTS

P. C. Barbe'*, L. Noristi*, and
M. A. Schexnayder**

*Himont Italia, Centro Ricerche G. NATTA, 44100 Ferrara (Italy)
**Himont Inc., 1313 N. Market Street, Wilmington, DE 19864

INTRODUCTION

The importance of Lewis bases in determining catalyst performance for polyolefin synthesis, above all when the catalyst is formed by $MgCl_2$-supported Titanium compounds is well known.[1] Most of the progress recently achieved in polyolefin synthesis is due undoubtedly to the fact that suitable Lewis bases have been used. The addition of the base to Alalkyl has several effects, in terms of catalyst stereospecificity, polymerization kinetics and polymer molecular weight.

That the base promotes increase in stereospecificity, and decrease in activity, is well known. In addition, it has been shown that polymer isotacticity increase is derived from a productivity decrease which is greater for the atactic polymer than for the isotactic one (Figure 1). The following effects have been attributed to the base:

- selective poisoning of atactic centers;
- transformation of atactic centers into isotactic ones;
- increase in K_p of isotactic centers.

Several hypotheses have been formulated with regard to the nature of the active centers and their interaction with the base in order to clarify these phenomena. Some hypotheses may be summarized as follows:

- the base poisoning effect is due to its reversible complexation with coordinatively unsaturated centers;[2,3,4]
- the base causes the Alalkyl concentration, and thus the Alkyl adsorption on the active centers, to decrease;[5]
- the base reacts irreversibly with the Ti-R bond;[6]
- the base adsorption poisons the active centers, whereas the base-alkyl complex adsorption transforms nonstereospecific centers into stereospecific ones.[7]

It is difficult to identify the most likely mechanism because the results available in the literature were obtained under different conditions.

With this work, we intended to verify the effect that chemically different bases could have on the performance of a single catalyst. We were

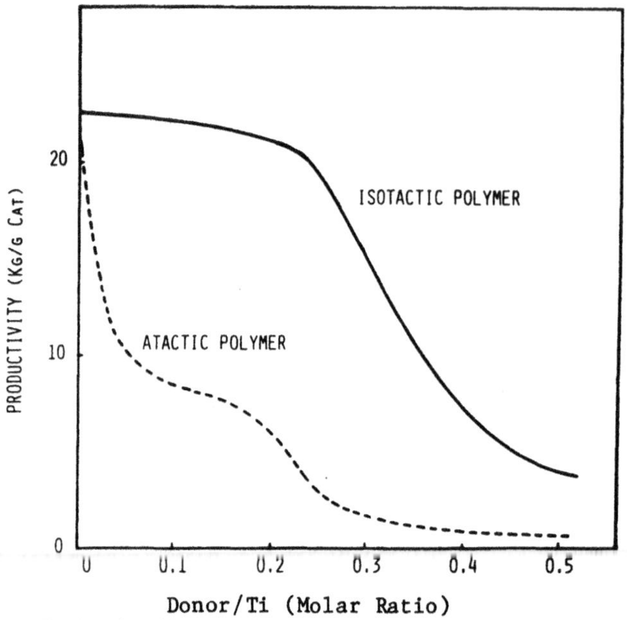

Fig. 1: Typical Effect of the Lewis Base on the Productivity of Isotactic and Atactic Polymer

spurred by the experimental verification that bases of different nature, such as an aromatic ester and a hindered amine, could give fundamentally different performance. As reported in Figure 2, the aromatic ester, gradually added to the alkyl, promoted a decrease in productivity, whereas the amine increased it. Such widely differing results could lead to very differentiated donor action mechanisms which could endow either the transformation of atactic centers into isotactic ones or could poison the atactic centers.

Experimental tests were thus planned to verify the role of different donors on the performance of a solid catalyst consisting of $MgCl_2$-supported $TiCl_4$. This was in order to avoid any major complication due to the presence of both an additional component and the consequent reaction products.

Three Lewis bases were investigated:

- an aromatic ester: methylparatoluate (MPT);
- a hindered amine: 2.2.6.6. tetramethylpiperidine (TMP); and
- a silanic ester: phenyltriethoxysilane (PES).

Polymerization tests were carried out at very low base concentration in order to better point out probable selective effects.

EXPERIMENTAL SECTION

The solid catalyst was prepared by a double treatment of "active" $MgCl_2$ with $TiCl_4$ (10 g.$MgCl_2$/100 ml $TiCl_4$) at 100°C for 2 hours, followed by washings with heptane at 70°C and drying under nitrogen.

The "active" $MgCl_2$ was obtained by the $MgCl_2$.EtOH reaction with $AlEt_3$ according to known procedures.

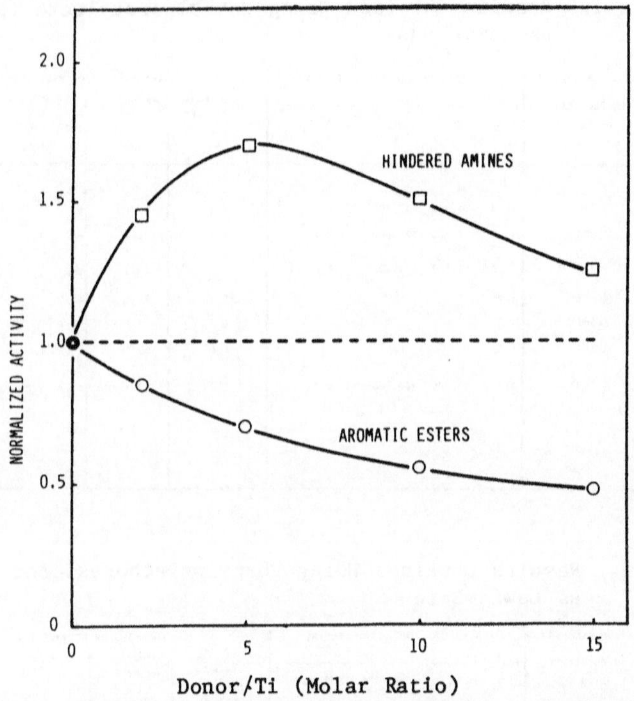

Fig. 2. Effect of Different Lewis Bases on the Propylene Polymerization with a Catalyst Prepared With a Base

Catalyst analysis gave: Ti 4.43% wt, Mg 18.91%, Cl 68.43%. Heptane-diluted $AlEt_3$ (10 g/100 ml heptane) was used as activator. Analyses of the Lewis bases gave:

- methylparatoluate (MPT) Montedison: 99% purity
- 2.2.6.6. tetramethylpiperidine (TMP) Fluka: 97% purity
- phenyltriethoxysilane (PES) Petrarch.: 98% purity

* * * *

Polymerization tests have been carried out in hexane under the following conditions: reactor volume 2.5 l - hexane 1 l - $AlEt_3$ 5 mmole - Al/Ti 50 molar - 50°C - Propylene 2 bar. Hydrogen was fed continuously so as to have a constant concentration (5% mol in the gas phase). Polymerization time 4 hours.

The following procedure was used: the diluent and then the catalyst system were fed into the reactor maintained at 25°C under propylene. Afterwards, both temperature and the amount of propylene were increased to the prefixed values.

The pressure was kept constant by continuous propylene feeding. The following analyses were made on the dried polymers:

- intrinsic viscosity in tetrahydronaphthalene at 135°C;
- xylene solubility at 25°C, after polymer complete dissolution at 135°C; the liquid phase was treated with excess of acetone and filtered. The solvent was evaporated to determine the "oil" fraction; and
- MWD by Gel Permeation Chromatography Waters mod. 150/C. Solvent Trichlorobenzene.

Table 1. Results Obtained Using Methylparatoluate (MPT) as Lewis Base

POLYMERIZATION CONDITIONS AND RESULTS (1)			ANALYSES ON THE TOTAL POLYMER				ANALYSES ON POLYMER FRACTIONS						
BASE/Ti	TIME	PRODUCTIVITY	I.V.	G. P. C.			Acetone soluble	Xylene soluble	Xylene insoluble				
										I.V.	G. P. C.		
m.r.	min.	g Pol/g Cat	dl/g	$\bar{M}_n$	$\bar{M}_w$	$\bar{M}_w/\bar{M}_n$	%wt	%wt	%wt	dl/g	$\bar{M}_n$	$\bar{M}_w$	$\bar{M}_w/\bar{M}_n$
0	240	3,390	1.48	18,400	157,300	8.5	5.3	27.0	67.7	1.48	29,200	208,100	7.1
0.19	240	4,000	1.17	12,600	123,600	9.7	4.7	30.4	64.9	1.42	22,400	155,900	6.9
0.37	240	3,860	1.17	12,800	137,400	10.7	6.3	28.2	65.5	1.40	23,200	185,900	8.0
0.82	240	3,890	1.38				6.1	24.6	68.9				
1.20	240	3,360	1.30				6.0	25.0	69.3				
2.34	240	2,850	1.52				4.4	20.2	75.4	1.89			
6.77	240	1,720	2.06	24,300	344,100	14.2	3.1	12.1	84.8	2.22	35,800	364,000	10.2

Table 2. Results Obtained Using Phenyltriethoxysilane (PES) as Lewis Base

POLYMERIZATION CONDITIONS AND RESULTS (1)			ANALYSES ON THE TOTAL POLYMER				ANALYSES ON POLYMER FRACTIONS						
BASE/Ti	TIME	PRODUCTIVITY	I.V.	G. P. C.			Acetone soluble	Xylene soluble	Xylene insoluble				
										I.V.	G. P. C.		
m.r.	min.	g Pol/g Cat	dl/g	$\bar{M}_n$	$\bar{M}_w$	$\bar{M}_w/\bar{M}_n$	%wt	%wt	%wt	dl/g	$\bar{M}_n$	$\bar{M}_w$	$\bar{M}_w/\bar{M}_n$
0	240	3,390	1.48	18,400	157,300	8.5	5.3	27.0	67.7	1.48	29,200	208,100	7.1
0.16	240	3,610	1.12				6.2	26.4	67.4				
0.34	240	3,860	1.40	17,700	165,500	9.3	5.7	25.9	70.6	1.55	27,500	205,800	7.5
0.65	240	3,390	1.45	23,600	198,200	8.4	3.6	23.3	73.5	1.62	30.000	255,000	8.5
0.96	240	2,460	1.61				3.7	18.3	77.5				
5.86	240	1,430	1.98	18,500	240,000	13.0	2.5	9.9	87.6	2.06	30.800	296,700	9.6

Table 3. Results Obtained Using 2,2,6,6 - Tetramethylpiperdine (TMP) as Lewis Base

POLYMERIZATION CONDITIONS AND RESULTS (1)			ANALYSES ON THE TOTAL POLYMER				ANALYSES ON POLYMER FRACTIONS						
BASE/Ti	TIME	PRODUCTIVITY	I.V.	G. P. C.			Acetone soluble	Xylene soluble	Xylene insoluble				
										I.V.	G. P. C.		
m.r.	min.	g Pol/g Cat	dl/g	$\bar{M}_n$	$\bar{M}_w$	$\bar{M}_w/\bar{M}_n$	%wt	%wt	%wt	dl/g	$\bar{M}_n$	$\bar{M}_w$	$\bar{M}_w/\bar{M}_n$
0	240	3,390	1.48	18,400	157,300	8.5	5.3	27.0	67.7	1.48	29,200	208,100	7.1
0.15	240	3,250	1.22				5.0	24.7	70.3				
0.32	240	3,260	1.27				6.3	23.8	69.9				
0.78	240	3,140	1.59	10,500	164,900	15.7			72.3	1.65	22,500	231,000	10.3
0.99	240	2,630	1.55				3.7	16.7	77.7				
7.5	240	1,930	1.95	11,900	218,700	18.4	3.2	12.8	84.0	2.10	22,300	286,000	12.8

The results are reported in Tables 1, 2 and 3.

RESULTS AND DISCUSSION

Figure 3 shows the effect of the gradual addition of the bases on the total productivity. For low donor concentration, MPT and PES exhibit an increase in productivity, while TMP shows constant values. Subsequent additions of base promote decreases in productivity. Figure 4 shows isotacticity behavior as a function of the donor/Ti ratio. With PES and TMP, a clear isotacticity increase is noticed, while with MPT this increase occurs only at higher donor concentrations.

For MPT, small amounts of donor cause a slight isotacticity decrease.

By using the data of the two previous Figures, we have calculated the isotactic polymer productivity as plotted in Figure 5. Low concentrations of the base causes the productivity to increase for MPT and PES but has a negligible effect for TMP. Greater amounts of base bring about productivity decrease. The relationship between atactic polymer productivity and base/Ti is reported in Figure 6. Again here the trends are very similar to those of the total productivity and clearly emphasize peaks for MPT and PES in combination with small additions of base. Trends relative to PES and TMP are very similar and distinct from those for MPT.

These Figures show two distinct effects as a function of the base concentration. At low concentrations for MPT and PES, an increase in catalyst activity occurs which is not selective as it involves both isotactic and atactic polymers. At greater concentrations, a highly selective decrease in activity is noticed, which is more marked for the atactic than for the isotactic polymer.

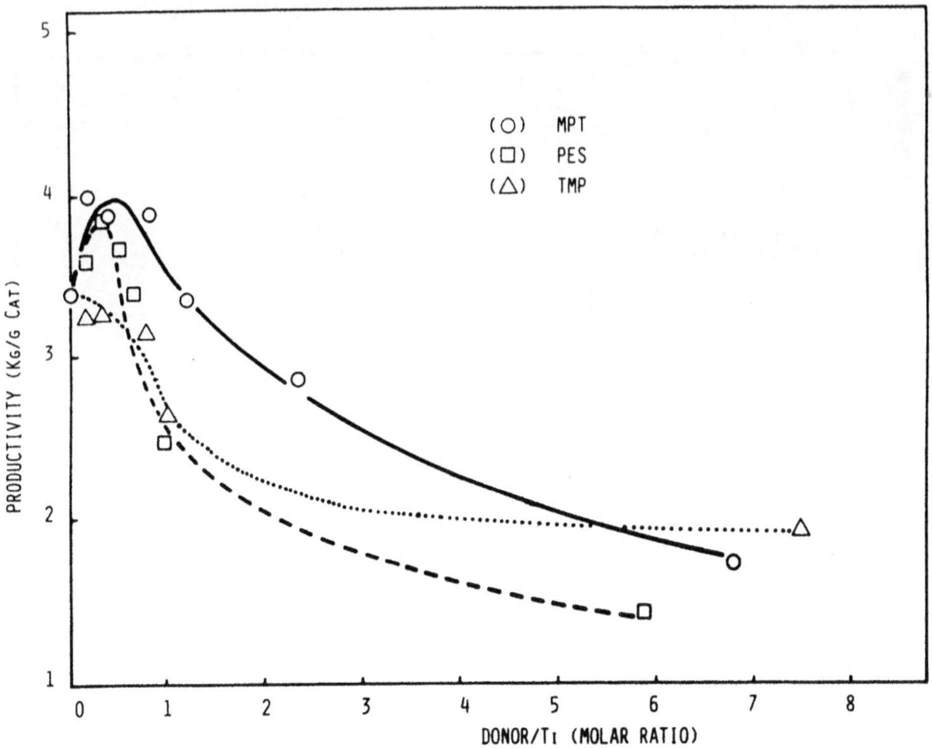

Fig. 3. Effect of the Donor/Ti Ratio on the Overall Productivity

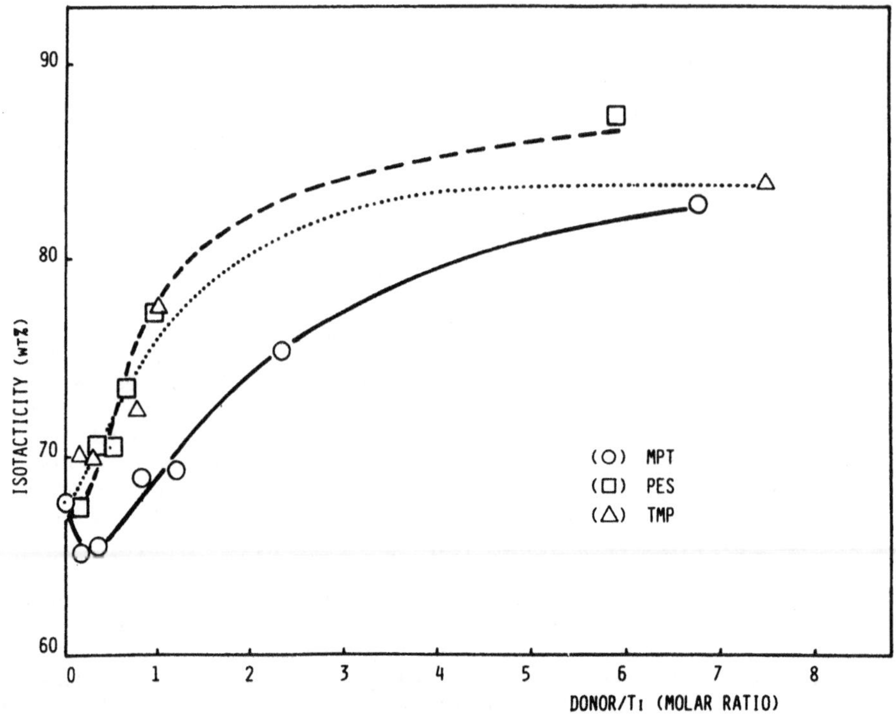

Fig. 4. Effect of the Donor/Ti Ratio on the Isotacticity

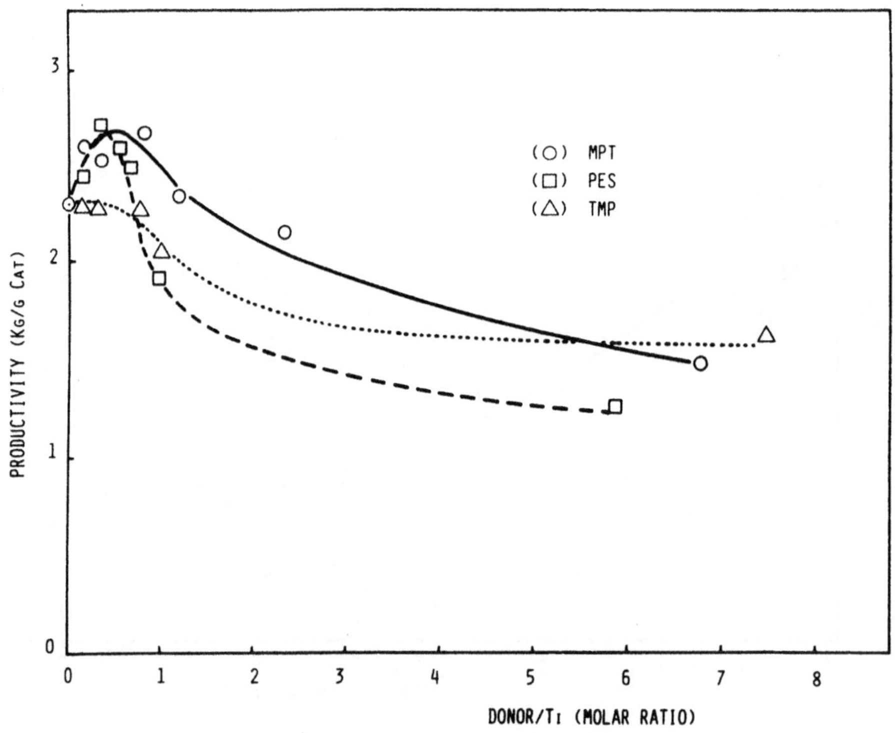

Fig. 5. Effect of the Donor/Ti Ratio on the Productivity of Isotactic Polymer

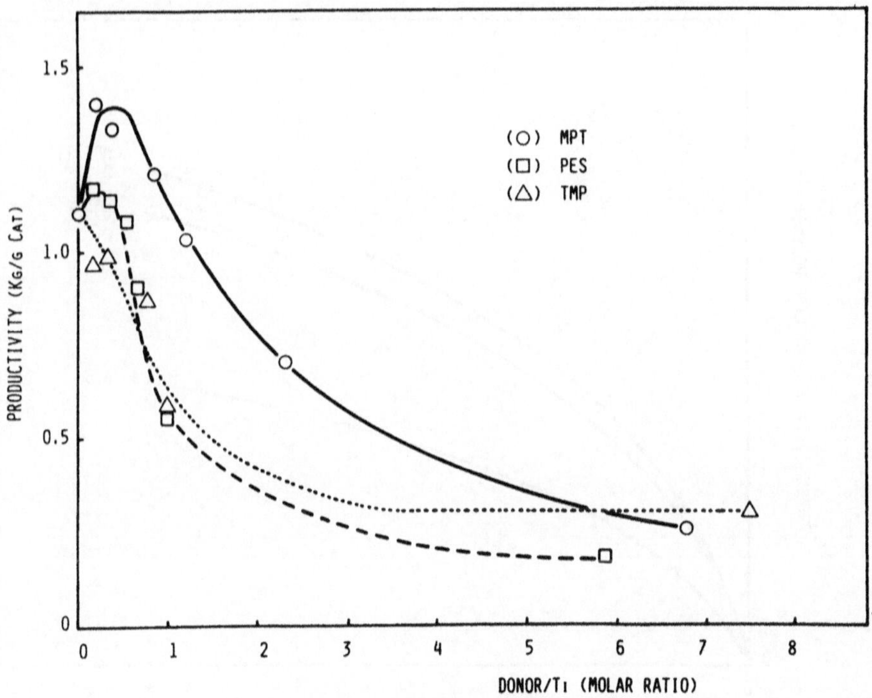

Fig. 6. Effect of the Donor/Ti Ratio on the Productivity or Atactic Polymer

Figure 7 presents polymerization kinetics obtained for different MPT concentrations and shows how the activity increase, which takes place with low base/Ti ratios, is connected to steadier kinetics, while the decrease which occurs at high ratios is a consequence of a lowering in the initial rate.

PES behaves similarly (Figure 8), while the curves obtained for TMP (Figure 9) in the absence of base and at low ratio are superimposed.

Such kinetic trends suggest that low base concentrations may have a stabilizing effect on the active centers, whereas higher concentrations inhibit the activity.

As for $\bar{M}n$, determined by GPC, two different effects can be distinguished as a function of the base/Ti ratio (Figure 10): i.e. at low ratios, a $\bar{M}n$ decrease, while at higher ratios, a $\bar{M}n$ increase which is high for MPT, low for PES and null for TMP.

Due to the experienced uncertainties of these determinations, we also analyzed other molecular parameters. Figure 11 reports $\bar{M}w$ with the more or less relevant decrease corresponding to low ratios, and a $\bar{M}w$ increase for higher ratios.

Figure 12 shows the behavior of the intrinsic viscosity which complies with those of the previous Figures.

$\bar{M}w/\bar{M}n$ polydispersity ratio is affected by the increasing concentration of base (Figure 13) which gives rise to a progressive increase up to 2-3 base/Ti ratio.

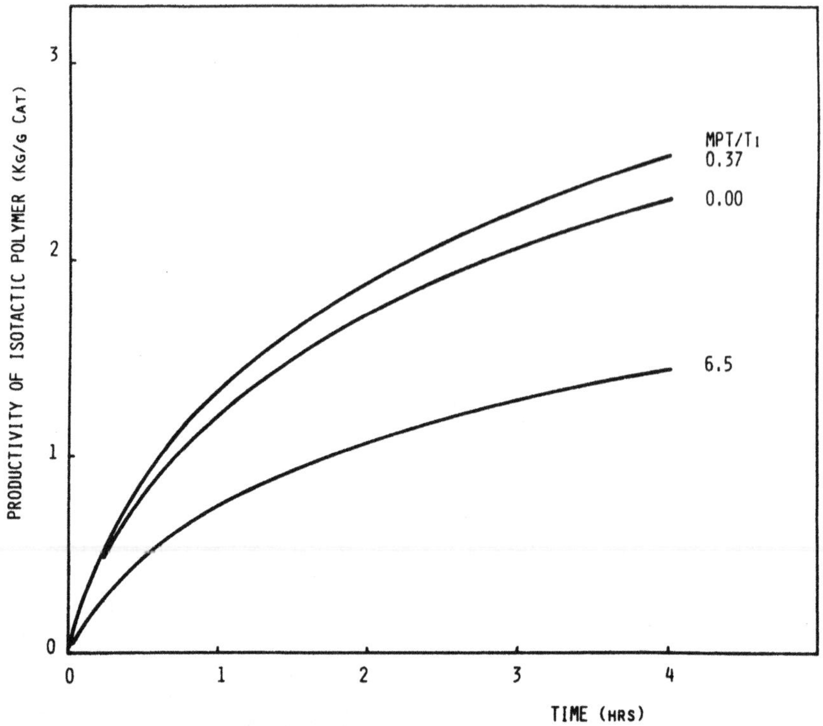

Fig. 7. Polymerization Kinetics at Different MPT/Ti Ratios

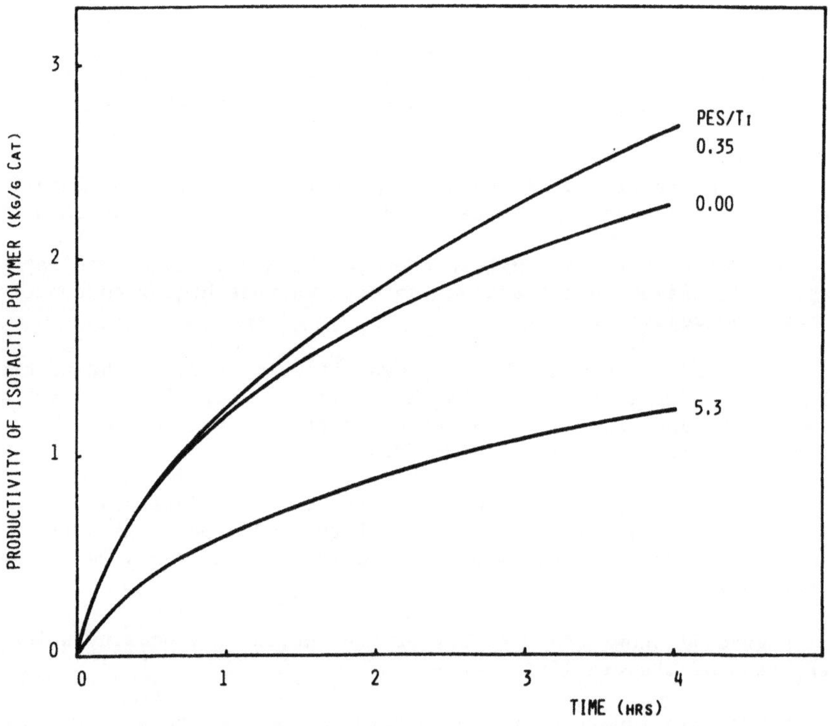

Fig. 8. Polymerization Kinetics at Different PES/Ti Ratios

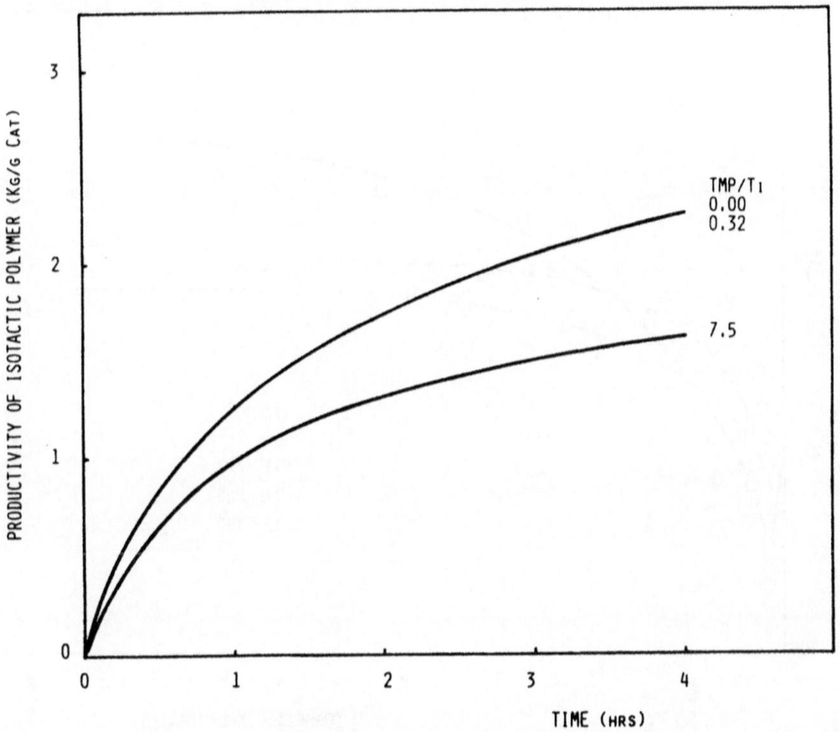

Fig. 9. Polymerization Kinetics at Different TMP/Ti Ratios

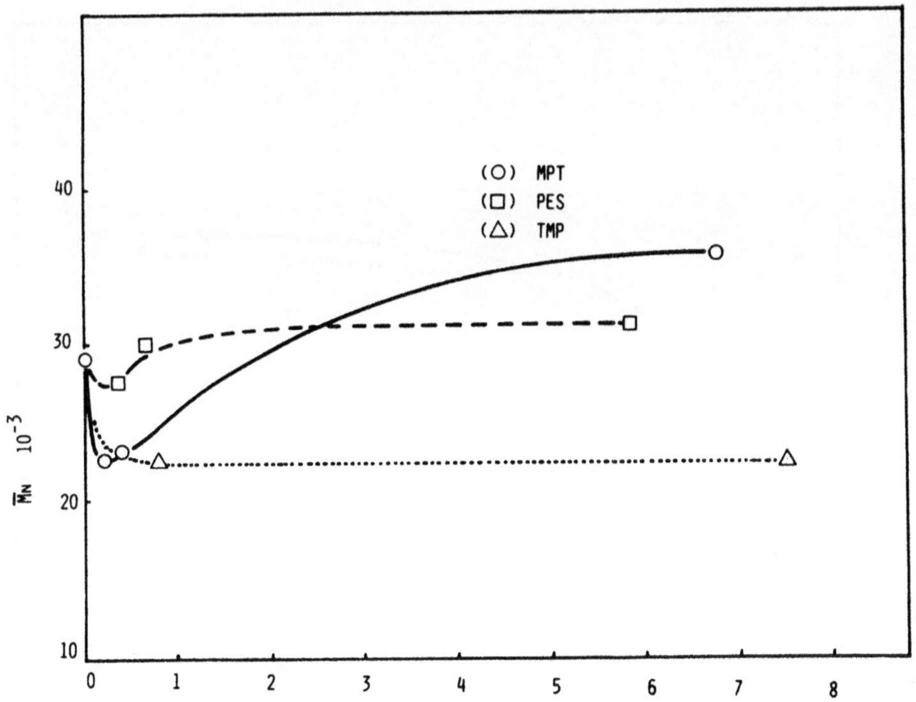

Fig. 10. Effect of the Donor/Ti Ratio on the $\overline{Mn}$ of the Isotactic Polymer

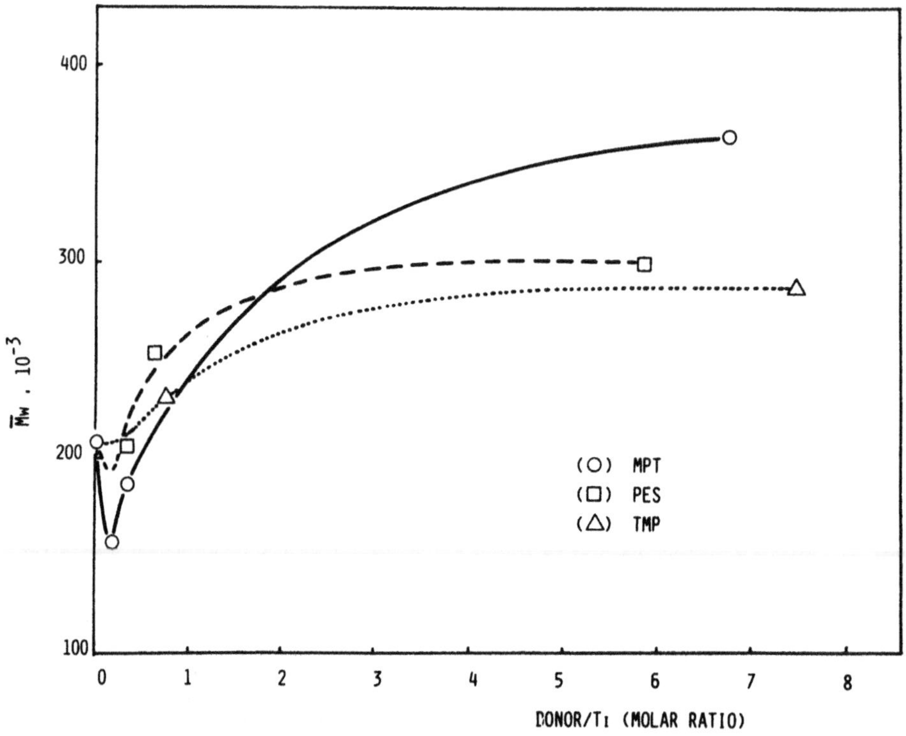

Fig. 11. Effect of the Donor/Ti Ratio on the $\overline{M}w$ of the Isotactic Polymer

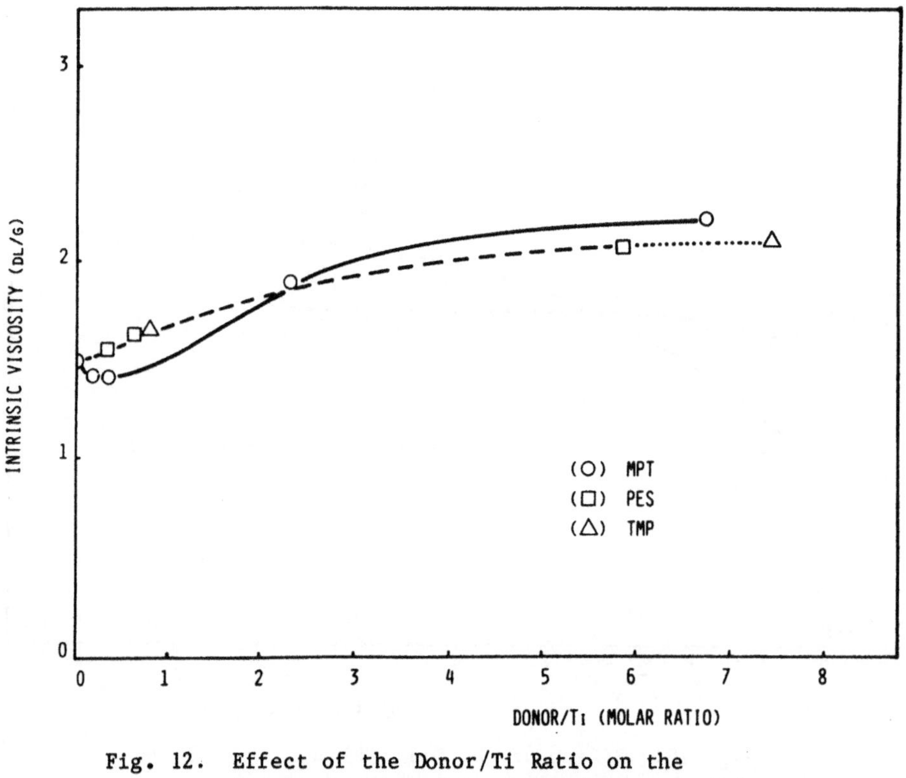

Fig. 12. Effect of the Donor/Ti Ratio on the Intrinsic Viscosity of the Isotactic Polymer

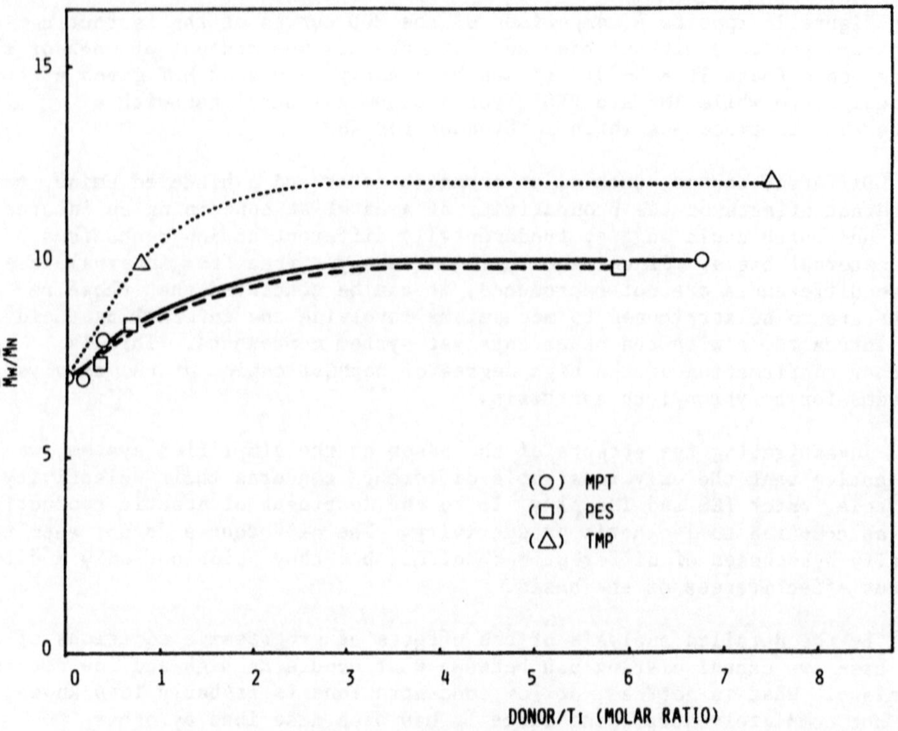

Fig. 13. Effect of the Donor/Ti Ratio on the MWD of the Isotactic Polymer

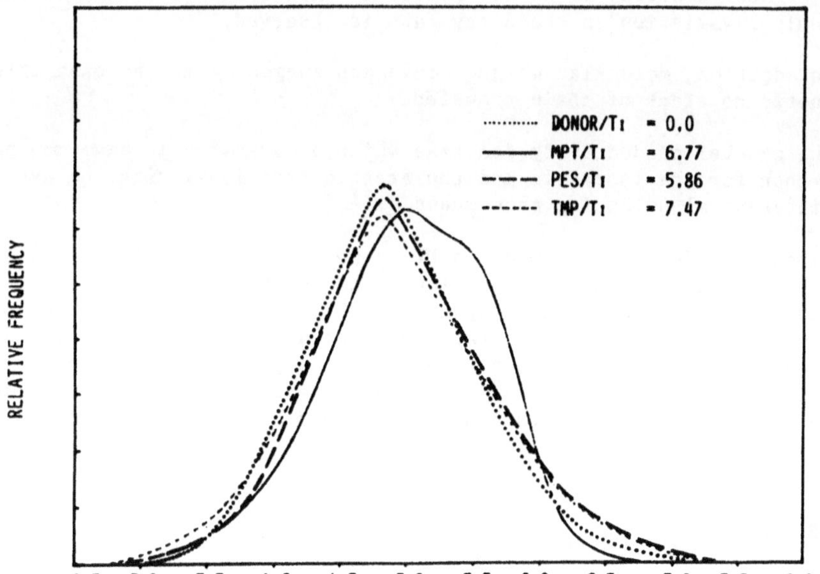

Fig. 14. Effect of Different Lewis Bases on the MWD of the Isotactic Polymer

Figure 14 reports a comparison of the MWD curves of the isotactic fractions produced without base and with the maximum content of each of the three bases (base/Ti = 5-7): it can be clearly seen that MPT gives a bimodal curve while TMP and PES gives a monomodal curve but with a difference in broadness which is broader for TMP.

Different donors, such as an aromatic ester and a hindered amine, have different effects on the productivity of a catalyst containing an internal base and which could suggest fundamentally different action mechanisms of the external bases. Since, when using a catalyst free from internal base those differences are not reproduced, it can be concluded that those results are to be attributed to mechanisms involving the internal base and its interactions with the other catalyst system components. This is another confirmation of the high degree of sophistication of the catalyst systems for polypropylene synthesis.

Investigating the effects of the bases on the simplified system, we can notice that the only remarkable difference concerns their selectivity. The role, which PES and TMP play, is to the detriment of atactic productivity as compared to isotactic productivity. The differences do not seem to justify hypotheses of different mechanisms; but they point out only a different effectiveness of the bases.

By the detailed analysis of the effects of progressive additions of the base, we cannot distingiush between what occurs at high and low concentrations. What is noticed, at low concentrations is probably less known, but not completely surprising since it has been described by other Authors.[8,9,10] In some cases, it was not emphasized because such an unusual concentration range was not taken into consideration.

It is our opinion that the productivity increase which takes place in this range is to be attributed to a stabilization of the active centers by the base rather than to an increase in their specific productivity or their number. If the productivity increase were caused by an increase in the number of active sites, we should notice an increase in the initial rate, while only a variation in the decay rate is observed.

In addition, molecular weight decreases suggest that the base affects the kinetic constant of chain transfer.

The general productivity decrease which occurs at high base concentrations, both for the isotactic and the atactic fractions, might be ascribed to a differential poisoning phenomenon.[2,11]

Also, an effect of the base on MW is noticed, even if it is opposite to that observed for low concentrations. In addition, with MPT, a bimodality in MWD arises which leads one to think that, beyond the poisoning, there is also an effect on the transfer constants. However, the appearance of bimodality does not allow one to exclude the presence of the phenomenon previously proposed by Kashiwa[12] according to which the base should contribute an increase in the propagation constant of some of the stereospecific sites.

CONCLUSIONS

The results suggest the following:

- by using a catalyst, free from internal base, no different productivity trends are noticed in quality as a function of the concentrations of chemically different external bases. The only difference is their effectiveness.

- Depending on their concentrations, all the bases seem to act according to two very different mechanisms, i.e., at low concentration, they seem to stabilize the active centers, whereas at higher concentration they poison the active centers selectively.
- Beyond these main effects, the bases affect also molecular weights and thus their effect on the kinetic constant of the active centers could be hypothesized.

REFERENCES

1. British Patent 1,387,890 to MONTEDISON (1971).
2. P. Pino and R. Mulhaupt, Angew. Chem. Int. Ed. Engl., 19, 857 (1980).
3. A. Caunt, J. A. Licchelli, I. W. Parsons, R. N. Harward, and R. Y. Al-Hillo, Polymer, 24, 121 (1983).
4. Ye. Ye. Vermel, V. A. Zakharov, Z. K. Bukatova, G. P. Shkmina, L. G. Yechevskay, E. M. Moroz, and S. V. Sudakova, Vysokomol. Soyed, A-22, 22 (1980).
5. T. Keii, Makromol. Chem., 183, 2285 (1982).
6. B. L. Goodall, Paper at the MMI Int. Symp. on "Transition Metal Catalyzed Polymerization: Unsolved Problems," Midland, MI (1981).
7. R. Spitz, J. L. Lacombe, and A. Guyot, J. Polym. Sci., Polym. Chem. Ed., 22, 2625 (1984).
8. K. Soga, T. Sano, K. Yamamoto, and T. Shiono, Chem. Lett., 425 (1982).
9. N. Kashiwa, Paper at the MMI Int. Symp. on "Transition Metal - Catalyzed Polymerization: Unsolved Problems," Midland, MI (1981).
10. P. J. T. Tait and A. E. Maher, Paper at the 9th Iberamericano Symp. of Catalysis, Lisbon, Portugal (1984).
11. A. W. Langer, T. J. Burkardt, J. J. Steger, Polym. Sci. Technol., 19, 225 (1983).
12. N. Kashiwa and J. Yoshitake, Pol. Bull., 12, 99 (1984).

POLYMERIZATION OF 4-METHYLPENTENE-1

WITH MAGNESIUM-CHLORIDE-SUPPORTED CATALYSTS

P. J. T. Tait, M. Ab Eid and
A. E. Enenmo

Department of Chemistry
UMIST, Sackville Street
Manchester, U.K., M60 1QD

INTRODUCTION

While it has been recognized for some time that the polymerization of ethylene by Ziegler-Natta catalysts was somewhat different from that of propylene and other α-olefins, it has often been assumed that the polymerization of all α-olefins proceeds more or less in the same manner, irrespective of the particular monomer under investigation. Recent research[1-4] however, indicates that this may not be so.

Tait and Abu Eid[1-3] have reported some important differences in kinetic behavior between the polymerization of 4-methylpentene-1 and propylene using highly active magnesium chloride-supported catalysts. Polymerization systems involving 4-methylpentene-1 showed more stable rate-time profiles and higher concentrations of active centers. Kashiwa and Yoshitake[4] have observed that the rate of copolymerization of ethylene and propylene was ten times higher than that for ethylene homopolymerization when using a magnesium chloride-supported titanium tetrachloride catalyst. This paper describes a systematic investigation of the polymerization of 4-methylpentene-1, using two types of $MgCl_2$/electron donor/$TiCl_4$ systems, viz., $MgCl_2$/ethyl benzoate/$TiCl_4$ and $MgCl_2$/diisobutyl phthalate/$TiCl_4$ along with the aluminum-alkyls, triethylaluminum, triisobutylaluminum, tri-m-hoxylaluminum and tri-n-dodecylaluminum. Ethyl benzoate and N,N,N',N'-tetra-methylethylenediamine (TMEDA) were used as cocatalyst donors.

EXPERIMENTAL

Preparation of Catalysts

The $MgCl_2$ support was dried under vacuum at a temperature of 400°C for 3 h before being placed in a stainless steel ball mill under an atmosphere of dried nitrogen. The $MgCl_2$ support was first treated with thionyl chloride at a molar ratio of $MgCl_2$:$SOCl_2$ of 1:0.05 and then ball milled for 1 h. Ethyl benzoate (EB) was then added so that $MgCl_2$: EB = 6.5 - 17:1 and ball milling was carried out for 72 h. The product from the treatment with $SOCl_2$ was ball milled with silicon oil ($MgCl_2$: silicon oil = 1:0.08) for 5 h before the diisobutyl phthalate (DIBP) was added

($MgCl_2$: DIBP = 6.5-17:1). Ball milling was then continued for 72 h. Finally the $MgCl_2$/electron donor support was treated with neat titanium tetrachloride at 115°C for 1.5 h with stirring under nitrogen and the resulting procatalyst product isolated by filtration and washed with n-heptane. The ethyl benzoate procatalyst was pale yellow in color and typically contained 1-3% Ti, while the phthalate procatalyst was pale grey in color and contained about 3% Ti. Results in the present study are reported for ethyl benzoate catalysts containing 1% and 2.6% Ti, and for a diisobutyl phthlate catalyst containing 3.2% Ti.

Polymerization Procedure

Polymerizations were carried out under rigorous experimental conditions using dilatometers.[5] All glassware used in the polymerization experiments was carefully cleaned, dried at 150°C and finally flamed out under vacuum. In order to obtain reproducible and comparable results, all polymerization mixtures were prepared in an identical manner and the components mixed in a prescribed order. In the absence of a cocatalyst donor, the order adopted was:

diluent/catalyst slurry/alkyl solution/monomer

When a cocatalyst donor was present, a different sequence was adopted, viz.,

diluent/catalyst slurry/monomer/donor solution/alkyl solution

Active Center Concentration

Active center concentrations were determined using a tritium quenching technique.[6,7] A molar ratio of [MeOH] : [AlR_3] of 15:1 was used in this study and extreme care was taken to remove all radioactive contamination products from the polymer; the procedure being checked by trial runs first quenched by inactive methanol followed by tritiated methanol addition after 10 min.

Tritium and ^{14}C contents of polymer samples were determined by means of a scintillation counting technique using a Packard Tricarb 300 C counter. For the calculation of active center concentrations, in the case of tritium quenching, allowance was made for the kinetic isotope effect and for chain transfer with aluminumalkyl.[6,7]

RESULTS AND DISCUSSION

General Kinetic Characteristics

Rate-Time Profiles. Valuable information about the physical nature of the catalyst matrix and about the stability of active centers in Ziegler-Natta polymerization, may be obtained from an examination of rate-time profiles.[13] A typical rate-time profile for the polymerization of 4-methylpentene-1 using a $MgCl_2/EB/TiCl_4$ - $Al(i\text{-}Bu)_3$ catalyst is shown in Figure 1. In this figure and in all other comparable figures, the rate values which are displayed have been corrected for the decrease in monomer concentration.

It can be seen that the polymerization reaction is characterized by an initial 'settling period,' during which, the overall rate of polymerization increases with the time of polymerization and reaches a maximum value, after which it decreases somewhat slowly with time. This behavior is different from that shown by propylene when polymerized by a similar catalyst

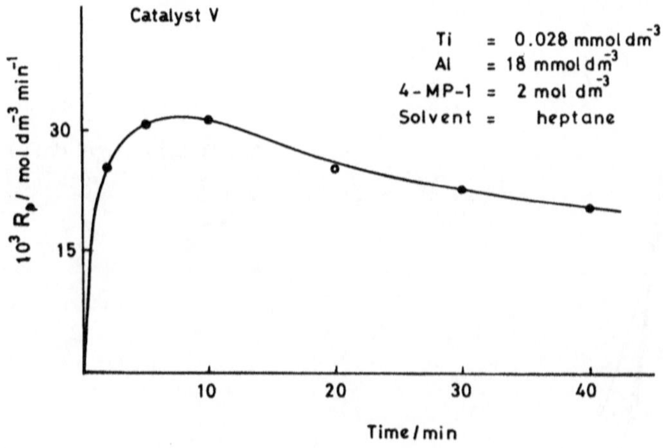

Fig. 1

when the overall rate of polymerization initially shows a maximum value which rapidly decreases with time as shown in Figure 2.

The behavior shown in Figure 2 is typical for the polymerization of ethylene and propylene by many catalyst systems of the type $MgCl_2/EB/TiCl_4 - AlR_3$.[10-15] However, the rate-time profile for the polymerization of 4-methylpentene-1 by the catalyst system $MgCl_2/EB/TiCl_4$ - Al(i-Bu)$_3$ resembles closely that shown for the polymerization of the same monomer by the first generation catalyst system, δ-$TiCl_3$.0.33 $AlCl_3$/donor-Al(i-Bu)$_3$, which is shown in Figure 3.

Examination of the corresponding rate-time profiles for the catalyst systems $MgCl_2/DIBP/TiCl_4$-AIR_3 shows that the rate of polymerization of 4-methylpentene-1 remains almost constant as a function of time. Figures 4 to 6 show typical rate-time profiles for polymerizations using triethylaluminum, tri-n-hexylaluminum and tri-n-dodecylaluminum, respectively. The same behavior was observed for triisiobutylaluminu. Throughout the present study $MgCl_2/DIBP/TiCl_4$ catalysts showed a greater stability, but a lower activity, than $MgCl_2/EB/TiCl_4$ catalysts.

Table 1 lists some comparative values for catalyst activities.

Dependence of Overall Rate of Polymerization on Aluminumalkyl Concentration. Increasing the concentration of aluminumalkyl while keeping the concentrations of monomer and catalyst constant has a profound effect on the observed rates of polymerization of the catalyst systems $MgCl_2$/DIBP/$TiCl_4$-AlR_3 and $MgCl_2/EB/TiCl_4$-Al(i-Bu)$_3$.

For $MgCl_2/DIBP/TiCl_4$-AlR_3 catalyst systems, the rate of polymerization initially increases with increase in AlR_3 concentration and then reaches a more or less constant value, as is shown in Figures 4, 5 and 6. This behavior is consistent with the greater stability of the phthalate-containing catalyst as indicated by the various rate-time profiles for this catalyst.

For $MgCl_2/EB/TiCl_4$ catalyst systems, the effect of increasing aluminumalkyl concentration is different as is shown in Figure 7. As the concentration of tri-iso-butylaluminum is increased, the rate of polymerization increases and reaches a maximum value and then decreases with further increase in aluminumalkyl concentration. This effect is shown more clearly

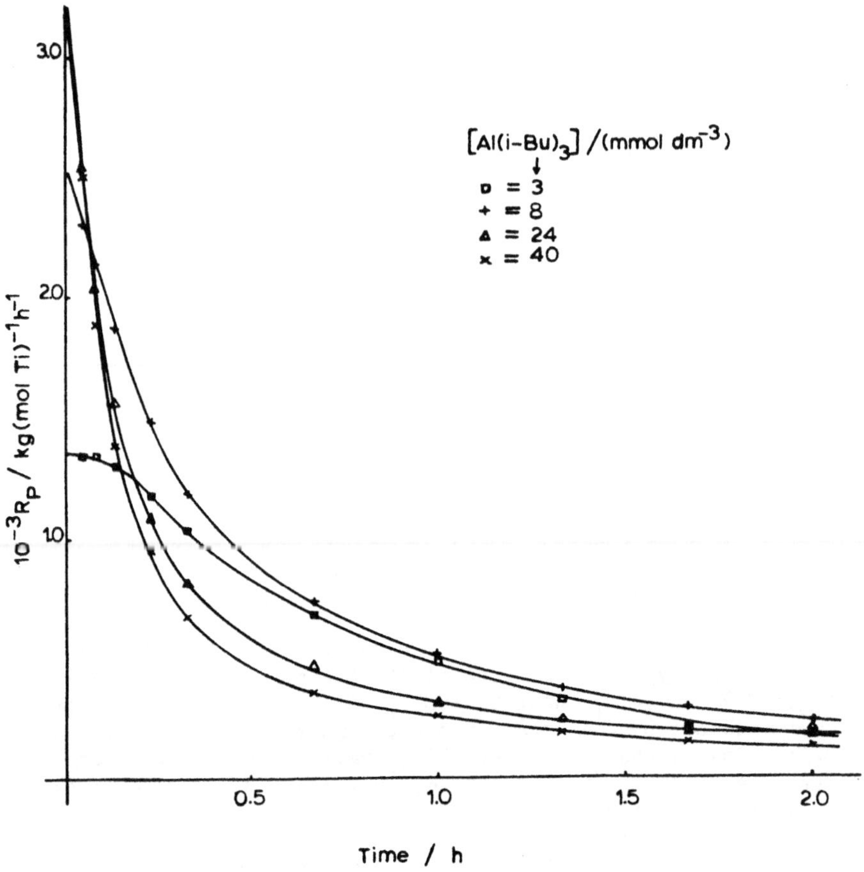
[Al(i-Bu)$_3$] /(mmol dm^{-3})
▫ = 3
+ = 8
▵ = 24
× = 40
$10^{-3}R_p$ / kg(mol Ti)$^{-1}$h^{-1}
3.0
2.0
1.0
0.5
1.0
1.5
2.0
Time / h

Fig. 2

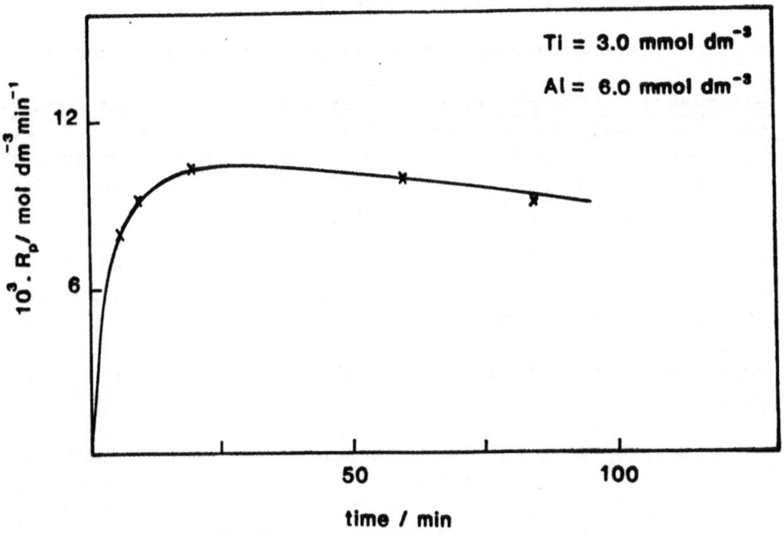
Ti = 3.0 mmol dm^{-3}
Al = 6.0 mmol dm^{-3}
$10^3 . R_p$/ mol dm^{-3} min^{-1}
12
6
50
100
time / min

Fig. 3

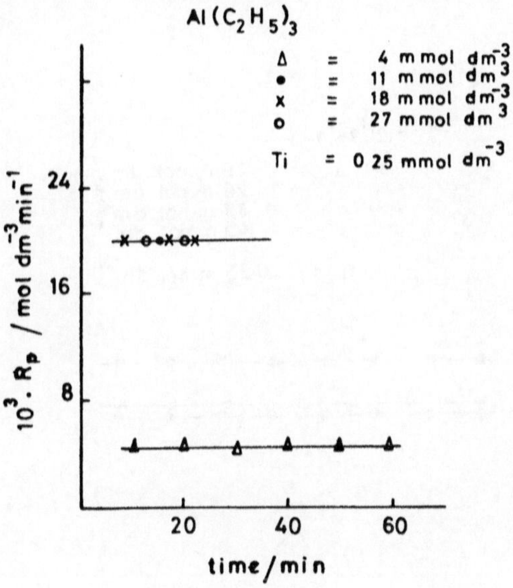

Al(C2H5)3
Δ = 4 m mol dm-3
● = 11 m mol dm-3
x = 18 m mol dm-3
o = 27 m mol dm-3
Ti = 0 25 mmol dm-3
10³. Rp / mol dm-3 min-1
24
16
8
20
40
60
time/min

Fig. 4

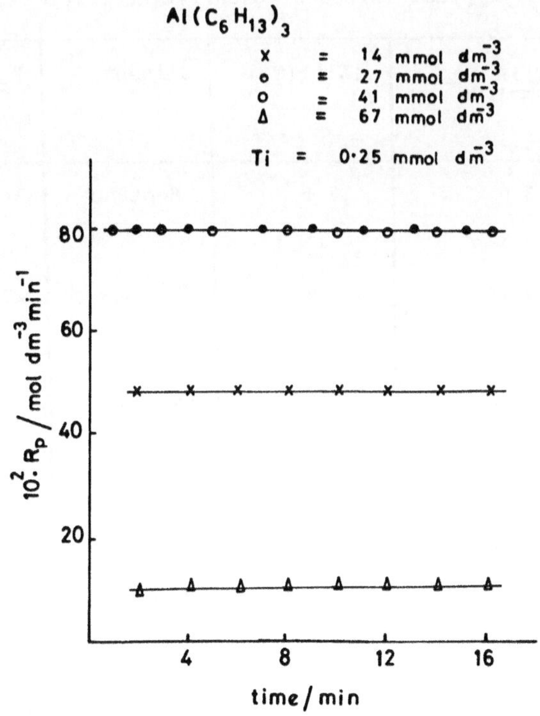

Al(C6H13)3
x = 14 mmol dm-3
● = 27 mmol dm-3
o = 41 mmol dm-3
Δ = 67 mmol dm-3
Ti = 0·25 mmol dm-3
10². Rp / mol dm-3 min-1
80
60
40
20
4
8
12
16
time/min

Fig. 5

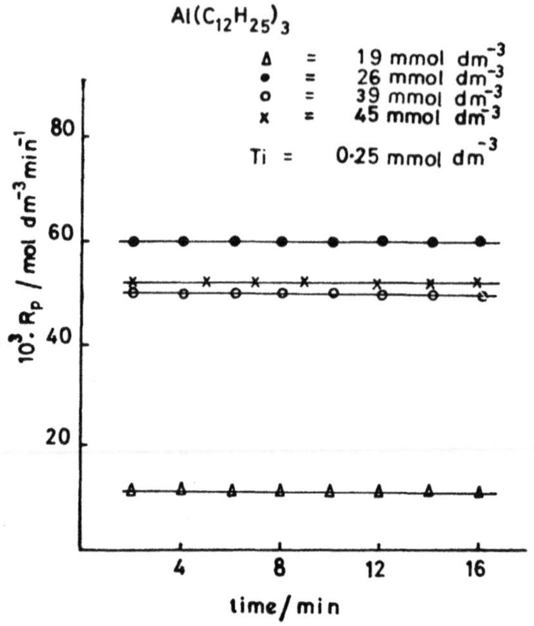

Fig. 6

Table 1. Some Typical Values of Catalyst Activities

Catalyst	[Ti]/ mmol dm^{-3}	[Al]:[Ti]	Diluent	R_p/[Ti] mol min^{-1}
δ-$TiCl_3$.0.33.$AlCl_3$* -Al(i-Bu)$_3$	3.0	2.0	Heptane	10.7
$MgCl_2$/DIBP/$TiCl_4$ -Al(i-Bu)$_3$	0.25	128	Toluene	208
$MgCl_2$/EB/$TiCl_4$ -Al(i-Bu)$_3$	0.028	643	Heptane	111

Temperature = 40 ^{o}C

[4-MP-1] = 2 mol dm^{-3}

*Stauffer 1.13

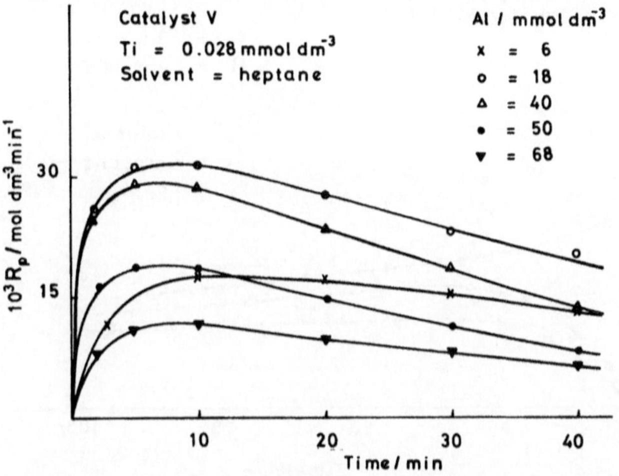

Fig. 7

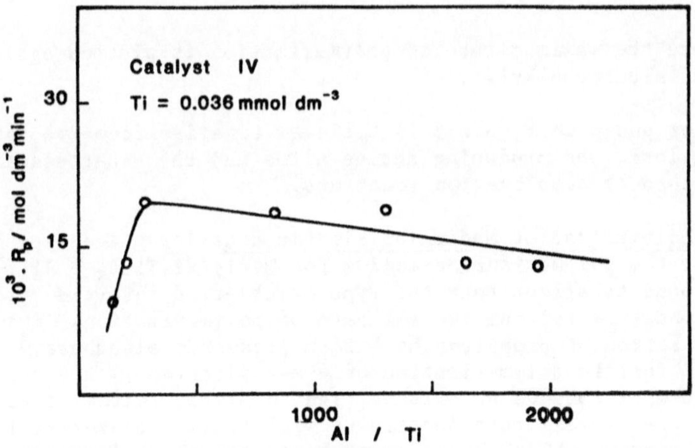

Fig. 8

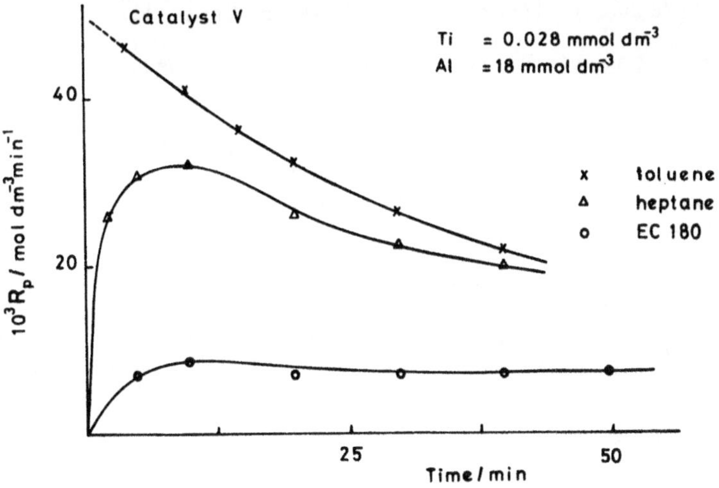

Fig. 9

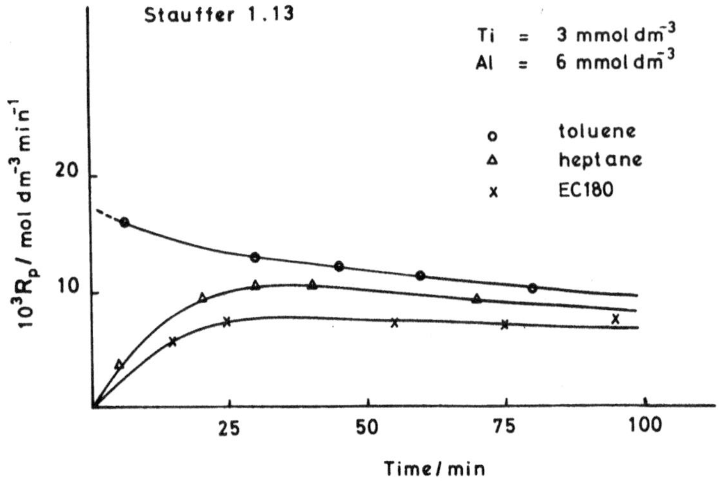

Fig. 10

in Figure 8 where the maximum rate of polymerization is plotted against the concentration of aluminumalkyl.

The behavior shown in Figure 8 is believed to arise from two sets of concurrent reactions, one producing active sites and the other deactivating active centers through complexation reactions.

Effect of Polymerization Medium on Kinetic Behavior. The type of diluent used for the polymerization medium for $MgCl_2/EB/TiCl_4$ - $Al(i\text{-}Bu)_3$ catalysts was found to affect both the type of observed rate-time profiles and the values obtained for the overall rate of polymerization. Results for the polymerization of propylene have been presented elsewhere.[3] The effects observed for the polymerization of 4-methylpentene-1 are shown in Figure 9. The use of toluene as diluent, led to higher initial rates of polymerization, the overall rate decreasing with time. Polymerizations in the hydrocarbon mixture, EC180, gave lower rates of polymerization, which after an initial "settling period," remained constant with time. A similar kinetic behavior was observed when a first generation type catalyst system was studied and comparable results for the use of the catalyst system δ-$TiCl_3.0.33AlCl_3$/donor - $Al(i\text{-}Bu)_3$, are shown in Figure 10.

The use of an aromatic diluent, such as toluene, leads in both cases to:

(1) the virtual elimination of the "settling period",
(2) higher overall rates of polymerization,
(3) a change in the type of rate-time profile, indicating the presence of destabilized polymerization centers.

While in the case of first generation titanium trichloride-matrix-type catalysts, the elimination or diminution of the "settling period" and the increased rates of polymerization observed when the toluene is used as a diluent, can be explained by enhanced catalyst matrix breakdown, it seems unlikely that similar arguments can be advanced for magnesium chloride-supported catalysts. The effects observed are considered more likely to be associated with the donor character of the π-electron system in toluene which can become of importance when high concentrations of toluene are used.

Active Center Concentrations in $MgCl_2/DIBP/TiCl_4-AlR_4/$ 4-MP-1 and $MgCl_2/EB/TiCl_4-AlR_3/4$-MP-1 Polymerization Systems

In the use of tritium quenching for the determination of active center concentrations, it is necessary to allow for both the kinetic isotope effect and for the accumulation of nonpropagation metal polymer bonds arising from chain transfer with aluminumalkyl.[6,7] Values for the kinetic isotope effect were determined experimentally[6] for each separate catalyst system involving the different aluminumalkyl cocatalysts. Values for active center concentrations (C*) were determined from metal polymer bond values using an extrapolation procedure.[6,7] A typical plot of $[MPB]_t$ versus time for polymerization system, $MgCl_2/DIBP/TiCl_4 - Al(C_{12}H_{25})_3/4$-MP-1 is shown in Figure 11.

$$\text{Since } [MPB]_t = C^* + \int_0^t R_{ta} dt \qquad (1)$$

where R_{ta} is the rate of chain transfer with metalalkyl, values of C* may be obtained from the intercept on the MPB-axis and values of R_{ta} may be obtained from the slope of the plot. Some values of C* for $MgCl_2$/DIPB/ $TiCl_4-AlR_3$ and $MgCl_2/EB/TiCl_4-AlR_3$ catalyst systems for the polymerization of 4-methylpentene-1 are listed in Table 2.

For diisobutyl phthalate containing catalysts in the polymerization of 4-methylpentene-1, it is apparent that:

(i) The numbers of active centers are still comparatively low and range from 1.2 - 9.6% mol/mol of the titanium content of the catalysts. These values are of the same order as those reported previously,[3] for the polymerization of propylene by the catalyst system $MgCl_2/EB/TiCl_4-Al(i\text{-}Bu)_3$ and also by other workers for similar catalysts.[4,8-13]

(ii) The numbers of active centers decrease as the alkyl groups in the aluminumalkyl increase in length and in size.

Nevertheless, it is evident that for ethyl benzoate-magnesium chloride catalysts that:

(i) Much higher numbers of active centers can be generated in some ethyl benzoate-magnesium chloride catalysts than in those containing diisobutyl phthalate.

(ii) As was observed for diisobutyl phthalate-magnesium chloride-

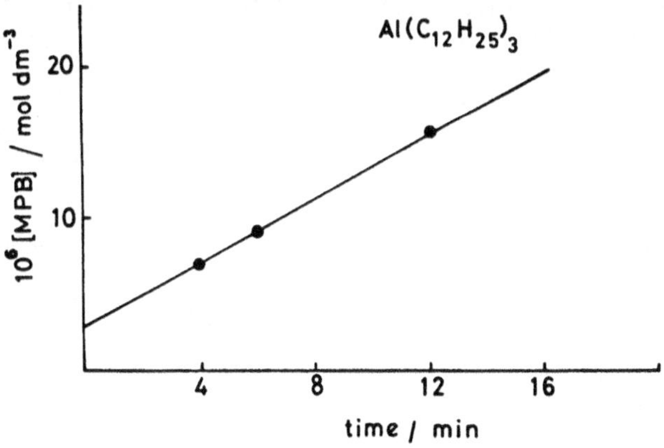

Fig. 11

Table 2. Comparative Data - Center Concentration - Tritium Quenching

Aluminium Alkyl	C*/mol/mol	
	$MgCl_2/DIBP/TiCl_4$*	$MgCl_2/EB/TiCl_4$**
$AlEt_3$	9.6×10^{-2}	-
$Al(i\text{-}Bu)_3$	8.0×10^{-2}	83×10^{-2}
$Al(n\text{-}Hex)_3$	5.0×10^{-2}	56×10^{-2}
$Al(n\text{-}Dodec)_3$	1.2×10^{-2}	11×10^{-2}

[4-MP-1] = 2 mol dm^{-3}

* [Ti] = 0.25 mmol dm^{-3}

** [Ti] = 0.036 mmol dm^{-3}

Temperature = 40 °C

containing catalysts the numbers of active centers decrease as the alkyl groups in the aluminumalkyl increase in length and size.

The decreased numbers of active centers observed, when using higher aluminumalkyls, is very likely associated with their reduced ability to alkylate exposed titanium atoms in these catalysts. The results indicate that titanium halide molecules, supported or chemisorbed on a $MgCl_2$/DIPB matrix, are less active towards alkylation than titanium halide molecules supported or chemisorbed on a $MgCl_2$/EB matrix, even when the relative titanium contents of the catalyst are the same.

Apparent values for the rate constants for chain propagation can be calculated from the simplified equation:

$$K_p = k_p C^* [M] \tag{2}$$

where k_p is the apparent rate constant. Values of k_p for both catalyst systems are listed in Table 3. Values of k_p for the $MgCl_2/EB/TiCl_4$-Al $(i\text{-}Bu)_3$ catalyst system are reasonably consistent with those reported previously for a different $MgCl_2/EB/TiCl_4$ catalyst.[2] It is apparent that the values of k_p are similar for both catalyst systems and that the major difference between the activities of two catalyst systems lies in the actual numbers of active centers.

The apparent increase in the values for the apparent propagation rate constant arises because of the decreased adsorption of the higher aluminumalkyls leading to higher values for the concentrations of adsorbed monomer.[16]

Adsorption processes, in general, are very important in heterogeneous catalysis for a proper understanding of the kinetic behavior of higher α-olefin polymerization. It is convenient to regard the propagation reaction as being a two-stage process, involving the adsorption of monomer, followed

Table 3. Comparative Data - Values of Propagation Rate Constants - Tritium Quenching

Aluminium-Alkyl	$k_p/dm^3mol^{-1}min^{-1}$	
	$MgCl_2/DIBP/TiCl_4$	$MgCl_2/EB/TiCl_4$**
$AlEt_3$	0.42×10^3	-
$Al(i\text{-}Bu)_3$	1.3×10^3	0.92×10^3
$Al(n\text{-}Hex)_3$	3.2×10^3	2.2×10^3
$Al(n\text{-}Dodec)_3$	10×10^3	8.1×10^3

[4-MP-1] = 2 mol dm^{-3}

* [Ti] = 0.25 mmol dm^{-3}

** [Ti] = 0.036 mmol dm^{-3}

Temperature = 40 °C

by an insertion reaction.[5] These adsorption processes become more important when dealing with higher olefins, since the sequence for increased capacity, to coordinate, will follow the sequence: ethylene < propylene < 4-methylpentene-1. Competitive adsorption of aluminumalkyl or chloroalkyl will also take place and consequently the overall rate of polymerization of 4-methylpentene-1 will be given by the equation

$$R_p = k_p C^* \theta_M \tag{3}$$

where θ_M is the fraction of active centers covered (or complexed by monomer and C* is the total concentration of active centers.[17]

The increased values of active center concentrations, obtained in higher α-olefin polymerization by magnesium chloride catalysts containing ethyl benzoate, is one of the more important features which have been established in the present study. Such a situation could arise from a number of possible situations.

(a) The monomer could be involved in the actual initiation step, and thus the relatively large concentrations of adsorbed higher - olefins could give rise to increased concentrations of active centers. However, in copolymerization reactions of ethylene and α-olefins, where enhanced rates of polymerization are observed, the values of the rates of copolymerization decrease, when the α-olefin is used up.[18] This would not be expected where centers had been established via the above mechanism.

(b) A large number of potentially active centers of different types and activity could be present, many of which are either inactive in their present form for chain propagation, or else are deactivated by irreversible complexation.

Considerable evidence has existed, for some time, for the existence of different types of active centers in heterogeneous Ziegler-Natta catalysts, and the following evidence can be considered to be relevant:

(i) The production of atactic and isotactic polymer in α-olefin polymerization using heterogeneous Ziegler-Natta catalysts indicates active centers with different stereoregulating ability. Different values for the propagation rate constants regulating the production of 'soluble' and 'insoluble' polymer have been established.[3,19]

(ii) Heterogeneity in copolymer composition in the copolymerization of α-olefins using heterogeneous catalysts indicating the presence of active centers located in different steric environments.[20,21]

(iii) Adsorption phenomena in allene adsorption studies, which indicated the presence of active centers with different reactivities, i.e. different k_p values.[3,7,19]

If the hypothesis, involving the presence of a number of different types of active centers, is accepted, then the increased values obtained for active center concentrations in higher α-olefin polymerizations and the observed rate enhancement experienced in ethylene-higher α-olefin copolymerizations, may be explained by one of two possible mechanisms:

(a) Complexation of an -olefin with a deactivated or inactive center leading to the formation of an active polymerization center.

A number of active centers are believed to exist in $MgCl_2$/EB/$TiCl_4$ catalysts, and Goodall[22] has described three particular types of active centers, viz., double vacancy corner sites, single vacancy corner sites, and single vacancy edge sites. In the case where coordination of an α-olefin can provide the necessary activation, an increased number of active polymerization centers will result. The presence of other donors, e.g. to toluene or electron donors, such as ethyl benzoate or diisobutyl phthalate or even aromatic compounds could have the same effect, and may explain the increased values of C* obtained when using $MgCl_2$/EB/$TiCl_4$ catalyst systems.

(b) Displacement of strongly adsorbed or complexed substances which have deactivated certain types of active centers leading to increased numbers of propagating centers. The general role of donors in this respect, has been discussed.[16,17,18] In both cases the increased numbers of active centers in higher α-olefin polymerization arises from adsorption or combination reactions between the higher α-olefin molecules and different types of potentially active centers.

In conclusion, while our knowledge of the true nature of heterogeneous catalysis and the structure of active polymerization centers is still somewhat uncertain and limited, some progress has been made in recent years towards a fuller understanding of the factors which control the behavior of these systems.

REFERENCES

1. P. J. T. Tait and M.Abu Eid, A.C.S. Meeting, Seattle, WA (1983).
2. P. J. T. Tait and M.Abu Eid, Proc. X, Iberoamerican Symposium on Catalysis, Lisbon, Portugal, Vol. 1, 163 (1984).
3. P. J. T. Tait, Future Aspects of Olefin Polymerization, Yokohama, Japan (July 4-6, 1985).
4. N. Kashiwa and J. Yoshitake, Makromol. Chem., 185, 1133 (1984).

5. I. D. McKenzie, P. J. T. Tait, and D. R. Burfield, Polymer, 13, 307 (1972).
6. D. R. Burford and P. J. T. Tait, Polymer, 13, 315 (1972).
7. P. J. T. Tait in: "Transition Metal Catalyzed Polymerizations: Alkenes and Dienes," R. P. Quirk et al. (Ed.) Harwood Academic Publishers, New York, p. 115 (1983).
8. G. D. Bukatov, S. H. Shepelev, V. A. Zakharov, S. A. Sergeev, and Y. I. Yermakov, Makromol. Chem., 183, 2657 (1982).
9. P. J. T. Tait, unpublished communication.
10. U. Giannini, Makromol. Chem. Suppl., 5, 216 (1981).
11. T. Keii, Makromol. Chem., 183, 2285 (1982).
12. N. Kashiwa and J. Yoshitake, Makromol. Chem., Rapid Commun., 4, 41 (1983).
13. J. C. Wu, C. I. Kuo, and J. C. W. Chien, Proc. IUPAC Macro 1982, 28th Macromol. Symposium, Amherst, MA (July 12-16, 1982).
14. N. F. Brockmeier and J. B. Rogan, Ind. Eng. Chem. Prod. Dev., 24, No. 2, 278 (1985).
15. P. Galli, Proc. IUPAC Macro, 1982, 28th Macromol. Symposium, Amherst, MA (July 12-16, 1982).
16. P. J. T. Tait and A. E. Enenmo (to be published).
17. P. J. T. Tait, A.C.S. Meetings, Miami Beach, FL (1985).
18. P. J. T. Tait and Y. Akinbami (to be published).
19. P. J. T. Tait in: "Preparation and Properties of Stereoregular Polymers," R. W. Lenz (Ed.), D. Reidel Publishing Co., Dordrecht, p. 85 (1980).
20. B. Baker and P. J. T. Tait, Polymer, 8, 225 (1967).
21. G. McKee, Ph.D. Thesis, University of Manchester (1967).
22. B. L. Goodall in: "Transition Metal Catalyzed Polymerizations: Alkenes and Dienes," R. P. Quirk et al. (Ed.), Harwood Academic Publishers, New York, p. 335 (1983).

POLYMERIZATION OF STYRENE VERSUS PROPYLENE WITH $MgCl_2$-SUPPORTED CATALYSTS

Norio Kashiwa, Akinori Toyota, and
Junichi Yoshitake

Mitsui Petrochemical Industries, Ltd.
Yamaguchi, Japan

INTRODUCTION

It is well known that $MgCl_2$-supported $TiCl_4$ ($MgCl_2/TiCl_4$) catalyst in combination with $AlEt_3$ shows much higher activity in propylene polymerization than the conventional $TiCl_3$ catalyst system,[1] but the stereospecificity of the obtained polymer is rather low, namely 20-40% as isotactic index (I.I.). In a study of kinetics on short time polymerization,[2] it was concluded that a cause of the low stereospecificity of this catalyst system was due to the high concentration of atactic active centers ($[C^*]_{ata}$ = 14-58%). We also reported that a catalyst system composed of $MgCl_2/TiCl_4$/ethyl benzoate- $AlEt_3$/ethyl benzoate showed very high stereospecificity in both propylene and styrene polymerization and the concentration of the isotactic active centers in both polymerizations was similar,[3,4] ($[C^*]_{iso(PP)}$ = 1.3%, $[C^*]_{iso(St)}$ = 3%, suggesting comparable isotactic active centers for both monomers.

In this report, we have investigated the behavior of styrene polymerization with $MgCl_2/TiCl_4$-Al-alkyl catalyst system, and discussed the nature of the active centers particularly in $MgCl_2/TiCl_4$-$AlEt_3$ catalyst system in comparison with propylene polymerization.

EXPERIMENTAL

Preparation of the Catalyst

$MgCl_2/TiCl_4$: Stainless steel balls (2.8 kg, 15mm in diameter) were put into a pot of internal volume, 0.8 l. The inside of the pot was purged fully with nitrogen and then 20 g of $MgCl_2$ was added, and the system was milled for sixty hours. Ten grams of milled products were suspended in 100 ml of $TiCl_4$ for two hours at 80°C. Solid products were filtrated and washed with decane. The said products contained 8 mg-Ti/g-catalyst.

Polymerization and Characterization

Two hundred ml of decane and 50 ml of styrene, or 100 ml of styrene, were added to a 500 ml-glass flask and Al-alkyl and Ti-catalyst were added. The system was stirred and stopped by the addition of methanol. The slurry was poured into 3 l of methanol and the polymer was recovered.

Isotactic Index (I.I.)

Polystyrene: One g of polymer was extracted with boiling methyl ethyl ketone by using a Sohxlet extractor.

Polypropylene: Five g of polymer were extracted with boiling n-heptane.

$\bar{M}_w, \bar{M}_n, \bar{M}_w/\bar{M}_n$: G.P.C. (Waters, model ALC/GPC 150), solvent, o-dichlorobenzene at 135°C.

RESULTS AND DISCUSSION

Activity

Table 1 shows the results of the polymerization of styrene with $MgCl_2/TiCl_4$ catalyst and $TiCl_3$ catalyst systems. The order of activity among these catalyst systems was as follows,

$TiCl_3/-AlEt_2Cl < TiCl_3-Ali\text{-}Bu_3 < TiCl_3-AlEt_3 < MgCl_2/TiCl_4-AlEt_3 <$

$MgCl_2/TiCl_4-AlEt_2Cl < MgCl_2/TiCl_4-Ali\text{-}Bu_3$

The activities of $MgCl_2/TiCl_4$ catalyst systems were 10-100 times higher than those of $TiCl_3$ catalyst systems.

Stereospecificity

The stereospecificity of polystyrene was evaluated by the extraction method with boiling methyl ethyl ketone.

Figure 1 shows the ^{13}C-NMR spectra of polystyrene with $MgCl_2/TiCl_4-AlEt_3$ catalyst system for whole polymer, boiling methyl ethyl ketone-insoluble fraction and soluble fraction. The ^{13}C-NMR spectra of polypropylene are shown in Figure 2 for the polymer, boiling n-heptane-insoluble fraction and soluble fraction obtained with the same catalyst system. From Figure 1, mmm tetrad sequences percent in methyl ethyl ketone-insoluble fraction was determined to be more than 90%, while that in the methyl ethyl ketone-soluble fraction was less than 30%, this indicated that methyl ethyl ketone (M.E.K.) extraction was suitable for the evaluation of the isotacticity of polystyrene.

Table 1 shows isotactic index (I.I.) of polystyrene with various catalyst systems by the M.E.K. extraction method. $MgCl_2/TiCl_4$ catalyst showed almost the same stereospecificity as $TiCl_3$ catalyst in conjunction with a common Al-alkyl, and both Ti-catalysts were more stereospecific in AlR_3 than with $AlEt_2Cl$ as a co-catalyst. (I.I. of polystyrene with AlR_3 or $AlEt_2Cl$ was 84-93% or about 50% respectively). $TiCl_3$-catalyzed polymerization of α-olefin such as propylene or butene-1, $TiCl_3-AlEt_2Cl$ catalyst system shows higher stereospecificity compared to $TiCl_3-AlR_3$ catalyst systems. Hatada et al.[5] explained the reason of the low stereospecificity of $TiCl_3-AlEt_2Cl$ catalyst system in styrene polymerization in the following way; "$AlEt_2Cl$ or $TiCl_3$ has an ability to polymerize styrene and to produce atactic polystyrene with a considerably high activity, by cationic polymerization mechanism, but this is not the case for propylene polymerization.

In styrene polymerization with $TiCl_3-AlEt_2Cl$ catalyst system, both Ziegler type coordination polymerization and cationic polymerization take place at the same time. Therefore, the stereospecificity or I.I. of

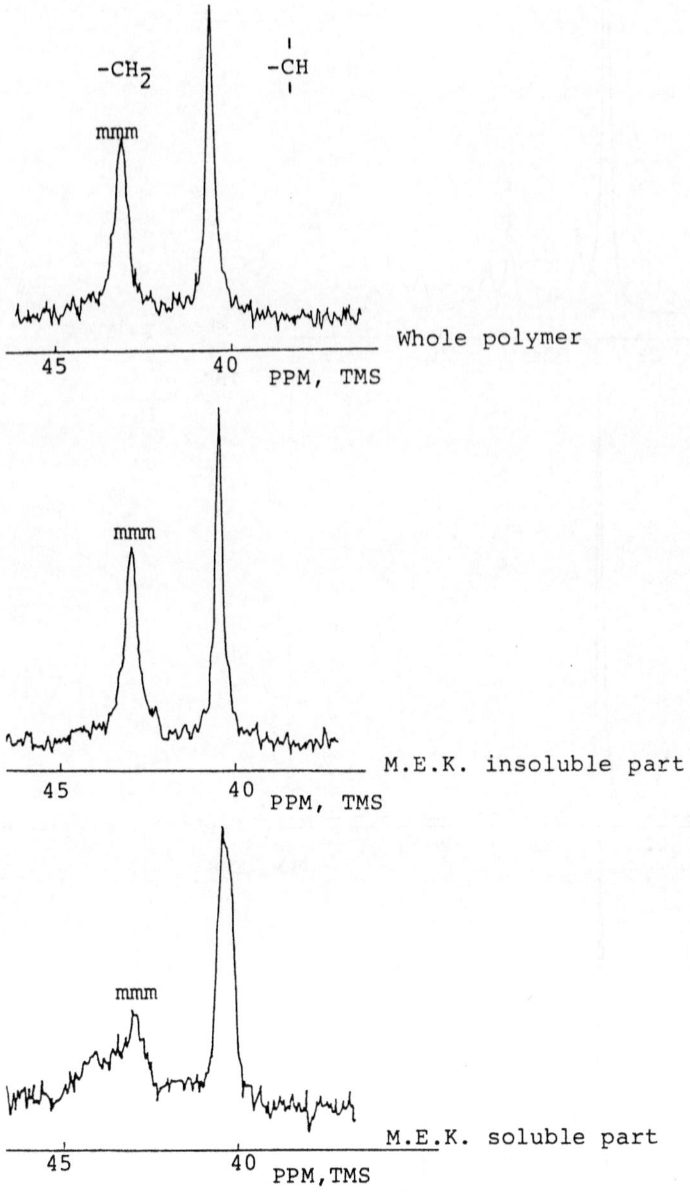

Fig. 1: ^{13}C NMR Spectra of Polystyrene Produced With $MgCl_2/TiCl_4$-$AlEt_3$ (Expanded view of methylene and methine carbon resonances)

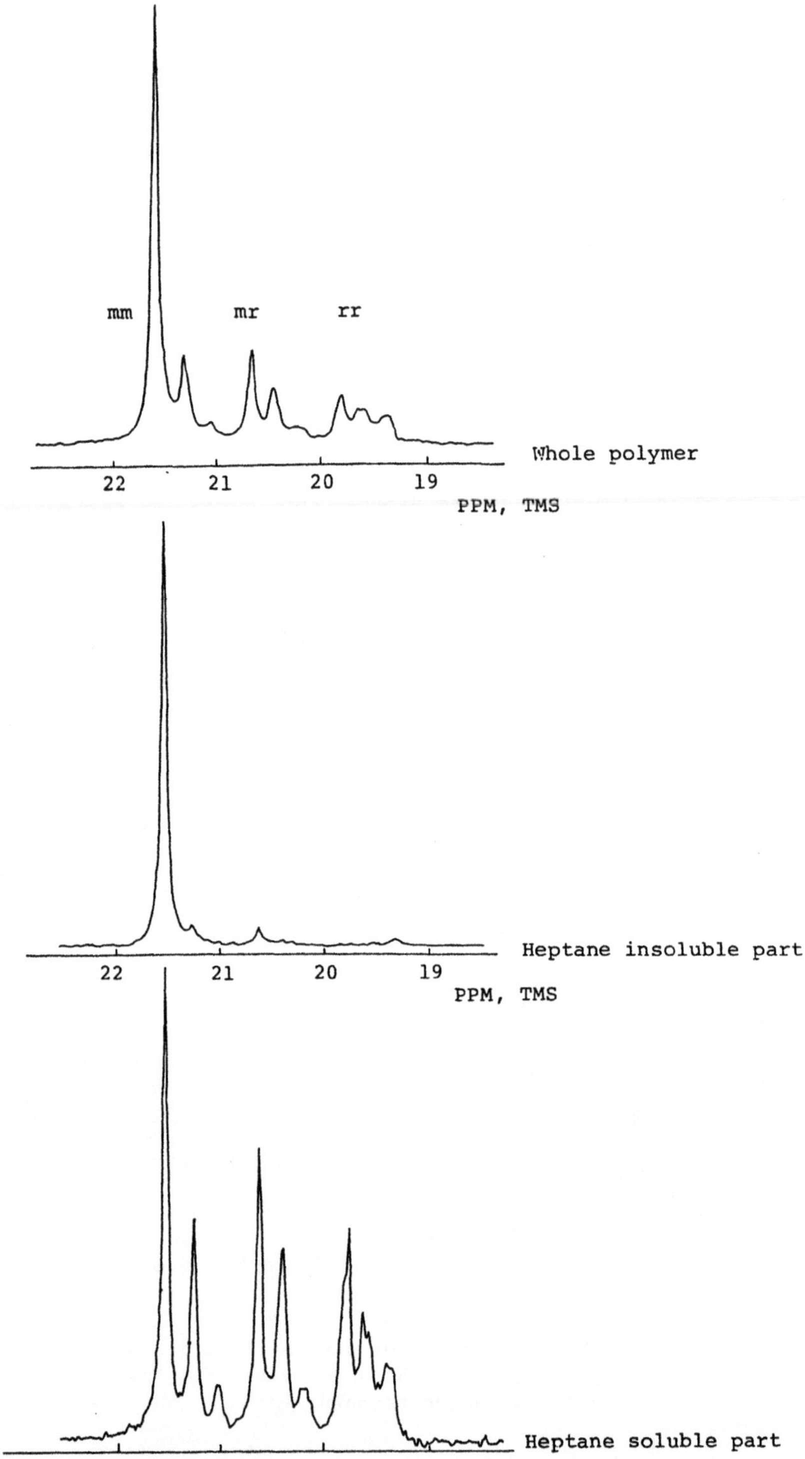

Fig. 2. ^{13}C NMR Spectra of Polypropylene Produced With $MgCl_2/TiCl_4$-$AlEt_3$ (Expanded view of the methyl carbon resonance)

Table 1. Styrene Polymerization Performance with Various Catalyst Systems

Polymerization conditions a)			Results		
Catalyst system	Al/Ti (mol/mol)	Time (min)	polymer yield (g)	Activity (g/mmolTi)	I.I. b) (wt.%)
$TiCl_3/1/3AlCl_3$ - $AlEt_2Cl$	2	30	3.7	0.37	51.9
$TiCl_3/1/3AlCl_3$ - $AlEt_3$	2	30	7.0	1.5	85.6
$TiCl_3/1/3AlCl_3$ - $AliBu_3$	8	120	1.1	0.48	83.6
$TiCl_3/1/3AlCl_3$	-	30	29.7	2.98	0
$MgCl_2/TiCl_4$ - $AlEt_2Cl$	10	30	8.1	23.8	52.9
$MgCl_2/TiCl_4$ - $AlEt_3$	10	30	6.9	20.2	84.2
$MgCl_2/TiCl_4$ - $AliBu_3$	20	30	11.6	68.2	92.9
$MgCl_2/TiCl_4$	-	30	3.9	29.3	0

a) The other polymerization conditions; solvent : n-decane 200ml, styrene 50ml, temp.:50°C.

b) Weight fractions insoluble in boiling methyl ethyl ketone

produced polystyrene is much lower. With $TiCl_3$-AlR_3 catalyst system, cationic polymerization is less important, because AlR_3 is not able to cause cationic styrene polymerization, and AlR_3 would inactivate the cationic active species in $TiCl_3$ catalyst by removing a small amount of water from the $TiCl_3$ catalyst. As a consequence, the $TiCl_3$-AlR_3 catalyst system shows higher stereospecificity than $TiCl_3$-$AlEt_2Cl$ in styrene polymerization." Therefore, in styrene polymerization, cationic polymerization is an important factor for understanding the stereospecificity of the catalyst system.

As shown in Table 1, $MgCl_2/TiCl_4$ catalyst itself showed an ability to polymerize styrene to atactic polystyrene. Therefore, Hatada's explanation for $TiCl_3$ catalyst system could be applicable to $MgCl_2/TiCl_4$ catalyst systems, and the observed order of the stereospecificity, $MgCl_2/TiCl_4$-$AlEt_3$ > $MgCl_2/TiCl_4$-$AlEt_2Cl$, was understandable.

Comparison Between Styrene and Propylene Homo-Polymerization with $MgCl_2/TiCl_4$-$AlEt_3$

Next, styrene and propylene homo-polymerizations for 30 min. with $MgCl_2/TiCl_4$-$AlEt_3$ catalyst system, were compared.

Table 2 shows the results obtained. The activity of the catalyst system in styrene polymerization was less than one-tenth as high as that in propylene polymerization. However, the I.I. of polystyrene was much higher (84%) than that of polypropylene (40%). The very poor stereospecificity in propylene polymerization with this catalyst system is due to the high concentration of atactic active centers in comparison to the isotactic active centers in the catalyst system.[2]

$$([C^*]_{ata(p)} = 14\text{-}58\%,\ [C^*]_{iso(p)} = 2\text{-}6\%)$$

It is very interesting that $MgCl_2/TiCl_4$ catalyst system shows the high stereospecificity in styrene polymerization in the presence of the high concentration of atactic active centers for propylene polymerization.

The Explanation for the High Stereospecificity in Styrene Polymerization with $MgCl_2/TiCl_4$-$AlEt_3$

We are proposing the following three explanations for the higher stereospecificity in styrene polymerization with $MgCl_2/TiCl_4$-$AlEt_3$;

Table 2. Styrene and Propylene Homo-Polymerizations With $MgCl_2/TiCl_4$-$AlEt_3$

Polymerization conditions[a)]			polymerization results	
Monomer	Temp. (°C)	Time (min)	Activity (g/mmolTi)	I.I. (wt.%)
Styrene	50	30	20.2	84.2[b)]
Propylene	50	30	360	39.5[c)]

a) The other polymerization conditions; Styrene polymerization was carried out in 200ml of n-decane and 50ml of styrene. $AlEt_3$(1.0mmol) and $MgCl_2/TiCl_4$(0.1mmol of Ti) were added in the said order, and then the polymerization was performed. Propylene polymerization was carried out under ambient pressure in n-decane(500ml).
b) boiling methyl ethyl ketone insoluble part
c) boiling heptane insoluble part

(1) Increase of the concentration of isotactic active centers for styrene polymerization.

There is a wider variety of active centers and some active centers which would be atacticspecific in propylene polymerization could act as isotactic active centers in styrene polymerization.

(2) Very small value of both propagation rate constant ($k_{p,ata}$) and $k_{p,ata}/k_{p,iso}$ ratio at active centers in styrene polymerization.

The catalyst activity can be considered to be proportional to the products of $k_p[C^*]$. Therefore, the proposed conditions would result in the low activity and the high stereospecificity even under the high $[C^*]_{ata}/[C^*]_{iso}$.

$$(I.I.(\%) \text{ of polymer} = 100k_{p,iso}[C^*]_{iso}/(k_{p,iso}[C^*]_{iso} + K_{p,ata}[C^*]_{ata})$$

$$= 100/(1 + (k_{p,ata}/k_{p,iso})([C^*]_{ata}/[C^*]_{iso}))$$

Styrene monomers could be coordinated at active centers, very stably and hardly inserted into the growing polymer chains, particularly at atactic active centers.

(3) A decrease of the concentration of atactic active centers for styrene polymerization.

In this explanation, atactic active centers are believed to have to be preferentially extracted from the solid catalyst component into the polymerization medium containing a large amount of styrene and this causes the inactivation of active centers. In this case, isotactic active centers should be relatively resistant to the extraction.

[C*] in Styrene Polymerization

At first, a possible explanation (1) was considered. A kinetic study was performed to determine the concentration of active centers, [C*], in styrene polymerization from the yield and molecular weight of polymers obtained by short time polymerization.

Polymerizations are shown in Table 3. The isotacticity of polymer increased with the lapse of time, particularly at an earlier stage. (I.I. = 39.9% at 5 sec., I.I. = 55.9% at 10 sec.)

Figure 3 shows the correlation between polymerization time and the polymer yield of isotactic fraction and atactic fraction. As for the isotactic fraction, the yield was increased linearly with polymerization time. Hence, the concentration of isotactic active centers are constant during the investigated period. $[C^*]_{iso(st)}$ was determined from the intercept of Figure 4 to be 1%, the yield of the atactic fraction was obtained within 20 sec. from the start of the polymerization, and $[C^*]_{ata(st)}$ could not be determined by the present study.

In contrast, in propylene polymerization with the same catalyst system, both the yields of isotactic and atactic polypropylene increased linearly with polymerization time in the same short time polymerization, and the concentrations of both active centers could be determined.[3,6] These results suggest that the production mechanism of atactic polymer was quite different in styrene polymerization and propylene polymerization.

Table 3. Results of the Styrene Polymerization With $MgCl_2/TiCl_4-AlEt_3$[a)]

Polymerization time (sec.)	Activity (g/mmolTi)	I.I.[b)] (wt.%)	$\bar{M}n\times10^{-4}$ [c)] (g/mol)	$\bar{M}w/\bar{M}n$ [c)]	$[N]\times10^2$ [c)] (mol/molTi)
5	2.9	39.9	6.26	7.30	1.84
10	3.2	55.9	5.95	11.5	3.01
15	5.6	46.4	7.66	8.18	3.39
20	6.5	50.7	5.95	9.03	5.53
25	6.5	60.1	6.05	14.4	6.46
40	7.9	64.8	6.80	14.2	7.53
60	11.8	68.5	8.47	9.73	9.54

a) Polymerization conditions; Temp.50°C, [Ti]=3.4mmol/ℓ styrene, solvent: styrene 100ml, Al/Ti=25 (mol/mol)
b) Weight fractions insoluble in boiling methyl ethyl ketone
c) Boiling methyl ethyl ketone insoluble part

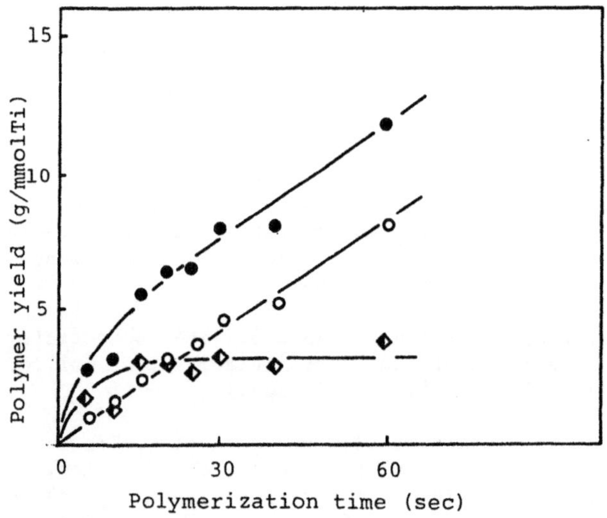

Fig. 3: Time Dependence of Polymer Yield in the Polymerization of Styrene With $MgCl_2/TiCl_4-AlEt_3$

(●): Whole polymer
(o): Methyl ethyl ketone insoluble part
(◆): Methyl ethyl ketone soluble part

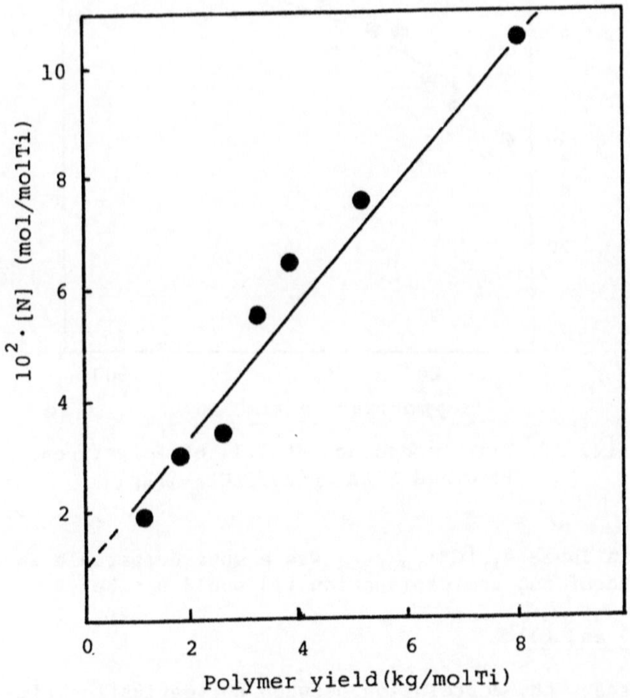

Fig. 4. Relation Between Polymer Yield and the Number of Polymer Chains Produced Per mol of Ti, [N]

Table 4. Comparison of [C*] in Propylene and Styrene Polymerization with $MgCl_2/TiCl_4$-$AlEt_3$ Catalyst System

Monomer	Polymerization Temperature °C	Fractions	[C*] mol %	Reference
Propylene	60	Iso.	2-6	2)
Styrene	50	Iso.	1	

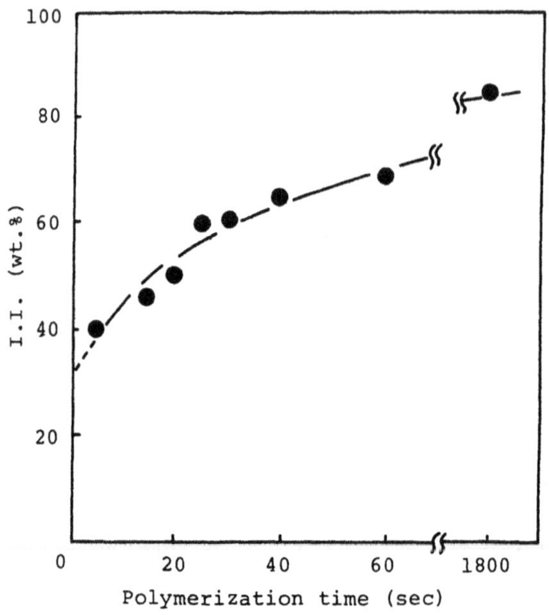

Fig. 5. Time Dependence of I.I. of Polystyrene Produced With $MgCl_2/TiCl_4$-$AlEt_3$

As shown in Table 4, $[C^*]_{iso(st)}$ was almost comparable to $[C^*]_{iso(p)}$. Therefore, the explanation (1) would not be valid.

Explanation (2) and (3)

Figure 5 shows the correlation between polymerization time and I.I. of polystyrene. I.I. was about 40% at 5 sec. and increased to 70% after 60 sec. and to 84% after 30 min. Therefore, the high stereospecificity of this catalyst system in styrene polymerization was observed for the polymerization longer than one min. This rapid increase of I.I. is attributable to the sharp decay of the activity to produce atactic polystyrene during the polymerization. Such a sharp decay is explained by (3), that is, $[C^*]_{ata(st)}$ decreased very rapidly with polymerization time, by the extraction of active centers in the polymerization medium and results in the inactivation. It is not clear why the extraction of atactic active centers should occur preferentially.

The explanation (2), that is, $k_{p,ata(st)}$ is quite small, could not explain a large decay in atactic polymer production during the polymerization. However, the cationic polymerization of styrene (2) could also explain the experimental facts.

G.P.C. Curves of the Obtained Polystyrene

Figure 6 shows the G.P.C. curves of polystyrene obtained at 5 sec., 1 min., and 30 min. in polymerization with $MgCl_2/TiCl_4$-$AlEt_3$ catalyst system. Each curve was composed of, at least, two separate peaks. The center of one peak was equivalent to molecular weight of 10^3-10^4 and the other was of 10^5-10^6. A lower molecular weight was predominant in the earlier polymerization stage, but with the time, it was overcome by formation of a polymer of higher molecular weight.

In this figure, the G.P.C. curve of boiling methyl ethyl ketone-insoluble part (isotactic polymer) of the polymer obtained in 30 min. polymeri-

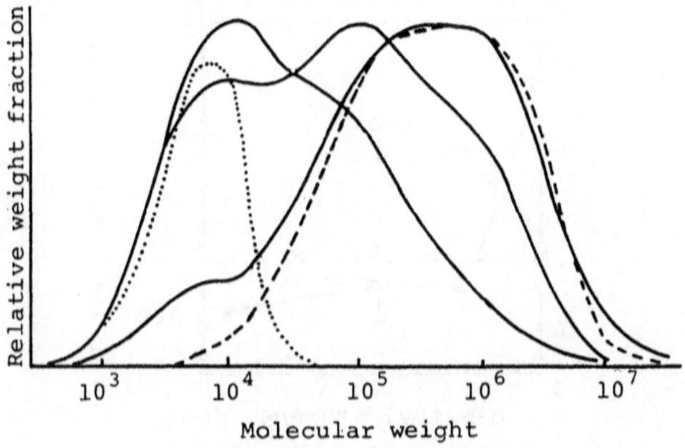

Fig. 6. G.P.C. Curves of Polystyrene

———— : Whole polymer produced with $MgCl_2/TiCl_4$-$AlEt_3$ in 5, 60 and 1800 sec.
- - - - -: Boiling methyl ethyl ketone insoluble part of polymer produced with $MgCl_2/TiCl_4$-$AlEt_3$ in 1800s.
··········: Whole polymer produced with $MgCl_2/TiCl_4$ (without $AlEt_3$)

zation was shown and it is consistent with the amount of high molecular weight polymer. Consequently, a lower molecular weight polymer should be atactic polymer, and the decrease of the contribution of the peak at lower molecular weight with polymerization time was coincident with the increase of I.I. with time in Figure 5.

Furthermore, the G.P.C. curve of atactic polystyrene obtained in cationic polymerization with $MgCl_2/TiCl_4$ catalyst and perhaps a trace amount of water contained in this Ti-catalyst (in absence of $AlEt_3$) was also exhibited in Figure 6. The G.P.C. curve of this atactic polymer was consistent with that of the atactic polymer part with $MgCl_2/TiCl_4$-$AlEt_3$ catalyst system.

In styrene polymerization, the presence of cationic active centers should be considered. Namely, in styrene polymerization with $MgCl_2/TiCl_4$-$AlEt_3$ catalyst system, there are two kinds of atactic centers regarding the so-called Ziegler type atactic active centers formed by the reaction of $TiCl_4$ with $AlEt_3$ and cationic atactic active centers formed by $MgCl_2/TiCl_4$ catalyst and water.

Modification of Explanation (2)

Here, we consider a case in which atactic polymer is produced in cationic polymerization and Ziegler type atactic active centers (coordination polymerization) would hardly work because of the very low $k_{p,ata}$ value. The observed polymerization behavior and the increase of I.I. with the polymerization time could be explained as follows.

"In the earlier stage, cationic polymerization produces atactic polymer with a considerably high activity and therefore, the I.I. of the polymer is very low. However, with the continued polymerization, active centers in cationic polymerization are inactivated. For example, as Hatada et al. pointed out, by the removal of water from the solid Ti-catalyst through the reaction of water with $AlEt_3$.

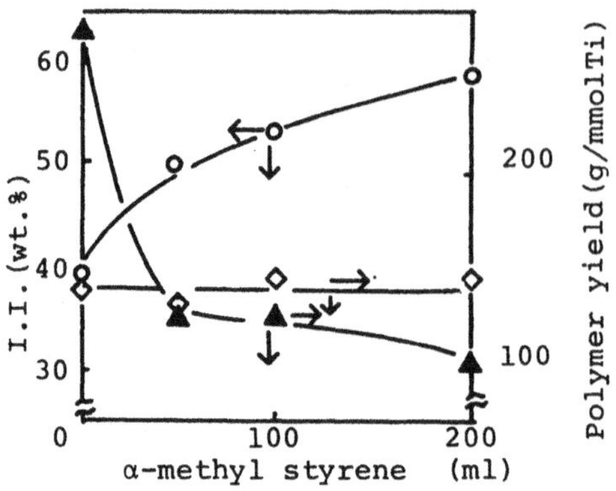

Fig. 7. Propylene Polymerization - Effects of Addition of -methyl styrene-

◇ : Boiling heptane insoluble part
▲ : Boiling heptane soluble part

Ziegler type active centers continue to produce the polymer with the high stereospecificity due to the very low value of $k_{p,ata}/k_{p,iso}$, and as a consequence, I.I. of polystyrene is increased with the polymerization time."

It seems that both explanation (2) and explanation (3) are valid, and we can not choose either one. Furthermore, the presence of cationic active centers which would disappear with polymerization time as well as Ziegler type atactic active centers is possible.[6]

Both explanations suggest that Ziegler type atactic active centers are partially inactivated by the presence of styrene monomer.

Effects of Addition of α-methyl Styrene to Propylene Polymerization System

Figure 7 shows the effects of addition of α-methyl styrene to propylene with $MgCl_2/TiCl_4$-$AlEt_3$ catalyst system. α-methyl styrene does not polymerize with this catalyst system. As shown in Figure 7, the yield of isotactic polypropylene was not affected by the addition of α-methyl styrene. α-methyl styrene was not found in the polymer by I.R. analyses, but the yield of atactic polypropylene was decreased considerably by the addition of α-methyl styrene. Consequently, I.I. was increased from 40% to 60%. This result shows that α-methyl styrene suppresses the production of atactic polymer in propylene polymerization, and supports the previous concept that styrene monomer inactivates Ziegler type atactic active centers.

We have discussed the increase of I.I. by addition of ethyl benzoate as an electron donor in propylene polymerization system and concluded that ethyl benzoate inactivates, very selectively, atactic active centers in the catalyst system.[7] Styrene monomer, or α-methyl styrene monomer could be regarded as a sort of electron donor with a benzene ring, like ethyl benzoate, and a similar effect on atactic active centers like ethyl benzoate is expected.

REFERENCES

1. N. Kashiwa, Polymer Journal, 12, 603 (1980).
2. N. Kashiwa, J. Yoshitake, Makromol. Chem., Rapid Commun., 4, 41 (1983).

3. N. Kashiwa, J. Yoshitake, Makromol. Chem., Rapid Commun., 3, 211 (1982).
4. N. Kashiwa, J. Yoshitake, Polymer Bulletin, 11, 485 (1984).
5. S. Murahashi, H. Yuki, K. Hatada, Kobunshikagaku, 22, 823 (1965).
6. N. Kashiwa, J. Yoshitake, Polymer Bulletin, 12, 99 (1984).
7. N. Kashiwa, Transition Metal Catalyzed Polymerizations, R. P. Quirk, Ed., Harwood Academic Publishers, New York, p. 379 (1981).

NEW CATALYSIS AND PROCESS FOR ETHYLENE POLYMERIZATION

F. J. Karol, B. E. Wagner, I. J. Levine,
G. L. Goeke, and A. Noshay

Unipol Systems Department
Union Carbide Corporation
Bound Brook, New Jersey 08805

INTRODUCTION

Olefin polymerization catalysis continues to be a fertile area of research with worldwide participation in both industrial and academic laboratories.[1-3] While much of this research has centered on methods to increase the productivity of catalysts, there has been and continues to be much active research on other features of olefin polymerization catalysis. The specific composition of the catalyst exerts an important effect on polymer molecular weight and molecular weight distribution (MWD), comonomer incorporation and copolymerization kinetics, and on the degree of stereoregularity. Moreover, the size, shape, and porosity (morphology) of the catalyst particle plays an important role in regulating the morphology of the resultant polymer. Development of low cost, reproducible processes for catalyst manufacture continues to be another important objective in catalyst research.[4] The focus of industrial research in olefin polymerization catalysis centers on the chemistry and technology necessary to obtain simultaneously favorable catalyst responses in all of the areas described above.

Union Carbide has developed a unique and versatile low-pressure, fluid-bed process for the production of HDPE and LLDPE (Figure 1). High-activity catalysts that are suitable for use in the UNIPOL process are a vital part of this technology. It was necessary throughout the development of high performance catalysts for the UNIPOL process to investigate all of the catalyst factors, to develop technology to improve performance in selected areas, and to combine specific, individual improvements into a final package of catalyst know-how (Table 1). This paper discusses some results and observations that were made during the course of investigations with different high-activity chromium and titanium catalysts.

POLYMERIZATION CONTROL WITH CHROMIUM-BASED CATALYSTS

Most workers have observed that productivity of the CrO_3/SiO_2 catalyst is sufficiently high that catalyst residues do not need to be removed or treated in the polymer.[5] Much of the research with chomium oxide catalysts has been directed to methods for controlling molecular weight (or melt index, MI) and molecular weight distribution (Table 2).[6-10] The

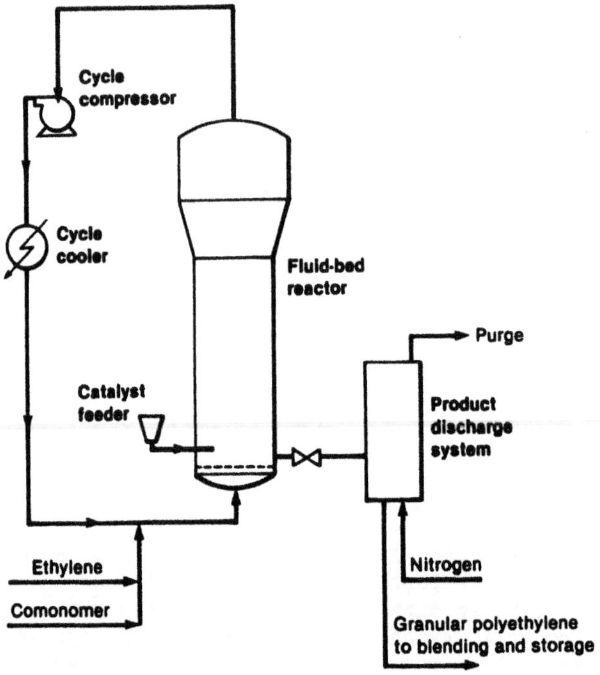

Fig. 1: Schematic of Low-Pressure, Fluid-Bed Process for Polyethylene Production

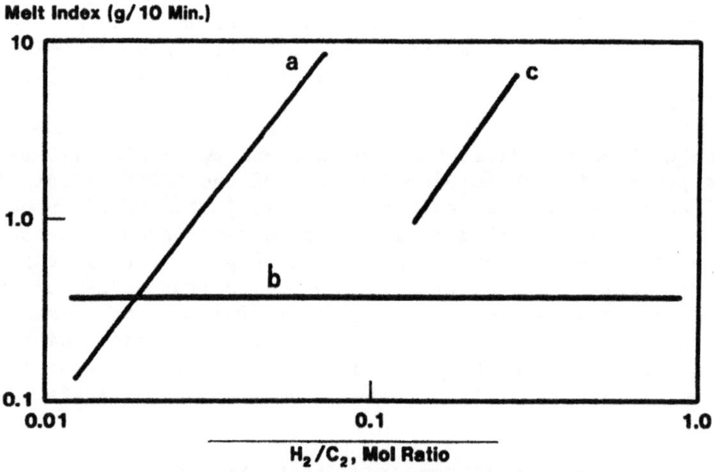

Fig. 2. Melt Index vs H_2/C_2 Ratio with Different Catalyst Families (A-C)

Table 1. Catalyst Requirements for UNIPOL Process

High Catalyst Productivity
Range of Molecular Weights
Range of Molecular Weight Distributions
Good Comonomer Incorporation
Attractive Particle Morphology
Low-Cost, Reproducible Catalyst Preparation

CrO_3/SiO_2 catalyst has very little, if any, response to hydrogen as a chain transfer agent (Figure 2, Curve B). Addition of organotitanates to the CrO_3/SiO_2 catalyst provides an attractive way to raise the polymer melt index.[7,8] Other methods include reduction and reoxidation of the chromium oxide catalyst using carbon monoxide or carbonyl sulfide, or pyrolysis of low-valent chromium salts.[12,13] Overall, these modifications of the CrO_3/SiO_2 catalyst provide wider control of polymer molecular weight.

Addition of inorganic fluorides such as ammonium silicofluoride [$(NH_4)_2SiF_6$] to the chromium oxide catalyst during catalyst preparation provides a means for production of polyethylenes with a more narrow molecular weight distribution.[6] Catalysts containing selected combinations of fluoride and titanate treatment show outstanding behavior for production of LLDPE.[11] Use of organoaluminum compounds such as diethylaluminum ethoxide (Et_2AlOEt) with the chromium oxide catalyst serves to broaden MWD.[14] All these catalyst modifications for controlling molecular weight or molecular weight distribution are effective because they provide a chemical modification of the silica surface in the vicinity of the active chromium center, or they alter the ligand environment or electron density at the active center.

Table 2. Chemical Modifications of CrO_3/SiO_2 Catalysts

Additive/Modifier	Typical Compound	Effect
Fluoride Compounds	$(NH_4)_2SiF_6$	Narrow
Organic Titanates	$(i\text{-}PrO)_4Ti$	Higher MI
Organoaluminum Compounds	Et_2AlOEt	Broaden MW
Reduction/Oxidation of Chromium	CO, COS	Higher MI
Pyrolysis of Low-Valent Chromium Cpd (0, 2, 3, 4) on Silica in N_2 and Reoxidation	Cr(III) Carboxylate	Higher MI, Broaden MWD

The use of π-organic ligands attached to an active chromium center can exert a significant effect on polymerization parameters.[15] In comprehensive studies, differences in hydrogen response, comonomer incorporation, and polymerization activity with these catalysts suggested that the nature of the active sites was different due, at least in part, to change in ligand environment. With π-bonded ligands it appears that at least one of these ligands (L) remains coordinated to the active chromium center during the polymerization process (Eq. 1).

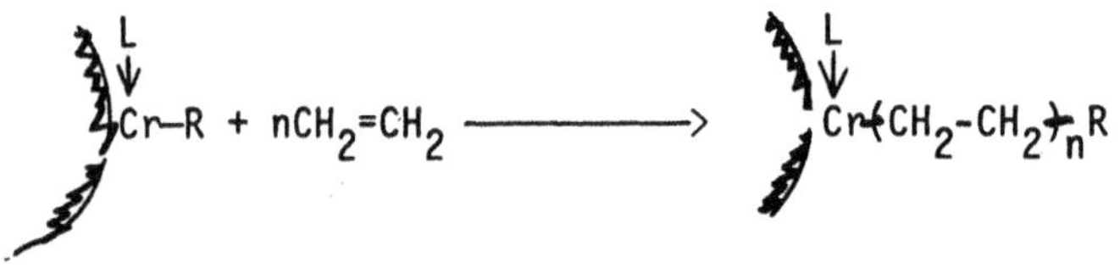

(1)

The presence of a cyclopentadienyl ligand on chromium [$(C_5H_5)_2Cr/SiO_2$ catalyst] significantly increases the hydrogen response of the catalyst (Curve A vs B in Figure 2).[16] As a result of this high hydrogen response, the catalyst produces a highly saturated polyethylene. However, polyethylenes produced with the CrO_3/SiO_2 catalyst usually have one double bond per molecule, indicating a different chain-transfer process. Thermal aging of the chromocene catalyst, $(C_5H_5)_2Cr/SiO_2$, leads to removal of the cyclopentadienyl ligand and loss of the high hydrogen response of the catalyst.[17] For comparison, Curve C in Figure 2 shows the hydrogen response of a magnesium chloride-titanium tetrachloride bimetallic catalyst (see below).

The overall studies with silica-supported, chromium catalysts illustrate that catalyst modifications can have a powerful and commercially significant effect on the response behavior of these systems.

POLYMERIZATION WITH CHEMICALLY-MODIFIED SILICA SUPPORTS

Role of Silica

Amorphous silicas of high surface area have been used commercially for many years with high activity chromium catalysts.[3,10] Silica supports play several vital roles in the polymerization process with chromium catalysts. Most importantly, silica participates in catalyst site formation by chemically anchoring chromium compounds to the surface silanol groups on silica. This chemical anchoring process isolates chromium sites on the high-surface-area silica thereby preventing destruction of active sites by mutual interaction.

Silica supports, containing the chemisorbed chromium compounds, determine the flow properties of the catalyst. Spheroidal silica supports display excellent flow properties and can be easily fluidized and conveyed in a dry state. Finally, the silica support provides a template for polymer particle growth. Spheroidal silicas with average particle sizes of 30-100 μm provide an ideal template for polymer particle growth.

Dehydroxylation of Silica Supports

Prior to heating at high temperatures, the silica surface is covered with silanol groups and adsorbed water. Thermal treatment at 200°C may remove all physically bound water, but hardly affects surface silanol groups. At elevated temperatures progressive elimination of water takes place by reaction of neighboring silanols. The surface chemistry of these heat-treated, amorphous silicas indicates that silanol and siloxane groups

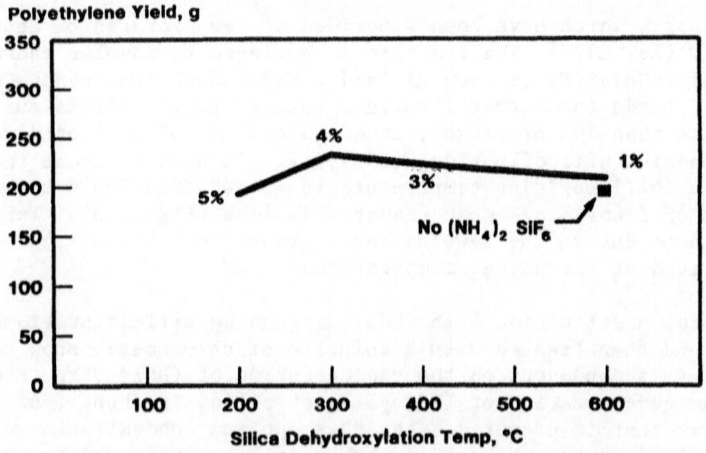

Fig. 3. Effect of $(NH_4)_2SiF_6$ Concentration and Silica Dehydroxylation Temperature on $(C_5H_5)_2Cr/SiO_2$ Catalysts

are present on the silica surface. With an increase in the dehydroxylation temperature of silica the fractional population of isolated silanol groups is increased. Many different techniques are available for determining the concentration of silanol groups on silica.[18,19]

Before chemical anchoring of organochromium compounds on silica it is necessary to dehydroxylate the silica support at temperatures in the range of about 500°C to about 800°C for periods of at least 4-8 hours. This dehydroxylation step provides isolated silanol groups for deposition of the chromium compounds.[20]

Polymerization with Fluoride-Modified Silica Supports

Chemical reaction at low temperatures of the surface silanol groups on silica offers an alternative to the high temperatures (500-800°C) used to prepare silicas for the $(C_5H_5)_2Cr/SiO_2$ catalysts. High-activity chromocene catalyst compositions for ethylene polymerization have been prepared with

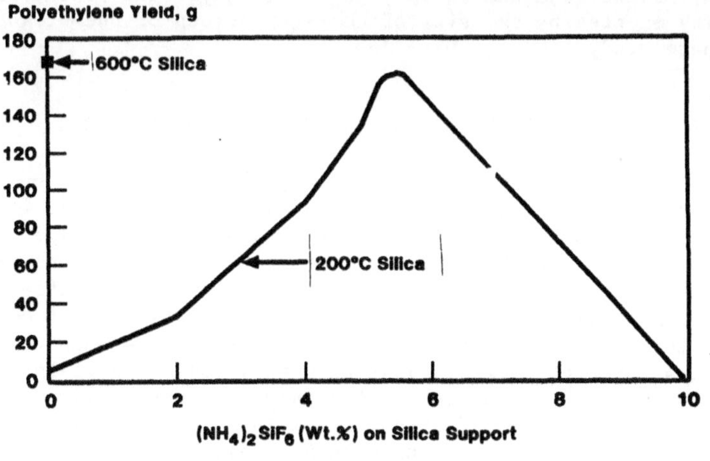

Fig. 4. Effect of $(NH_4)_2SiF_6$ Concentration on $(C_5H_5)_2Cr/SiO_2$ Catalysts

silica supports which have been fluorided at temperatures of about 200°C to about 500°C (Eq. 2).[21] The reaction is believed to involve thermal decomposition of $(NH_4)_2SiF_6$ to form HF (and SiF_4), which then reacts with silica to form SiF bonds (and other fluorided species no the silica surface). Usually less than 10% by weight, based on silica, of a fluoriding agent such as ammonium silicofluoride, $(NH_4)_2SiF_6$, is used to treat the silica. As the fluoriding temperature is raised from 200°C toward 600°C the amount of fluoriding agent required is less (Figure 3). This effect is believed to be due to the fewer silanol groups that are present on silicas dehydroxylated at the higher temperatures.

Chemical modification with $(NH_4)_2SiF_6$ using silicas previously dried at 200°C, and then treated with a solution of chromocene, show that catalytic activity is dependent on the concentration of $(NH_4)_2SiF_6$ (Figure 4). The optimum concentration of $(NH_4)_2SiF_6$ on silica is about 5-6% by weight. The catalyst that is prepared using this optimum concentration of $(NH_4)_2$ SiF_6 is equivalent to unfluorided chromocene catalysts which require silica dehydroxylation temperatures near 600°C. Treatment of silica supports with higher levels of $(NH_4)_2SiF_6$ (10-20 wt %) led to catalysts with little or no activity. Presumably too many of the silanol groups are removed by fluoridation, and not enough sites are available for anchoring of chromocene to form an active catalyst.

Polymerization with Silane-Modified Silica Supports

Another method for treating the silica support to remove silanol groups involves the use of hydrosilane compounds in the presence of base (Eq. 3).[22] This treatment is carried out prior to deposition of the chromocene compound on the support. Such a treatment provides a modified support to which the hydrosilane is chemically anchored. For example, butylsilane reacts with silanol groups in the presence of triethylamine to form butyl hydrosiloxy groups and hydrogen.

The butylsilane/triethylamine modification of dried silica provided $(C_5H_5)_2Cr/SiO_2$ catalysts with productivity, hydrogen response, and MWD behavior equivalent to that exhibited by $(C_5H_5)_2Cr/SiO_2$ catalysts based on high temperature silicas. The effectiveness of the hydrosilane treatment of silica is related to the number of silicon hydride bonds in the hydrosilane compound. Amylsilane was similar to butylsilane in providing highly active catalysts. Diethylsilane treatment gave only fair activity and triethylsilane resulted in very poor activity. The effect on catalyst productivity exerted by the size of the alkyl group of the silane compounds requires more study.

COPOLYMERIZATION WITH DIFFERENT MAGNESIUM-TITANIUM CATALYSTS

The efficiency of titanium polymerization catalysts can be improved using magnesium chloride by several methods.[13] These methods include formation of discrete magnesium-titanium complexes, and ball-milling which results in insertion of titanium compounds into defects of the magnesium chloride substrate. Some other approaches involve the formation of high surface area magnesium-titanium sponges, and the formation of $MgCl_2TiCl_e3$ solid solution, by coprecipitation. The operational methods used to improve catalyst productivities appear to have a common basis for their effectiveness. Those techniques that provide the opportunity to expose and isolate the maximum amount of transition metal compound, either in a solid matrix or a bimetallic complex, have been highly effective.

Some informative features of high activity magnesium-titanium catalysts may be established by contrasting two catalyst families, one based on

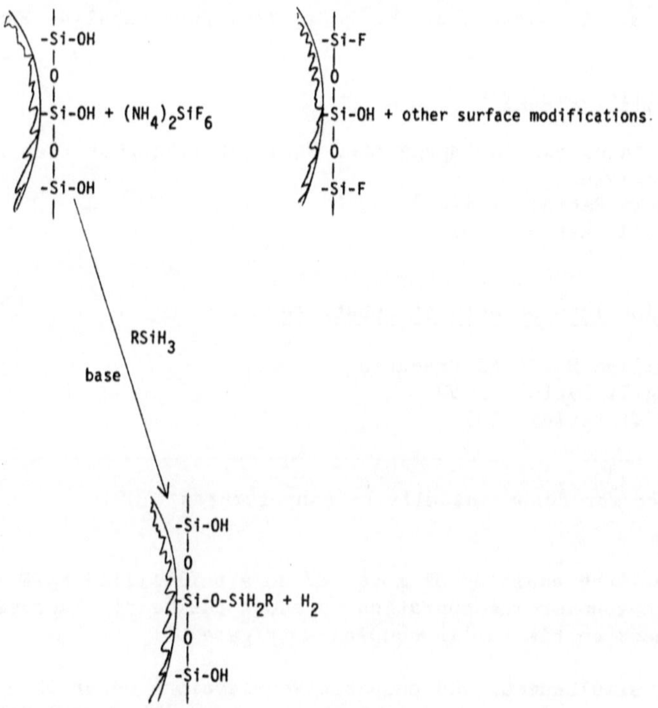

(2)

(3)

discrete magnesium-titanium complexes,[23,24] and the other based on ball-milled compositions (Table 3).[27] The two catalyst families show distinctly different copolymerization behavior (Figure 5). Lower comonomer ratios (butene/ethylene) in the gas stream are required in order to obtain polymers of equivalent density using ball-milled catalyst compositions. It was of interest to determine features in these catalysts that influence comonomer incorporation.

Treatment of bimetallic complexes of the type outlined in Table 3 with an alkylaluminum dihalide, prior to activation with a trialkylaluminum compound, improves comonomer incorporation of the resultant catalyst.[28]

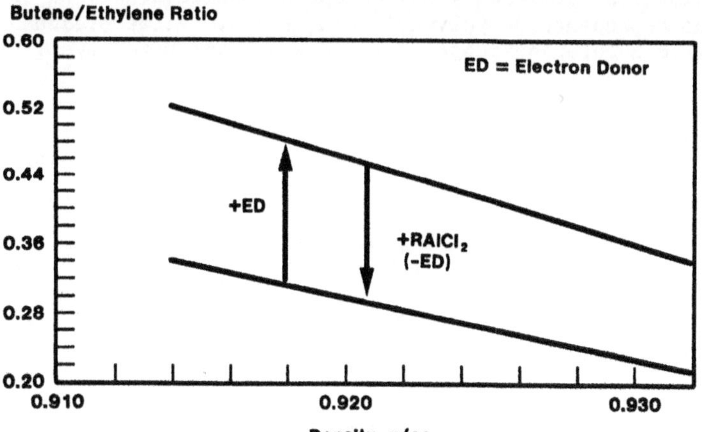

Fig. 5. Effect of Butene/Ethylene Ratio on Polymer Density for Different Catalyst Families

Table 3. Comparison of Two $MgCl_2/TiCl_4/ED^1$ Catalyst Families

Bimetallic Complexes	
Precipitated/Silica-Impregnated Mg/Ti/ED Precursor Prepared in ED Solvent	
Low Mg/Ti Ratio:	1-6
High ED/Ti Ratio:	7-10
Insertion into Defects of Substrate	
Ball-Milled Mg/Ti/ED Precursor	
High Mg/Ti Ratio:	> 50
Low ED/Ti Ratio:	1-2

1 = electron donor, usually tetrahydrofuran (THF)

Moreover, the addition of more THF to a ball-milled Mg/Ti/THF precursor lowered comonomer incorporation to about the level observed with catalysts based on bimetallic complexes (Figure 5).

Several simultaneous and competitive reactions occur when a $MgCl_2/TiCl_4/THF$ precursor is treated with Lewis acids such as $EtAlCl_2$. Infrared analysis has shown that this treatment removes about 80% of the THF from the precursor with no THF complexed to magnesium. Apparently redistribution of the THF occurred, with THF complexed to the strongest Lewis acids ($TiCl_4$ and $EtAlCl_2$) available. Lewis-acid treatment increased the porosity of the precursor particle. The three-to fourfold increase in catalyst activity after Lewis-acid treatment can in part be attributed to opening-up of coordination sites at the titanium centers.

COMPOSITIONAL HETEROGENEITY OF LOW-PRESSURE, ETHYLENE COPOLYMERS

Low-pressure, ethylene copolymers produced with high activity catalysts are compositionally heterogeneous in short chain branching. This heterogeneity can affect the physical properties of copolymers. The extent of compositional heterogeneity may depend on molecular weight or show little or no dependence on polymer chain length. Films produced from vanadium-based homogeneous copolymers show a lower haze level, higher

Table 4. Influence of $(C_2H_5)AlCl_2$ on Copolymerization with Magnesium-Titanium-Electron Donor (THF) Complexes

$EtAlCl_2$/THF	C_4/C_2, Mol Ratio in Reactor	Polymer Density, g/cc
0	0.402	0.925
0.9	0.370	0.920
1.6	0.326	0.925
1.7	0.325	0.925

impact strength, and a better balance of physical properties than films produced with heterogeneous copolymers.[29]

With homogeneous copolymers, fewer comonomer units in the polymer chain are required to reach a specific polymer density. The critical factors influencing compositional heterogeneity require more study. The chemical factors affecting the breadth of molecular weight distribution do not have a corresponding effect on compositional heterogeneity. Low density polyethylenes of narrow MWD can be more compositionally heterogeneous than copolymers of broad MWD. Ethylene copolymers produced in a solution process can also be compositionally heterogeneous.

KINETIC PROFILES OF HIGH ACTIVITY CHROMIUM AND TITANIUM CATALYSTS

High activity chromium and titanium catalysts usually show different kinetic profiles during ethylene polymerization. Moreover, even with catalysts based on a specific transition metal, such as chromium or titanium, the kinetic profile can be different depending on the ligand environment and oxidation state of the active transition metal species. The profiles illustrated in Figure 6 portray the type of behavior that can be observed with different catalysts. Polymerization rates are shown on a total catalyst basis. Changes in the concentration of transition metal compound on the catalyst will affect the total amount of polymer produced, but usually not the shape of the kinetic profile.

The rate of polymerization with a typical CrO_3/SiO_2 silica catalyst (Curve B) shows a characteristic induction period after which the rate of polymerization is fairly constant with little or no loss in catalyst activity after several hours of polymerization.[30] The catalyst based on triephenylsilyl chromate $[(\phi_3SiO)_2CrO_2/SiO_2]$ with alkylaluminum alkoxides (Curve D)[31] shows a fairly constant rate profile. No significant induction period is observed with this catalyst.

The kinetic behavior of the $(C_5H_5)_2/SiO_2$ catalyst is different (Curve A).[16] This catalyst shows very high catalytic activity during the first hour of polymerization, but rapidly decays with little polymerization after two hours. Clearly a deactivation mechanism is occurring which depends upon factors associated with this specific catalyst. The deactivation is

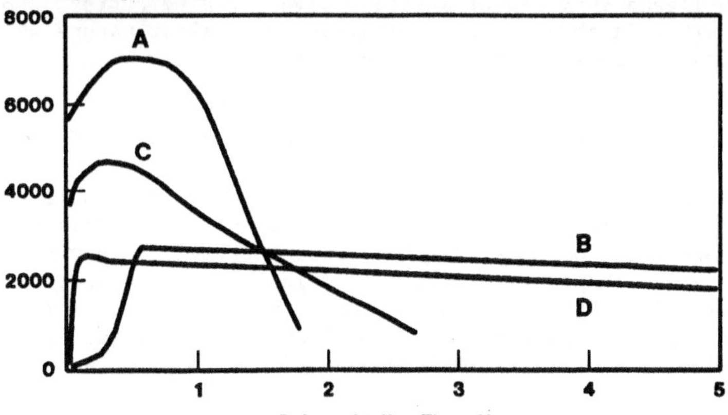

Fig. 6. Kinetic Profiles in Ethylene Polymerization with Different Catalyst Families

not related to the total amount of polymer that is produced. The CrO_3/SiO_2 catalyst, for example, under some conditions can produce more polymer per chromium atom than the $(C_5H_5)_2Cr/SiO_2$ catalyst and yet still retain a high level of catalyst activity. A likely mechanism for decay with the $(C_5H_5)_2Cr/SiO_2$ catalyst involves decomposition of the active site or reaction of the active site with surface groups on the silica support. Kinetic evidence suggests that the active site may have some mobility on the silica surface during the polymerization process.

The rate curve illustrated by Curve C shows the type of rate profile that is observed with several high-activity magnesium-titanium catalysts of the type described in Table 3. Initial catalyst activity is high with no induction period. Catalyst decay became measurable fairly early in the polymerization process. Decay with catalysts of this type is believed to be associated with catalyst decomposition through processes such as bimolecular coupling of the catalyst or deactivation by reaction with alkylaluminum halides formed in the polymerization process.

CONTROL OF MOLECULAR WEIGHT DISTRIBUTION

Considerable experimental data support the view that a diversity of chemically distinct active species is primarily responsible for the range of molecular weight distributions measured in ethylene polymerization. Changes in molecular weight distribution brought about by diffusion limitations do not provide an adequate explanation for the broad range of molecular weight distributions encountered with different catalyst systems.[3,32] The transition metal with the surrounding ligand environment can have a profound influence on polymer molecular weight distribution. Although research into the factors controlling MWD has received much attention, no comprehensive theory has evolved that explains or predicts the molecular weight distribution of an olefin polymer with any level of certainty.

Under comparable reaction conditions using polymerization temperatures near 90°C, many high activity catalysts based on titanium produce polyethylenes of relatively narrow MWD ($\bar{M}_w/\bar{M}_n$) = 2.5-4.5).[23-26] On the other hand, high activity, chromium-based catalysts provide polyethylenes of intermediate and broad MWD (Table 5). The breadth of MWD depends, among other factors, upon the particular chromium compound and the method of catalyst preparation. Ligand effects on MWD seem to be much less significant with titanium catalysts. With many titanium catalysts the active center is likely surrounded by chloride ligands, whose presence appears to control the ligand environment at the active titanium center. Possibly, the active sites in chromium catalysis have a covalent nature while those based on titanium show a more ionic character.

EFFECTS OF CATALYST MORPHOLOGY ON POLYMER MORPHOLOGY

Heterogeneous olefin polymerization catalysts are capable of replicating their morphology into the morphology of polymer particles. Because of this observation, workers have been able to regulate polymer particle morphology (size, shape, porosity) by regulating the morphology of the catalyst particle. The catalyst particle acts as a template for initial growth of the polymer particle. For high activity catalysts in olefin polymerization, the average particle size of the polymer is 15-20 times larger than the catalyst particle size.[3] Although the polymer particle replicates the morphology of the catalyst particle, experimental data show that the catalyst particle breaks down or shatters during the polymerization process and the fragments become dispersed throughout the polymer particle.[33]

Table 5. Molecular Weight Distribution with Chromium and Titanium Catalysts

Catalyst	Cocatalyst	MWD	Typical $\bar{M}_w\bar{M}_n$
$MgCl_2/TiCl_4/$ Electron Donor	R_3Al	Narrow	2.5-4.5
$(C_5H_5)_2Cr/SiO_2$	--	Intermediate	4-10
CrO_3/SiO_2	—	Intermediate	8-12
$(\phi_3SiO)_2CrO_2$	R_2AlOR'	Intermediate-Broad	10->20

Scanning electron microscopy has been widely used in the study of catalyst morphology and the replication effect in the resultant polymer particle.[34] Microtoming of catalysts and polymers to examine internal structure, and visual observations of ongoing polymerizations under a video recorder-equipped optical microscope have also proven to be powerful analytical tools.

Needle-shaped precursors usually form needle-shaped polymers. Transformation of the same precursor into spheres via shaping methods such as controlled precipitation or spray-drying permits spherical polymers to be formed. When a catalyst precursor is impregnated in or chemisorbed on a support, the external morphology of the polymer replicates the support, not the morphology of the catalyst precursor within the support.

The physical manifestation of polymer morphology--resin bulk density and flow properties--are crucial parameters which can determine the success or failure of commercial catalysts (Figure 7). The packing of resin particles (interparticle porosity) and the fullness of particles (intraparticle porosity) determine the resin bulk density (Figure 8). Resin packing properties and flow properties are related to particle roundness and particle size distribution. The significant factors influencing polymer morphology include the activity and kinetic profile of the catalyst, and the features of the catalyst support. The polymer growth conditions, and

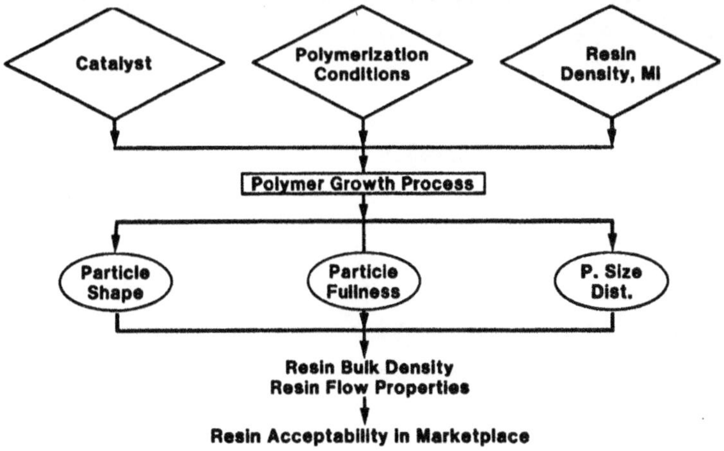

Fig. 7: Components of Polymer Morphology

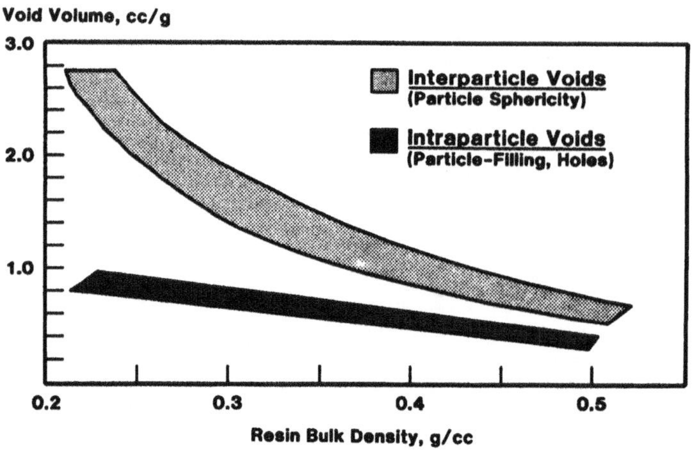

Fig. 8: Bulk Density vs Void Volume

the extent to which the nascent polymer particles can accommodate the stresses of the growth process, contribute to the resultant morphology (Table 6).

The growing polymer particle must be able to resist the forces that threaten to pull it apart. Experimental data support the following view of the growth of the polymer particle. In the initial stages of polymerization, the diameter of the growing particle expands rapidly. Since olefin polymerization is a highly exothermic process, the temperature within the particle increases rapidly, and may exceed the softening or melting point of the resin. The morphology of the final particle will depend on the balance between annealing and distorting forces during particle growth.[35]

In the early stages of polymerization the "hydraulic pressure" inside the particle is relieved by rapid, radially-outward motion of the entire mass within the particle (Figure 9). At excessive initial rates of polymerization the stress on the growing polymer particle is high and the particle becomes hot and soft. The expanding interior catches up with the outer shell of the particle creating a thick outer skin and a poorly-filled interior (Figure 10).

Table 6. Factors Influencing Polymer Growth

Catalyst
Activity
Distribution in Support
Support
Size, Shape, Porosity of Silica Particle
Size of Microparticles Used to Form Silica Particle
Fragility
Kinetic Profile of Polymerization
Nascent Polymer Viscosity

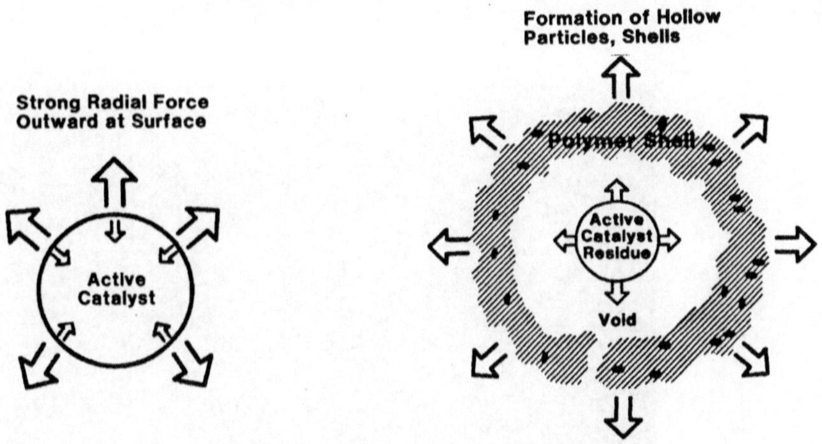

Polymer Particle Loses Cohesiveness

Fig. 9. Effect of Fast Induction or Effect of Uneven Catalyst Distribution

Under less vigorous polymerization conditions, when polymerization rates are moderate enough to allow the nascent polymer to take a semi-liquid form, a beneficial annealing process can occur. The exterior surface becomes smoother and more spherical and the polymer beads being produced by the shattered subparticles are fused into a solid, coherent mass within the interior of the particle (Figure 11).

Finally, if the initial rate of polymerization is not sufficiently high as to soften the resin during the early stages of polymerization, then a highly irregular, fragile and sponge-like polymer particle is obtained (Figure 12). This model of slow internal growth assumes that the morphology of the polymer particle arises from independent polymerization of the shattered subparticles and from the inability of the resultant polymer beads to fuse into one cohesive particle.

This discussion emphasizes that the polymer particle growth process is complex and the behavior of the catalyst is very important in the growth process. Control of the catalyst features responsible for the growth process represents another challenge in polymerization catalysis and catalyst design.

OUTLOOK FOR THE FUTURE

Worldwide attention devoted to the study of olefin polymerization catalysts will continue to reap substantial benefits. Catalyst characterization studies using a wide range of modern analytical techniques will continue to provide greater insight into the structure and composition of these catalysts. More attention will be focused on the factors affecting catalyst decay and on ways to provide more stable catalyst compositions. Control of molecular weight distribution and compositional heterogeneity of ethylene copolymers will receive much attention, but significant results in the near future will be based on empirical approaches. Comparative studies with different transition metal compounds will provide the basis of a greater understanding of the important factors in olefin polymerization and provide the basis for attractive new catalysts. More microscopy studies, coupled with kinetic profile studies of catalyst behavior, are needed if more progress in polymer particle growth mechanisms is to be made.

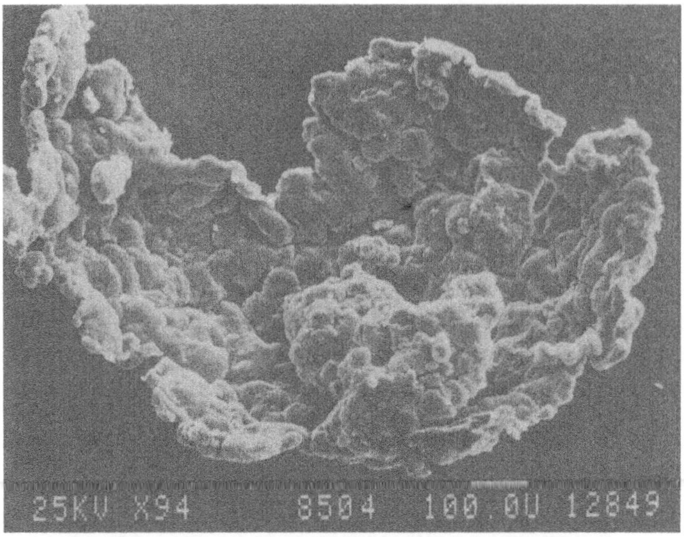

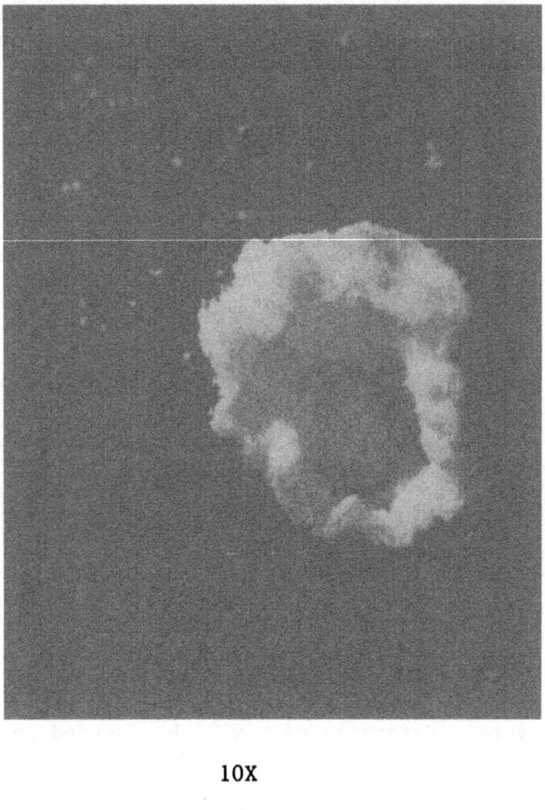

10X

Fig. 10. Morphology of Polyethylene Particles Produced at Very High Initial Polymerization Rates

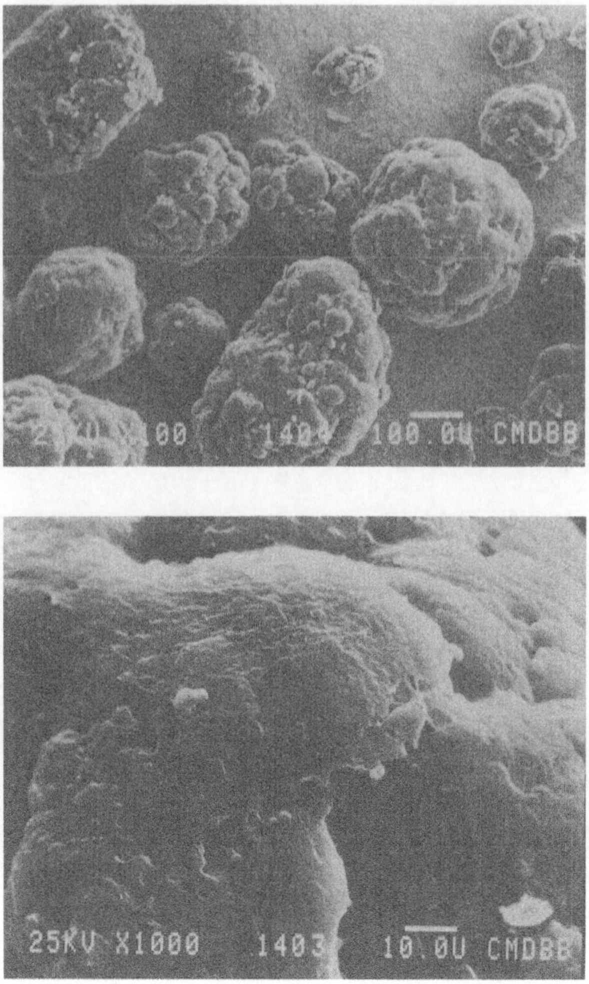

Fig. 11. Morphology of Polyethylene Particles Produced Under Favorable Polymerization Conditions

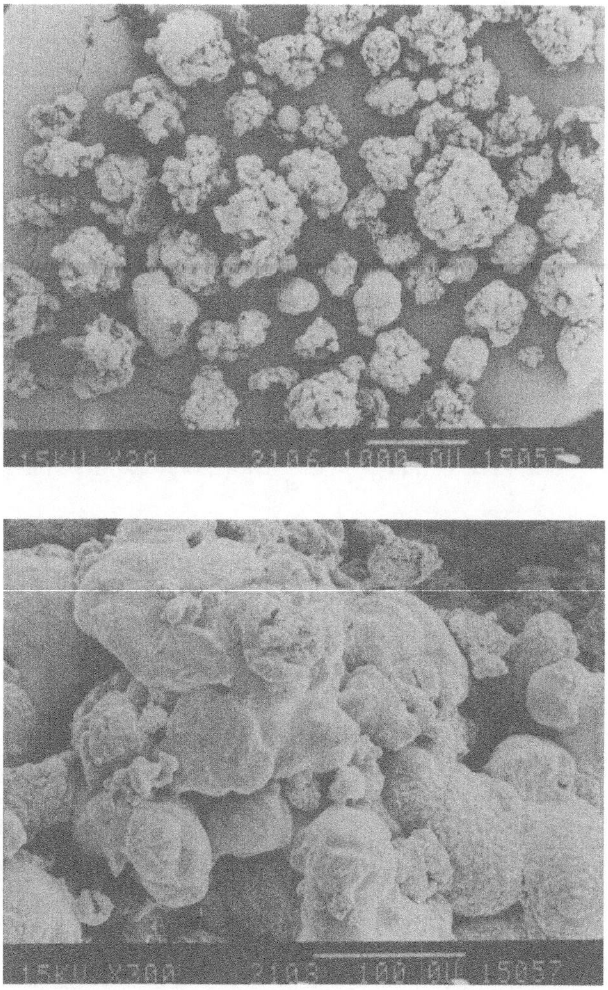

Fig. 12. Morphology of Polyethylene Particles Produced Under Ethylene-Starved Polymerization Conditions

REFERENCES

1. R. D. Quirk, Ed., Transition Metal Catalyzed Polymerizations, MMI Press Symposium Series, Vol. 4, Parts A and B (1983).
2. P. Pino and R. Mulhaupt, Angew. Chem., Int. Ed. Engl., 19, 857-875 (1980).
3. F. J. Karol, Catal. Rev.-Sci. Eng., 26 (3 & 4), 557-595 (1984).
4. L. Bohm, Chem-Ing.-Tech., 56, No. 9,674-684 (1984).
5. J. N. Short, see ref. 1, Part B, 651-669 (1983).
6. J. P. Hogan, U. S. Patent 3,130,188 (1964).
7. J. P. Hogan and D. R. Witt, U.S. Patent 3,622,521 (1971).
8. M. P. McDaniel, M. B. Welch, and M. J. Dreiling, J. Catal., 82, 118-126 (1983).
9. T. J. Pullukat, M. Shida, and R. E. Hoff, see ref. 1, Part B, 697-712 (1983).
10. H. L. Hsieh, Catal. Rev.-Sci. Eng., 26 (3 & 4), 631-651 (1984).
11. I. J. Levine and F. J. Karol, U.S. Patent 4,011,382 (1977).
12. M. P. McDaniel and M. B. Welch, U.S. Patents 4,152,122 (1979), 4,177,162 (1979), 4,182,815 (1980),4,247,421 (1981), 4,277,587 (1981).
13. W. Kirch and P. A. Thompson, U.S. Patents 4,184,979 (1980) and 4,224,428 (1980).
14. J. P. Hogan, U.S. Patent 3,878,179 (1975).
15. See ref. 3, 575-577.
16. F. J. Karol, G. L. Brown, and J. M. Davison, J. Polym. Sci., Polym. Chem. Ed., 11, No. 2, 413-424 (1973).
17. F. J. Karol and C. Wu, J. Polym. Sci., Polym. Chem. Ed., 12, 1549-1558 (1974).
18. L. H. Little, Infrared Spectra of Adsorbed Species, Academic Press, New York (1966).
19. B. E. Wagner, J. N. Helbert, E. H. Poindexter, and R. D. Bates, Jr., Surface Sci., 67, 251 (1977).
20. F. J. Karol, G. L. Karapinka, C. Wu, A. W. Dow, R. N. Johnson and W. L. Carrick, J. Polym. Sci., Part A-1, 10, 2621-2637 (1972).
21. A. Noshay and F. J. Karol, U.S. Patents 4,077,904 (1978) and 4,100,337 (1978).
22. A. Noshay and F. J. Karol, Canadian Patent, 1,087,595 (1980).
23. G. L. Goeke, B. E. Wagner, and F. J. Karol, U.S. Patent 4,302,565 (1981).
24. F. J. Karol, G. L. Goeke, B. E. Wagner, W. A. Fraser, R. J. Jorgensen, and N. Friis, U.S. Patent 4,302,566 (1981).
25. U. Giannini, E. Albizatti, S. Parodi, and F. Pirinoli, U.S. Patents 4,124,532 (1978) and 4,174,429 (1979).
26. A. Greco, G. Bertolini, and S. Cesca, J. Appl. Polym. Sci., 25, 2045-2061 (1980).
27. A. Noshay, F. J. Karol, and R. J. Jorgensen, U.S. Patent 4,482,687 (1984).
28. G. L. Goeke, B. E. Wagner, and F. J. Karol, U.S. Patent 4,354,009 (1982).
29. C. T. Elston, U.S. Patent 3,645,992 (1972).
30. J. P. Hogan, B. E. Nasser, and R. T. Werkman, Preprints of XXII International Congress on Pure and Applied Chemistry, Boston Vol. II, 703-710 (1971).
31. W. L. Carrick, R. J. Turbett, F. J. Karol, G. L. Karapinka, A. S. Fox, and R. N. Johnson, J. Polym. Sci., Polym. Chem. Ed., 10, 2609-2620 (1972).
32. U. Zucchini and G. Cecchin, Adv. Polym. Sci., 5, 101-153 (1983).
33. M. P. McDaniel, J. Polym. Sci., Polym. Chem. Ed., 19, 1967-1976 (1981).
34. J. Boor, Jr., Ziegler-Natta Catalysts and Polymerizations, Academic, New York, Chap. 8 (1979).

35. M. G. Chiovetta, Heat and Mass Transfer During the Polymerization of Alpha-Olefins From the Gas Phase, University Microfilms International, B 1983, 44 (4), 1183, Ann Arbor, Michigan (1983).

NEW CATALYSIS FOR POLYPROPYLENE POLYMERIZATION

Steven P. Sawin and Gene W. Powers

Union Carbide Corp.
P. O. Box 670
Bound Brook, New Jersey 08805 and
Shell Development Co.
P. O. Box 1380
Houston, Texas 77001

INTRODUCTION

A gas-phase, fluidized-bed process was developed initially for high density polyethylene, HDPE, in the 1960's and was commercialized first in 1968 at Union Carbide's Seadrift, Texas plant. The simplicity, operational stability, cost effectiveness, and versatility of the process led to the extension of the process to linear low density polyethylene, LLDPE, which was produced initially on a commercial reactor in 1975. The technology has been licensed with more than 25 licensees operating over 40 reactors in fifteen countries. By 1986 the combined capacity of these reactors will be sufficient to supply 25% of the world's total demand for low density polyethylene.

The first world-scale plant utilizing this technology for the production of polypropylene started up on schedule in May of this year at Union Carbide's Seadrift, Texas plant. This process combines Union Carbide's process for polyolefin polymerization with Shell Chemical's high-activity catalyst to produce a full range of homopolymers, random and impact (in-situ) copolymers.

In the PP process (see Figure 1), catalyst, propylene and other reactants are fed continuously to the reaction system consisting of a fluidized-bed reactor, a cycle gas blower and a cycle gas cooler. The reactant is a vertical fluidized-bed containing granular polypropylene and a small amount of catalyst. The blower continuously circulates reaction gas, mainly propylene, through the reactor and heat exchanger. The gas provides agitation and absorbs the heat of reaction. Unlike other gas-phase processes for making polypropylene, no mechanical agitation is required to keep the reactor well-mixed.

Polypropylene is removed as a granular product, directly from the reactor, via a product discharge system where the resin is separated from the reaction gas, which is returned to the reactor. The resin is then transferred to a purge bin where residual hydrocarbons are removed, prior to pelletization and blending of the product.

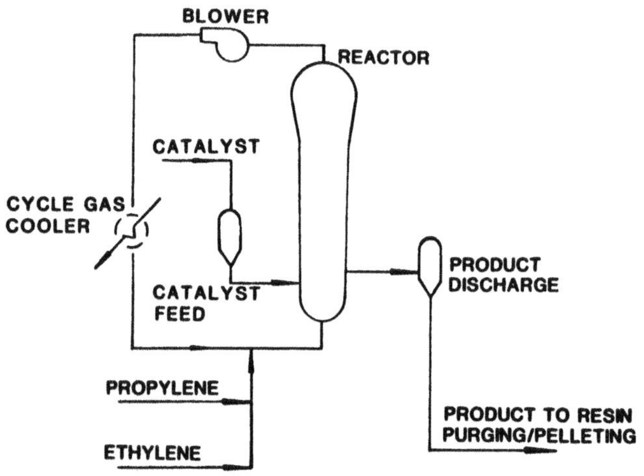

Fig. 1

A single reaction system as shown in Figure 1 is used to make homopolymers and random copolymers. For the production of in-situ impact copolymers, two reactors are operated in series as shown in Figure 2. Resin containing active catalyst is transferred from the homopolymer reaction system to a second, smaller reaction system where the ethylene-propylene rubber fraction of the product is produced. The resin is then discharged from the reactor, purged, pelletized, and blended as for homopolymer. With the PP process, only two reactors in series are required to make the full range of medium to super-high impact products.

The PP process has many advantages. Investment costs are 10-15% lower than the other processes because only a few major items and no exotic equipment or materials of construction are required.[1,2] Operating costs are estimated to be 10-15% lower than competitive processes because of efficient use of monomer, low energy consumption, and modest requirements for both operating and maintenance skills. Since this is a gas phase process, there are no large volumes of liquid hydrocarbon that could pose a fire hazard. There are no liquid effluents and gas and solid effluents are minimal.

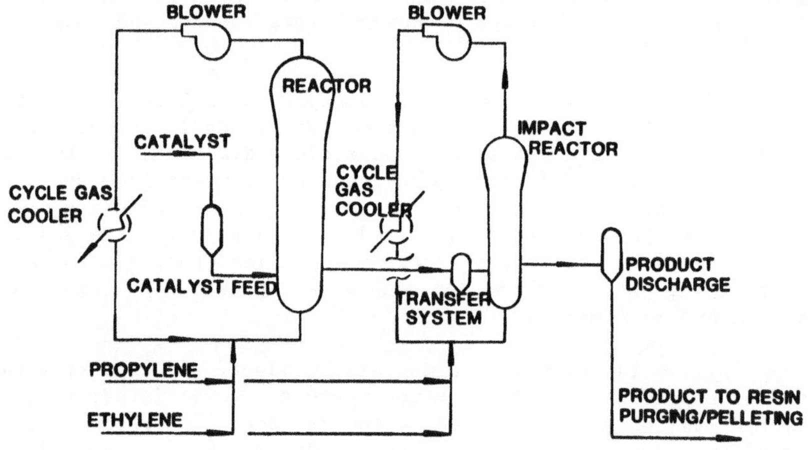

Fig. 2

CATALYST REQUIREMENTS

The catalyst which is a key factor in the success of any polymerization process, must meet certain requirements to work successfully. It must have high productivity so that catalyst removal is not necessary and advantage can be taken of the simplicity of the process. The catalyst must produce the proper molecular weight distribution, must allow for the control of molecular weight and must also produce a polymer with high and controllable stereoregularity, so that removal of an atactic fraction is not necessary. The resin must have a morphology which permits good fluidization in the reactor. And, the catalyst must yield reproducible catalyst batches.

Shell developed a high-activity catalyst for use in liquid propylene in the late 70's.[3] This "new generation" polypropylene catalyst with high selectivity and high productivity was adapted successfully for operation under gas phase reactor conditions where it produces the same excellent products as in the liquid process. This important capability of this catalyst was demonstrated first with products produced in the PP pilot plant reactors, and it has been confirmed by the successful start-up and market acceptance of products from the Seadrift commercial reactor.

SCALABILITY OF PP TECHNOLOGY

The PP process, with its back-mixed reactor and heat removal external to the reactor, theoretically should be an ideal process from a scalability viewpoint. This has been confirmed by the excellent translation of technology developed in the PP pilot plant to the commercial PP facility. The first scale-up of polypropylene to a commercial reactor was made in 1983 on a converted polyethylene reactor at Seadrift.

As shown in Figure 3, the catalyst-productivity/resin-stereoregularity relationship developed in the pilot plant was duplicated almost exactly on the commercial reactor. The response to changes in process variables was not only similar in slope, but nearly the same magnitude in the commercial unit as in the pilot scale unit. This is especially pleasing considering the capacity of the Seadrift reactor is 175 million lbs/yr or about 1000 times that of the pilot plant reactor. Figure 4 is a picture of the Shell/Union Carbide UNIPOL PP reactors at Seadrift, Texas.

OPERATION OF THE SEADRIFT PP PLANT

The first world-scale PP reactor started up very smoothly and easily on May 27, 1985 at Seadrift, Texas. Initially, a series of homopolymer products were produced which have been and are being qualified by customers. Less than two months after start-up, the unit began producing in-situ impact copolymers with excellent properties and with very stable operation of the two reactors in series. Very-high impact copolymer was produced at nameplate capacity three days after the copolymer run began.

PRODUCT SCALE-UP AND PERFORMANCE

Since plant start-up, a number of products have been made for specific end use markets. Among the homopolymer products are included grades for: injection molding, oriented polypropylene film, tape/carpet backing, and fine fiber multifilament. The results of initial commercial evaluations in these applications have been very gratifying and are summarized in Table 1.

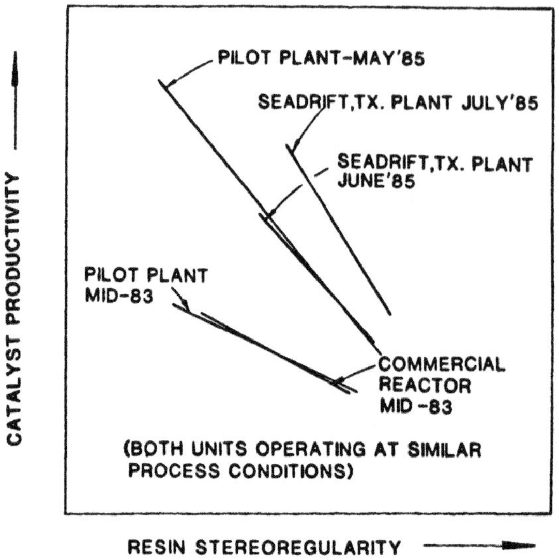

PILOT PLANT-MAY'85
SEADRIFT,TX. PLANT JULY'85
SEADRIFT,TX. PLANT JUNE'85
PILOT PLANT MID-83
COMMERCIAL REACTOR MID-83
(BOTH UNITS OPERATING AT SIMILAR PROCESS CONDITIONS)
CATALYST PRODUCTIVITY
RESIN STEREOREGULARITY

Fig. 3

Fig. 4

Table 1. PP Homopolymer Products - Commercial Performance

MOLDING

- Matched competitive product in molding performance
- Parts appearance met customers' requirements

FILM

- Extruded and oriented readily at standard conditions
- Film visual appearance and optical properties acceptable

TEXTILES

- Carpet-backing yarn made at maximum line speed (750 fpm)
- Displayed low water carryover
- Properties met yarn specifications
- Multifilament, 3dpf (22μm), met application requirements

Several impact copolymers have been made using the two reactors in series. These copolymers span the range from medium impact (10 wt. % of product made in the second reactor) to very-high impact (25 wt. % made in second reactor). Optimum performance with these types of products is only attained if compositional parameters are controlled closely. The amounts of hydrogen and propylene entering the second reactor with the resin from the first reactor can be easily controlled and facilitates control of product properties.

Polymerization conditions have been changed rapidly to produce products with different melt flows and different copolymer compositions. These easy transitions occur because the amounts of reagents in the reactors are low relative to their rate of consumption as compared with liquid polymerization reactors.

PROCESS VERSATILITY

A unique feature of the PP process is its versatility. Products are produced covering the entire commercial range of melt flows including the ultra-high flow, stereospecificity (isotactic index), comonomer content (in random copolymers) and impact (medium to super-high). The range of products is summarized in Table 2.

As seen from Table 2, the PP process can make homopolymers over a wide range of melt flows, including fractional melt flow (<1) grades for extrusion, filament and film grades (2-5 melt flows), fine denier fiber grades (8-50 melt flow), and ultra high flow grades (> 100 melt flow). Random copolymers with more than 5% ethylene have been made and higher α-olefin co- and terpolymers have been made.

The full range of medium, high and super-high impact copolymers made with PP exhibit an excellent balance between low-temperature impact strength and stiffness. Microscopic structural analysis indicates that the morphology of these products is identical to polymers made in complex reaction systems having more stages.

Table 2. PP Product Capabilities

HOMOPOLYMERS
° High, controllable stereoregularity (II from 93-98%)
° Broad MW range (< 0.1 to > 100 MF)
RANDOM COPOLYMERS
° Broad range of comonomer content (up to 7 wt%)
° Not limited to ethylene
° Broad MW range (< 0.1 to > 100 MF)
IMPACT (IN-SITU) COPOLYMERS
° Excellent impact/stiffness balance
° Competitive to products made utilizing more complex technology
° Broad range of total ethylene content (up to 25% elastomer)
° Not limited to ethylene
° Broad MW range (< 0.1 to > 50 MF)

An exciting feature in this technology is its potential for making products, beyond today's commercial range. This is made possible by the nature of the gas-phase, fluidized-bed process, which allows the reactor to operate at a wide variety of conditions. For example, high melt flow resins can be produced without the problems associated with high hydrogen concentrations in liquid phase processes.

REFERENCES

1. S. P. Sawin and C. J. Bass, Chem. Eng., 92, 42-43 (May 27, 1985).
2. H. Short, Chem. Eng., 92, (July 8, 1985).
3. R. D. Quirk, Ed., Transition Metal Catalyzed Polymerizations, MMI Press Symposium Series, Vol. 4, Part A (1983).

POLYMERIZATION OF OLEFINS WITH A HOMOGENEOUS ZIRCONIUM/ALUMOXANE CATALYST

W. Kaminsky and H. Hahnsen

Institute of Technical and Macromolecular Chemistry
University of Hamburg
Bundesstr. 45, D-2000 Hamburg 13

INTRODUCTION

Using the soluble catalytic system formed from bis(cyclopentadienyl) zirconium compounds and the coactivator methylalumoxane with the structure $(Al(CH_3)O)_{6-20}$) activities up to $5 \cdot 10^6$ g polyethylene/g Zr·h·bar were found (Table 1).[1-3]

For the given ethylene pressure of 8 bar, there is an hourly production of $40 \cdot 10^6$ g polyethylene per gram of zirconium component, which is much more when "optimized" supported catalysts can be provided. The average molecular weight of the polyethylene obtained is 150 000, i.e., a productivity, where at least 24 000 macromolecules per zirconiumatom an hour are formed. Hence, the rate of growth of one macromolecule is 0.15 sec and the insertion of one ethylene molecule in the growing polymer chain is 2.8×10^{-5} sec. These are turnover times which usually can only be found with radical or ionic polymerizations. These times would be even shorter, if each zirconium complex produces to build a new active center.

Table 1.

Transition metal (mmol)		Cocatalyst	Activity (g PE/g M·h·bar)	T (°C)	Molecular weight
$Cp_2Ti(CH_3)_2$	$(3 \cdot 10^{-6})$		5 000	20	520 000
$Cp_2Ti(CH_3)Cl$	$(3 \cdot 10^{-6})$		50 000	20	450 000
Cp_2TiCl_2	$(3 \cdot 10^{-6})$	Methylalumoxane	90 000	20	430 000
$Cp_2Zr(CH_3)_2$	(10^{-7})		9 000	20	730 000
$Cp_2Zr(CH_3)_2$	(10^{-7})		70 000	70	190 000
$Cp_2Zr(CH_3)_2$	(10^{-8})		3 100 000	90	106 000
Cp_2ZrCl_2	(10^{-8})		5 000 000	90	122 000
$Cp_2Hf(CH_3)_2$	$(3 \cdot 10^{-8})$		60 000	70	441 000
Cp_2HfCl_2	$(3 \cdot 10^{-8})$		65 000	70	490 000
Cp_2ZrCl_2	$(1,5 \cdot 10^{-6})$	Ethylalumoxane	23 000	60	>500 000
$Cp_2Zr(CH_3)_2$	$(3 \cdot 10^{-6})$	Tetraisobutyldi-alumoxane	175 000	70	>500 000
$TiCl_4$ auf R MgCl [29]		Aluminumtrimethyl	200 000	70	>500 000

Table 1 shows that the zirconium catalysts are more active than analogous titanium or hafnium systems. As a cocatalyst, methylaluminoxane is better suited than ethyl- or the dimer isobutylaluminoxane. Bis(cyclopentadienyl) compounds, containing chloride ligands, possess higher activities, especially at lower polymerization temperatures than "chloride free" systems.

Total substitution of all hydrogens of the Cp-ligands by methyl groups leads to a graphic decrease of polymerization activities, but the molecular weight of the polyethylene obtained is higher under similar conditions.

The titanium system is not applicable at temperatures over 50°C because of rapid aging and subsequent deactivation.

The density of the polymer can be lowered by using hexene-1 or butene-1 as an additional monomer in the forming of LLDPE. Incorporation of 1 mole percent hexene-1 yields a density of less than 0.95 g/ml for the molten polymer. With 5 mole percent of hexene-1, a density of 0.92 g/ml is obtained. This value decreases to 0.89 g/ml with a hexene concentration of 9 mole percent.[4,5]

The polymerization activity is influenced by the hexene content. At a concentration of 65 mole percent hexene, the activity reaches its maximum value which is higher than that for the ethylene polymerization. This suggests that a double bond in the solvation of the active center increases the activity. The activation energy was found to be 47,5 kJ/mol.

In contrast, the hexene is incorporated into the polymer chain at a slower rate than ethylene. By the method of Fineman and Ross, the copolymerization parameters were measured r_1 (ethylene) = 55 and r_2 (1-hexene) = 0,005 (20°C). The molecular weight of the polymer is lowered by incorporation of 50 mole percent hexene from 200,000 to 100,000 at 60°C.

The zirconium catalyst provides ethylene/propylene copolymers (EPM). Changing the solvent, for example toluene or liquid propylene and the ethylene pressure, is a means by which the desired ratio of propylene to ethylene in the polymer can be regulated. As a result, the properties of the copolymerization products can be influenced considerably. Copolymerization products with 10 to 90% of propylene have been produced.

In the same way, EPDM terpolymer with ethylidene norbornene as the diene monomer was prepared. The overall activities lie between 100 and 1.000 kg EPDM/mole Zr·h·bar. This polymerization could be carried out at zirconium concentrations as low as $5 \cdot 10^{-7}$ mol/l. Propylene sequences produced by the bis(cyclopentadienyl)zirconium compounds are totally atactic as was observed in the case of polypropylene. In the polymers which were produced, $r_1 \cdot r_2$ = 0,3 was obtained for the EPDM samples, and $r_1 \cdot r_2$ = 0,25 was obtained for the ethylene/propylene copolymers. This corresponds to a random distribution of ethylene and propylene. Molecular weight distributions (M_w/M_n) obtained by GPC-curves have an average value of only 1.7. This narrow distribution implies that only one active center is predominantly present.

Biomass, like starch, cellulose and lignin or inorganic substances, like $CaCO_3$, $CaSO_4 \cdot 0.5H_2O$ and Al_2O_3 can be covered with small amounts of the catalyst.[6]

Particles, treated in this manner, have considerable polymerization activity when they are suspended in toluene and the transition metal compound is added. It is possible to cover them with polyethylene or other polyolefins. The polyethylene forms scaly fibers on the surface. Polyolefin formation occurs only on the surface of the activated particle. There is no decrease in polymerization activity in π solvents. By varying the polymerization time, the desired weight ratio between filling materials and synthetic polymers can be regulated over a wide range.

ATACTIC POLYOLEFINS

Olefins can be converted into atactic or isotactic polymers by using catalysts, whose ligands are transition metals. Bis(cyclopentadienyl) compounds of zirconium and titanium yield pure atactic polypropylene (Table 2). The concentration of transition metals are in the range of 10^{-7} and 10^{-6} mol/l.

After passing a maximum, a constant rate of polymerization is observed.[7] Olefins can be polymerized in solution as well as in liquid propane. Under similar conditions, the zirconium compound becomes more active than the titanium compound. Methyl substitution at the Cp-ligand also decreases the activity. Atactic polymers with molecular weights up to 590 000 are obtained. However, polymerization of 1-butene produces only low-molecular atactic polybutene. All atactic polymers show a very small molecular weight distribution between 1.6 and 2.5.

ISOTACTIC POLYOLEFINS

Chiral, soluble zirconium compounds have been reported. If those compounds or analogous titanium structures, together with methylaluminoxane, are used, e.g., rac-ethylene (bisindenyl)zirconiumdichloride,[8,9] propene is polymerized to highly isotactic polypropylene (Table 3).[10]

The (indenyl)zirconocene is more active than the (tetrahydroidenyl) zirconocene. The activity increases with increasing polymerization temperature and reaches a value of 16 000 kg isotactic polypropylene per mol of Zr per h.

1-Butene can also be polymerized with the zirconocene/methylalumoxane catalyst. Thus, polymerization activity is 2640 kg of polybutene per hour

Table 2.

Catalyst	Al (mol)	Toluene (ml)	Propene (ml)	T (°C)	Time (h)	Yield (g)	Activity (g PP/g M·h)	$\bar{M}$
$Cp_2Ti(CH_3)_2$	$1,7\cdot10^{-3}$	50	200	50	48	13,5	481	58 000
$Cp_2Ti(CH_3)_2$	$1,8\cdot10^{-3}$	150	200	20	48	15,1	533	100 000
$Cp_2Ti(CH_3)_2$	$1,5\cdot10^{-3}$	-	250	10	76	18,0	430	93 000
$Cp_2Zr(CH_3)_2$	$5,0\cdot10^{-3}$	-	330	5	24	48,5	13 300	106 000
$Cp_2Zr(CH_3)_2$	$5,0\cdot10^{-3}$	5	340	5	24	120,0	15 700	106 000
$Cp_2Zr(CH_3)_2$	$5,4\cdot10^{-3}$	5	340	-10	12	32,0	8 400	191 000
$Cp_2Zr(CH_3)_2$	$5,9\cdot10^{-3}$	50	250	-30	50	202,0	1 300	84 000
$Cp_2Zr(CH_3)_2$	$3,4\cdot10^{-3}$	50	100	-70	58	8,0	75	590 000
Cp_2ZrCl_2	$1,8\cdot10^{-3}$	120	210	20	24	5,6	17 600	75 600
$Cp(CpMe_5)ZrCl_2$	$1,8\cdot10^{-3}$	120	210	20	24	2,6	8 600	180 000
$(CpMe_5)_2ZrCl_2$	$1,8\cdot10^{-3}$	120	210	20	24	0,7	2 300	147 000

Table 3. Polymerization of 70 ml Propylene or Butene with $7 \cdot 10^{-7}$ mol/l rac-Et(Ind)$_2$ZrCl$_2$(I) or $8,4 \cdot 10^{-6}$ mol/l rac-Et(TH-Ind)$_2$ZrCl$_2$(II) and $1,6 \cdot 10^{-2}$ mol Al/l Methylaluminoxane in 330 ml toluene

Monomer	T (°C)	Time (min)	Yield (g)	Activity (kg PP/mol Zr·h)
Propene I	15	620	3,6	1,500
Propene I	21	500	31	16,000
Propene I	32	20	31	12,000
Propene II	-20	360	1,5	80
Propene II	-10	270	4,5	300
Propene II	0	255	12,5	880
Propene II	8	180	13,0	1 300
Propene II	15	170	26,7	2 900
Propene II	20	120	31,3	4 750
Propene II	60	90	38,7	7 700
Butene II	-10	330	9,1	500
Butene II	+20	200	29,2	2 640

per mole of Zr at 20°C. Although the polymers have molecular weights of 150,000 or 50,000, they are soluble in toluene. The product is wax-like to crystalline, but not clear and viscous like atactic polybutene, which was produced with bis(cyclopentadienyl)zirconium compounds and characterized by ^{13}C-NMR spectroscopically.

The properties of the polypropylene obtained are shown in Table 4.

The toluene-soluble fraction consists of less than 0,2 weight percent and therefore lies below that of isotactic polypropylene produced with supported catalysts.

The IR-spectroscopically-determined isotacticity index of 86 to 91% is also very high. The crystallinity of the highly isotactic polypropylene is largely determined by the stereochemical uniformity of the polymerization. High-resolution ^{13}C-NMR spectroscopy affords an insight into the pattern of the sequences in the polymer chain. Isotactic sequences (mm triads) are found to be present to the extent of 95%, heterotactic sequences (mr triads) are present to the extent of 3,2%, and syndiotactic sequences (rr triads) are present to the extent of 0,9%. Notable is the very small amount of rr triads compared to mr triads.

A profound analysis of the number and intensity of the methyl peaks in the nmr-spectrum shows the distribution between isotactic percentage I, syndiotactic percentages S and iso- and syndiotactic sequential lengths

Table 4. Properties of Isotactic Polypropylene Obtained with the Zirconium Catalyst

T (°C)	$\bar{M}_w$	M_w/M_n	Soluble Prod.	Isotact. Ind.
-10	305 000	2,6	0,25 %	90,0
0	144 000	2,4	0,2 %	88,1
15	62 000	2,0	0,7 %	87,3
20	45 000	1,9	1,0 %	86,0

n_{iso} and n_{syn}, respectively. Typical values are as following: 97,5% I; 2,5% S; 61 n_{iso}. Thus, it can be stated that polypropylene produced via the ansa-zirconocene compound consistent of an extraordinarily high degree of isotacticity. Only every 40th propene unit has the opposite configuration from the foregoing. The molecular weight distribution M_w/M_n is small 2 to 2,5. This is a characteristic value for homogeneous catalysts with only one type of active centers.[11] The means molecular weights are strongly dependent on reaction temperature.

By variation of reaction temperature in the range from -20°C to 60°C, molecular weights from 300,000 to 1,500 are obtained.

The racemic mixture of the zirconocene compound was resolved into optically active pure enantiomeric forms, leading for the first time to optically active polypropylene. Table 5 represents different values for the specific rotations of polymer suspensions found in various solvents and after different pretreatment. Since it is bound to the optically active center, the polymer chain is formed dominantly in only one screw sense.

After getting rid of the metal center, the so formed helix (3/1) remains stable, unless heated at higher temperatures or being completely dissolved. In these cases, the optical activity will be destroyed by racemization into helices of both screw senses. A transparent foil of app. 0.1 mm thickness from the optically active polypropylene has an angle of rotation of -65°C.

KINETICS

Catalysts based on bis(cyclopentadienyl)zirconium compounds[12] together with methylaluminoxane are deactivated at temperatures above 90°C (Figure 1). Yet, at 70°C and below, while the rate of polymerization is nearly constant. In all cases, the rate of polymerization increases after addition of the olefin and goes through a maximum at higher temperatures. This indicates, that the highly active center is not created until the olefin is added.

The activation energy is approximately 34 kJ/mol. The rate of polymerization is determined by the ethylene and zirconium concentration and is first order. This indicates a bimolecular rate determining step between growth center and monomer molecule. For alumoxane concentration, a kinetic of 0.4 till 0.6 order, depending on the conditions, is found, resulting in an overall rate of reaction as follows:

Table 5. Measurements of Specific Rotations.
Instrument: Perkin Elmer 243; employed wavelength: Na-D-line

(589 nm); specific rotation: $\alpha_{\lambda}^{T} = \frac{\alpha_{meas.} \cdot 100}{d \cdot l}$;

l = length of sample tube 1 dm; d = concentration in $\frac{g}{100\ ml}$

specific rotation of the employed catalyst: $\alpha_{436}^{RT} = +297°$

Sample	Solvent	d	$\alpha_{meas.}$	α_{589}^{T}
atactic PP	decalin	1,775	-0,025	-1,4
isotactic PP	decalin	0,285	-0,350	-123
isotactic PP	o-xylene	0,280	-0,700	-250
isotactic PP	toluene	53,0	± 0	-
isotactic PP	dcca*	0,210	± 0	-
isotactic polybutene	decalin	0,114	±0,150	+130

*dichloroacetic acid, helix-destroying solvent

$$rp = Kp \quad [Monomer] \quad [Zr] \quad [Alox]^{0.5}$$

At temperatures below 0°C, a macromolecule is produced within a few minutes, but little transfer or inhibition reactions occur.

From the transition metal concentration, the polymer yield and the average molecular weight, the number C* of mols of active transition metals

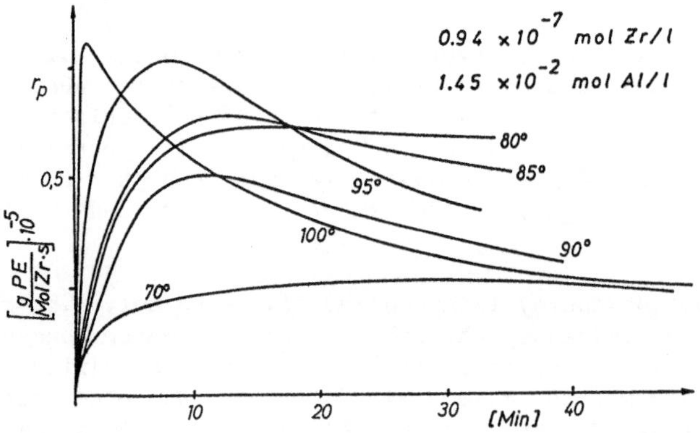

Fig. 1

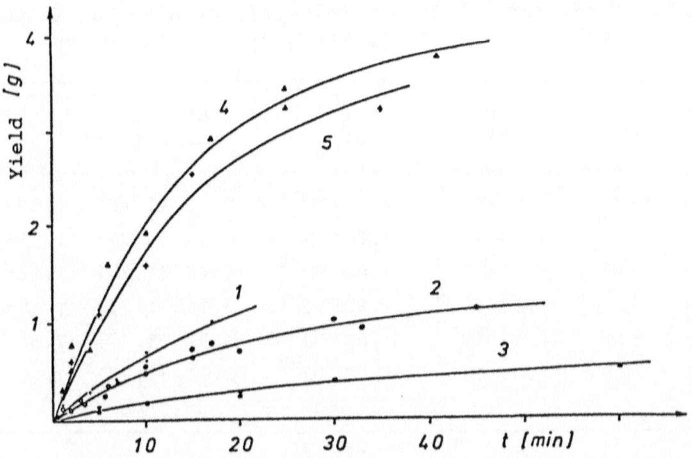

Fig. 2

versus the total inserted amount, can be estimated. Figure 2 shows the yield of polyethylene for different temperature conditions.

The conditions are shown in Table 6. The resulting average molecular weights depend on the time of polymerization as shown in Figure 3.

The rates of reaction decrease. The polymerization is strongly dependent on the Al/Zr ratio. Hence, the exact determination of C* is hindered, especially at higher zirconium concentrations, where bimolecular molecular weight distributions are present. This indicates at least two different active centers. The estimated data for C* are shown as follows:

No	Zr mol/l	T °C	C* $\left(\frac{\text{Zr active}}{\text{Zr total}}\right)$
1	10^{-5}	-10	0.45
2	10^{-5}	-25	0.40
3	10^{-5}	-40	0.30
4	10^{-4}	-25	0.15
5	10^{-3}	-25	0.05

For low zirconium concentrations, C* is increasing from 0.30 at -40°C to 0.45 at -10°C. For the optimal polymerization temperature between 70°C and 90°C and a zirconium concentration of approximately 10^{-6} to 10^{-7} mol/l, the contribution will increase and approach a value 1.0.

MECHANISM

The oligomeric alumoxanes appear as dimers and trimers which dissociate to form an UV-spectroscopically observed complex.[13] The following set of reactions are proposed for the polymerization mechanism.

Table 6. Conditions of Ethene Polymerization by Low Temperatures with the Catalyst $Cp_2Zr(CH_3)_2$/Methylalumoxan

No	T °C	Zr mol/l	Al mol/l	Ethene mol/l (bar)	Al/Zr
1	-10	$1.03\cdot10^{-5}$	$1.60\cdot10^{-2}$	0.13 (1.40)	$1.55\cdot10^{3}$
2	-25	$1.03\cdot10^{-5}$	$1.60\cdot10^{-2}$	0.31 (1.77)	$1.55\cdot10^{3}$
3	-40	$1.03\cdot10^{-5}$	$1.60\cdot10^{-2}$	0.48 (1.77)	$1.55\cdot10^{3}$
4	-25	$1.03\cdot10^{-4}$	$1.60\cdot10^{-2}$	0.31 (1.77)	$1.55\cdot10^{2}$
5	-25	$1.03\cdot10^{-3}$	$1.60\cdot10^{-2}$	0.31 (1.77)	$1.55\cdot10^{1}$
6	-25	$1.03\cdot10^{-6}$	$1.60\cdot10^{-2}$	0.31 (1.77)	$1.55\cdot10^{4}$

Polymerization of Olefins
$Cp_2\ Zr\ R_2$ + Aluminoxane

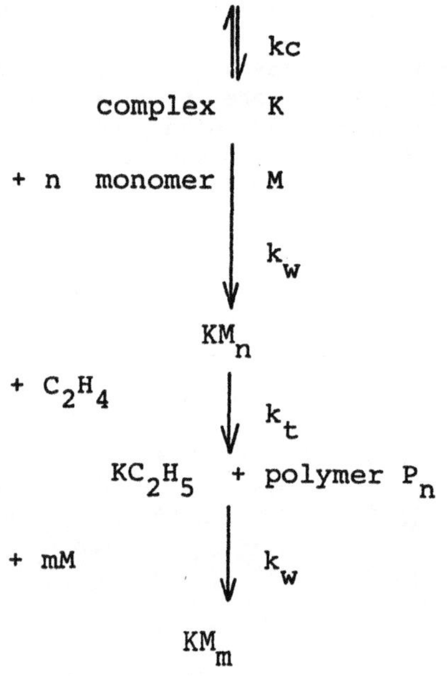

The active complex K inserts n monomer units. The transfer of hydrogen takes place to an ethylene unit and the polymer Pn is formed.[14] The complex $K{-}C_2H_5$ is able to insert new monomers.

Additional steps are necessary to form the active complex. The reaction between zirconium compound and aluminoxane occurs after separation of methane.

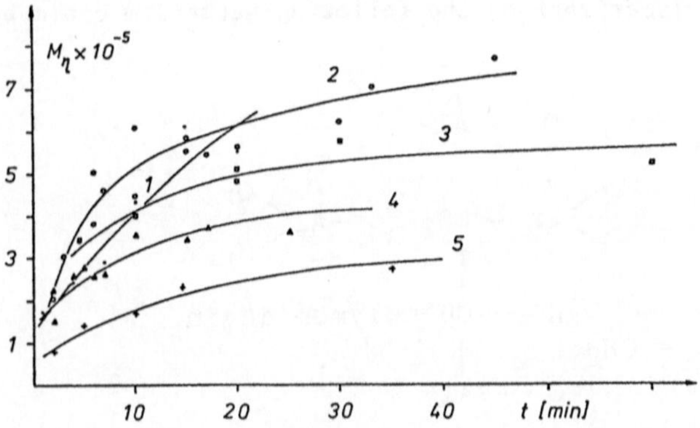

Fig. 3

Because exchange is also possible for halogen compounds, Zr-O-Al-structures are observed.

In a Zr-O-Al bond, electron density is withdrawn from the zirconium atom, hence the appearance of a hydrogen bond to the zirconium atom occurs as a prestep for β-splitting.[15]

In the presence of ethylene, it can become bonded as a π-complex. The Zr-CH_2-bonds is released with the aid of an acidic center, as shown:

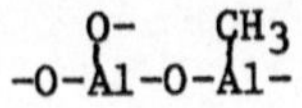

or
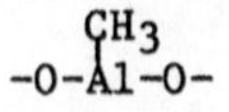

in alumoxane.

Insertion then takes place and the initial condition under chain extension is completed again.

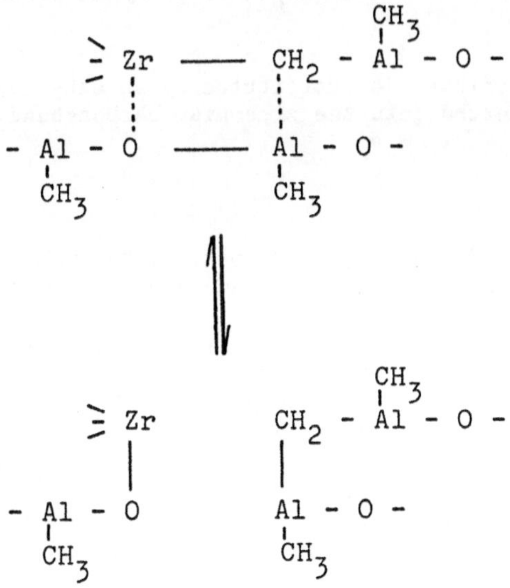

This model is in accord with hypotheses, where the chain growth in systems of bis(cyclopentadienyl)titanium compounds and aluminiumalkyl halogenes is explained by the existence of electron deficient compounds in five times coordinated bimetal complexes.[16,17]

For the polymerization, the following mechanism could be possible:

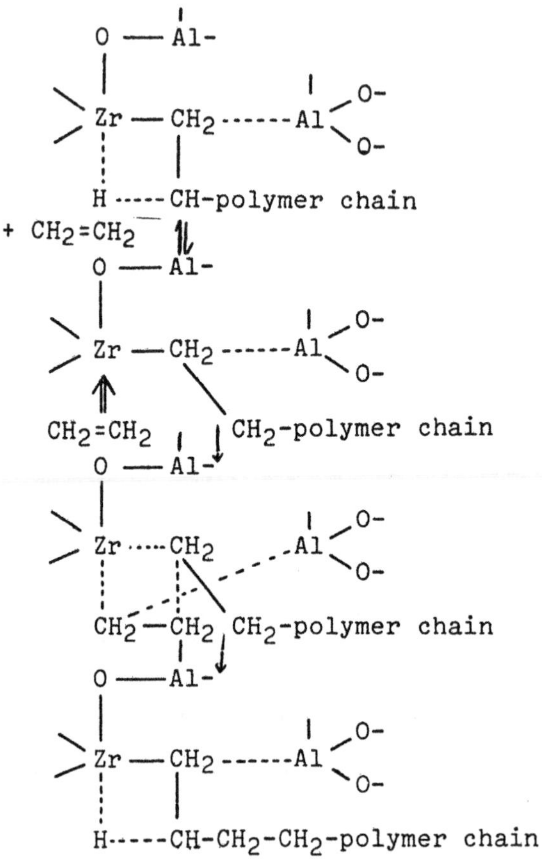

The bridging hydrogen is substituted by an ethylene which is first π-bonded and then inserted into the zirconium-carbon-bond.

REFERENCES

1. A. Andresen, H.-G. Cordes, J. Herwig, W. Kaminsky, A. Merck, R. Mottweiler, J. Pein, H. Sinn, and H.-J. Vollmer, Angew. Chem. Int. Ed. Engl., 15, 630 (1976).
2. H. Sinn, W. Kaminsky, and H.-J. Vollmer, and R. Woldt, Angew. Chem. Int. Ed. Engl., 19, 390 (1980).
3. H. Sinn and W. Kaminsky, Adv. Organomet. Chem., 18, 99 (1980).
4. J. Herwig and W. Kaminsky, Polymer Bull., 9, 464 (1983).
5. W. Kaminsky, M. Miri, H. Sinn, and R. Woldt, Makromol. Chem. Rapid Commun., 4, 464 (1983).
6. J. Dutschke, W. Kaminsky, and H. Lüker, in: "Polymer Reaction Engineering," Ed. K. H. Reichert a. W. Geiseler, Hanser Publishers, Munchen, p. 207 (1983).
7. W. Kaminsky, Nachr. Chem. Techn. Lab., 29, 373 (1981).
8. F. R. W. P. Wild, L. Zsolnai, G. Huttner, and H. H. Brintzinger, J. Organomet. Chem., 232, 233 (1982).
9. F. R. W. P. Wild, M. Wasiacionek, G. Guttner, and H. H. Brintzinger, J. Organomet. Chem., in Press.

10. W. Kaminsky, K. Külper, H. H. Brintzinger, F. R. W. P. Wild, Angew. Chem., 97, 507 (1985).
11. K. Soga, K. Izumi, S. Ikeda, and T. Keii, Makromol. Chem., 178, 337 (1977).
12. W. Kaminsky, H.-J. Vollmer, E. Heins, and H. Sinn, Makromol. Chem., 175, 443 (1974).
13. W. Kaminsky in MMI Press, Symposium Series Vol. 4 in: "Transition Metal Catalyzed Polymerization," R. P. Quirk Ed., Harwood Academic Press Publishers, New York, p. 225.
14. W. Kaminsky and H. Lüker, Makromol. Chem. Rapid Commun., 5, 225 (1984).
15. H. Takaya, M. Yamakawa, and K. Mashima, J. Chem. Soc. Commun., p. 1283 (1983).
16. F. Patat and H. Sinn, Angew. Chem., 70, 496 (1958).
17. J. W. Lauher and R. Hoffmann, J. Am. Chem. Soc., 98, 1729 (1976).

CRYSTALLINITY AND MORPHOLOGY OF

ETHYLENE/ α-OLEFIN COPOLYMERS

P. Schouterden, G. Groeninckx, and H. Reynaers

Laboratory of Macromolecular Structure Chemistry
Department of Chemistry
KULeuven
Celestijnenlaan 200 F
B-3030 Leuven

INTRODUCTION

Linear low density polyethylenes (LLDPE) are copolymers of ethylene and α-olefins, such as propene, 1-butene, 1-hexene, and 1-octene. The chain microstructure of LLDPE depends on the type, concentration and distribution of the short chain branches. Besides the molecular weight and molecular weight distribution, these factors influence the crystallization and molecular segregation behavior, the ultimate crystallinity and the semi-crystalline morphology of the copolymer. In this report, the influence of different types and contents of short chain branches on the thermal behavior and the morphology of LLDPE and the incorporation of the short chain branches in the crystal lattice for different types of LLDPE are reported.

EXPERIMENTAL

Differential scanning calorimeter (DSC) scans were performed on a Perkin-Elmer DSC-2C. The degree of branching of the various LLDPE samples was determined by 13C-NMR spectroscopy (NMR Bruker WM250 Cryospec, PW=60°, RD=10s), according to the 13C chemical shifts established by Randall.[1] Wide-angle (WAXS) X-ray diffraction spectra were recorded in the reflection mode using a Philips vertical diffractometer. Small-angle X-ray (SAXS) patterns were measured using a Kratky-camera. Absolute intensities were obtained by normalization using the lupolen standard 13/6 supplied by Prof. Kratky. Both WAXS and SAXS apparatus were equipped with a proportional counter and a pulse-height analyzer. The wavelength of the Ni-filtered CuK_{α} radiation was $\lambda = 1.54$ A.

Isothermal crystallization of LLDPE from the melt was carried out in a nitrogen atmosphere at various temperatures (105°C $\leqslant$ Tc $\leqslant$ 125°C). The crystallization time (17h) was sufficient to insure complete crystallization.

T_c (°C)	X_t (%)	cryst. segr. phase: X_s(%)	phase cryst. at T_c: X_c(%)
105	32.2	4.0 (12.3/100)	28.2 (87.7/100)
110	27.8	7.5 (26.8/100)	20.3 (73.2/100)
115	29.0	11.1 (38.2/100)	17.9 (61.8/100)
120	28.7	17.2 (60.0/100)	11.5 (40.0/100)

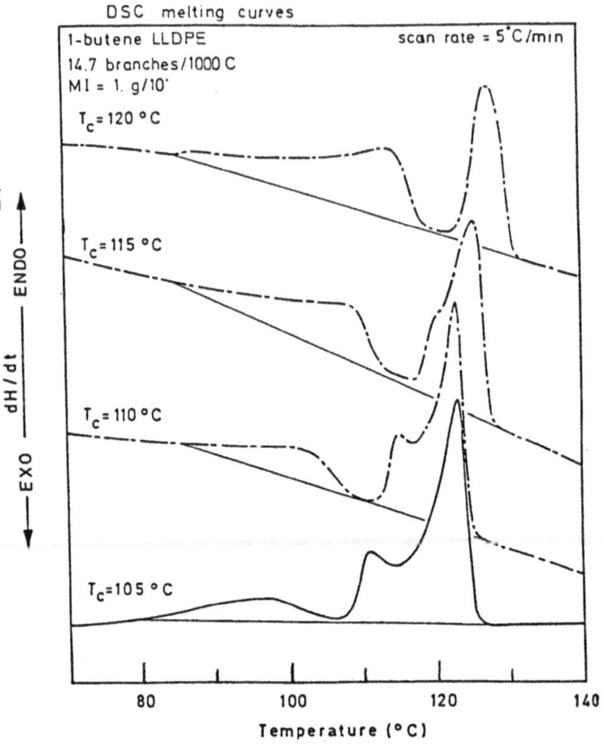

Fig. 1

RESULTS

Molecular Segregation Process During Isothermal Crystallization From the Melt

LLDPE specimens with propene, 1-butene and 1-octene as comonomer exhibit segregation phenomena on isothermal crystallization in the temperature range of 105-125°C. A certain fraction of the copolymer (segregated fraction) is not able to crystallize in this temperature interval. Part of this segregated material, however, crystallizes at temperatures markedly below the original crystallization temperature Tc. During a subsequent DSC heating scan (Figure 1), the crystallized segregated fraction melts over a broad temperature range just below Tc (endotherm I); the crystalline structure formed at Tc gives rise to multiple melting endotherms above Tc (endotherms II and III).

The effect of the crystallization temperature on the amount of crystallized segregated material and on the amount of originally formed crystals is very pronounced: the amount of segregated material increases (Figure 1 and 2), and the amount of originally formed crystalline phase decreases strongly (Figure 1 and 3). The total crystallinity, however, decreases only very slightly (Figure 1).

The comonomer concentration has an important influence on the thermal behavior: the total crystallinity decreases when the comonomer content increases, both for the 1-octene (Figure 4) and for the propene comonomer (Figure 5). Also the relative amount of originally-formed crystals drops, and the comonomer content (Figure 4 and 5).

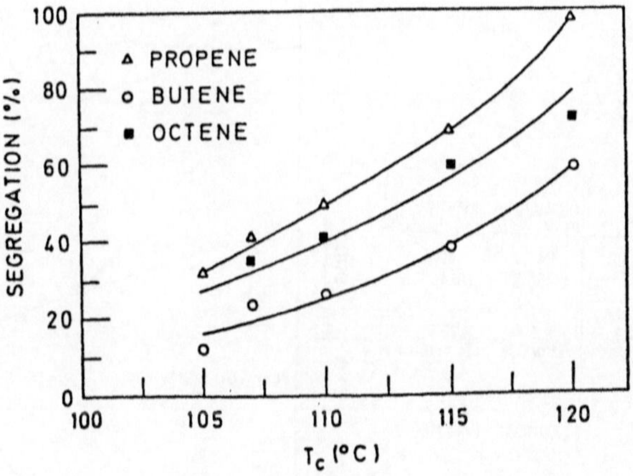

SEGREGATION (%)
100
80
60
40
20
0
PROPENE
BUTENE
OCTENE
100
105
110
115
120
T_c (°C)

Fig. 2

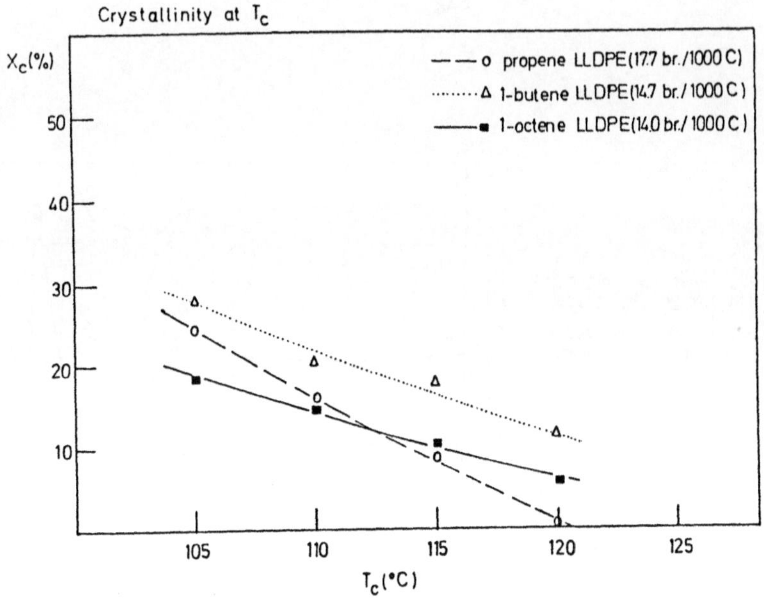

Crystallinity at T_c
X_c(%)
50
40
30
20
10
o propene LLDPE(17.7 br./1000 C)
Δ 1-butene LLDPE(14.7 br./1000 C)
■ 1-octene LLDPE(14.0 br./1000 C)
105
110
115
120
125
T_c(°C)

Fig. 3

	n_b/1000C	MI (g/10')	X_t (%)	cryst segr. phase: X_S(%)	phase cryst. at T_c: X_c(%)
A	6.8	1.0	39.8	9.4 (236/100)	30.4 (76.4/100)
B	10.3	1.0	32.1	8.2 (25.6/100)	23.9 (74.4/100)
C	14.0	2.3	25.2	10.3 (409/100)	14.9 (591/100)

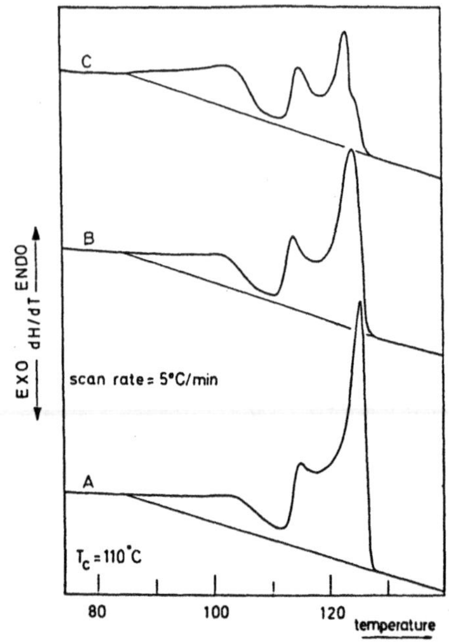

Fig. 4

	branches /1000 C	MI (g/10')	X_t (%)	cryst. segr. phase: X_S(%)	phase cryst. at T_c: X_c(%)
A	4.2	1.0	52.3	5.8 (11.0/100)	46.5 (89.0/100)
B	17.7	1.0	39.5	15.0 (38.0/100)	24.5 (62.0/100)
C	28.4	1.0	21.7	14.8 (684/100)	6.9 (31.6/100)

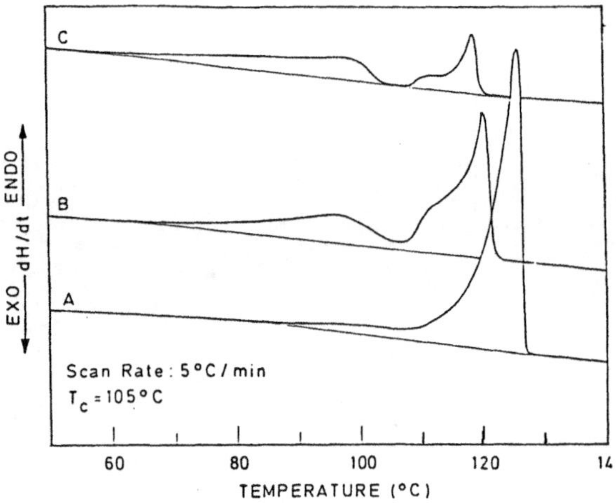

Fig. 5

The complex melting behavior above Tc (multiple endotherms) reflects the existence of various kinds of crystalline lamellae differing in size and perfection. This phenomenon is not due to reorganization (partial melting and recrystallization) during heating in the calorimeter but is the result of the heterogeneous distribution of both intramolecular and intermolecular short chain branches.[2,3,4] This will, of course, give rise to a broad ethylene sequence length distribution.

Semi-Crystalline Morphology Using SAXS

From optical microscopic observations, it is clear that the spherulites are impinging and space-filling at Tc. Hence, the crystallizable segregated phase has to be located intraspherulitically. SAXS measurements were performed both at room temperature and at the crystallization temperature Tc=107°C. All SAXS spectra were background corrected, desmeared and Lorentz-geometrically corrected. Figure 6 illustrates the SAXS patterns obtained with LLDPE samples containing propene, 1-butene and 1-octene as comonomer, but with about the same number of side branches (17.7 methyl-, 14.7 ethyl-, 14.0 hexyl branches per 1000 C atoms in the backbone, resp.), crystallized at Tc=107°C. The long period, as calculated by application of Bragg's equation to the position of the first peak maximum, is markedly different for these samples and depends strongly on the type of comonomer (Table 1).

Moreover, the difference in long spacing at Tc and room temperature (RT) points out that at least part of the copolymer is segregated and crystallizes between the original crystalline lamellae.

The lamellar thickness, both at room temperature (l_{RT}) and at Tc (l_{Tc}), was calculated using the equation (5):

$$l = X_c \cdot L$$

and is shown in Table 1. The difference in lamellar thickness at Tc and RT indicates a smaller lamellar thickness of the crystallized segregated lamellae in comparison with the thickness of the original lamellae, which is one reason for the lower melting endotherm (I) in the DSC scans.

Study of the Crystalline Perfection for Different LLDPE Types

In the case that all short chain branches are rejected from the crystals into the amorphous regions during the crystallization process, the dimensions of the lattice vectors of LLDPE will be exactly the same[6] as those of linear PE (HDPE). However, when at least a fraction of the side chain branches can be incorporated in the crystal lattice, the unit cell

Table 1. Long Spacings From SAXS Measurements for Different LLDPE Types

Type	L_{RT} (A)	L_{Tc} (A)	l_{RT} (A)	l_{Tc} (A)
Propene	224	411	81	86
1-Butene	202	341	59	81
1-Octene	153	362	42	63

must necessarily expand, which results in larger lattice vectors. So measuring the unit cell vectors of LLDPE provides information on whether the side groups are rejected from the crystal or not.

In order to study this rejection process, LLDPE specimens with about the same amount of side branches (See 2.) and HDPE were compared. A WAXS spectrum was recorded for each specimen at room temperature and at Tc = 107°C, and the unit cell parameters were calculated using the following equation for the orthorombic unit cell of LLDPE:[7]

$$2\frac{1}{\sqrt{\frac{h^2}{a^2}+\frac{k^2}{b^2}+\frac{l^2}{c^2}}}\qquad \sin\theta = n\lambda$$

The unit cell parameters a, b, and the cell volume are listed in Table 2. The c parameter, which is the direction along the chain axis, appeared to be constant: c = 2.547 Å.

Inspection of Table 2 shows that all LLDPE specimens have larger unit cell dimensions than HDPE. This indicates that a fraction of the side branches are incorporated into the crystal lattice. At Tc=107°C, the unit cell is expanded due to thermal expansion. At room temperature the unit cell volume is a weighted average of the volume of the crystals formed at Tc and the volume of the crystallized segregated phase, which is different for the three LLDPE types (40.9/100 for propene, 23.3/100 for 1-butene and 34.8/100 for 1-octene).

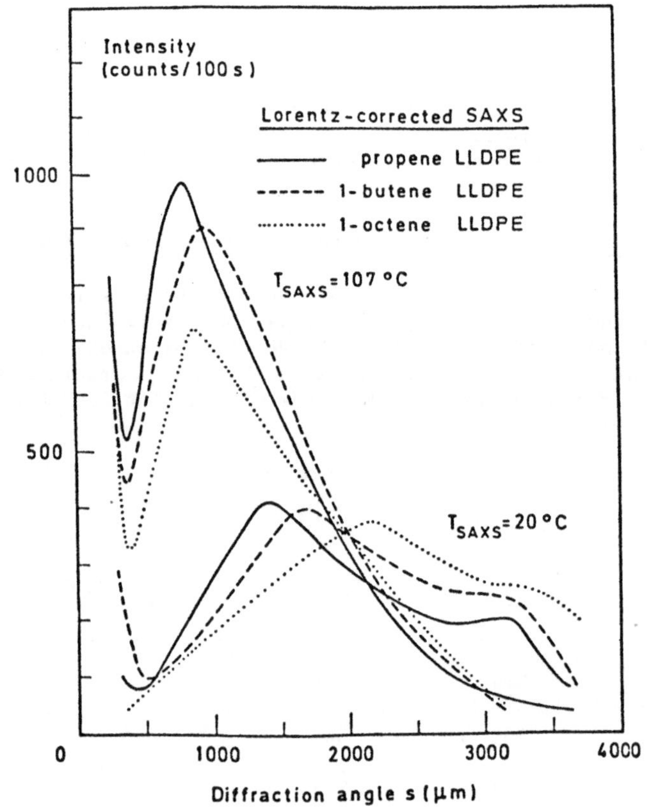

Fig. 6

Table 2. Unit Cell Parameters a, b, and Unit Cell Volume V for HDPE and Different LLDPE Types, Measured at Room Temperature and at the Crystallization Temperature

Specimen Type	Room Temperature			Tc=107°C		
	a (Å)	b (Å)	V (Å^3)	a (Å)	b (Å)	V (Å^3)
HDPE	7.4418	4.9645	94.098	7.6358	4.9777	96.808
Propene	7.5838	4.9923	96.431	7.7178	4.9725	97.746
1-Butene	7.5018	4.9988	95.521	7.7000	4.9940	97.942
1-Octene	7.5118	5.0308	96.252	7.7059	4.9756	97.656

The difference in cell volume between RT and Tc is merely due to thermal expansion, because the HDPE specimen does not show any crystallized segregated fraction at Tc. This thermal expansion accounts for 2.89% enlargement of its unit cell. The fact that the cell volume at Tc for all three LLDPE specimens is smaller than the cell volume at RT, enlarged for correction of the thermal expansion, proves that the crystallized segregated phase incorporates more short side chain branches than the crystals originally formed at Tc.

CONCLUSION

A molecular segregation process during crystallization from the melt in LLDPE containing propylene, 1-butene and 1-octene is reported. It was found that the amount of crystallized segregated polymer increases with increasing crystallization temperature and comonomer content. The amount of originally formed crystals drops with increasing crystallization temperature and comonomer concentration. The total crystallinity is very little affected by Tc, but decreases strongly with increasing comonomer content.

The semi-crystalline morphology was studied for different LLDPE types using SAXS at room temperature and at Tc. It has been pointed out that at least part of the crystallized segregated phase is located between the originally formed lamellae. The lamellar thickness of the crystallized segregated phase is smaller than the thickness of the crystals formed at Tc.

The internal crystalline perfection was studied for different types of LLDPE using WAXS at room temperature and at Tc. All LLDPE types incorporate part of the side branches. The crystals formed at Tc contain fewer side branches than the crystallized segregated phase.

REFERENCES

1. J. C. Randall, "Characterization of long chain branching in polyethylenes using high-field carbon 13 NMR," in: "Polymer characterization by ESR and NMR," p. 94-118.
2. K. Shirayama, T. Okada, and S. Kita, J. Polym. Sci., Part A, 3, 907 (1965).
3. R. Florin, R. Spitz, A. Douillard, A. Guyot, R. F. Boyer, L. R. Denny, and P. L. Kumler, Europ. Polym. J., 16, 1079 (1980).
4. V. F. F. Mathot and M. J. Pijpers, Polymer Bulletin, 11, 297 (1984).

5. Kavesh and Schultz, J. Polym. Sci., A2, 8, 243 (1970).
6. P. Zugenmaier and H. Cantor, Koll. Z. Z. Polym., 230, 229 (1969).
7. L. Azaroff and M. Buerger, "The powder method," McGraw-Hill, New York, p. 83 (1958).

PHOTOANTIOXIDANTS IN POLYOLEFINS

Gerald Scott

Department of Molecular Sciences
University of Aston
Birmingham B47ET

INTRODUCTION

The term "antioxidant" is normally understood by polymer technologists to refer to the stabilization of rubbers and plastics against the effects of oxygen at high temperatures. Plastics technologists have adopted the term "stabilization" to describe the protection of plastics against light, and until recently, UV stabilizers for plastics were commonly believed to act by quite different mechanisms. This technological classification has tended to obscure the basic mechanisms common to the behavior of antioxidants in many polymers and under a wide range of environments.

CHAIN-BREAKING ANTIOXIDANTS

The mechanism of action of the most widely recognized group of antioxidants, the chain-breaking (CB) antioxidants was elucidated about 30 years ago. It was seen to involve the deactivation of chain-carrying alkylperoxyl radicals by electron donors, notably hindered phenols and aromatic amines, of which the antioxidants shown in Table 1 are widely used in the polymer industries.[2] None of these compounds is an effective light stabilizer for polymers and some of them actually catalyze photooxidation.[3] This explains why thermal and light stabilizers were originally thought to act by different mechanisms. Recently, however, it has been recognized that some phenols containing the same antioxidant functional group as BHT and 1076 are considerably more effective as light stabilizers in polyolefins.[2] Some typical examples are given in Table 2. The feature that distinguishes these CB-D antioxidants from the conventional hindered phenols is their greater stability to UV light. This permits them to fulfill their antioxidant function for much longer times under UV irradiation conditions. Pospisil has shown[4] that BHT readily undergoes photolysis to ring-rearranged products, see Scheme 1, and the formation of reactive intermediate radicals accounts for its pro-oxidant behavior. The UV absorbing group in the modified hindered phenols shown in Table 2 may be considered to be autoprotective.

The long established UV absorbers, see Table 3, may act similarly. That is, they do not simply screen the polymer from the effects of light (although this is certainly one of their functions), since they are relatively ineffective when used as screening agents in a "sandwich" struc-

Table 1. Typical Chain-Breaking Donor (CB-D) Antioxidants

STRUCTURE	NAME	USE
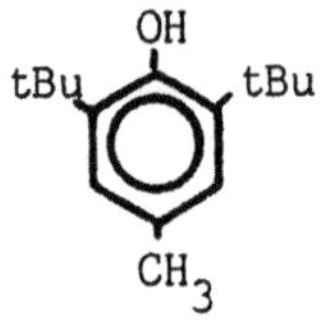	BHT IONOL	PROCESSING STABILISER (THERMOPLASTICS)
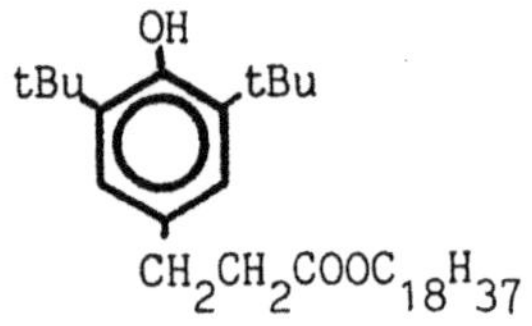	IRGANOX 1076	HEAT STABILISER (THERMOPLASTICS)
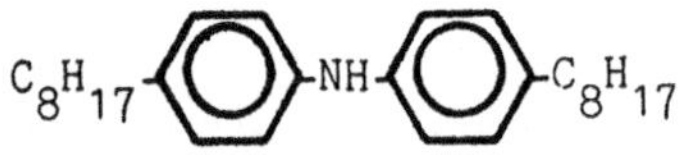	NONOX OD	HEAT STABILISER (RUBBERS)
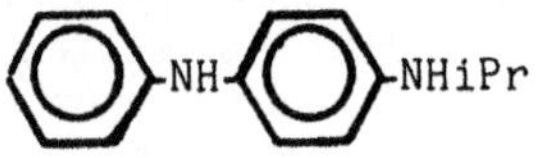	SANTOFLEX IP	ANTIFATIGUE AGENT, ANTI-OZONANT (RUBBER)

Table 2. UV Stabilizers for Polypropylene Containing a CD-D Function

STRUCTURE	TRADE NAME
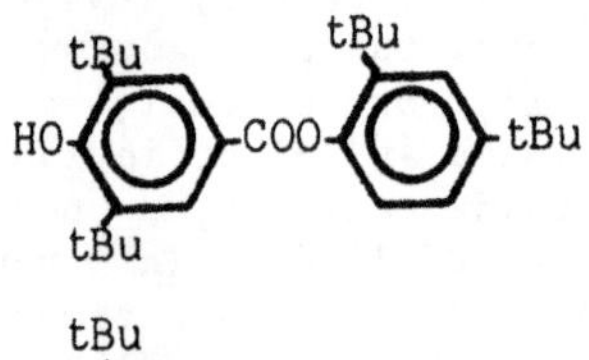	TINUVIN 121
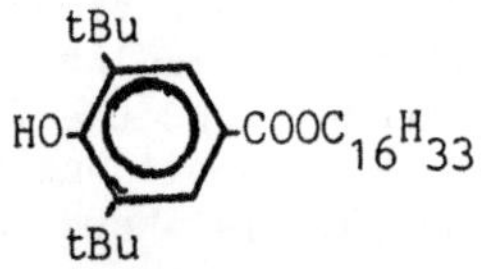	CYASORB 2908
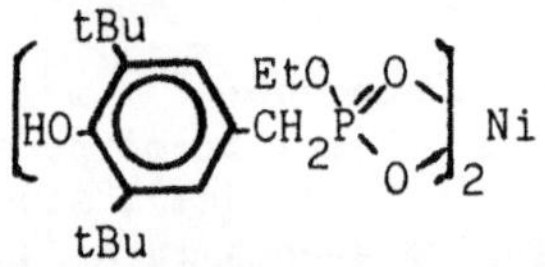	IRGANOX 2002

Table 3. Mechanisms of Action of "UV Absorbers"

STRUCTURE	TRADE NAME	MECHANISM
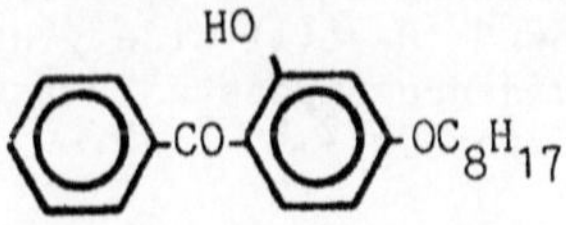	CYASORB UV 531	UVA, CB-D
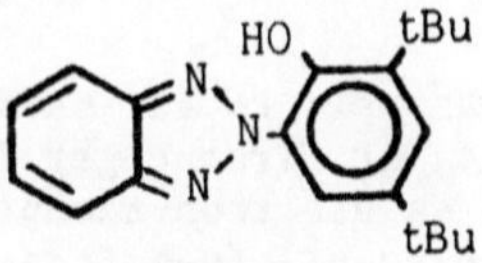	TINUVIN 326	UVA, CB-D
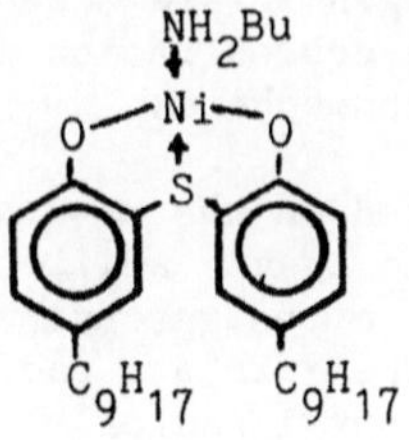	CYASORB 1084	UVA,CB-D, (PD-C)
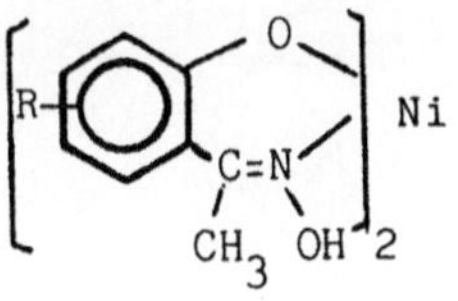	NEGOPEX*	UVA,CB-D, PD-S

*No longer available commercially

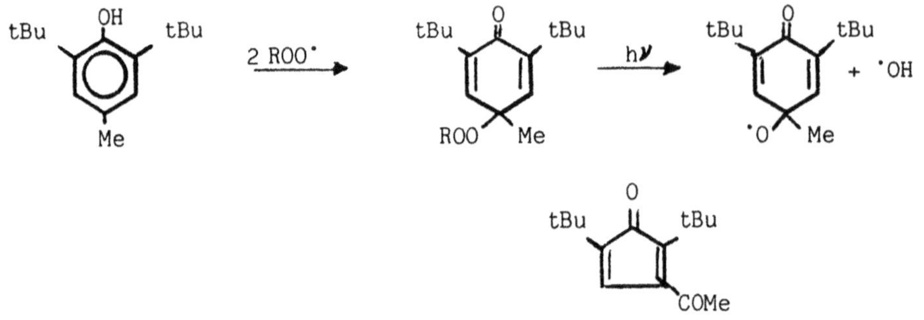

Scheme 1. Photo-Oxidation of BHT

ture.[5,6] They have all been shown to have at least one additional antioxidant function, and in some cases, more than one (see Table 3). In the case of the nickel complexes, they were at one time thought to act as excited state quenchers,[12,18] and the term "quencher" is still widely used in the technological literature. More recent evidence suggests[6-8] that they act primarily as light stable peroxide decomposers.[2,8-17,19,20]

CATALYTIC CHAIN-BREAKING ANTIOXIDANTS

Until quite recently, very little attention was paid to antioxidants that act by oxidizing intermediate radical species. At first sight it may seem rather unlikely that oxidizing agents can act as electron acceptors (CB-A antioxidants) in competition with molecular oxygen[21] (see Scheme 2) but it must be remembered that the equilibrium oxygen concentration in a polymer matrix is very much lower than it is in the atmosphere.

Nitroxyl radicals that are fully alkyl substituted on the α-carbon atoms are highly effective CB-A antioxidants in Scheme 2 and oxidize macroalkyl radicals to macro-olefins during the processing of polypropylene. However, under these conditions, they do not deactivate just one radical, but by alternating reduction and oxidation of the nitroxyl radical-hydroxylamine couple (see Scheme 3), many chain-propagating radicals can be removed from the system.[2,22-28]

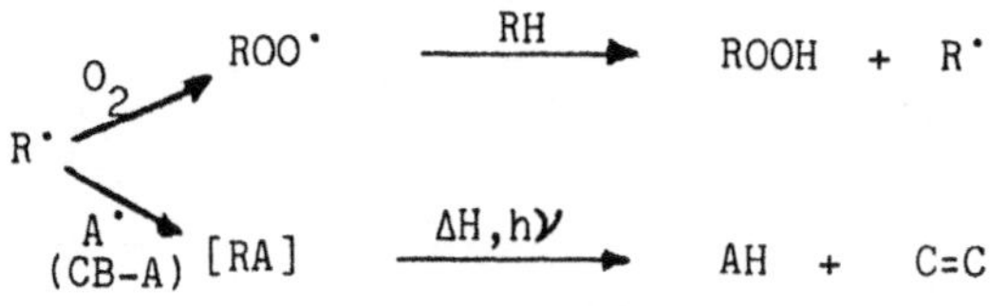

Scheme 2. Competition Between Oxygen and a CB-A Antioxidant for Alkyl

Scheme 4 summarizes the photo-oxidation process and indicates the ways in which it can be interrupted by the removal of chain carrying species or of potential radical generating species. It is clear that the removal of both the chain propagating radicals (R• and ROO•) must constitute a very powerful antioxidant mechanism provided that the oxidized form of the antioxidant has photo-oxidatively stable and can compete with oxygen for alkyl radicals. In a commercial processing operation, air is essentially excluded and the only oxygen present is that trapped between the polymer granules or dissolved in the polymer under ambient conditions. These are

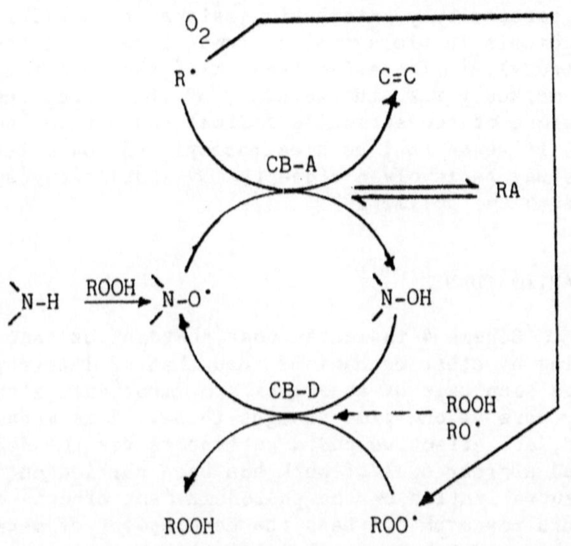

Scheme 3. Catalytic Mechanism of "Stable" Nitroxyls

the ideal requirements for the operation of the CB-A/CB-D cycle. It is perhaps somewhat more surprising that hindered nitroxyls are able to compete with oxygen in thin polymer films where oxygen saturation might be expected. Nevertheless, this appears to be the case and the hindered piperidinoxyls and their cognate hydroxylamines are among the most effective photoantioxidants so far discovered for polypropylene (see Table 4).

In commercial practice, the hindered piperidines are normally used as additives rather than the nitroxyls or hydroxylamines, and they are partially converted to the nitroxyl during processing.[24] However, their effectiveness is very dependent upon the severity of the processing operation. Very severe processing leads to the oxidative destruction of the nitroxyl radical.[27]

Table 5 lists other precursors of nitroxyl radicals that also behave as photoantioxidants in polypropylene. They are, in general, as effective as or superior to the hindered piperidines.[29,30] The spin trapping agent

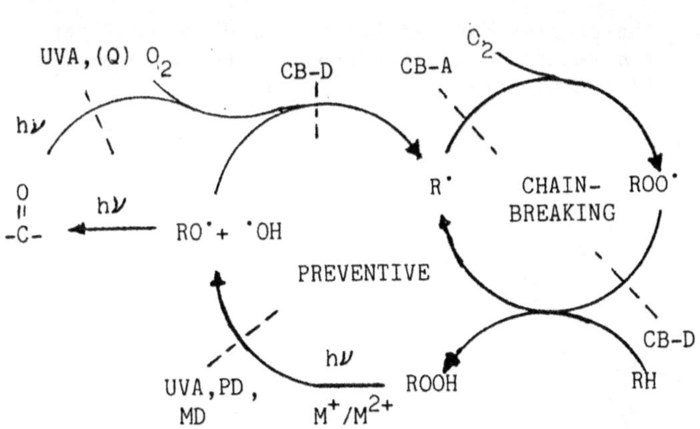

CB-A,CHAIN-BREAKING ACCEPTOR, CB-D,CHAIN-BREAKING DONOR, UVA,UV ABSORBER, PD,PEROXIDE DECOMPOSER, Q,QUENCHER

Scheme 4. Potential Photoantioxidant Mechanisms

agent, NTB, is particularly interesting, since it is widely used to detect and identify radicals in biological systems, because of the stability of the derived nitroxyl.[31] The effectiveness of the spin traps listed in Table 5 varies markedly with the severity of the processing operation and although the nature of the attacking radical has not yet been identified with certainty, it seems that both macroalkyl and low molecular weight "oxyl" radicals may be involved since the UV stabilizing species is partially attached to the polymer.[30]

PEROXIDOLYTIC ANTIOXIDANTS

Reference to Scheme 4 indicates that photoantioxidants are potentially capable of acting by other mechanisms than that of interrupting kinetic chains. We have seen that UV absorbers are important, although not as important as they were at one time thought to be. Some metal ions, notably iron and cobalt, are effective photo-activators for the degradation of polyolefins, and a great deal of work has been carried out in recent years to study the neutralization of the photodegradant effects of metal ions. A byproduct of this research has been the development of metal complexes which confer a limited but controlled lifetime on polymers used out-of-doors.[32]

However, by far the most important class of preventive antioxidants is the peroxide decomposers, since by removing hydroperoxides in a process which does not involve the formation of free radicals, they thereby remove the most important indigenous initiator for photooxidation.[2,19,20,33-37]

Some peroxidolytic antioxidants, which are highly effective heat stabilizers, (e.g., DLTP and ZnDRC in Table 6), are, like the hindered phenols, not UV stable and are consequently poor UV stabilizers. However, others and in particular the nickel thiolate complexes, are not only effective heat stabilizers, but some of the more polymer-soluble members of this class have photoantioxidant activities similar to or better than the hindered amines, particularly when used in combination with UV absorbers. Table 7[38] shows that combinations of the higher alkyl dithiocarbamates are particularly effective in combination with UV 531. A considerable advantage of the three component antioxidant system shown in Table 7 is its all-round effectiveness as a melt stabilizer, thermal antioxidant and photoantioxidant.[38-40]

Table 4. Photoantioxidant Activity of a Hindered Piperidine and Related Oxidation Products in Polypropylene (All Concentrations 6×10^{-4} mol/100g)

STRUCTURE	TIME TO EMBRITTLEMENT, h*		
X =	H	O˙	OH
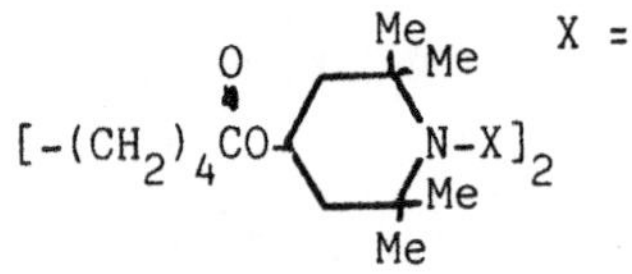	750	960	-
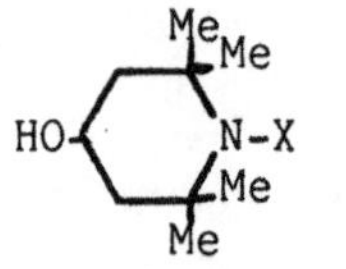	-	920	1040

*Control without Stabilizer, 90 hours

Table 5. Nitroxyl Precursors as Photoantioxidants in Polypropylene (Concentrations 10^{-3} mol/100g)

PRECURSOR	CODE	EMBRITTLEMENT TIME,h
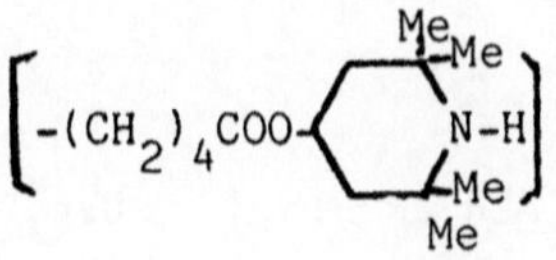	PS-H (TINUVIN 770)	760
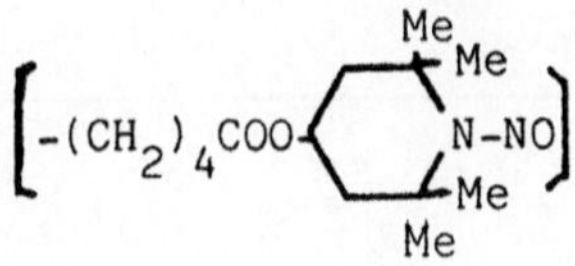	PS-NO	1580
tBuNO	NTB	875
tOctNO	NTO	690

Table 6. **Peroxidolytic (PD) Antioxidants as UV Stabilizers and Synergists**

ANTIOXIDANT	CODE	USE	MECHANISM
$P(OR)_3$	-	SYNERGIST WITH UVA	PD-S
$(C_{12}H_{25}OCOCH_2CH_2)_2S$	DLTP	SYNERGIST WITH UVA AND CB-D	PD-C
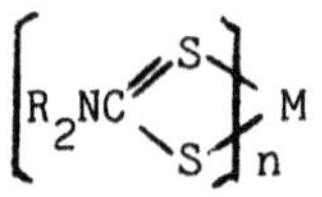			
M= Zn, n=2	ZnDRC	SYNERGIST WITH UVA	PD-C
M=Ni, n=2	NiDRC	UV STABILISER IN ITS OWN RIGHT	UVA,PD-C
M=Fe, n=3	FeDRC	DELAYED ACTION PHOTO-ACTIVATOR	PD-C,UVA
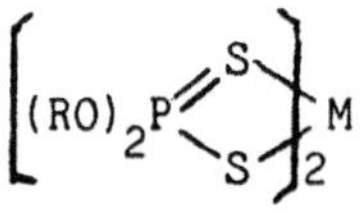			
M=Ni, n=2	NiDRP	UV STABILISER IN ITS OWN RIGHT	UVA,PD-C

Table 7. **Synergism Between Peroxidolytic Antioxidants and a "UV Absorber"**

ANTIOXIDANT	CONCENTRATION (g/100g)	UV EMBRITTLEMENT TIME,h	
		ALONE	WITH UV 531 (0.2g/100g)
ZnDNC	0.2	385	1400
NiDNC	0.2	1160	2250
ZnDNC + NiDNC	0.1 + 0.1	-	2600
TINUVIN 770	0.2	640	1780
CONTROL	-	70	400

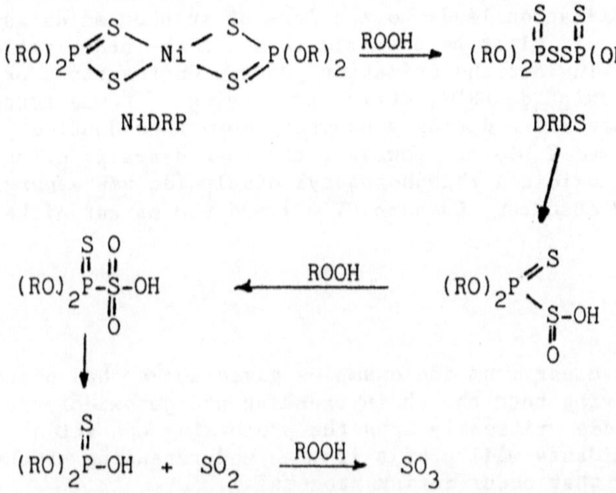

Scheme 5. Antioxidant Mechanisms of the Nickel Dialkyl Dithiophosphates

EFFECT OF OXIDATIVE PROCESSING ON THE ACTIVITY OF PEROXIDOLYTIC ANTIOXIDANTS

It was shown many years ago that the metal thiolate antioxidants shown in Table 6 are converted to sulphur acids during use by hydroperoxides present in the polymer in a radical generating step. The latter gives rise to initial pro-oxidant activity. More detailed studies have recently shown[34] that in the case of the nickel dialkyl dithiophosphates (NiDRP),

Table 8. Effects of Oxidative Processing and Synergism on Peroxidolytic Antioxidants in Polypropylene

ANTIOXIDANT	CONCENTRATION (g/100g)	UV EMBRITTLEMENT TIME, h: NORMAL COMPOUNDING	UV EMBRITTLEMENT TIME, h: TBH ADDED (0.1g/100g)
DBDS	0.2	300	700
DBDS + UV 531	0.2 + 0.1	1300	2050
DBDS + UV 531 + NiDBP	0.2 + 0.1 + 0.1	1400	5000
CONTROL	-	90	-

DBDS = $[(BuO)_2\overset{S}{\overset{\|}{P}}S]_2$, NiDBP = $[(BuO)_2\overset{S}{\overset{\|}{P}}S]_2Ni$, TBH = tBuOOH

hydroperoxide oxidation leads to the loss of sulphur acids as shown in Scheme 5. Thus, it might be possible to avoid the undesirable pro-oxidant effects by carrying out the oxidation process on the first oxidation product, the disulphide DRDS, during processing. It was found that incorporating hydroperoxides during processing more than doubled photoantioxidant activity, see Table 8. However, the most dramatic effects were observed when the oxidized thiophosphoryl disulphide was synergised with a conventional UV absorber, Cyasorb UV 531 and the parent nickel dithiophosphate.[41]

CONCLUSION

It will be clear from the examples given above that photoantioxidant activity, involving both the chain-breaking and peroxidolytic mechanisms, frequently depends critically upon the processing operation. The design of new photoantioxidants will profit from an understanding of the chemical transformations that occur during processing.

ACKNOWLEDGEMENTS

I am grateful to my coworkers, Dr. K. B. Chakraborty and Dr. S. Al-Malaika who have helped to put the theories of antioxidant action on a firmer foundation. Also to my former research students, Drs. R. P. R. Ranaweera, R. Bagheri, H. Yaghmour, E. O. Omikorede and Mr. M. Coker who contributed to the studies reported.

REFERENCES

1. G. Scott, Atmospheric Oxidation and Antioxidants, Elsevier, London and New York, Chapter 4 (1965).
2. N. Grassie and G. Scott, Degradation and Stabilization of Polymers, Cambridge University Press, Chapter 5 (1985).
3. Ref. 1, p. 400, et seq.
4. J. Pospisil, Developments in Polymer Stabilization-1, G. Scott Ed., Appled Science Publishers, London, p. 1 (1979).
5. D. J. Carlsson and D. M. Wiles, J. Macromol. Chem., C14, 155 (1976).
6. K. B. Chakraborty and G. Scott, Europ. Polym. J., 15, 35 (1979).
7. D. K. Hodgeman, J. Polym. Sci., Polym. Lett. Ed., 16, 161 (1978).
8. G. Scott, ACS Symposium Series, 25, 340 (1976).
9. G. Scott, Pure and Appl. Chem., 52, 365 (1976).
10. R. P. R. Ranaweera and G. Scott, Europ. Polym. J., 12, 591 (1976).
11. R. P. R. Ranaweera and G. Scott, Europ. Polym. J., 12, 825 (1976).
12. P. J. Briggs and J. F. McKellar, J. App. Polym. Sci., 12, 1825 (1969).
13. A. M. Trozolo and F. H. Winslow, Macromol., 1, 1529 (1968).
14. J. P. Guillory and C. F. Cook, J. Polym. Sci., A1, 9, 1529 (1971).
15. B. Ranby and J. F. Rabek, Photodegradation, Photo-oxidation and Photostabilization of Polymers, London and New York, p. 379 et seq. (1975).
16. M. L. Kaplan and P. E. Kelleher, J. Polym. Sci., 139, 565 (1981).
17. A. Adamczik and F. Wilkinson, J. Chem. Soc., Farad. II, 68, 2031 (1972).
18. D. J. Carlsson, T. Suprunchuk, and D. M. Wiles, Macromol., 5, 659 (1972).
19. G. Scott, Brit. Polym. J., 16, 271 (1984).
20. G. Scott, Polym. Deg. and Stab., 10, 97 (1985).
21. Ref. 1, p. 161.
22. D. J. Carlsson and D. M. Wiles, J. Macromol. Chem., C14, 65 (1986).

23. D. J. Carlsson, A Garton, and D. M. Wiles, Developments in Polymer Stabilization-1, G. Scott Ed., Applied Science Publishers, London, p. 219 (1979).
24. K. B. Chakraborty and G. Scott, Chem. and Ind., 239 (1978), Polymer, 21, 252 (1980).
25. R. Bagheri, K. B. Chakraborty and G. Scott, Polym. Deg. and Stab., 4, 1 (1982).
26. V. Ya. Shlyapintokh and V. B. Ivanov, Developments in Polymer Stabilization-5, G. Scott Ed., Applied Science Publishers, London, p. 41 (1982).
27. G. Scott, Developments in Polymer Stabilization-7, G. Scott, Ed., Elsevier, Applied Science Publishers, London, Chapter 2 (1984).
28. S. Al-Malaika, E. O. Omikorede and G. Scott, J. Polym. Sci. in press.
29. K. B. Chakraborty, G. Scott, and H. Yaghmour, Polym. Deg. and Stab., 10, 221 (1985).
30. K. B. Chakraborty, G. Scott, and H. Yaghmour, J. App. Polym. Sci., 30, 189 (1985).
31. E. G. Janzen and B. J. Blackburn, J. Am. Chem. Soc., 91, 4481 (1969).
32. D. Gilead and G. Scott, Developments in Polymer Stabilization-5, G. Scott Ed., Applied Science Publishers, London, p. 71 (1983).
33. G. Scott, Developments in Polymer Degradation-1, N. Grassie Ed., Applied Science Publishers, London, p. 205 (1978).
34. S. Al-Malaika, K. B. Chakraborty, and G. Scott, Developments in Polymer Stabilization-6, G. Scott Ed., Applied Science Publishers, London, p. 73 (1983).
35. S. Al-Malaika and G. Scott, Degradation and Stabilization of Polyolefins, N. S. Allen Ed., Applied Science Publishers, London, p. 283 (1983).
36. K. B. Chakraborty and G. Scott, Europ. Polym. J., 13, 1007 (1977).
37. K. B. Chakraborty and G. Scott, Polym. Deg. and Stab., 1, 37 (1979).
38. K. B. Chakraborty, G. Scott, and W. R. Poyner, Plast. and Rubb. Proc. and App., 3, 59 (1983).
39. J. D. Holdsworth, G. Scott, and D. Williams, J. Chem. Soc., 4692 (1964).
40. S. Al-Malaika and G. Scott, Polym. Comm., 24, 24 (1983); Europ. Polym. J., 19, 235, 241 (1983).
41. S. Al-Malaika and G. Scott, Brit. Pat., 2,117,779A (1983).

^{13}C NMR SPECTRA OF POLYOLEFINS

H. N. Cheng and M. A. Bennett

Hercules Incorporated
Research Center
Wilmington, Delaware 19894

INTRODUCTION

^{13}C NMR spectroscopy has been used successfully to study ethylene and propylene polymers. Some common polymer systems are summarized as follows:

C_3 ⟶ isotactic C_3 ⟶ irregular C_3

C_2 ⟶ stereoregular C_2C_3 ⟶ EPR ⟶ ethylene-propylene oil

where polymer microstructure becomes increasingly complex as one moves from left to right. The ^{13}C NMR spectral features correspondingly increase in complexity from left to right. C_2 and C_3 refer to ethylene and propylene respectively.

Isotactic and atactic polypropylenes were among the first polymers studied by ^{13}C NMR.[1] Much work has been done on these polymers[2-5] and the assignments are now well established. In contrast, irregular polypropylene (with head-to-head and tail-to-tail structures) gives considerably more complex spectra and has been studied less. Although the main features of the spectrum have been assigned,[6-8] it was only recently that a complete interpretation was accomplished.[9]

For copolymers, much of the attention[10-14] has concentrated on stereoregular poly(propylene-co-ethylene), and the ^{13}C spectra have been completely elucidated.[10] Ethylene-propylene rubbers (EPR) have also been studied,[15-18] but the spectra were only recently[18] fully interpreted. It was shown that the complexity of the ^{13}C NMR spectrum could be interpreted as due to the interplay of ethylene-propylene sequence placements, propylene tacticity, and propylene inversion. On this basis, it is possible to provide detailed spectral assignments for EPR's.[18] For ethylene-propylene oils the chain ends must be taken into account; again the assignments have been made recently.[19]

Owing to the complexity of the spectra and the various combinations and permutations different structural elements can undergo, it would be desirable to determine the rules governing the ^{13}C shift behavior in these systems. An advantage (if such rules exist) is that the ^{13}C shifts of all subsequent homo- or co-polymers of ethylene and propylene can be predicted, thereby simplifying the task of interpretation of any unknown ethylene-propylene systems. It is the purpose of this work to examine all the existing ^{13}C shift data for polyolefins and low-molecular-weight model compounds. Indeed, detailed additive rules can be devised for all the carbons. The rules are general, and apply to all homopolymers and copolymers of ethylene and propylene, as well as low molecular weight analogs.

Because of the comprehensive nature of these rules, they require a fair amount of computation. To simplify the calculations and to minimize arithmetical errors, we have automated these rules. The resulting computer program (CALMOD) can readily reproduce the predicted ^{13}C shift values for any methyl substituted alkanes or ethylene/propylene polymers.

EXPERIMENTAL

The program CALMOD was written in the BASIC language for the Nicolet 1280 computer. The simulated spectra were produced on a Zeta plotter. All ^{13}C shifts were referenced to tetramethylsilane at 0 ppm.

The samples used as examples in this work were all experimental polymers. These were run as 20% (w/w) solutions in 1,2,4-trichlorobenzene with d_6-benzene as the lock material. The spectra were taken on a Nicolet NT 360 WB spectrometer at 90.55 MHz and 120°C. The free induction decays were stored in 8K memory addresses, zero-filled upon processing, with a spectral window of 6000 Hz. A pulse angle of 45° was used with a 4-s. delay. All ^{13}C shifts were referenced to tetramethylsilane at 0 ppm.

RESULTS AND DISCUSSION

Rules for Spectral Prediction

In our earlier efforts on spectral prediction,[20,21] ^{13}C shifts were computed by empirical additive rules corresponding to the α, β, γ, δ, and ε effects. For detailed spectral assignments of ethylene-propylene copolymers, these rules must be modified in order to account for the meso/racemic configuration of propylene placements, and (in low molecular weight materials) the presence of chain ends.

The chemical shift rule for the methyl group is taken from the work of Zambelli and Gatti.[22] They have shown that the following equation approximately holds for a large number of model compounds:

$$\delta_{CH_3} = 19.99 + \sum_i A_i + \sum_{i,j} B_{ij} \tag{1}$$

where A_i is the substituent parameter for position i, and B_{ij} is the correction term corresponding to the configurations of the groups at i and j. These parameters are reproduced in Table 1. It was previously found[18] that this rule agreed well with out assignments for the ethylene-propylene rubbers.

For the methine and methylene carbons, similar rules do not exist and must be devised. These were accomplished by examining the chemical shifts of similar polymers or model compounds. The equations we devised are given below;

$$\delta_{CH} = 33.26 + \sum_i A_i + \sum_{i,j} B_{ij} \tag{2}$$

Table 1. Parameters for the Estimation of ^{13}C Shifts of Methyl Group (Equation 1)

A_i	γ	δ	ϵ	ζ
m	-3.05	0.77	0.16	0.05
r	-4.83	0.22	0.06	-0.03

i	j	$B_{\delta\zeta}$	$B_{\delta\epsilon}$	$B_{\gamma\delta}$	$B_{\gamma\epsilon}$	$B_{\epsilon\zeta}$	$C_{\gamma\gamma}$
m	m	0.02	0.17	0.75	-0.12	0	-1.7
m	r	-0.10	-0.06	0.75	-0.40	0	0
r	m	-0.15	-0.33	0.05	-0.14	0	0
r	r	-0.20	-0.39	0.05	-0.52	0	1.1

$$\delta_{CH_2} = 29.9 + \sum_i A_i + \sum_{i,j} B_{ij} + \sum_{i,j} C_{ij} + \sum_k D_k \tag{3}$$

where A_i corresponds to the substituent parameter for position i, B_{ij} is the correction term for methyls at positions i and j on the same side of the methylene group, and C_{ij} is to be used for methyls as positions i and j on the opposite sides of the methylene group in question. After a careful study of the published literature, consistent sets of values are obtained (Tables 2 and 3). Most of the parameters are taken from the assignments of irregular polypropylene,[6-9] ethylene-propylene copolymers[10-14] and ethylene-propylene rubbers.[15-19] The results of Zambelli et al.,[3,22-25] Tonelli and Schilling,[26-27] Freche et al.,[28] and Moeller and Cantow[29] are also used. The term D_k is specifically designed to take into account the effect of polypropylene tacticity on methylene carbons:

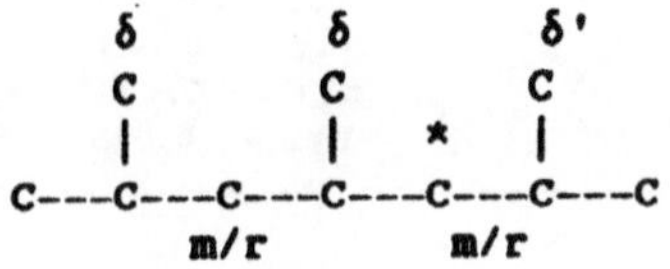

$D_{\delta\beta\beta'}$

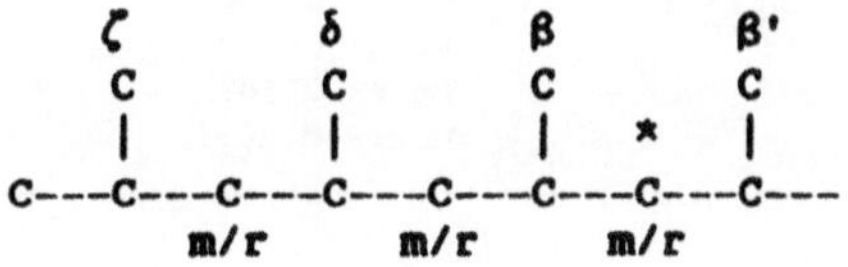

$D_{\zeta\delta\beta\beta'}$

Thus, for the (mmmmm) hexad, the values of 2 $D_{\delta\beta\beta'}$ (mm) + 2 $D_{\zeta\delta\beta\beta'}$ (mmx) should be added.

Equations 1-3 can be used directly for carbons in the middle of a large molecule (e.g. for polymers). For small molecules or for carbons near the chain ends, the surplus carbons beyond the chain ends must be truncated. Correction factors must then be introduced for these carbons near the chain ends.

$$\delta_{end} = \delta_{chain} - n_\alpha E_\alpha - n_\beta E_\beta - n_\gamma E_\gamma - n_\delta E_\delta - n_\epsilon E_\epsilon \tag{4}$$

where E_i is the correction factor and n_i is the number of carbons in the ith position to be truncated. For better accuracy, again separate parameter values are given for methyls, methines, and methylenes (Table 4).

Table 2. Empirical Parameters for the ^{13}C Shifts of Methines (Equation 2)

A_i	β	γ	δ	ε
m	4.9	-2.38	0.36	0.04
r	4.2	-2.46	0.27	0.00

i	j	$B_{\gamma\varepsilon}$	$B_{\gamma\delta}$	$B_{\beta\gamma}$	$B_{\beta\delta}$	$B_{\delta\varepsilon}$	$B_{\beta\varepsilon}$	$C_{\beta\beta}$
m	m	0	0	-0.8	0	0	0	0.8
m	r	0	0	-1.4	0	0	0	-0.5
r	m	-0.14	0	-1.0	0	0	0	-0.5
r	r	0	0	-1.5	0	0	0	-1.2

Table 3. Empirical Parameters for the Estimation of ^{13}C Shifts of Methylene Group (Equation 3)

A_i	β	γ	δ	ε	ζ
	7.59	-2.56	0.40	0.05	0

	$C_{\beta\beta}$	$C_{\beta\gamma}$	$C_{\beta\delta}$	$C_{\beta\varepsilon}$	$C_{\beta\zeta}$	$C_{\gamma\gamma}$	$C_{\gamma\delta}$
m	0.54	-0.001	-0.04	-0.107	0	-0.023	-0.017
r	0.34	-0.059	0	-0.072	0	0	0

	$B_{\beta\gamma}$	$B_{\beta\delta}$	$B_{\beta\varepsilon}$	$B_{\gamma\delta}$	$B_{\gamma\varepsilon}$	$B_{\delta\varepsilon}$	$B_{\delta\zeta}$
m	-1.8	-0.34	-0.25	0.5	-0.220	0	0.15
r	0	0.51	0	0	-0.115	0	0

$D_{\delta\beta\beta'}$:		$D_{\zeta\delta\beta\beta'}$:	
	mm = 0.390		mmx = -0.257
	mγ = 0.119		γmx = 0.043
	γm = -0.149		mγx = 0.006
	γγ = -0.178		γγx = 0.045

Table 4. Additive Parameters for the Estimation of Base Values in Chain Ends

	Base	E_α	E_β	E_γ	E_δ	E_ε
Methyl	19.9	---	8.8	-3.6	0.59	0.1
Methylene*	29.99	8.5	9.5	-2.46	0.35	0.05
Methine	29.99	4.9	6.4	-1.9	0.35	0.05

*A steric correction factor of +1.0 ppm is added for the terminal carbon in this structure: $...CH_2CH_2CH(CH_3)_2$.

Program CALMOD

The rules delineated above have been completely encoded in the program CALMOD. Our basic goal for this program is that it should be user-friendly and fast. As examples, we show several structural patterns that can arise from ethylene and propylene polymer systems.

A. Stereoregular Sequences

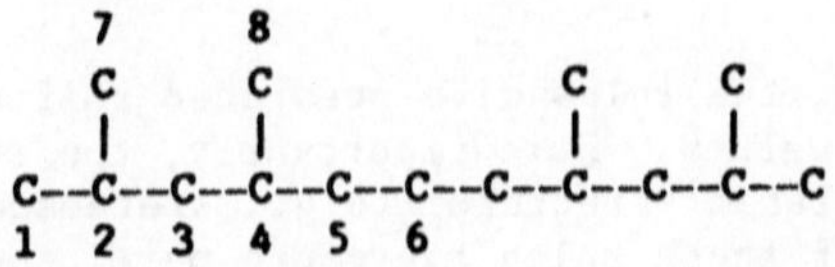

Carbon	Observed[25]	Predicted
1	22.80	22.68
2	26.04	25.98
3	47.63	47.73
4	31.22	30.52
5	38.37	38.06
6	24.66	24.53
7	23.55	23.67
8	20.32	20.71

B. Stereoirregular Sequence

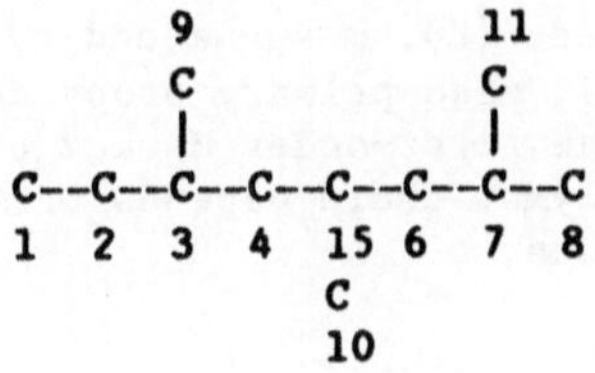

Carbon	Observed[24]	Predicted
1	11.27	11.43
2	30.77	31.03
3	32.52	32.20
4	45.63	45.71
5	28.93	27.89
6	48.47	48.22
7	25.98	25.80
8	23.51	23.53
9	19.53	19.43
10	20.40	20.35
11	22.86	23.03

C. Regioirregular Sequence

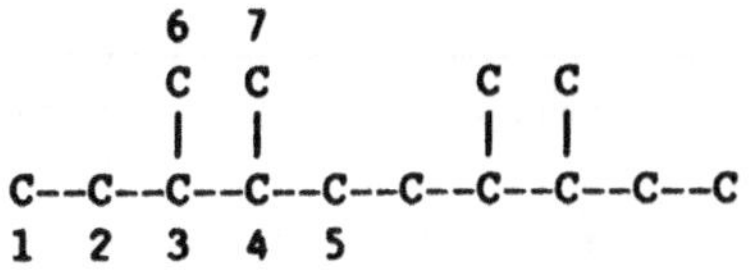

Carbon	Observed[29]	Predicted
1	12.29	12.29
2	25.88	26.06
3	39.93	39.70
4	38.18	38.16
5	31.36	31.65
6	16.21	16.30
7	16.69	17.05

In most cases examined, the rules give predicted shifts that are within 0.3 ppm of the observed values. More importantly, the rules give accurate relative shifts for related structures (e.g., stereosequences and regiosequences). The use of these rules therefore permits detailed assignments of ^{13}C NMR spectra.

Spectral Simulation

In an earlier work,[21] we proposed a spectral simulation approach for the interpretation of vinyl and vinylidene copolymers. Although it is desirable to superimpose the empirical rules obtained here to the spectral simulation approach, much work is needed to accomplish this. A major consideration is the complexity of the copolymerization.

The propylene monomer can add either tail-first onto the polymer chains (secondary insertion), or head-first (primary insertion). We shall designate the propylene undergoing secondary insertion as P, and the propylene undergoing primary insertion as Q. Inclusion of tacticity then gives the following five comonomers: ethylene (E), meso-secondary propylene (P_m), racemic-secondary propylene (P_r), meso-primary propylene (Q_m), and racemic-primary propylene (Q_r). In the first-order Markov process, the addition of a monomer to a growing polymer chain depends on the identity of the last unit of the polymer chain, thus,

--E* + E ---> --EE* k_{11}

--E* + P_m ---> --EP_m* k_{12}

--E* + P_r ---> --EE_r* k_{13}

--E* + Q_m ---> --EQ_m* k_{14}

--E* + Q_r ---> --EQ_r* k_{15}

The asterisk indicates the position of the metal atom of the catalyst complexes. Twenty more equations can be written for the other cases. The reactivity ratios can be defined as $r_{ij} = k_{ii}/k_{ij}$, where i,j = E, P_m, P_r, Q_m, or Q_r (identified as 1,2,3,4,5 respectively). The reaction probabilities can be related to the reactivity ratios and the comonomer feed concentrations. Typical equations are given below:

$$P_{ij} = \frac{k_{ij}[j]}{k_{i1}[E] + k_{i2}[P_m] + k_{i3}[P_r] + k_{i4}[Q_m] + k_{i5}[Q_r]}$$

where the notation P_{ij} refers to the probability of a polymer chain terminating in unit i adding to monomer j.

When the molecular weight is low, the likelihood of chains being ter-

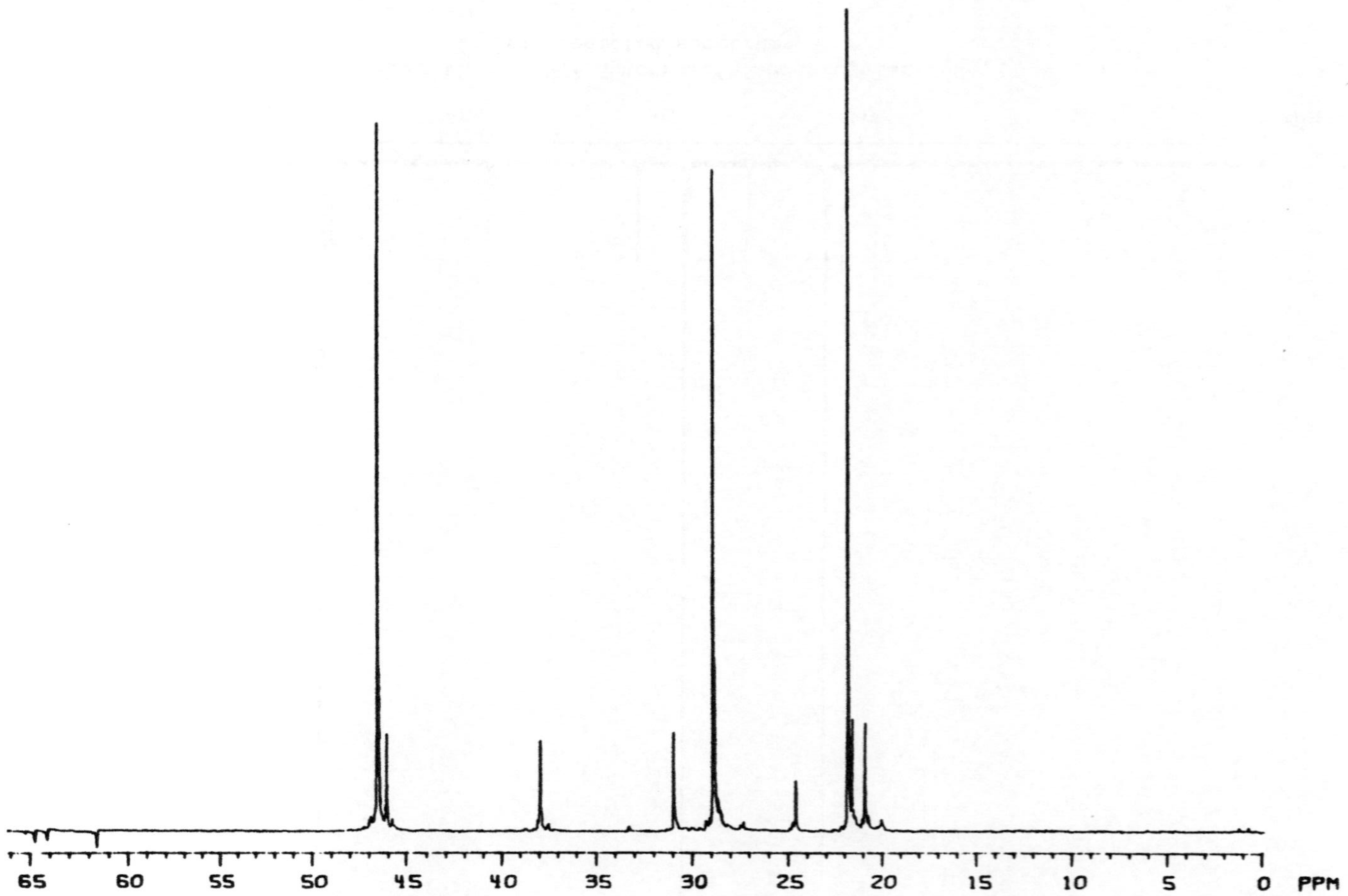

Fig. 1. ^{13}C NMR Spectra of Stereoregular C_2C_3; (a) observed

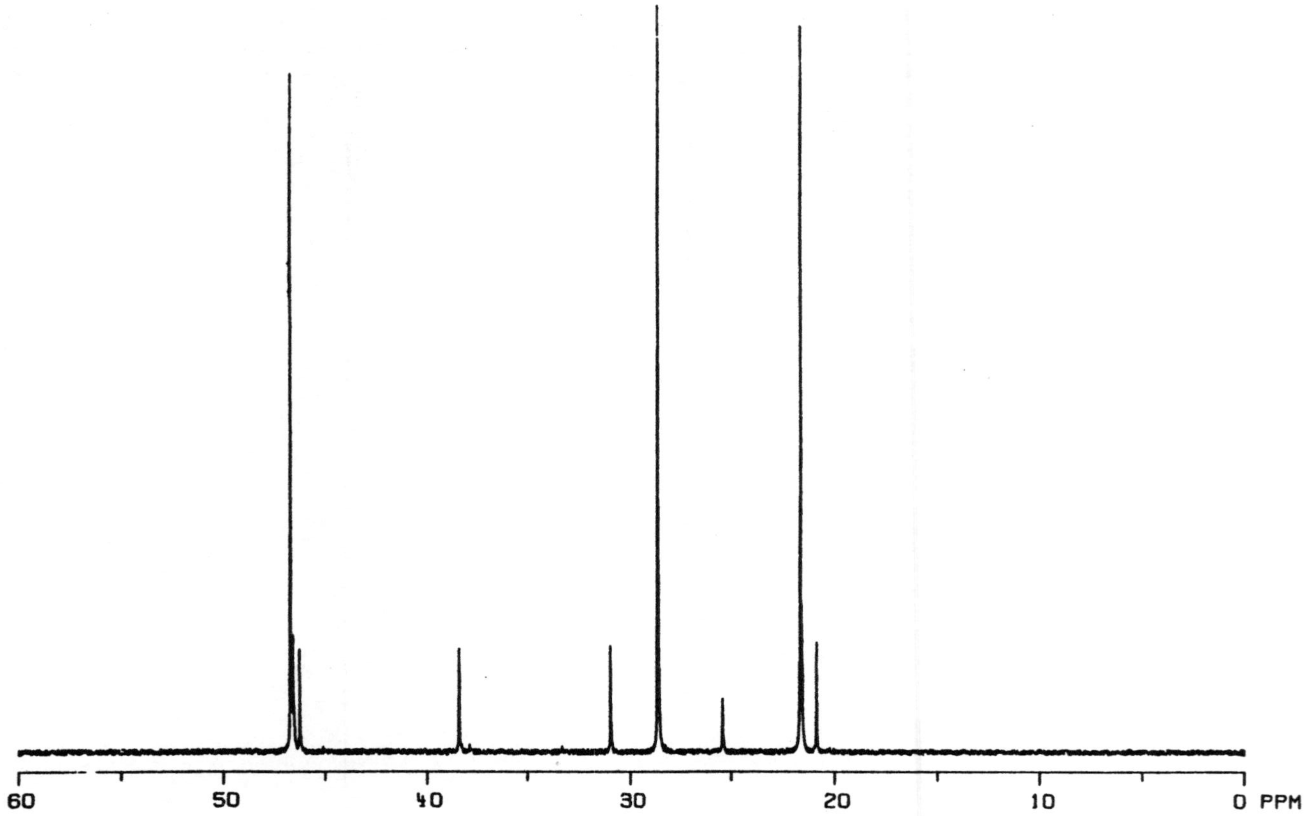

Fig. 1. ^{13}C NMR Spectra of Stereoregular C_2C_3;
(b) predicted spectrum

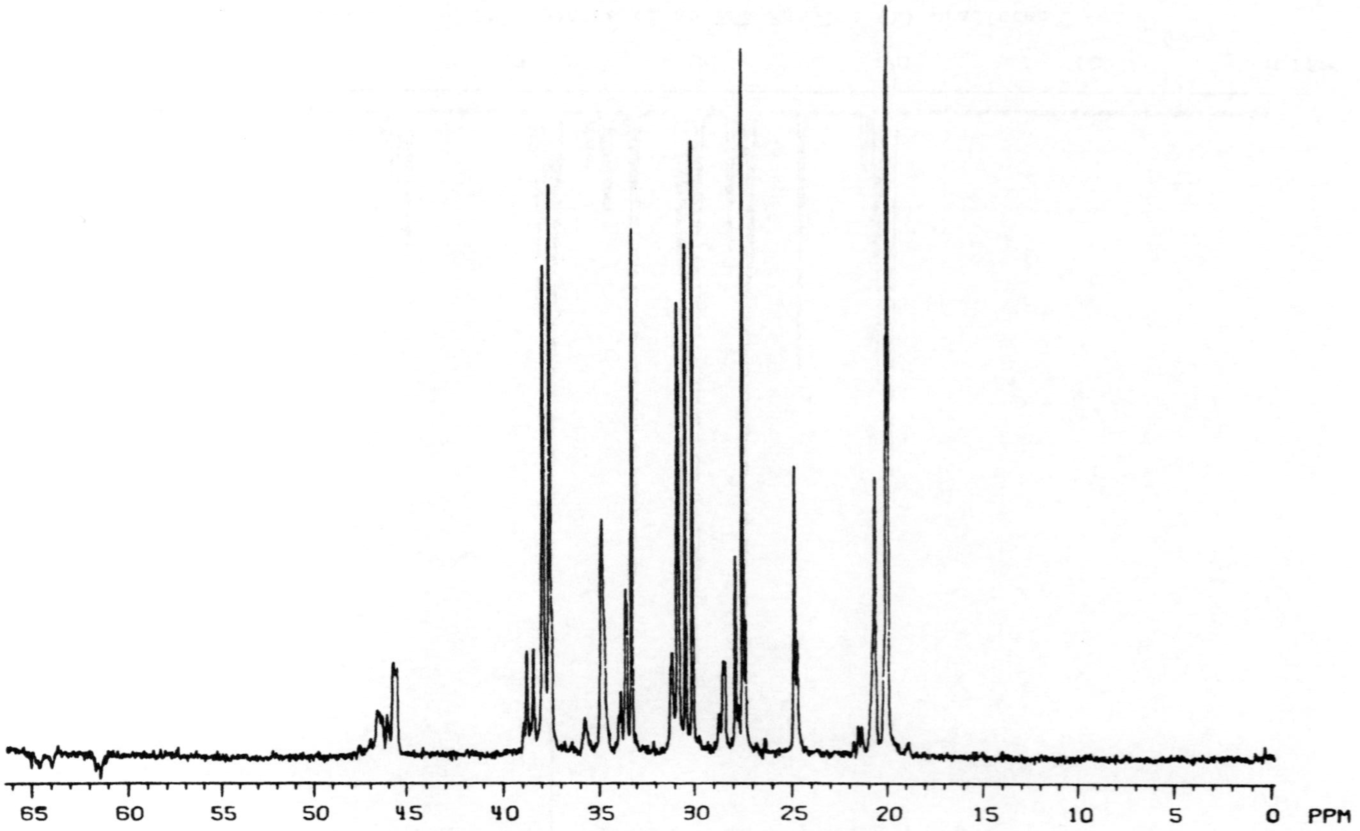

Fig. 2. ^{13}C NMR Spectra of an EPR Sample; (a) observed

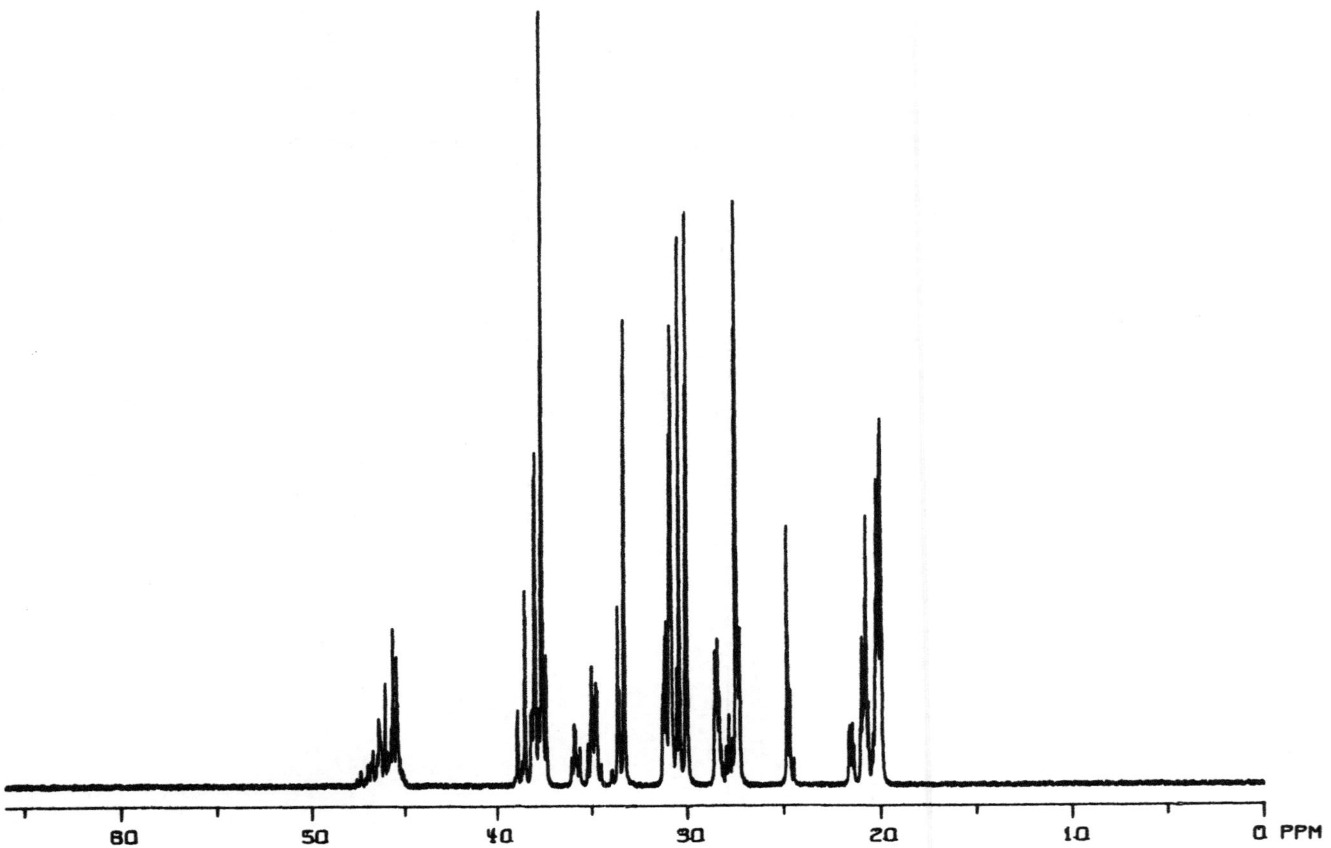

Fig. 2. ^{13}C NMR Spectra of an EPR Sample; (b) predicted spectrum

minated is high. Heuristically, such a system can be regarded as a 6-component copolymer, with the termination step as the 6th equation:

$$--E^* + X --> --EX \qquad k_{16}$$

with similar equations for k_{i6} (i = 2, 3, 4, 5). The group X may be hydrogen, carbon monoxide, water, methanol or other chain transfer or termination species.

The use of the spectral simulation approach can produce predicted (simulated) spectra of any propylene homopolymers or ethylene-propylene copolymers of variable molecular weights. The following example can be shown to illustrate this approach.

The ^{13}C NMR spectrum of a stereoregular ethylene-propylene copolymer is given in Figure 1a. This spectrum can be readily simulated (Figure 1b). This particular copolymer appears to give a reasonable fit with a first-order Markovian statistics, with P_{41} = 0.055 and P_{14} = 0.955.

In Figure 2a is shown the ^{13}C NMR spectrum of an ethylene-propylene rubber. Compared to Figure 1a, the spectrum is considerably more complex. After some experimentation, the following first-order Markovian reaction probabilities are found to adequately describe the system:

P_{14} = 0.186	P_{21} = 0.310	P_{31} = 0.310	P_{41} = 0.608	P_{51} = 0.608
P_{15} = 0.344	P_{23} = 0.448	P_{32} = 0.242	P_{42} = 0.047	P_{52} = 0.047
P_{12} = 0.002	P_{24} = 0	P_{34} = 0	P_{43} = 0.087	P_{53} = 0.087
P_{13} = 0.004	P_{25} = 0	P_{35} = 0	P_{45} = 0.168	P_{54} = 0.090

The simulated spectrum is given in Figure 2b. The good agreement between the two spectra (Figures 2a and b) is a measure of the essential correctness of the present approach. The reaction possibilities thus obtained can then be used to determine copolymer composition, comonomer sequence distribution, and information on stereoregularity and regiospecificity.

CONCLUSION

In this work, we have shown how the ^{13}C NMR shifts of ethylene/propylene polymers can be reduced to several additive rules. The rules give accurate ^{13}C shifts for stereoisomers and regioisomers, and are applicable to polymers and low-molecular-weight model compounds. With these rules, one can make detailed assignments of ethylene/propylene polymer spectra, as well as predict the spectral features of any ethylene/propylene homopolymers or copolymers that have not yet been synthesized.

REFERENCES

1. F. A. Bovey, Polymer Preprints, 10 (1), 9 (1969).
2. A. Zambelli, P. Locatelli, G. Bajo, and F. A. Bovey, Macromolecules, 8, 687 (1978).
3. A. Zambelli, P. Locatelli, A. Provasoli, and D. R. Ferro, Macromolecules, 13, 267 (1980).
4. F. C. Schilling and A. E. Tonelli, Macromolecules, 13, 270 (1980).
5. H. N. Cheng and G. H. Lee, Polymer Bull., 13, 549 (1985).

6. T. Asakura, I. Ando, A. Nishioka, Y. Doi, and T. Keii, Makromol. Chem., 178, 791 (1977).
7. Y. Doi, Macromolecules, 12, 248 (1979).
8. Y. Doi, Makromol. Chem., 180, 2447 (1979).
9. H. N. Cheng, Polymer Bull., in press.
10. G. J. Ray, P. E. Johnson, and J. R. Knox, Macromolecules, 10, 773 (1977).
11. J. M. Sanders and R. A. Komoroski, Macromolecules, 10, 1214 (1977).
12. J. R. R. Paxson and J. C. Randall, Anal. Chem., 50, 1977 (1978).
13. E. G. Brame, Jr. and H. E. Holmquist, Rubber Chem. Technol., 52, 1 (1979).
14. H. N. Cheng, Anal. Chem., 54, 1828 (1982).
15. C. J. Carman, R. A. Harrington, and C. E. Wilkes, Macromolecules, 10, 536 (1977).
16. J. C. Randall, Macromolecules, 11, 33 (1978).
17. W. V. Smith, J. Polymer Sci., Polymer Phys., Ed., 18, 1587 (1980).
18. H. N. Cheng, Macromolecules, 17, 1950 (1984).
19. H. N. Cheng and D. A. Smith, Paper presented at the National ACS meeting in Washington, D.C., 1983; manuscript in preparation.
20. H. N. Cheng and S. J. Ellingsen, J. Chem. Inf. Computer Sci., 23, 197 (1983).
21. H. N. Cheng and M. A. Bennett, Anal. Chem., 56, 2320 (1984).
22. A. Zambelli and G. Gatti, Macromolecules, 11, 485 (1978).
23. A. Zambelli, G. Bajo, and E. Rigamonti, Makromol. Chem., 179, 1249 (1978).
24. A. Zambelli, P. Locatelli, and G. Bajo, Macromolecules, 12, 154 (1979).
25. A. Zambelli, M. C. Sacchi, and P. Locatelli, Macromolecules, 12, 783 (1979).
26. A. E. Tonelli, Macromolecules, 11, 634 (1978).
27. F. C. Schilling and A. E. Tonelli, Accts. Chem. Res., 14, 233 (1981), and references cited therein.
28. P. Freche, M.-F. Grenier-Loustalot, F. Metras, and A. Gascoin, Makromol. Chem., 184, 569 (1983).
29. M. Moeller and H.-J. Cantor, Macromolecules, 17, 733 (1984).

AMBIENT-TEMPERATURE AND IR LASER-INDUCED CHEMILUMINESCENCE FROM γ -IRRADIATED POLYPROPYLENE

G. D. Mendenhall and H. Byun

Department of Chemistry and Chemical Engineering
Michigan Technological University
Houghton, Michigan 49931

J. M. Cooke

Himont U.S.A., Inc.
800 Greenbank Road
Wilmington, Delaware 19808

INTRODUCTION

For some years we and other workers[1-3] have attempted to find ways to apply chemiluminescence to the study of polymeric materials. For certain substrates such as polypropylene, there is a fairly good correspondence between the rate of oxidation and the chemiluminescence emission, and this correspondence has been used to rank antioxidants. Most of this work has been carried out with samples heated at 100-150°C. While the results are relevant to oxidation of the polymer during melt extrusion or other high-temperature processing, or for comparison with oven aging of stabilized polymers at those temperatures, the questionable validity of any subsequent extrapolation of test data to ambient temperature is generally acknowledged.

A year ago we reported[1] that samples of polypropylene, stabilized with various combinations of phenols and thioethers and irradiated with electron beams, showed striking differences in stability toward impact strength when aged at 60°C compared with samples stored at ambient temperatures. The chemiluminescence from the samples, although measured in somewhat different ways, could be correlated qualitatively with the impact strengths of polymers stored at both ambient and elevated temperatures.

In the present study we wish to report two approaches toward evaluating the γ -induced oxidation of polypropylene. First, we have directly measured the chemiluminescence from samples at ambient temperature after irradiation, and we have studied the enhancement of this emission by added fluorescers. Second, we have used an infrared laser to heat small polymer samples in an evacuated tube to very high temperatures, in order to obtain chemiluminescence from peroxidic groups which have accumulated in the sample as a result of the oxidation process. The former approach is nondestructive and is a qualitative measure of the oxidation process in real time. The latter one is destructive and is a qualitative measure of a peroxide value.

EXPERIMENTAL

Polypropylene powder from Hercules, Inc. was extracted with refluxing isopropanol or dichloromethane to remove stabilizers, and then was coevaporated with a solution of Irganox 1010 (CIBA-GEIGY Corporation) in methylene chloride. 9,10-Dibromoanthracene (Aldrich, Inc., DBA) was recrystallized from ethanol and coevaporated onto the stabilized polypropylene in the same way.

The γ-irradiations were carried out under ambient conditions with a ^{60}Co source at a dose rate of 0.0057 MR/hr. The source was calibrated with ferrous sulfate solution as described.[4] The irradiated powder was stored at ambient temperature in the dark, and examined by placing a glass vial containing the powder directly in the well of a chemiluminescence apparatus with an automated counting function (Turner Designs, Inc., Mountain View, CA, Model 20e Luminometer).

The apparatus for IR heating of the polypropylene was constructed from a standard 1 3/4" steel pipe cross with an extra port for the laser beam drilled into the center perpendicular to the others (Figure 1).

The laser beam passed through a 3/4" germanium flat (Janos Optical Co., antireflection coated) held in place with two O-rings and a threaded aluminum cap. The fiber-optic cable (Dolan-Jenner Model BXT 424, 1/4" fiber bundle) was screwed into a threaded adapter in one of the threaded ports. The other end of the fiber optic cable was attached through an ordinary rubber stopper to the sample well of the Turner luminometer. A small, blue color filter (CS 4-72) was taped to the bottom of the stopper to eliminate black-body emission at the red end of the spectrum.

A plug in the top port was drilled to accommodate a threaded steel cylinder with a 1/16" hole into which a capillary tube was inserted with the sample. The tube was held in place with clay in the expanded part of the hole in the cylinder. The viewing port was closed with a conventional pipe cap before carrying out an experiment.

Samples of polypropylene powder were placed in weighed capillary tubes, and tapped against a hard surface to pack the material at the sealed end. The weight of the sample was determined to ± 0.1 mg, and the tube was then evacuated (oil pump) and sealed off with a small flame.

A temperature stabilized, 5w CO_2 laser (Laakmann Electro-optics, Inc. Model RF 22) emitting in the 0-0 mode at 10.6 μ, was placed 2' from the apparatus in order to broaden the beam slightly, since preliminary experiments showed that the beam caused localized melting of the glass at closer positions.

The experiments were carried out by positioning the capillary tube so that the laser would intersect it initially above the polymer sample. After taking background readings with the laser off, it was allowed to heat the capillary while slowly drawing it through the beam during the third minute of observation (Table 1).

Determination of POV in polymer samples was carried out spectrophotometrically after a 60 minute reflux period as described.[5]

Analysis of samples for Irganox 1010 content was carried out on extracts by liquid chromatography with a 25 x 4.6 cm column of 5 μ alumina

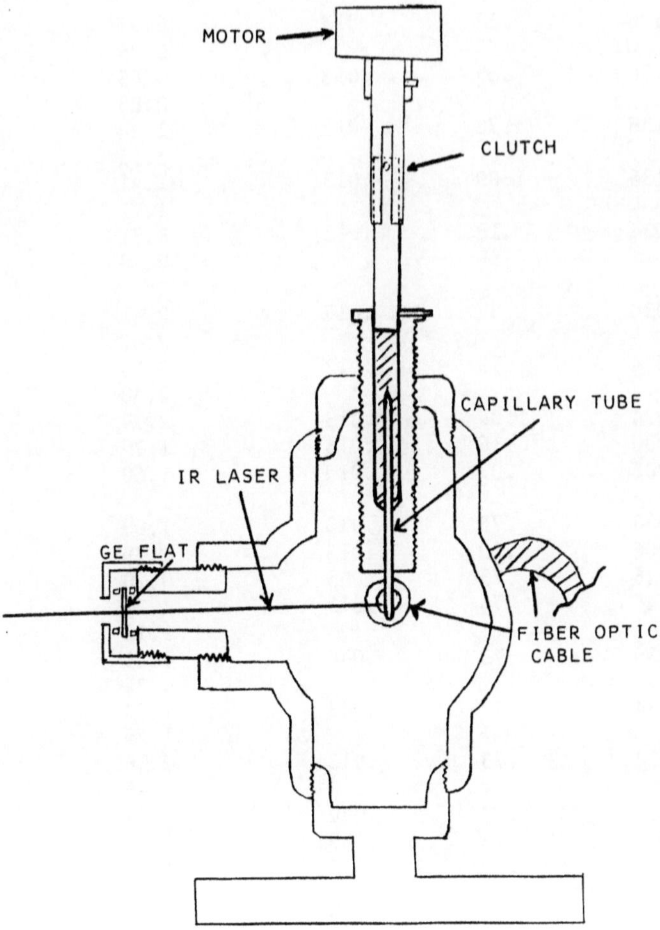

Fig. 1. Apparatus for IR Laser Heating of Polypropylene

Table 1. First-Order Decay of Chemiluminescence at 22°C from Irradiated Polypropylene Samples

Sample	Dose (MR)	DBA[a] (%)	IRG[b] (%)	$10^4 k_{obs}$, sec.$^{-1}$	I_o[c] (cpm)
1	.136	.18	.013	0.97	178.0
				0.85	173.0
2	.136	.37	.013	0.85	116.0
				0.85	120.0
3	.136	.73	.013	1.24	106.0
				1.13	102.0
4	.136	1.09	.013	1.41	82.0
				1.21	82.3
5	.034	.18	.013	2.31	157.0
				2.25	152.0
6	.034	.37	.013	2.61	109.0
				2.47	108.0
7	.034	.73	.013	2.56	74.8
				2.48	75.9
8	.034	1.09	.013	2.77	47.2
9	.006	.18	.013	1.77	38.3
10	.006	.37	.013	1.68	24.8
11	.006	.73	.013	1.98	22.1
12	.006	1.09	.013	2.39	17.5
13	.136	.73	.005	0.20	198.0
14	.136	.73	.013	1.24	106.0
				1.13	102.0
15	.136	.73	.040	1.21	105.0
				0.32	210.0
16	.034	.73	.005	1.24	123.0
				1.25	128.0
17	.034	.73	.013	2.56	74.8
				2.48	75.9
18	.034	.73	.040	2.57	67.4
19	.006	.73	.005	0.66	40.7
20	.006	.73	.013	1.98	22.1
21	.006	.73	.040	3.32	21.4

[a]9,10-Dibromoanthracene

[b]Irganox 1010 [tetrakis [methylene-3-(3',5'-di-tert-butyl-4'-hydroxphenyl)propionate]methane]

[c]Intercept of semilogorithmic plot of decay curve, in counts/minute.

at 40°C, 50μL injection volume, 280nm detection, and gradiant elution from 100% CH_2Cl_2, 0.3mL/m, at 0m to 4% MeOH in CH_2Cl_2, 3mL/m, at 25m. The retention time was 13.1m.

RESULTS

Ambient-Temperature Chemiluminescence

In the absence of phenolic stabilizer, γ-irradiated PP showed chemiluminescence with a hyperbolic decay curve, so that a double-log plot of emission rate versus time was linear (Figure 2). In the presence of the phenol, the decay curves became exponential with a half-time for decay that depended on dosage, stabilizer concentration, and to a lesser extent the DBA concentration (Table 1). Representative plots of data appear in Figures 3 and 4, and plots of the first-order decay versus radiation dose and concentrations of additives appear in Figures 5-8. The effect of increasing DBA concentration in a given case is to decrease the initial chemiluminescence intensity, but with a small effect on the slope (i.e. half-life) of the decay. It should be noted, however, that the amounts of DBA in the samples are much higher than are usual for enhancement of chemiluminescence. The values of the first-order rate constants for decay of chemiluminescence after irradiation do not change in a uniform way with antioxidant concentration or radiation dose (Table 1, Figures 5-8).

In Figures 4 and 8, the value of k_{obs} is seen to be identical for the samples with 0.013 and 0.040% stabilizer at the two higher radiation doses. Since the concentrations of stabilizer were small, a saturation effect of this sort was not expected. The result suggested that the antioxidant was largely confined to the sample surface. The following experiments confirmed this hypothesis. A sample of polypropylene containing 0.73% DBA and 0.040% Irganox 1010, prepared in the usual way, was melted above 160°C in a nitrogen atmosphere, cooled to room temperature, ground to a powder under liquid nitrogen in a clean mortar and pestle, and irradiated to 0.136 mrads as before. The chemiluminescence decay followed a power function (Figure 9) just as with an extracted sample with no antioxidant after irradiation (Figure 2). When a sample of polypropylene was prepared in the same manner with 0.40% Irganox 1010, however, a tenfold increase, the decay of chemiluminescence was once again exponential with a half-life of 150 minutes (Figure 10).

Peroxide Value

Plots of POV versus time for γ-irradiated samples appear in Figures 11-16, and the data are also listed in Table 2. The curves all increase linearly with time during the two weeks' examination, although the total increases are not very large. The presence of DBA reduces the rate of increase of POV, as does the phenolic antioxidant at a lower concentration range. The initial POV after irradiation is inversely proportional to the amount of additive. The rate of increase of POV in the samples is greatest with the smallest amount of DBA or phenol, while the other curves in each figure, with the larger amounts of additives, have slopes that are respectively similar.

The POV of polypropylene samples five days after irradiation, is plotted against their respective radiation doses in Figures 17 and 18. At each level of irradiation, the POV decreased monotonically with increasing concentrations of additives.

Table 2. Peroxide and Laser-Induced Chemiluminescence from Irradiated Polypropylene

Sample	Dose(MR)	DBA[a](%)	IRG[b](%)	CL[e](counts)	CL/POV[f]x10[13]	POV[e,g]
1	.006	.18	.013	24±1	0.46	0.80±0.2
2	.034			67±1	1.03	0.99±0.2
3	.136			242±13	2.27	1.63±0.3
$a^c = 1687$, $a'^d = 6.36 \times 10^{-3}$, $a/a' = 2.65 \times 10^{-5}$						
4	.006	.37	.013	36±1	0.72	0.77±0.7
5	.034			83±4	1.41	0.90±0.3
6	.136			249±10	2.76	1.38±0.4
$a = 1635$, $a' = 4.70 \times 10^{-3}$, $a/a' = 3.48 \times 10^{-5}$						
7	.006	.73	.013	45±10	0.94	0.73±0.1
8	.034			104±9	1.94	0.82±0.4
9	.136			339±5	4.35	1.19±0.2
$a = 2281$, $a' = 3.56 \times 10^{-3}$, $a/a' = 6.41 \times 10^{-5}$						
10	.006	1.09	.013	73±8	2.94	0.38±0.4
11	.034			147±1	4.16	0.54±0.6
12	.136			419±3	5.67	1.13±0.2
$a = 2663$, $a' = 5.77 \times 10^{-3}$, $a/a' = 4.62 \times 10^{-5}$						
13	.006	.73	.005	56±1	0.69	1.24±0.2
14				97±12	1.08	1.37±0.3
15	.136			196±3	1.61	1.86±0.5
$a = 1048$, $a' = 4.78 \times 10^{-3}$, $a/a' = 2.19 \times 10^{-5}$						
16	.006	.73	.040	76±6	1.81	0.64±0.3
17	.034			103±1	2.25	0.70±0.2
18	.136			224±3	3.36	1.02±0.2
$a = 1151$, $a' = 2.98 \times 10^{-3}$, $a/a' = 3.86 \times 10^{-5}$						

[a]9,10 Dibromoanthracene

[b]Irganox 1010

[c]Slope of CL(counts) vs. dose (Mrad) for samples 1, 2 and 3; 4, 5 and 6; etc.

[d]Slope of POV (moles peroxide/L polypropylene) vs. dose (Mrad) for samples 1,2, and 3, etc.

[e]Chemiluminescence (CL) per milligram of sample, measured 5 days after -irradiation. Average of 3 values.

[f]Einsteins of photons/moles of peroxide.

[g]Peroxide value (millimoles of peroxide/L of polypropylene). Average of 3 values, measured 5 days after irradiation.

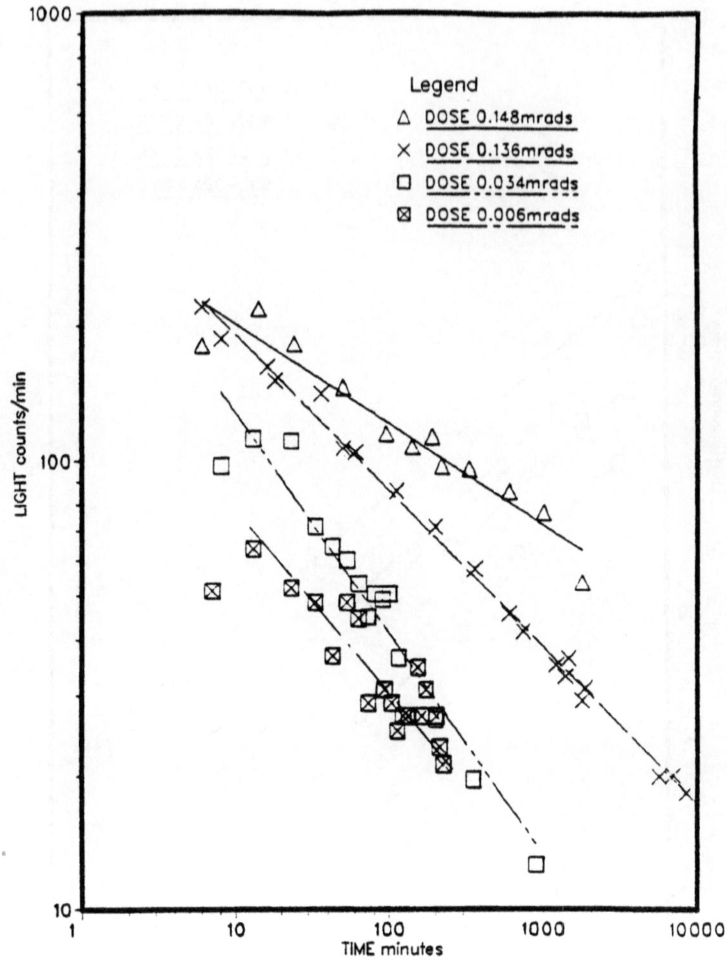

Fig. 2. CL at 22°C from Polypropylene Powder with 9,10-DBA (0.73%) after Irradiation

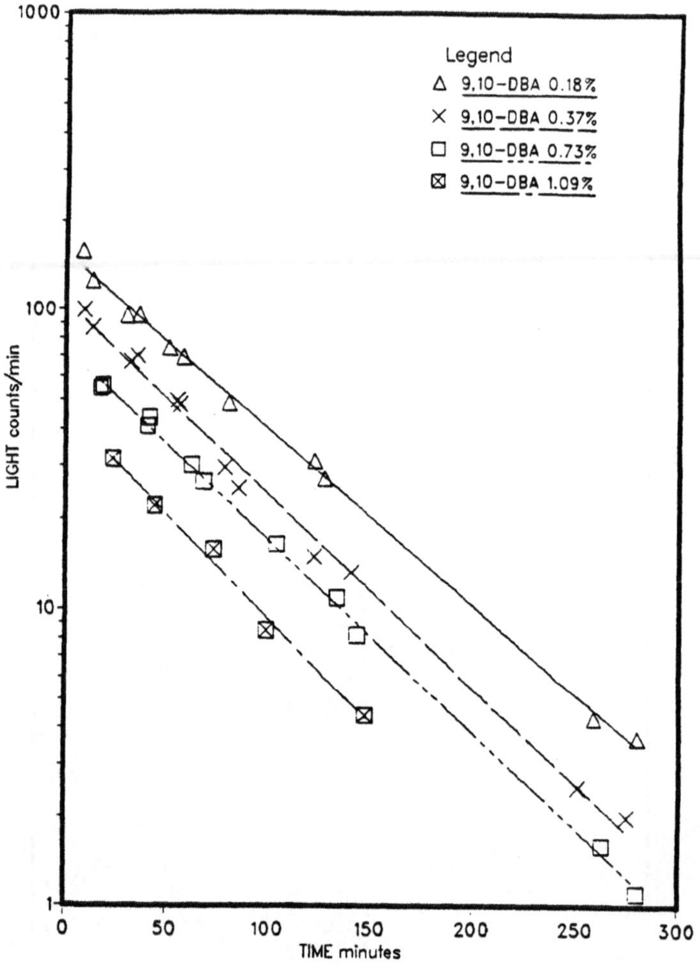

Fig. 3. CL at 22°C from Polypropylene Powder with Irganox 1010 (0.013%) and Various 9,10-DBA Concentrations after Irradiation (0.034 mrad)

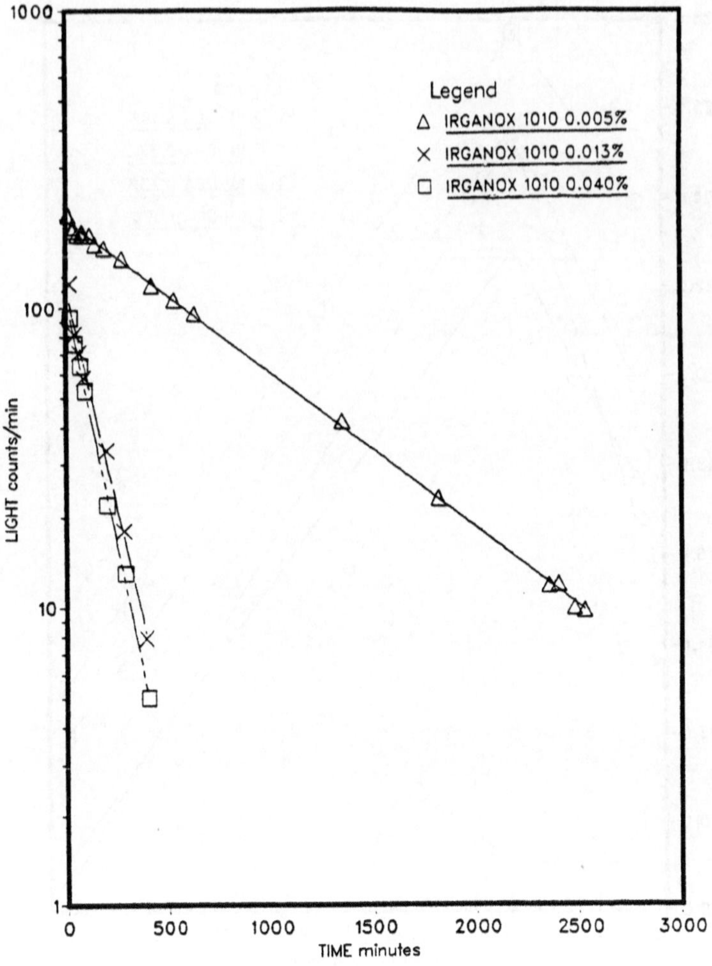

Fig. 4. CL at 22°C of Polypropylene Powder with 9,10-DBA (0.73%) and Various Irganox Concentrations after Irradiation (0.136 mrad)

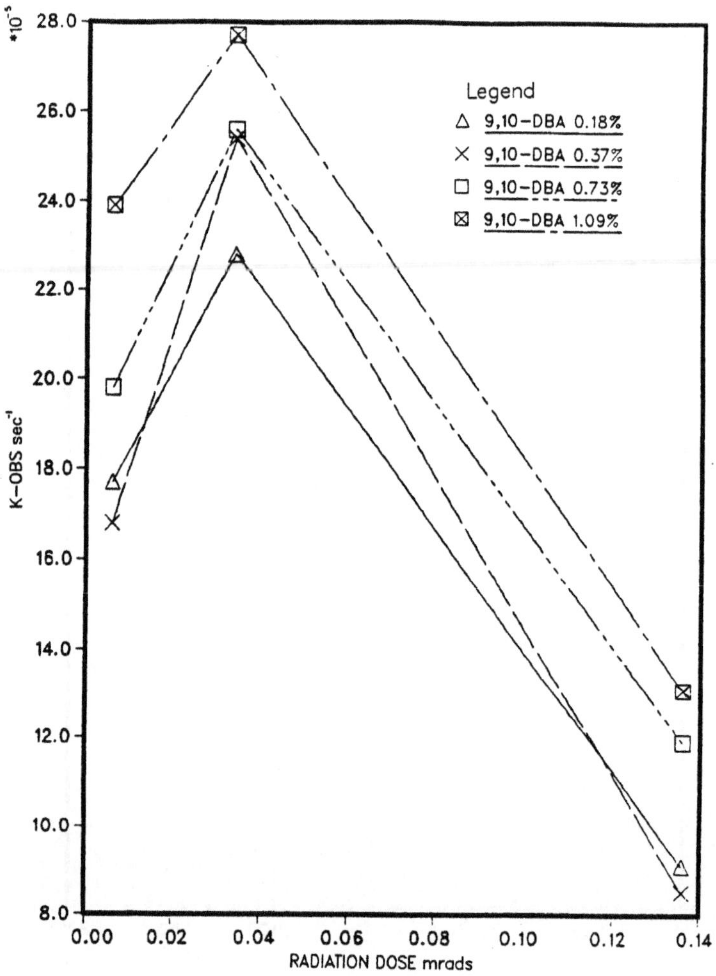

Fig. 5. Values of k_{obs} vs. Radiation Dose for Different Concentrations of DBA with Irganox 1010 (0.013%) in Polypropylene

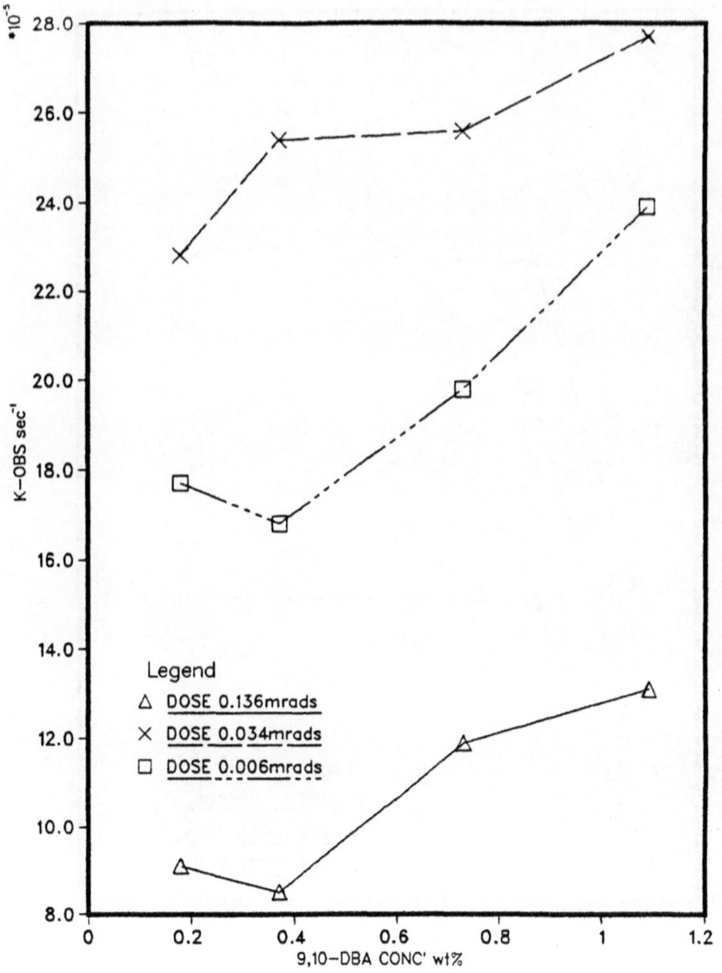

Fig. 6:- Values of k_{obs} vs. Concentration of 9,10-DBA for Different Radiation Doses in Polypropylene with Irganox 1010 (0.040%)

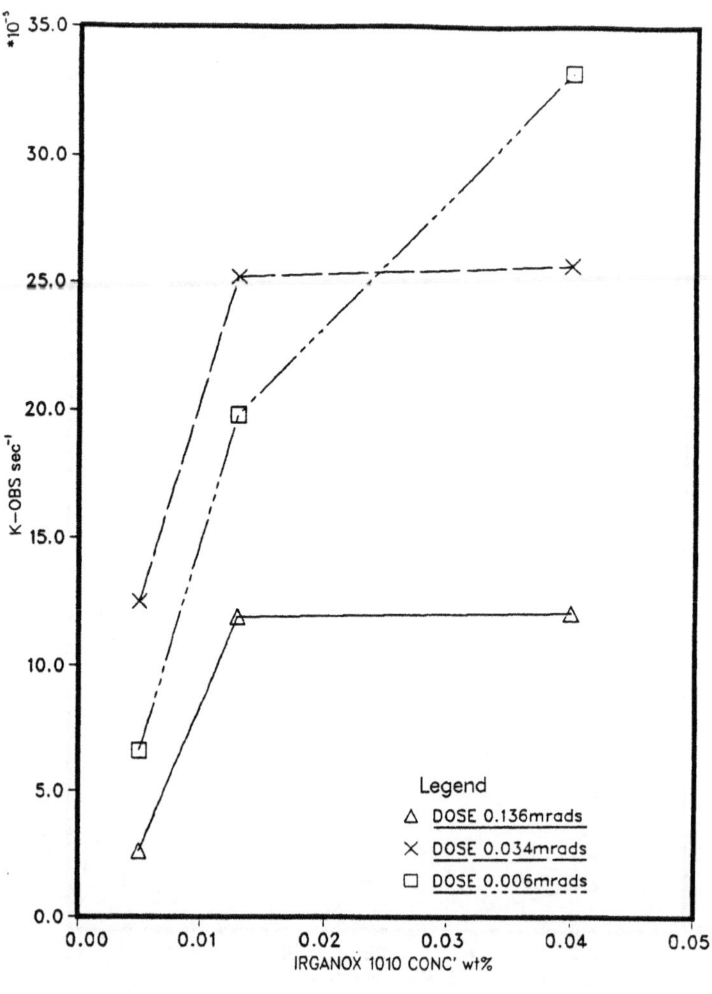

Fig. 7. Values of k_{obs} vs. Irganox 1010 Concentrations for Different Radiation Doses in Polypropylene with 9,10-DBA (0.73%)

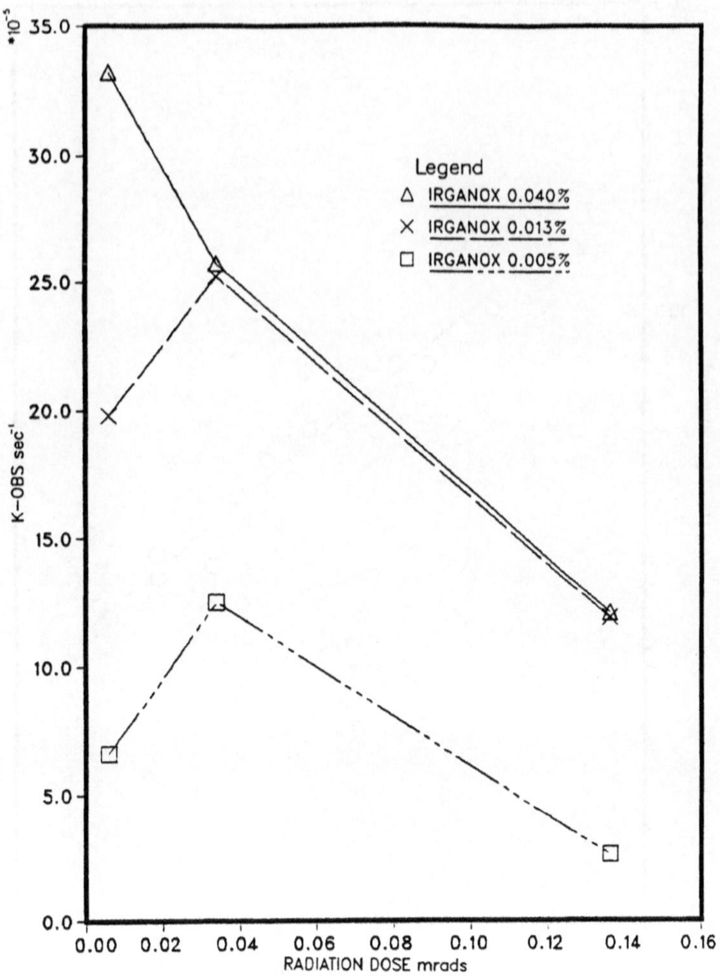

Fig. 8. Values of k_{obs} vs. Radiation Dose for Different Concentrations of Irganox 1010 in Polypropylene with 9,10-DBA (0.73%)

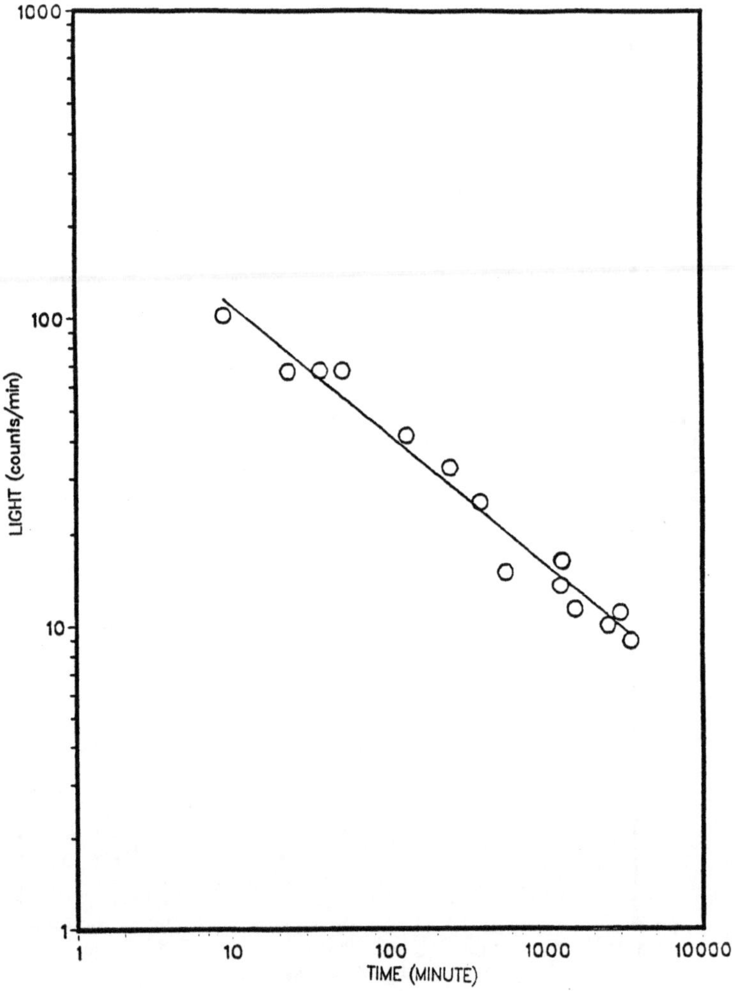

Fig. 9. Chemiluminescence at 22°C after Irradiation (0.136 Mrads) from Polypropylene Powder Melted with 9,10-DBA (0.73%) and Irganox 1010 (0.040%)

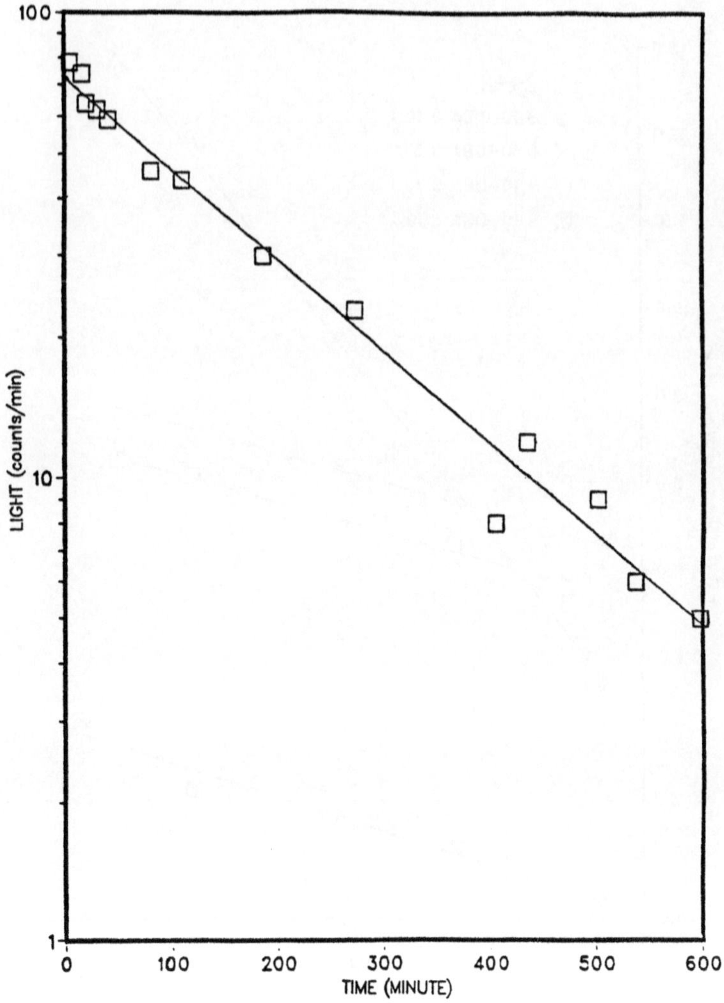

Fig. 10. Chemiluminescence at 22°C after Irradiation (0.136 Mrads) from Polypropylene Powder Melted with 9,10-DBA (0.73%) and Irganox 1010 (0.40%)

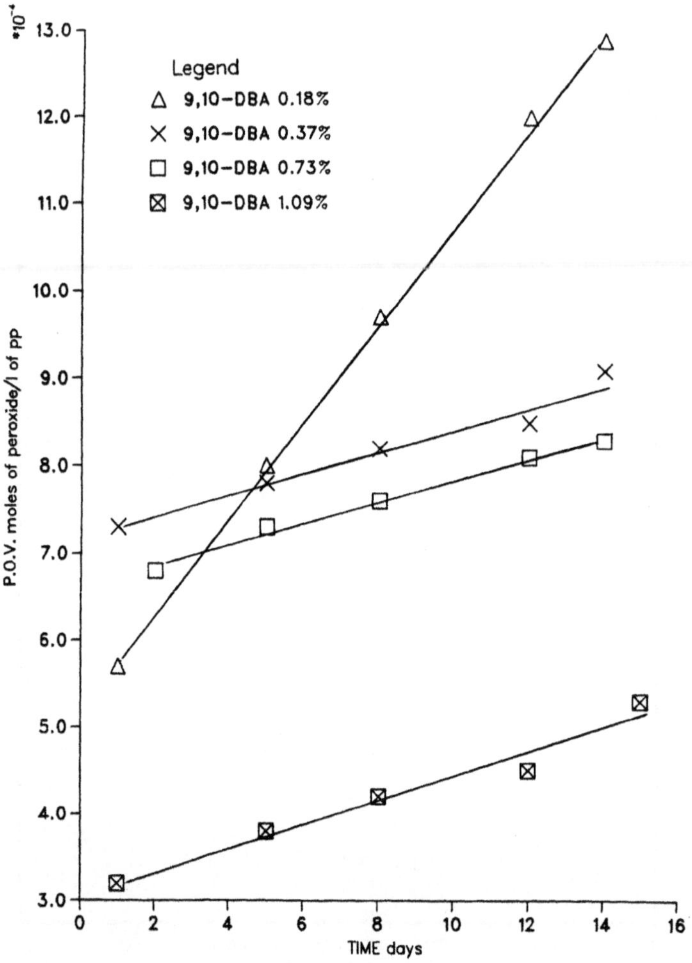

Fig. 11. Effect of 9,10-DBA Concentrations on POV as a Function of Time, in Polypropylene Powder with Irganox 1010 (0.013%) after Irradiation (0.006 mrad)

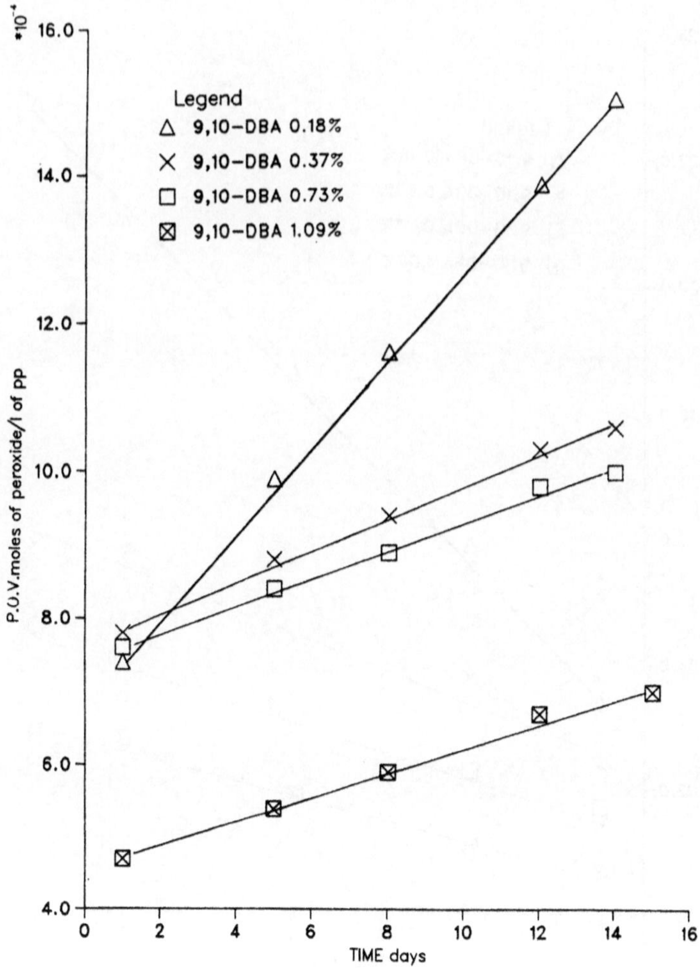

Fig. 12. Effect of 9,10-DBA Concentrations on POV as a Function of Time, in Polypropylene Powder with Irganox 1010 (0.013%) after Irradiation (0.034 mrad)

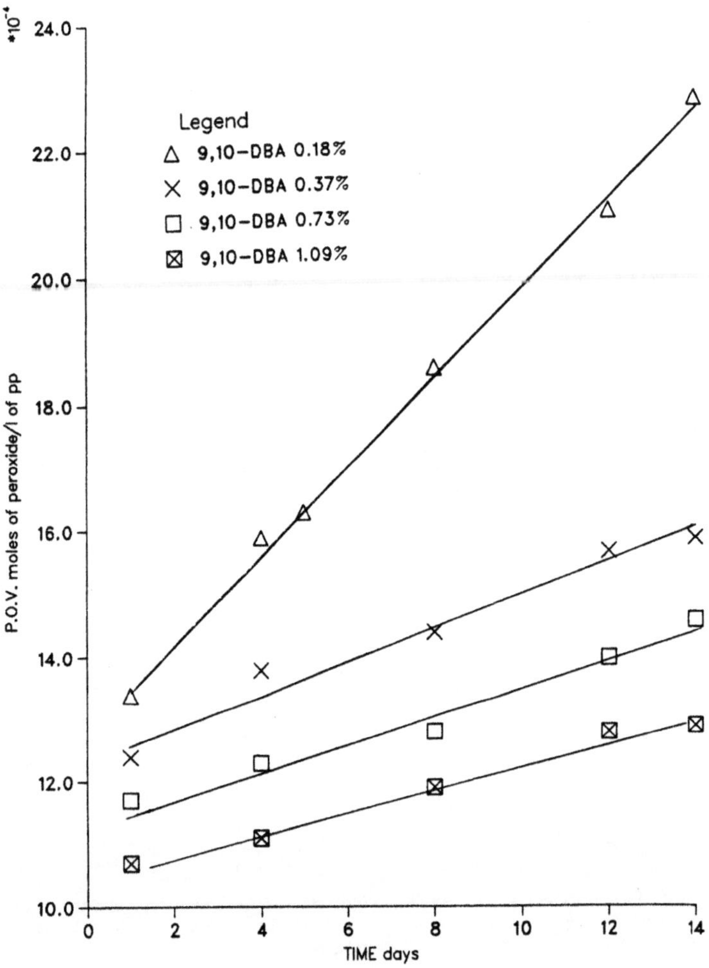

Fig. 13. Effect of 9,10-DBA Concentrations on POV as a Function of Time, in Polypropylene Powder with Irganox 1010 (0.013%) after Irradiation (0.136 mrad)

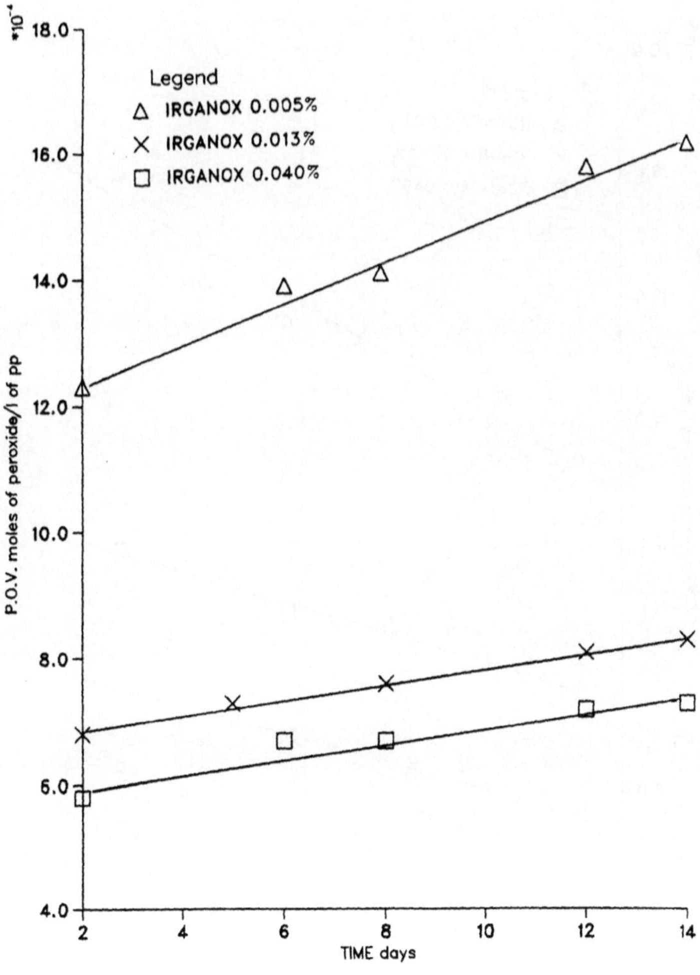

Fig. 14. Effect of Irganox 1010 Concentrations on POV as a Function of Time, in Polypropylene Powder with 9,10-DBA (0.73%) after Irradiation (0.006 mrad)

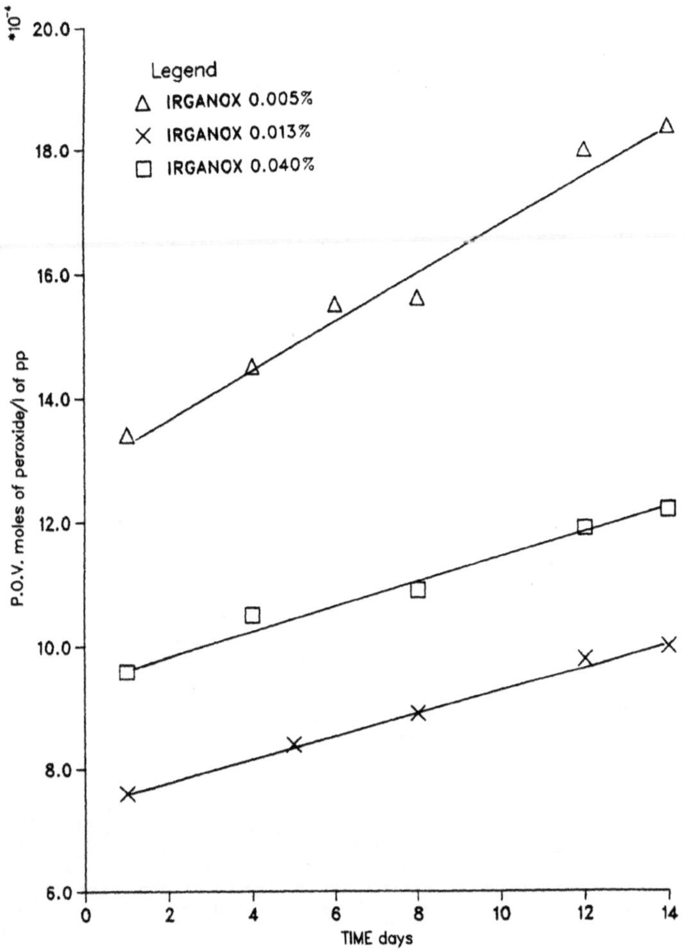

Fig. 15. Effect of Irganox 1010 Concentrations on POV as a Function of Time, in Polypropylene Powder with 9,10-DBA (0.73%) after Irradiation (0.034 mrad)

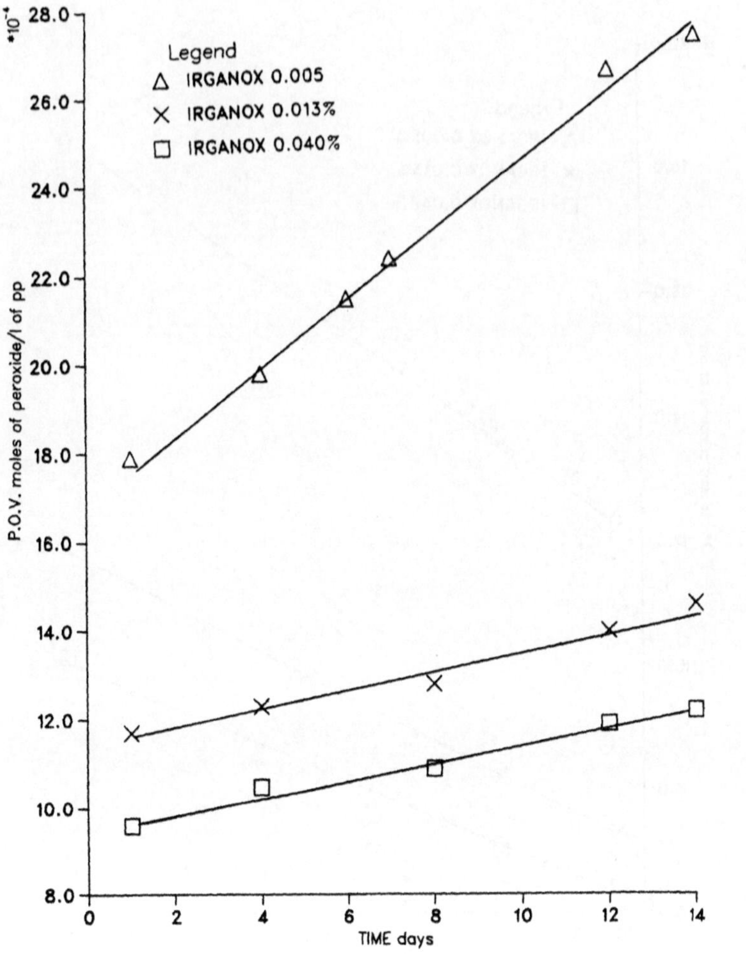

Fig. 16: Effect of Irganox 1010 Concentrations on POV as a Function of Time, in Polypropylene Powder with 9,10-DBA (0.73%) after Irradiation (0.136 mrad)

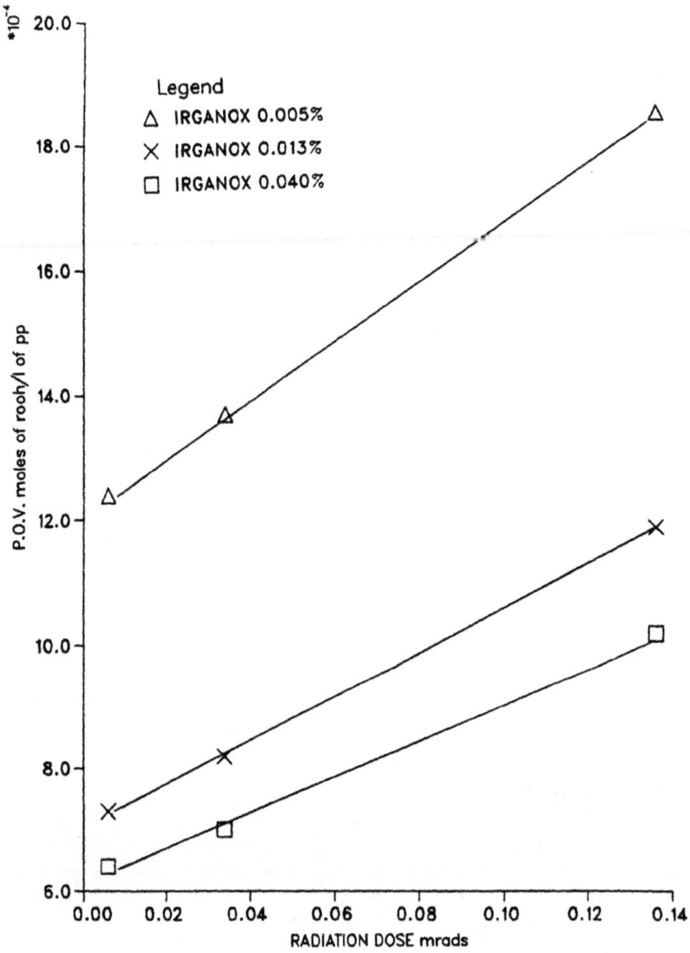

Fig. 17. POV vs. Radiation Dose for Polypropylene Containing 9,10-DBA (0.73%) and Different Irganox 1010 Concentrations

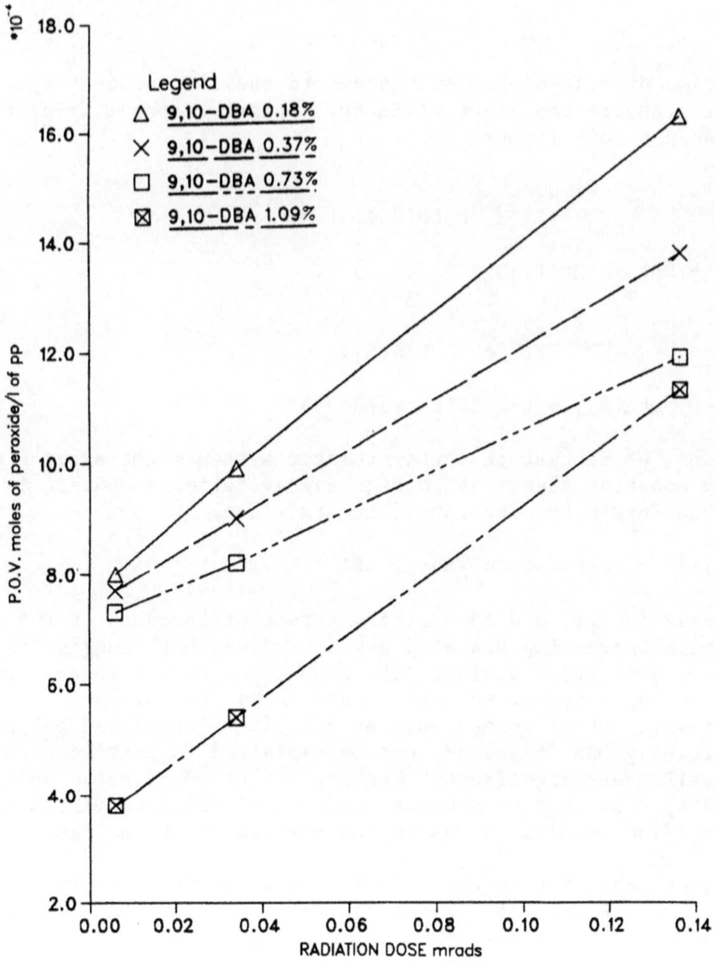

Fig. 18. POV vs. Radiation Dose for Polypropylene Containing Irganox 1010 (0.013%) and Different Concentrations of 9,10-DBA

IR Laser-Induced Chemiluminescence

Normalized values of the photons emitted in the laser experiments appear in Table 2. Plots of the counts are given in Figures 19 and 20 as a function of dose, for differing DBA concentrations with constant phenolic stabilizer, and for differing amounts of phenol with constant DBA. Unlike the emission at 22°C after irradiation, the counts increase with increasing DBA. At each additive concentration, the chemiluminescence increase with dose appears to be nearly linear.

DBA

The addition of 9,10-dibromoanthracene to chemiluminescent systems is a common way to enhance the light yield by trapping carbonyl triplets, which otherwise not emit light:

$$R_2CO(T_1) \xrightarrow{\text{i.s.c.}} R_2CO(S_o) \qquad (1)$$

$$R_2CO(T_1) + DBA \rightarrow DBA(T_2) + R_2CO(S_o) \qquad (2)$$

$$DBA(T_2) \xrightarrow{\text{i.s.c.}} DBA(S_1) \qquad (3)$$

$$CBA(S_1) \rightarrow DBA(S_o) + h\nu \text{ (fluorescence)} \qquad (4)$$

In addition, we thought that the aromatic system might exert a protective effect by donating electrons to high-energy cationic centers in the polymer that are formed by electron-γ interactions:

$$[\text{Polymer}]\cdot^{+} + DBA \rightarrow \text{Polymer} + DBA\cdot^{+} \qquad (5)$$

The decrease in POV, and the decreased rate of increase of POV after irradiation, with increasing DBA at a given γ-dose, both suggest the presence of some protective action. The amounts of DBA, however, are so high that all of the material was not dissolved in the polymer. The observation of decreasing chemiluminescence at 22°C from irradiated polypropylene with increasing DBA (Figure 3) can be explained in part in terms of a screening or self-quenching effect. Examination of DBA-treated polypropylene, prepared by the standard methods, under a fluoresence microscope revealed crystalline needles of DBA on the surface of the polymer.

On the other hand, the increase in IR laser-induced chemiluminescence with increasing DBA (Figure 19) indicates that the light-enhancing action of the fluorescer outweighs self-absorption. This is a reasonable result, since the laser melts the polymer sample (i.e. T > 160°C). The DBA will be more soluble in the polymer at the elevated temperatures, and the increased diffusional rate of molecular species and segments in the melt should further increase the efficiency of excited state trapping.

A survey of various samples of polypropylene was made with the laser technique (Figure 21), which showed that the enhancement in chemiluminescence by DBA was actually quite modest, and that the aromatic hydrocarbon perylene displayed the largest enhancement of any fluorescer tested. We were also intrigued by the large amount of chemiluminescence from polypropylene in the presence of triethylene diamine, which is usually regarded as a quencher for excited states.

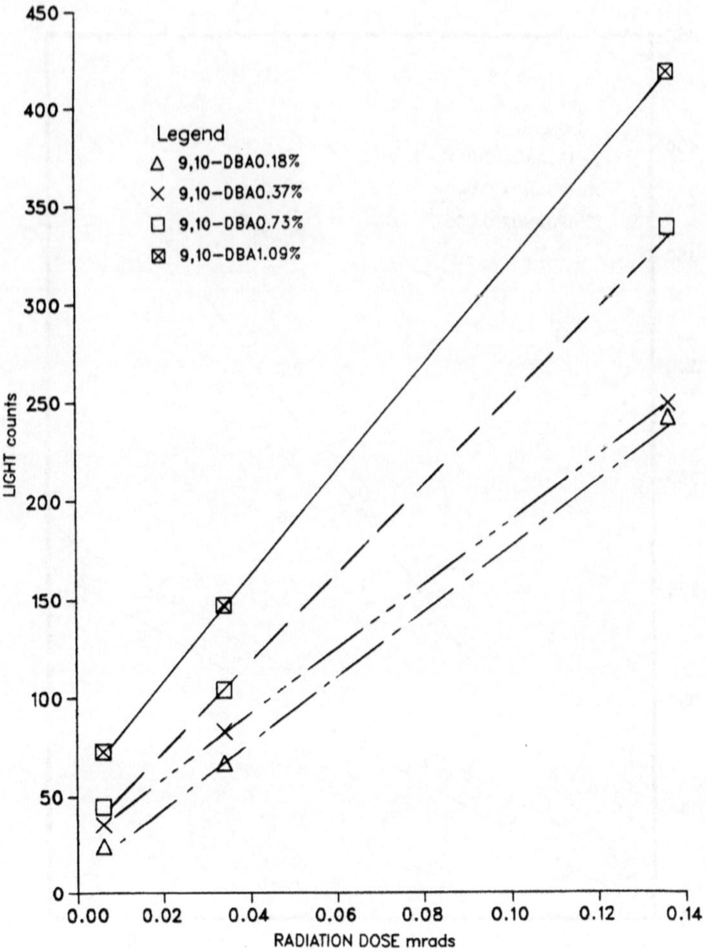

Fig. 19. Laser-Induced Chemiluminescence vs. Radiation Dose for Polypropylene Containing Irganox 1010 (0.013%) and Different Concentrations of 9,10-DBA

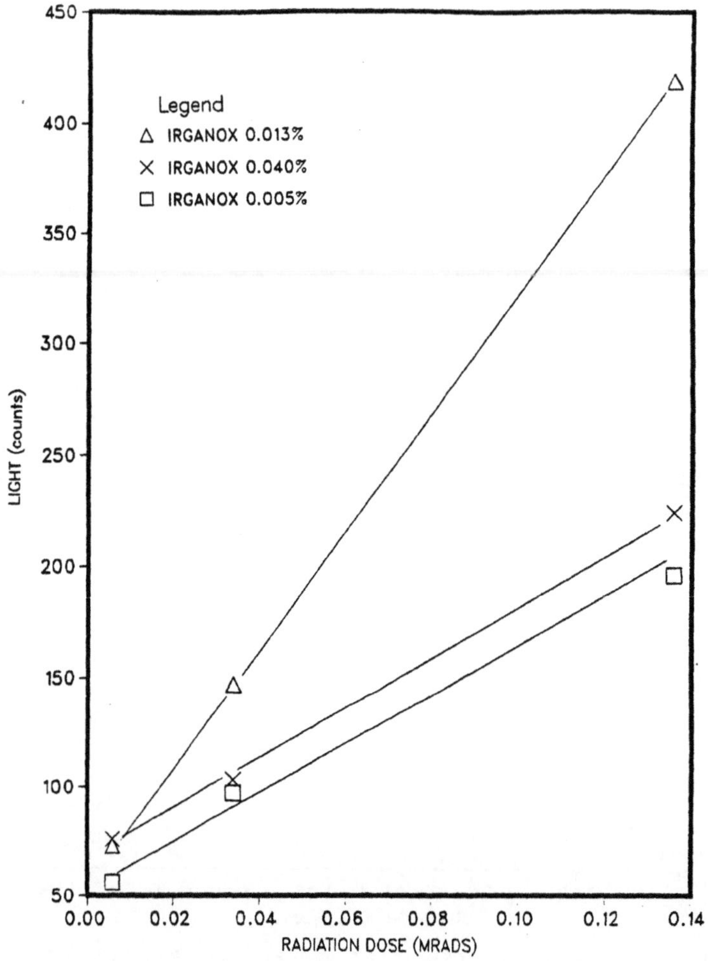

Fig. 20. Laser-Induced Chemiluminescence vs. Radiation Dose for Polypropylene Containing 9,10-DBA (0.73%) and Different Concentrations of Irganox 1010. Experiments Conducted 10 Days after Irradiation.

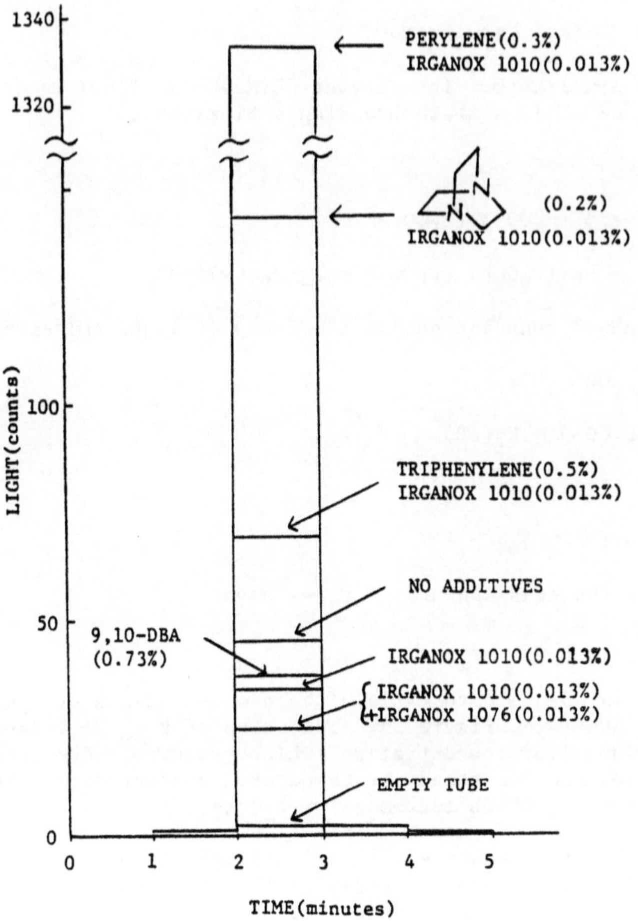

Fig. 21. Counts Per Milligram of Chemiluminescence from Laser-Heated Polypropylene and from Polypropylene Containing Various Additives

Semilogarithmic Decay at 22°C

The first-order kinetics of chemiluminescence at ambient temperature (Table 1) can be explained quite easily in terms of a rapid decay of primary or secondary peroxyl radicals, which can give excited states on reaction with another peroxyl, in the presence of a nearly constant concentration of tertiary peroxyls and phenolic inhibitor. Thus;

$$P_pOO\cdot + P_tOO\cdot \xrightarrow{k_1} h\nu \text{ and termination} \quad (6)$$

$$P_pOO\cdot + AH \xrightarrow{k_2} P_pOOH + A\cdot \quad (7)$$

$$A\cdot + P_tOO\cdot \xrightarrow{} \text{fast termination} \quad (8)$$

where $P_pOO\cdot$ is polymer-$CH_2OO\cdot$ (or polymer-CH(Me)$OO\cdot$), $P_tOO\cdot$ is (polymer)$_2C(CH_3)OO\cdot$, and AH is a chain-breaking antioxidant.

Since

$$dh\nu/dt \equiv I = \phi\,(P_pOO\cdot)\,(P_tOO\cdot) \quad (9)$$

$$d(P_pOO\cdot)/dt = -k_1(P_pOO\cdot)\,(P_tOO\cdot) = k_2(P_pOO\cdot)(AH) \quad (10)$$

If $(P_tOO\cdot)$ and (AH) $\cong$ constant during the decay of light emission,

$$(P_pOO\cdot) = (P_pOO\cdot)_o e^{-k't} \quad (11)$$

where k' and $k_1(P_tOO\cdot) + k_2(AH)$

Substituting, we obtain

$$I = \phi(P_tOO\cdot)\,(P_pOO\cdot)_o e^{-k't} \quad (12)$$

Identifying I_o as the emission at t = 0, we have

$$I = I_o e^{-k't} \quad (13)$$

This scheme is adequate to explain the odd dependence of k_{obs} (=k') in some of the plots (e.g. Figure 5). The value of k_{obs} will depend both on the phenolic inhibitor concentration, which presumably decreases with increasing γ-dose, and the steady-state peroxyl concentration after the end of the irradiation, which increases with dose.

We caution that the sequence 5-7 is a minimal set of reactions sufficient to explain the exponential decay of the chemiluminescence. In particular, the sequence ignores reversibility (e.g. regeneration of $P_pOO\cdot$ from P_pOOH by H-transfer to $P_tOO\cdot$), which can lead to extremely complicated kinetic expressions.[6] We found that, however, the ESR signal of sample 8 in Table 1 decreased only about 10% in the time for 90% of the decay of the ambient chemiluminescence emission from the same sample. This observation supports one of the assumptions in the derivation of equation (13).

The other assumption in the derivation, that the phenolic antioxidant concentration remained about constant, was tested by analyzing a sample of polypropylene, containing (nominally) 0.040% of Irganox 1010 and irradiated with a dose of 0.14 Mrad. The irradiated samples showed 0.031 and 0.030% antioxidant 1 and 6 days after irradiation, respectively. Non-irradiated controls, analyzed at these times, both showed concentrations of 0.039%. Since the antioxidant was inhomogeneously distributed in these samples, it is apparently not able to prevent the continued oxidation that is evident in Figures 11-18.

Peroxide Value and Laser Chemiluminescence

The dichotomy between non-tertiary and tertiary peroxides is also presumably sufficient to explain the absence of any direct correlation between POV and photons emitted during the laser heating. Tertiary peroxides and hydroperoxides contribute to POV but not, at least directly, to the chemiluminescence, since they lack abstractable α-hydrogens. The laser technique thus gives results which may be dependent on the mix of various peroxidic species present. We are encouraged that the laser results, unlike the first-order decay constants discussed above, appear monotonically related to γ-dose (Table 2, Figures 19-20).

CONCLUSIONS

1. Inhibited polypropylene at ambient temperature shows a short-lived chemiluminescence with an exponential decay after γ-irradiation. These observations can be reconciled with the production of excited states from a transient subpopulation of alkylperoxyl radicals that disappears with pseudo-first order kinetics.

2. Rapid heating of polypropylene samples with an IR-laser gives chemiluminescence emission that is dependent on the γ-dose and which can be related to POV within a given formulation.

These conclusions are limited by the scope of this study, which were restricted to one type of antioxidant, low γ-doses compared with those in commercial operations, powdered samples, concentrations of DBA which were above the solubility limit at ambient temperature, and antioxidant in most samples confined to the surface layer. In spite of these shortcomings, we suspect that the basic features we have observed probably will be similar in samples in which different variables are chosen.

Our results show the potential utility of two chemiluminescence methods to monitor radiation damage in formulated polypropylene.

ACKNOWLEDGEMENT

The authors thank the Himont U.S.A. for a graduate fellowship to Hongsik Byun, and Mark D. Nachbor for preparation of some samples. The authors express their appreciation to Dr. Jeffrey W. C. Kuo for helpful comments, and to Dr. Gary Gimmestad for loan of the IR laser.

REFERENCES

1. G. D. Mendenhall, H. K. Agarwal, J. M. Cooke, and T. S. Dziemianowicz, in: "Polymer Stabilization and Degradation," P. P. Klemchuk, ed., ACS Symposium Ser. 280, pp. 373-385 (1985).
2. G. A. George, Develop. Poly. Degrad., 3, 173 (1981) and references therein.
3. W. K. Fisher, J. Indust. Irrad. Tech., Tech., 3, 167 (1985).
4. W. J. Price, "Nuclear Radiation Detection," McGraw HIll, pp. 295-7 (1964).
5. D. J. Carlsson and D. M. Wiles, Macromolecules, 2, 597 (1969).
6. L. R. Mahoney, Ang. Chem. Int. Ed., 8, 547 (1969).

CHARACTERIZATION OF POLYETHYLENE IN FILM AND FIBER

David Cook,[a] Alfred Rudin,[a,b]
Henry P. Schreiber[c] and Joel Young[a]

[a]Guelph-Waterloo Centre for
Graduate Work in Chemistry
University of Waterloo
Waterloo, Ontario, Canada N2L 3G1

[b]Address communications to this author

[c]Ecole Polytechnique
Department of Chemical Engineering
P. O. Box 6079, Station A
Montreal, Quebec, Canada H3C 3A7

INTRODUCTION

The range of polyethylenes, available to film producers, has been greatly increased with the large scale production of linear low density (LLDPE) polymers and the use of high density polyethylene (HDPE) in tubular film extrusion. These polymers are being used alone and in blends with high pressure low density polyethylenes (LDPE) in many film applications.

In general terms, the LLDPE's that are currently available have the advantage of lower price, strength and the ability to be drawn into thin films. Their drawbacks include high melt viscosities and a tendency for melt processing problems, like cyclic melt fracture.

Both deficiencies can be counteracted, to some extent, by increasing the melt temperature. This, however, promotes bubble instability in tubular film extrusion. One remedy for this problem is blending with LDPE. Film producers have found that LDPE/LLDPE mixtures can be employed to stabilize and strengthen the bubble or sheet in tubular or cast film extrusions, respectively. Conventional LDPE's for film applications are known to provide strong, stable melts, in general, but to be limited in their ability to yield very thin films.

Film producers are faced with choices between a wide range of polyethylenes and blends, as well as a variety of suggested modifications to the film extrusion process. Examination of these alternatives, in film extrusion trials, can be very tedious and expensive. In this report, we describe a simple, small scale procedure that can be used to screen polymer blends and some extrusion variables for certain important aspects of film extrusion behavior.

The experimental method is based on the extrusion and stretching of a monofilament while measuring its diameter along the drawline and the tension at a fixed point. In a slight variation, the ultimate extensibility of the fibre can be determined, under given experimental conditions.

Others have reported on various aspects of isothermal and non-isothermal treadline spinning.[1-3] Mention has been made of the similarity of film and fibre extrusions in that both involve stretching flows of thermoplastics.[4]

The particular contribution of this work is to illustrate the application of the method, described to the solution of certain current problems in film production and to point out a simple procedure for condensing the experimental data.

The method requires no particularly special apparatus. It yields a relation between parameters, with some similarity to elongational viscosity and shear rates. We call these the fictive elongational viscosity and shear rate. They have the proper units but are based on an analysis, which is rheologically incorrect, because the threadline is not isothermal and the axial stress and elongation are actually position-dependent.

The calculated parameters apply to the whole process of tensile deformation of a particular cooling polymer melt. They do not characterize the polymer alone but also reflect the cooling and stretching processes. They are useful for comparing different polymers or blends in a given process of this type or to examine the effects of variations in processing conditions. The results illustrate how this basic experiment anticipates some important experiences in film extrusion of current commercial polyethylenes.

EXPERIMENTAL

The extrusion and drawdown apparatus are shown in Figure 1. The extruder employed a 1.2 cm diameter 27:1 L/D screw with a 3:1 compression ratio. The orifice was a flat entry die with 0.19 cm diameter, 1.71 cm length and L/D = 9. The screw speed was set at 5 rpm (by timing shaft revolutions), unless noted otherwise.

An air coil (3.2 cm diameter x 10 cm length) was used to cool the extruded monofilament. The air ring was constructed of 0.64 cm copper tubing, with 0.32 x 0.15 cm slit air holes spaced every 0.9 cm along the interior of the 12 turns of the tubing.

Air flow to the cooling coil was kept constant at 1.5 L/s. The air temperature was 17°C. The heat transfer coefficient, $h(W/m^2\ °C)$ was determined, along the coil centerline, using a thermometer cool-down technique and lumped heat transfer calculation. The results are shown graphically in Figure 2. With the coil configuration used, the highest cooling rate was in the high draw region of the extruded monofilament.

The extrudate was wrapped 5 times around a variable speed pair of godet rolls each with 30.5 cm circumference and then taken up by a constant torque winder. Measurements were generally made at 30, 60 and 100 rpm godet speeds. Some polymers, however, could not be drawn at the highest godet speed without breaking.

A spring tensiometer was installed between the godet and extruder about 20 cm from the die face. The tensiometer was calibrated using weights attached to a line as shown in Figure 3.

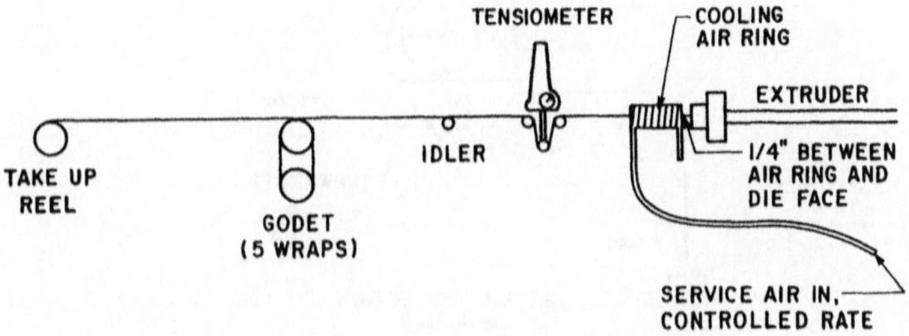

Fig. 1. Experimental Arrangement

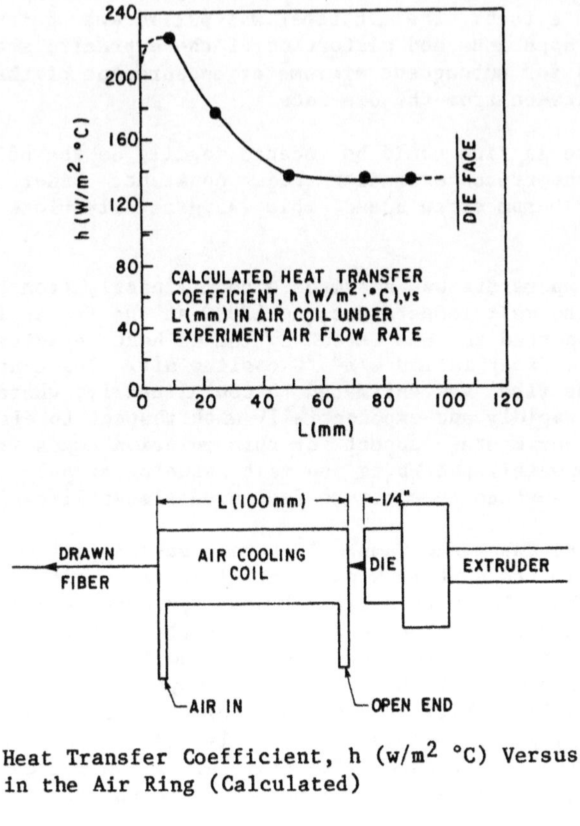

Fig. 2. Heat Transfer Coefficient, h (w/m^2 °C) Versus Position in the Air Ring (Calculated)

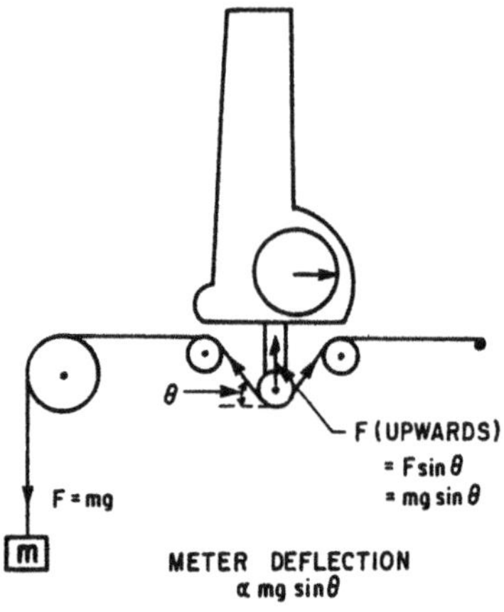

Fig. 3. Tensiometer Calibration

Samples for estimates of the fictive elongational viscosity were obtained in the following manner. The die pressure, die temperature and tension (in dynes) were recorded. The fibre was sheared off at the die face with a sharpened stapula while it was clamped stationary about 20 inches from the die face. The cut fiber was pulled away quickly to prevent contact with the apparatus and distortion of the extrudate shape. Filaments were stored for subsequent micrometer measurement of the diameter as a function of distance from the die face.

The fibre freeze line could be located readily as the point at which the diameter of the frozen extrudate became constant. Under our experimental conditions of 5 rpm screw speed, this was generally close to 5 cm from the die face.

The fibre temperature was assumed to cool linearly from the die face, where it was at the melt temperature, to 120°C at the freeze line. This assumption is supported to some extent by lumped heat capacity-temperature calculations for a fiber in 130 w/m^2 °C cooling air. The centerline temperature of the fiber was estimated to cool linearly, whereas the surface cooled more rapidly and exponentially with respect to distance from the die face. Experimental support for this relation comes from the measurements of Spruiell and White[5] on melt spinning of polypropylene fibres, where the surface temperature decreased almost linearly.

More recently, Huang and Campbell[6] have reported surface temperature profiles that exhibited a shallow exponential decrease with distance from the die face in the tubular film extrusion of low density and linear low density polyethylenes. The choice of 120°C as the freezing temperature is obviously only an approximation for different samples. We assume, however, that errors in assignment of the freeze line temperature will not affect the relative comparisons between different polymers at processing conditions. In any event, allowance is easily made for a different or variable freeze-line temperature. The diameter of the molten extrudate was calculated from its frozen dimensions according to the following procedure.

The initial radius (r_o) taken from the die swell value measured on the extrudate without tension. The radius (r_t) at a distance h from the die face was measured as described above. All frozen radii were corrected for the difference in densities of the solid and molten polymers by using known density-temperature relations.[7]

The experimental data can be used with the analysis of Cogswell[8] to provide estimates of a fictive elongational viscosity versus a fictive strain rate.

In this case:

$$\nu = \frac{Th}{Q \ln \left(\frac{r_o}{r_t}\right)^2} \quad (1)$$

and

$$\dot{\varepsilon} = \frac{Q \ln \left(\frac{r_o}{r_t}\right)^2}{\pi r_t^2 h} \quad (2)$$

Here ν is a parameter with units of viscosity (here poises), T is the measured tension on the fibre (dynes), h is the distance from the die face (cm), Q is the volumetric flow rate of the polymer (cm^3/s), r_o (cm) is the initial radius, r_t (cm) is the radius at distance h from the die face and $\dot{\varepsilon}$ has the units (s^{-1}) of a deformation rate.

The values of ν and $\dot{\varepsilon}$ do not characterize the polymer alone. They also reflect the stretching process in the cooling melt. It is suggested therefore that these figures may be useful in comparing different polymers and process conditions in extrusions that involve stretching of the melt, as in fiber formation and tubular or cast film production.

The filament extension in a particular experiment was calculated as follows: The cross-sectional area of the molten monofilament at the die face was calculated. The volumetric flow rate, Q, was known from the measured mass flow rate and melt density and hence the linear speed of the fibre at the die face could be calculated from the ratio of volumetric flow rate to the cross-sectional area of the molten fibre. The extension ratio is determined from the linear speed of the fibre at the die face and the godet surface speed at break.

The ability of the fibres to be drawn to thin diameters was assessed in separate experiments. In this case, godet speed was slowly increased until the fibre broke in the melt zone. The tensiometer was not in place in these experiments. The data are reported in terms of an ultimate extension ratio.

RESULTS AND DISCUSSION

Generally, 4-6 replicate experiments were conducted and fibre diameters were averaged as function of distance from the die face (h). Typical data are shown in Figure 4.

Five polymers were studied. Their specifications are given in Table 1. Table 2 records ultimate elongations at various melt temperatures. As expected, the LLDPE's exhibit much higher drawdown ability than LDPE's, when comparisons are made as usual at more or less the same melt index. These polymers can be drawn to thinner gauges with higher melt temperatures.

LDPE B is recommended by the manufacturer for blending with LLDPE A.

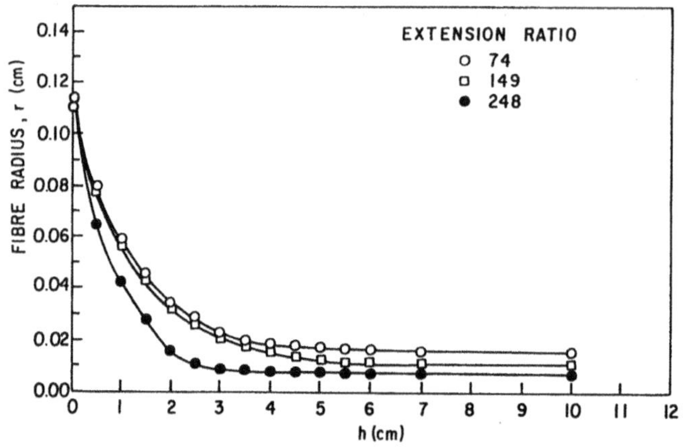

Fig. 4. Filament Radius (cm) Versus Distance from Die Face h (cm) for LLDPE A at 200°C Melt Temperature and 5 rpm Extruder Screw Speed. Total Extension Ratios: o 74, 149, 248

Table 1.

Polymer	Type	Density	Melt Index
A	LLDPE ethylene-butene copolymer	0.922	1.0
B	LDPE polyethylene	--	3.0
C	LLDPE ethylene-octene copolymer	0.920	1.0
D	LDPE 2.5% vinyl acetate-autoclave	0.925	2.2
E	LDPE 1.5% vinyl acetate-autoclave product	0.923	0.8

Table 2.

Polymer	Melt Temp. (°C)	Filament Linear Speed at Die (cm/s)	Extension Ratio at Break
A (LLDPE)	190	0.20	660
	200	0.20	740
	220	0.23	750
A(70%) + B(30%)	200	0.14	630
C (LLDPE)	200	0.19	1320
	220	0.20	no break
D (LDPE)	200	0.20	190
	220	0.17	280
E (LDPE)	190	0.60	80
	200	0.58	90
	220	0.42	160

Table 3.

Polymer	Melt Temperature	Extruder Speed (rpm)	Pressure (psi)	Line Tension (dynes)
A	190	5	550	5500
	200	5	495	4900
	220	5	390	3300
A(70%) + B(30%)	200	5	300	7900
C	200	5	465	3600
	220	5	365	2300
D	200	5	225	5500
	220	5	165	1700

Addition of LDPE B at a 30 weight percent level resulted in a moderate decrease in the drawdown of the host LLDPE, as seen in Table 2. It also produced a substantial increase in melt tension.

With all polymers, except the low melt index LDPE E, melt tensions were not sensitive to extension rates. Measured line tensions were essentially constant at godet speeds of 30, 60 and 100 rpm, with a fixed extruder screw speed of 5 rpm and set melt temperature. This is because the increased extension rates resulted primarily in enhanced thinning in the hottest melt zones, where the elongational viscosity of the material was low (cf. Figure 4). The data for samples A-D and the A/B blend are given in Table 3. As shown in Table 3, the particular extrusion conditions were at slow speed, with low die pressures.

The benefits of blending LDPE with LLDPE are visible in these results. Addition of 30 weight percent LDPE to LLDPE A results in a moderate decrease in drawdown (Table 2) and a very significant increase in melt tension (and hence in bubble or cast sheet stability).

The lower melt index LDPE, polymer E, differed from the other materials studied in that melt tension increased with higher extension ratios. These results are shown in Figure 5. Melt tension is much higher for this LDPE and its ultimate drawability is comparatively low.

The drawdown data presented to this point are consistent with practical performance of these polymers. Polymers A and C are high drawdown LLDPE film grade resins. Polyethylene D is used for thin gauge packaging as in bread bags and polymer E is for packaging where some toughness is required. The commercial gauge range is down to 25 μm for D, where E cannot be used to produce film thinner than about 40 μm.

Figure 6 shows the measured diameter-die distance plots for the poor drawdown LDPE E. In this case the extruder screw speed was set at 10 rpm, which is double the setting for LLDPE A in Figure 4. There is, however, no clear distinction between the behavior of these two polymers from the filament diameter-draw distance plots. The differences are apparent, however, when fictive elongational viscosities are calculated as functions of apparent shear rates.

Figure 7 shows elongational "viscosity" versus elongational "shear rate" plots for LLDPE A at 190°C melt temperature and 3 different extension

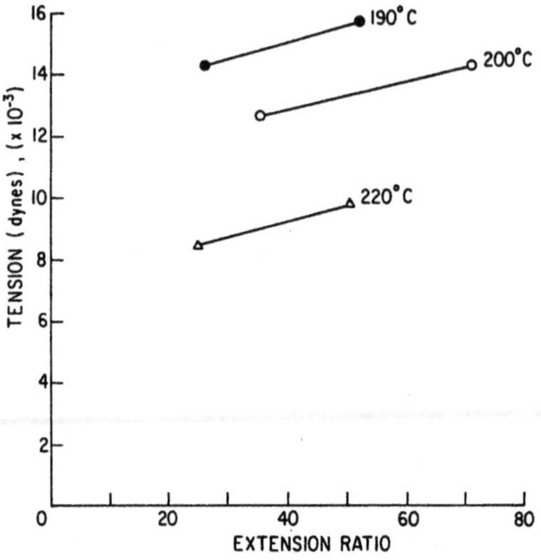

Fig. 5. Fiber Tension Versus Extension Ratio for LLDPE E

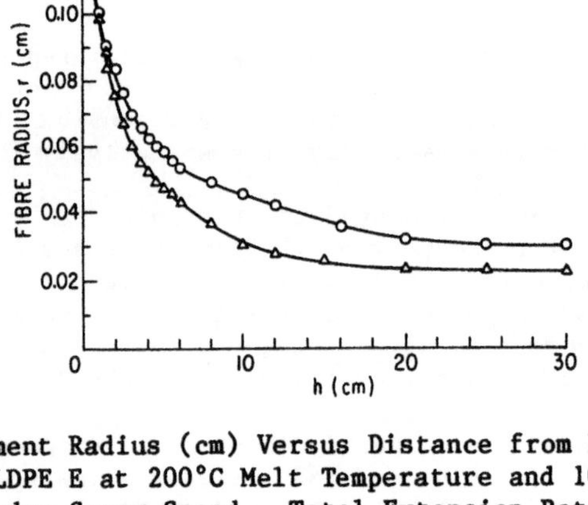

Fig. 6. Filament Radius (cm) Versus Distance from Die Face (cm) for LDPE E at 200°C Melt Temperature and 10 rpm Extruder Screw Speed. Total Extension Ratios: o 36, x 71.

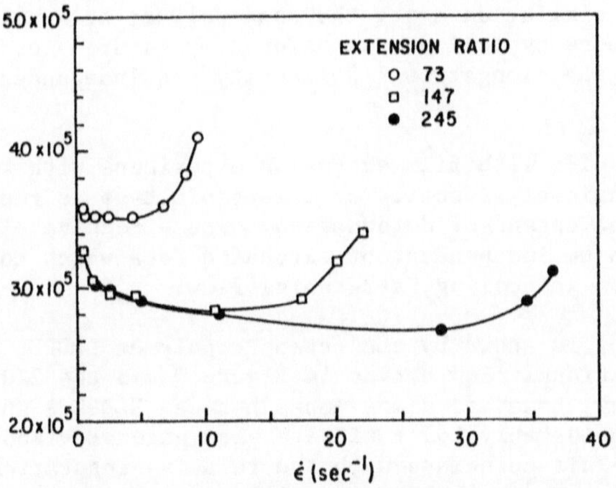

Fig. 7. Fictive Elongational Viscosity Versus Fictive Elongational Shear Rate for LLDPE A at 190°C Melt Temperature and 3 Extension Ratios: o 73, 147, 245

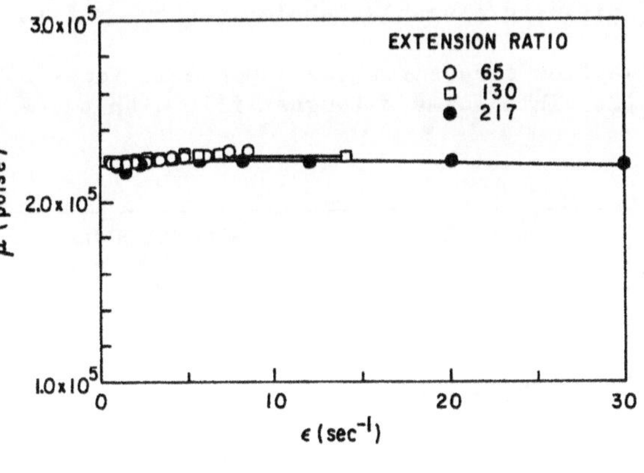

Fig. 8. Fictive Elongational Viscosity Versus Fictive Elongational Shear Rate for LLDPE A at 220°C Melt Temperature. Extension Ratios are: o 65, 130, 217

ratios. At this melt temperature, which is lower than that used commercially with this polymer, the melt is apparently shear-thinning at low shear rates. The fictive elongational strain rates are apparently higher than those achieved in tubular film extrusion where values less than 5 s^{-1} seem to be common. The range of $\dot{\varepsilon}$ could be compressed in our experiments by increasing the cooling rate in the air ring.

Figure 8 shows similar data for the same polymer at a melt temperature of 220°C. This is more typical of extrusion temperatures for this polymer. It can be seen that the elongational "viscosity" is independent of elongational rate.

These data coincide with film extrusion experience with this polymer. The isothermal extensional viscosity of this resin type is reported to be almost independent of extent of deformation. The effective elongational viscosity appears to be independent of extension rate which corresponds to extent of deformation in cooling, stretching flow.

Similar behavior is shown by the octene copolymer LLDPE, resin C, the 200°C fictive elongational flow curves in Figure 9 and the 220°C plots in Figure 10. There are, however, differences between LLDPE A and LLDPE C. The latter has a considerably lower fictive elongational viscosity (cf. Figures 8 and 10). This coincides with the relative reputations of the two polymers in the trade, where LLDPE C is regarded as being more suitable for very thin gauge applications.

Figure 11 records the fictive elongational flow curves for the higher melt index LDPE, resin D, at 200°C. Data points are included for replicate trials at the higher extension ratio, to illustrate the precision of these experiments. This LDPE has relatively good drawdown properties for a low density polyethylene. It is decidedly inferior to the lower melt index LDPE's, however, as shown in Table 2. The elongational behavior of this LDPE is very different from that of the LLDPE's. The fictive elongational viscosity of polyethylene D increases with increasing elongation rate. In cooling, stretching flow, the melt will become very viscous with increasing degree of extension, yielding a stable bubble in tubular film extrusion.

The 200°C fictive flow curves for the lower melt index LDPE, resin E, are shown in Figure 12. This forms a tougher film with poorer drawdown

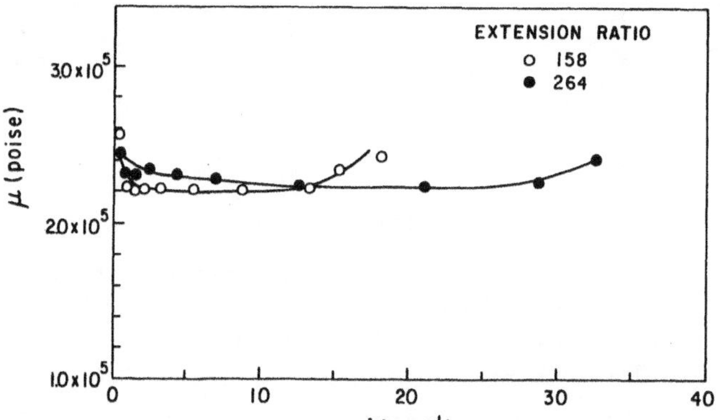

Fig. 9. Fictive Elongational Viscosity Versus Fictive Elongational Shear Rate for LLDPE C at 200°C Melt Temperature. Extension Ratios: o 158, 264

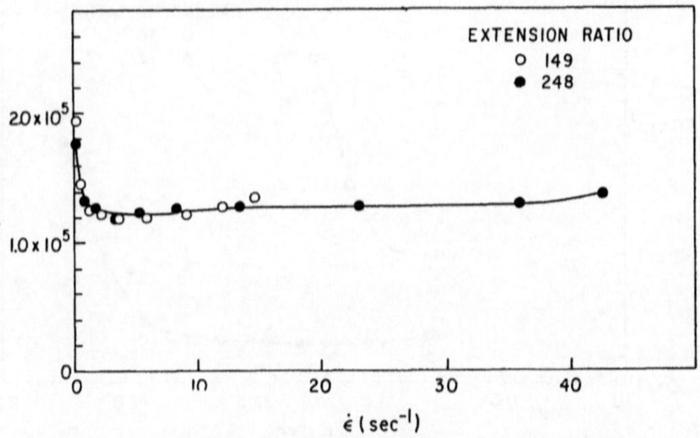

Fig. 10. Fictive Elongational Flow Curves for LLDPE C at 220°C. Extension Ratios: o 149, 248

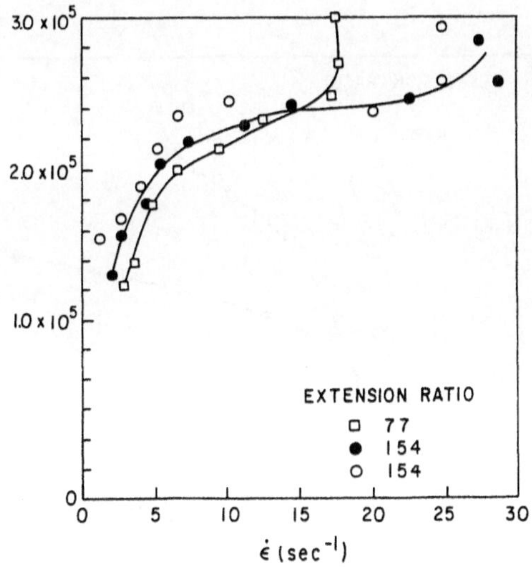

Fig. 11. Fictive Elongational Flow Curves for LDPE D at 200°C. Extension Ratios: 177, o 154

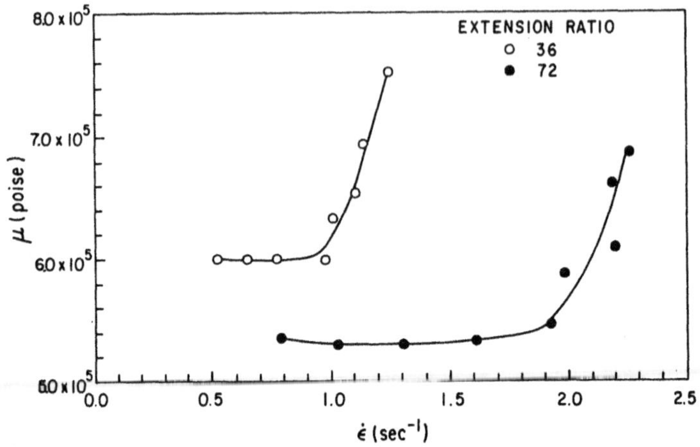

Fig. 12: Fictive Elongational Viscosity Versus Fictive Elongational Shear Rate for LDPE at 200°C. Extension Ratios: o 36, 72

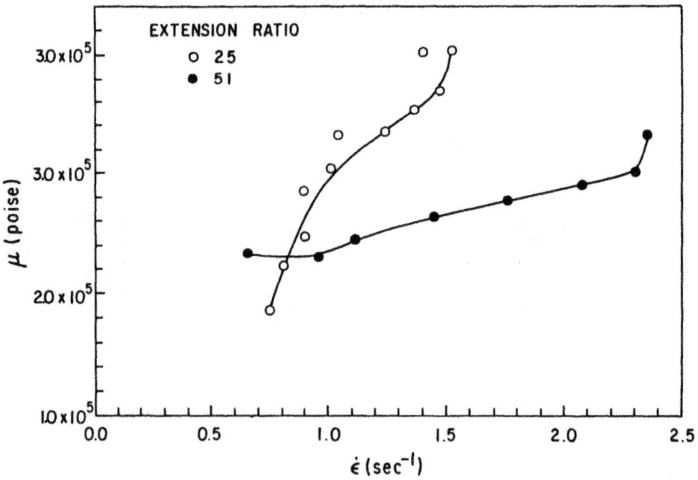

Fig. 13. Fictive Elongational Flow Curves for LDPE E at 220°C. Extension Ratios: o 25, 51

properties than LDPE D. The fictive elongational flow curves in Figure 12 indicate that strain hardening occurs at low fictive elongational rates, with the condition for the onset of this phenomenon dependent on extrusion conditions. The apparent elongational viscosity of polymer E is much higher than that of D, at the same elongational rate.

The fictive flow curves for resin E at 220°C are shown in Figure 13. General strain hardening is observed.

CONCLUSIONS

Certain aspects of the behavior of polyethylenes in tubular film extrusion can be simulated in a simple, relatively fast melt spinning experiment. Both processes involve stretching of the polymer melt while it is being cooled. There are differences, of course, since biaxial extension is involved in tubular film production while melt spinning is characterized by uniaxial elongation. Nevertheless, the effects of various polymers and process changes can be evaluated relatively rapidly in the simple procedure described here.

The ability to be downgauged is an important characteristic of commercial polyethylenes. Total extension to break in the filament extrusion, ranks different polymers reliably and indicates the probable effects of changes in such process variables, as melt temperature, cooling air temperature or flow rate. These data cannot be used, however, to predict minimum running gauges.

The measured threadline profile at different extrusion conditions also provides useful predictions of the relative bubble stability and drawdown behavior of polyethylenes, as well as of the effects of changes in extrusion conditions. The differences between polymers are most clearly seen when the raw data are transformed into corresponding parameters that represent a fictive elongational viscosity and elongational rate under the particular non-isothermal experimental conditions. These characteristics correlate well with the full scale performance of different polyethylenes. The procedure described does not provide fundamental rheological parameters. It has been designed solely to aid in the evaluation of the effects of variations in polymer structure or film extrusion conditions.

ACKNOWLEDGEMENT

This research was supported by a PRAI grant from the Natural Sciences and Engineering Council of Canada.

REFERENCES

1. A. Acierno, J. N. Dalton, J. M. Rodriquez, and J. M. White, J. Appl. Polym. Sci., 15, 2395 (1971).
2. C. D. Han and R. R. Lamonte, Trans. Soc. Rheol., 16, 447 (1972).
3. I.-J. Chen, G. E. Hagler, L. E. Abbott, D. C. Bogue, and J. L. White, Trans. Soc. Rheol., 16, 473 (1972).
4. C. D. Han, SPE Antec Papers, 33, 193 (1975).
5. J. E. Spruiell and J. L. White, Polym. Eng. Sci., 15, 660 (1975).
6. T. A. Huang and G. A. Campbell, SPE Antec Papers, 43, 84 (1985).
7. A. Rudin, K. K. Chee, and J. H. Shaw, J. Polym. Sci. C., 30, 415 (1970).
8. F. N. Cogswell, Rheol. Acta, 8, 187 (1969).

CHARACTERIZATION OF HIGHLY DRAWN POLYPROPYLENE FILMS

Francis M. Mirabella, Jr.

NORCHEM, Inc.
Technical Center, 8805 North Tabler Road
Morris, Illinois 60450

INTRODUCTION

The production of fibers and films with high strength and modulus through extension is common in the plastics industry. The relationship between the draw ratio and the degree of macromolecular orientation of oriented polypropylene is an area of increasing interest. This interest is directed toward the achievement of the highest possible strength and modulus.

The maximum theoretical values of tensile strength and Young's modulus have been approached but never achieved experimentally. This study was undertaken to correlate the draw ratio and mechanical properties with the degree of macromolecular orientation.

The systems studied include polypropylene films with draw ratios up to about 50X which were produced by gelation/crystallization of very high molecular weight polypropylene from dilute solution. The dried gel films can be drawn to remarkably high extension ratios.[1-3]

The reported mechanical properties[2,3] of these highly drawn polypropylene films approach the theoretical maximum values and make the study of the macromolecular orientation of such films particularly interesting. The results of polarization-modulated infrared dichroism measurements on these highly drawn polypropylene films are presented in this report. Similar measurements are presented on melt crystallized and drawn polypropylene films for comparison.

EXPERIMENTAL

Materials

The highly extended polypropylene films used in this study were prepared by a method described previously.[2,3] It consists of gelation/crystallization of very high molecular weight polypropylene (ca. 3.4×10^6) from dilute decalin solution. The dried gel is pressed into flat sheets and drawn on an Instron Tensile testing machine at 2.0 cm/min cross-head speed at 150°C. The highly drawn films have draw ratios of 8.3X, 20X, 36.5X and 47.5X. The lower extension range was covered with melt crystallized films which were pressed into 5 mil plaques at 200°C and drawn on an Instron Tensile testing machine at 2.0 cm/min cross-head speed at 150°C. The melt crystallized films had draw ratios from 1X to 9X.

Techniques

The infrared measurements were made on a Nicolet 6000 FTIR. The polarization was modulated with a germanium polarizer (Harrick Scientific Corp., PSD-J1R). Background and sample interferograms were collected at each polarization and ratioed. Signal-averaging of 640 scans in each case yielded excellent signal-to-noise ratio. The drawn films were in the form of long (machine direction, MD), narrow (transverse direction, TD) and thin (normal direction, WD) specimens. These specimens were mounted vertically (along MD) on a sample holder with scored lines to facilitate accurate mounting. Absorption spectra for each film were obtained with the infrared electric vector parallel to the MD and parallel to the TD by rotation of the polarizer. The reproducibility of these measurements was checked by repeating the measurements with the film mounted horizontally (along TD) on the sample holder. The reproducibility was within ± 10%. This showed that the infrared beam was not vignetted differently when the polarizer was rotated from 0° to 90° settings. Spectral coaddition was done on the Nicolet computer.

RESULTS AND DISCUSSION

The average orientation of macromolecular chains relative to the fabrication extension direction can be expressed in terms of the Hermans orientation function

$$f_p = \frac{3\cos^2\theta_p - 1}{2} \tag{1}$$

where f_p (p = c, crystalline phase; AM, amorphous phase; and AV, average phase) is the orientation function for a particular phase, and θ_p is the average angle between the extension direction and the helix chain axis for polypropylene in that phase p. The Hermans orientation function can be related to the dichroic ratio obtained from infrared measurements of the principal absorption coefficients for radiation parallel (k_π) and perpendicular (k_σ) to the orientation axis ($D = k_\pi/k_\sigma$) by

$$f_p = \frac{(D-1)}{(D+2)}\,\frac{(D_o+2)}{(D_o-1)} \tag{2}$$

where $D_o = 2\cot^2\phi$ and ϕ is the transition moment angle, between the helix chain axis and infrared transition moment. If suitable infrared bands are observed for each phase, then the Hermans orientation function can be calculated for macromolecular chains in each phase and the average angle θ, formed by such chains with the fabrication extension direction, can be determined for each phase.

The infrared bands chosen were a 1220 cm^{-1} due to crystalline phase and θ = 90°,[4] 2725 cm^{-1} due to the amorphous phase and ϕ = 90°[4] and 841 cm^{-1} due to the average phase and ϕ = 5°.[5] The 1220 cm^{-1} band disappears on melting while the 2725 cm^{-1} band remains essentially unchanged. The dichroic ratios D, Hermans orientation functions f and the average angle of orientation θ were calculated from measurements of the above infrared bands for each drawn polypropylene film.

The results for the highly extended films are collected in Table 1. It can be observed in Table 1 that the dichroic ratios range from very small to very large values depending on the orientation of the infrared electric vector relative to the MD and TD. This presents unique problems in determining the dichroic ratios for such highly drawn films. These ratios are formed by a large absorbance and a small absorbance, so that the experimental error is potentially serious. This requires extraordinary care in order to obtain precise values of D. The dichroic ratios for other infrared bands follows similar patterns as exhibited in Table 1, but these bands were chosen for the convenient intensity, position in the spectrum and transition moment angle and phase assignments that these possess.

The Hermans orientation function is plotted for the low (melt crystallized) and high (gel crystallized) draw ranges against the draw ratio in Figure 1. It can be readily observed in Figure 1 that the crystalline orientation (f_c) rapidly reaches a maximum value at low draw ratios while the amorphous orientation (f_{AM}) increases nearly linearly over the entire extension range. Thus, the average orientation (f_{AV}) reflects the average between these two extremes. This same orientation behavior is reflected in the calculated θ values in Table 1.

The modulus for the highly drawn (gel crystallized) films is plotted against the draw ratio in Figure 2 as determined by Peguy and Manley.[2] The modulus increases linearly with the draw ratio. The modulus is plotted against the three Hermans orientation functions in Figure 3. It is apparent that increases in modulus are not correlated with the crystalline orientation function f_c. However, the modulus exhibits a roughly linear increase as a function of the amorphous orientation function f_{AM}. Again, the correlation of modulus with the average orientation function f_{AV} exhibits a behavior which is between these two extremes.

Before considering the significance of Figure 3, it should be noted that since the films were drawn at 150°C there is an opportunity for significant variations in crystallinity (and density) among the films. The direct measurement of crystallinity in drawn polypropylene is not possible with current techniques. The usual x-ray diffraction and differential scanning calorimetric techniques yield fallaciously high values due to the strain imposed on the crystallites by the taut amorphous chains. The density method often yields fallaciously low values due to voiding in the extended polymer. If voiding can be minimized the density method is the best approach to measurement of crystallinity.

Peguy and Manley reported[2] that the densities range from 0.908 to 0.912 g/cc for these highly drawn films. The range of crystallinities can be estimated from the equation ($V_c = (d-d_{AM})/d_c - d_{AM}$) where d is the density and $d_c = 0.936$ g/cc and $d_{AM} = 0.853$ g/cc are the theoretical crystalline and amorphous densities, respectively. This equation yields a range of crystallinities of 66.3% to 71.1%.

The densities can likewise be used to estimate the range of moduli expected for polypropylene. Using the data of Frank,[6] the range of moduli corresponding to this density range is 3.5 GPa to 6.4 GPa. Another approach to the estimation of the crystallinity of the drawn films is through the infrared data. As was stated previously, the 1220cm^{-1} and 2725 cm^{-1} bands are due to the crystalline and amorphous phases, respectively. Thus, changes in the relative intensities of these bands are expected with changes in crystallinity.

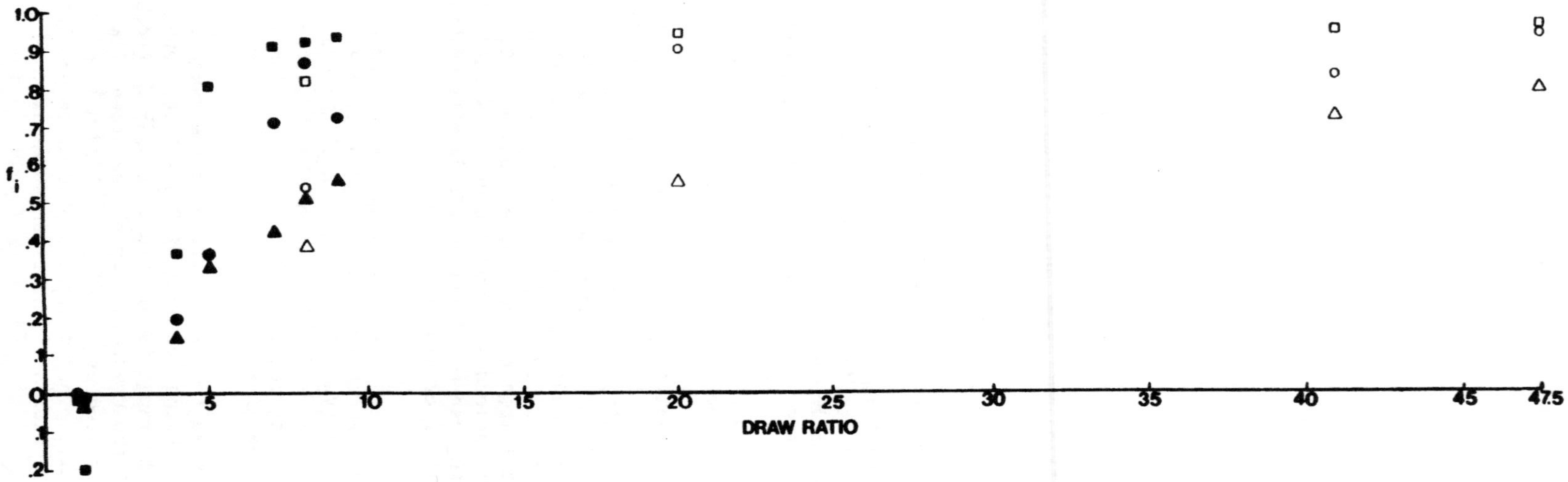

Fig. 1. Hermans Orientation Functions for Melt and Gel Crystallized Polypropylene Versus Draw Ratio.

	f_c	f_{AM}	f_{AV}
Gel Crystallized	□	△	○
Melt Crystallized	■	▲	●

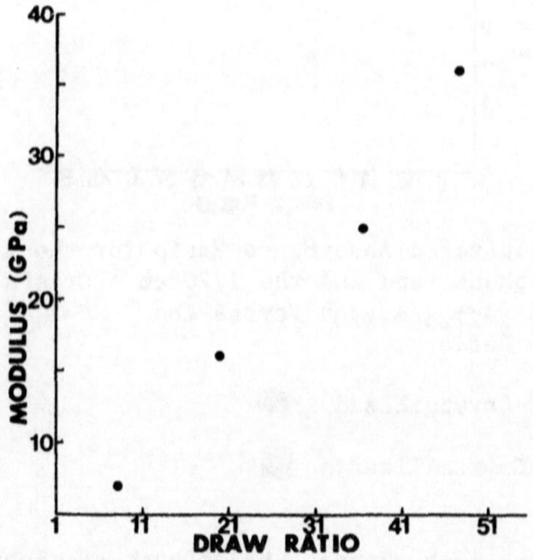

Fig. 2. Modulus Versus Draw Ratio for a Series of Ultradrawn Polypropylene Films

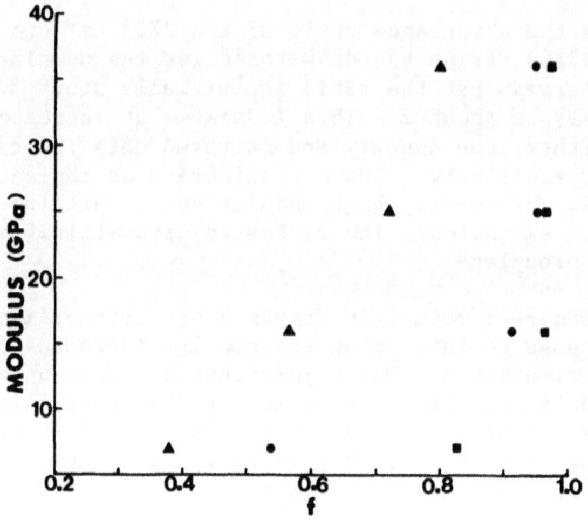

Fig. 3. Modulus Versus Hermans Orientation Functions f for a Series of Ultradrawn Polypropylene Films:

f_C ■

f_{AM} ▲

f_{AV} ●

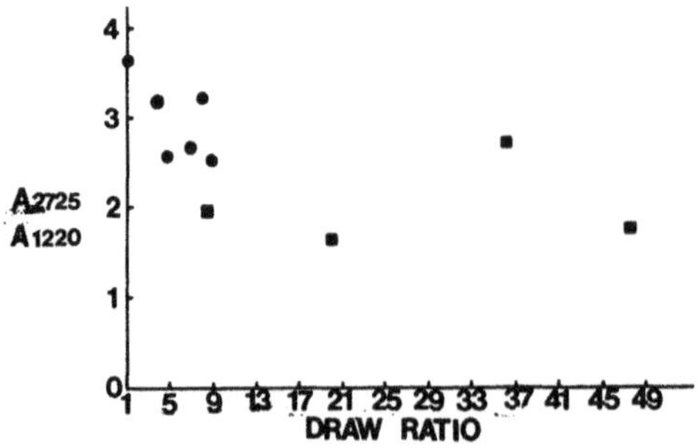

Fig. 4. The Infrared Absorbance Ratio for the 2725 cm^{-1} Amorphous Band and the 1220 cm^{-1} Crystalline Band (A_{2725}/A_{1220}) Versus the Draw Ratio

Melt Crystallized ●

Gel Crystallized ■

In order to observe such changes the effects of dichroism must be eliminated. This was done by coadding the spectra of each specimen taken parallel to the MD and parallel to the TD to produce a truly unpolarized spectrum. The coadding procedure yields a truly unpolarized spectrum, while a spectrum recorded in the FTIR, without a polarizer, may be partially polarized due to partial polarization of the infrared radiation by the mirrors in the FTIR.

Figure 4 shows the absorbance ratio of the 2725 cm^{-1} to the 1220 cm^{-1} band (A2725/A1220) versus the draw ratio for the unpolarized spectra. It can be observed that the ratio is initially about 3.5 and decreases progressively to about 2. This indicates an increase in crystallinity. Taken together, the density and infrared data indicate a significant increase in crystallinity. These considerations indicate that these films are not of extraordinarily high modulus due to attainment of high crystallinity during extension. The estimated crystallinities and moduli are typical of polypropylene.

However, the observed moduli in Figure 2 are extraordinary. The above considerations are used to infer that the modulus increases primarily due to macromolecular orientation. The significant conclusion that can be drawn from Figure 3 is that the orientation of the crystalline phase is not responsible for the dramatic increase in modulus over this extension range. Instead, the modulus increases roughly linearly with the amorphous orientation function f_{AM} over this extension range.

This indicates that the sequence of events during extension begins with reorientation of the crystallites with the c-axis, parallel to the fabrication draw direction. This reorientation is completed at relatively low draw ratio, as evidenced by the rapid rise of f_c at the lower draw ratios. As extension continues fibrillation occurs. Fibrillation was observed to be well developed even at the lower draw ratios, and this is in agreement with the observations of Cannon[1] on similarly drawn polypropylene films in which fibrillation was observed in 6X drawn films. Thus, the major orientation which occurs between 8.3X and 47.5X is in the amorphous phase.

This orientation is responsible for the linear increase in modulus over this extension range. The ability of the amorphous phase to achieve higher orientation perfection in the gel-crystallized specimens was apparently due to the minimization of entanglements in the undrawn gel-crystallized starting material. The higher level of entanglements in the melt crystallized starting material was apparently the cause of the lower level of amorphous orientation perfection and lower modulus achievable in these specimens.

The highest modulus achieved in the gel-crystallized films was about 36 GPa.[2] The theoretical maximum value of modulus for polypropylene determined by x-ray range from 35 GPa[7] to 41 GPa.[8] However, Fanconi and Rabolt[9] have recently shown that the uniform stress assumption used in the x-ray analysis may not be justified and that the resulting values of longitudinal crystal moduli may be too low. The uniform stress assumption used in the x-ray analysis is that the applied macroscopic stress is uniformly distributed throughout the crystalline and amorphous phases.

Fanconi and Rabolt[9] presented higher values of modulus, which were determined by longitudinal acoustic modes observed by Raman spectroscopy and by neutron scattering techniques, which were in good agreement for several polymers. The new modulus value determined from Raman longitudinal acoustic modes for polypropylene was 88 GPa.[9] The value from neutron scattering has not yet been determined for polypropylene.

This new theoretical maximum modulus value for polypropylene refers to 100% crystalline material. This much higher value appears to be reasonable since a value of 36 GPa was achieved with polypropylene which exhibited a Hermans amorphous orientation function (f_{AM}) of 0.8 (see Table 1) and an apparent crystallinity of much lower than 100%. The major conclusion of this work is that a pretreatment to increase the crystallinity of polypropylene to near 100%, followed by a second pretreatment (e.g. gel-crystallization) to minimize the amorphous entanglements, should yield a material that can approach the new maximum modulus "benchmark" of 88 GPa upon drawing.

ACKNOWLEDGEMENT

The author would like to express his appreciation to Dr. R. St. John Manley and Saroj K. Roy for supplying the highly extended polypropylene films used in this study, and to Daniel Schleck for preparing the melt crystallized specimens.

REFERENCES

1. C. G. Cannon, Polymer, 23(8), 1123 (1982).
2. A. Peguy and R. St. John Manley, Polym. Commun., 25, 39 (1984).
3. S. K. Roy and R. St. John Manley, ACS Preprints, Div. Polym. Mater. Sci. Eng., 52, 53 (1985).
4. R. J. Samuels, Makromol. Chem. Suppl., 4, 241 (1981).
5. R. J. Samuels, Private Communication.
6. H. P. Frank, Polypropylene, Gordon and Breach Science Publishers, New York (1968).
7. I. Sakurada and K. Kaji, J. Polym. Sci., Polym. Symp., 1, 31, 57 (1970).
8. I. Sakurada, T. Ito, and K. Nakamae, J. Polym. Sci., Polym. Symp., C, 15, 75 (1966).
9. B. Fanconi and J. F. Rabolt, J. Polym. Sci., Polym. Phys. Ed., 23, 120 (1985).

A STUDY OF POLYOLEFIN FILMS AND YARNS UNDER CORONA DISCHARGE

B. Catoire, P. Bouriot, and M. Sotton

Institut Textile de France

C. Chabert, J. P. Soulier, and D. Sage

Laboratoire d'Etudes des Matériaux
Plastiques et des Biomatériaux, Université Lyon I

INTRODUCTION

It is well known[1,2,3] that if one wants to increase their adhesivity, polyolefins have to be subjected to a pretreatment entailing substantial modifications of their surfaces. Among the usual activation processes, we chose electrical discharge; such a treatment is very specific and therefore is widely used in industry. Corona discharge occurs in the gap between two electrodes, one of which is at a high potential and the other is grounded.[4] This discharge in presence of air entails substantial physico-chemical modifications such as formation of carbonyl groups, ethylenic double bonds and alteration of external appearance.[5]

These modifications strongly enhance the autohesion between two sheets of treated polymer. Although the corona discharge is widely used, the basic knowledge underlying the modifications and evolution of the treated material and its influence on adhesion is still lacking and the relevant literature is controversial.[6] Therefore we chose to study the corona discharge treated polyolefin films and yarns, ESR spectroscopy, ATR and transmission IR, DSC, tonometry, GPC and heat sealing and stripping.

EXPERIMENTAL PART

An unstretched (crystallinity = 30%) LDPE film was studied as well as biaxially stretched (crystallinity = 70%) PP film and 660 dtex PP yarns with a relatively high content of stabilizers.

The corona discharge reactor used is shown in Figure 1. The ground electrode B is an iron rotating roller overlapped with two dielectrics (mica covered with a polyester film). The second electrode is supplied with an alternative tension of 13 KV. Nearly all of the experiments were carried out in ambient air. The maximum main power was 1 KW.

The optical density of each characteristic band of ATR and transmission IR measurements was referred to that of a reference band at 1380 cm^{-1} for LDPE and at 2840 cm^{-1} for PP. The IR experiments were carried out on

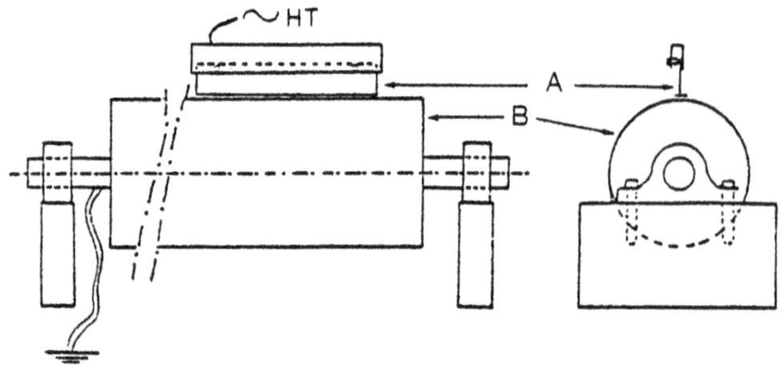

Fig. 1. Electrical Corona Discharge Cell

a PERKIN ELMER 983 apparatus. The ESR spectra were obtained with a BRUKER 100D ESR spectrometer working in X band. The DSC experiments were carried out on a DUPONT 990 thermal analyzer (heating speed 20°C/mn).

The samples were treated, under a 90 s discharge. Although this timelag is unusually high as compared to classical industrial treatments, we observed results of the same nature with treatment times close to 1 s.

RESULTS

Characterization of the Modified Superficial Layer

Chemical Modifications.

ESR. When the samples are corona-treated in the presence of air, no ESR spectrum can be detected, even if they are immediately immersed into liquid nitrogen. Conversely, if the discharge is carried out in an oxygen-free, argon medium, an ESR spectrum is observed. The ESR spectrum of a treated PP film in argon is depicted on the Figure 2. The line shape with no hyperfine structure is quite asymmetric and characteristic of peroxy macroradicals.[7,8] Probably hydrogen atoms on the polymer surface are abstracted, under the electrical discharge, giving rise to carbon macroradicals. These radicals react easily with atmospheric oxygen and give stable peroxy radicals.

When the discharge is carried out in the presence of air, no ESR signal is observed. This result indicates that nonparamagnetic species such as carbonyl groups, or hydroperoxides are produced. The radicals, induced by thermal decomposition of hydroperoxides are generally too short lived species to be observed in ESR, unless previously stabilized by other molecules, called spin traps.[9]

ATR. Figure 3 shows the ATR IR spectra of an unstretched LDPE film, of a corona discharge treated-film and of the superficial layer, stuck on a KRS 5 Crystal, after removing of the bulk of the treated sample. One can see that, on the latter spectrum, the intensities of the double band characteristic of methylenic groups at 2860-2920 cm^{-1} is highly attenuated, while the bands at 1720 cm^{-1} characteristic of carbonyl groups and at 1630 cm^{-1}, characteristic of ethylenic bonds are still present. The same phenomena were observed on PP film and PP yarn.

Tonometry and GPC. The value of the number average molecular weight of the LDPE superficial layer, dissolved in methanol, was obtained by tonometry.[10] The number average molecular weight and the weight average molecular weight of this layer were also determined by GPC. Infrared spectroscopy showed that, the spectrum of the dissolved material in methanol,

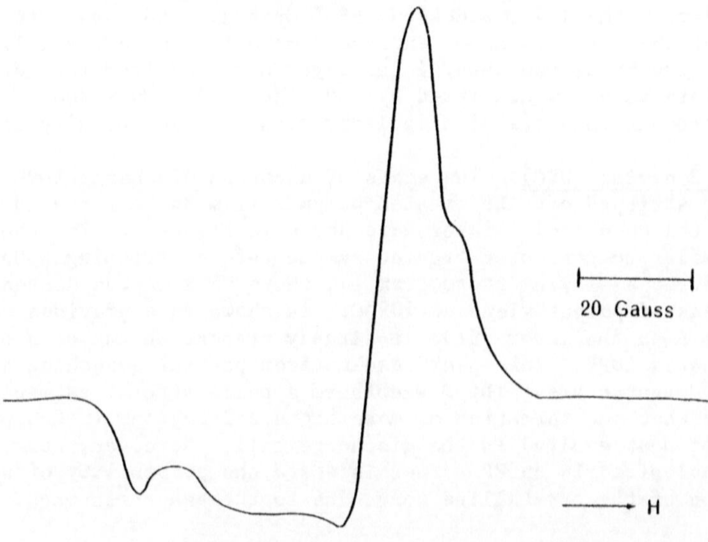

Fig. 2. ESR Spectrum of a PP Film Corona Treated in an Argon Atmosphere and Exposed to Atmospheric Oxygen

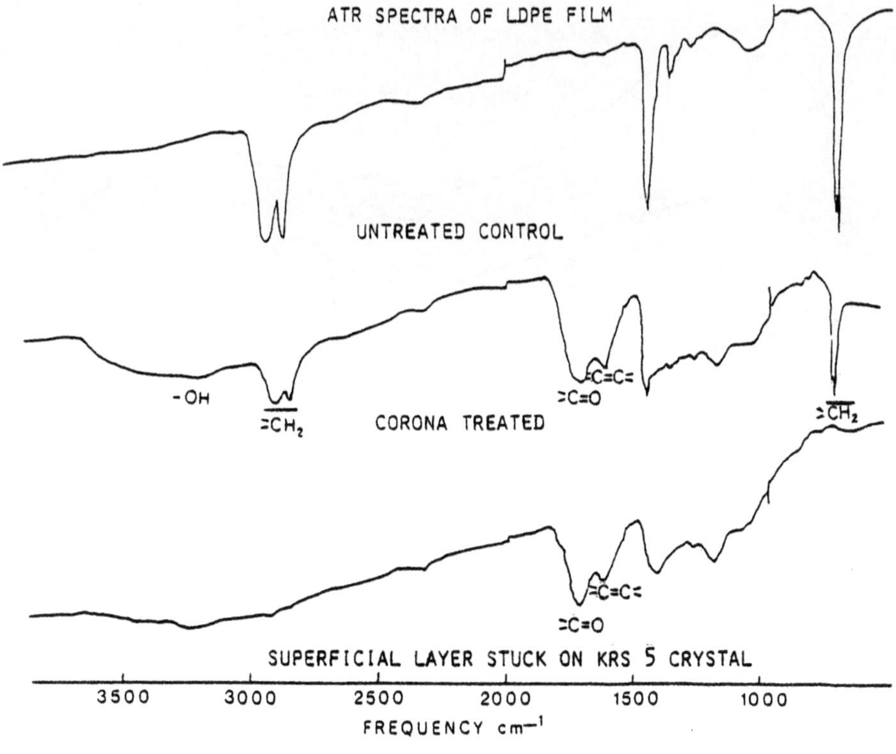

Fig. 3. ATR IR Spectra of LDPE Film

is very similar to the layer stuck on KRS 5 Crystal. Moreover, it has to be pointed out that the amount of extracted material is very small.[11] Nevertheless, the $\overline{Mn}$ of the superficial layer obtained from tonometry is # 200 - 250. This value is confirmed by GPC ($\overline{Mn}$ = 180, $\overline{Mw}$ = 700). This superficial product consists of oligomers, with a polydispersity of 4.

Thermal Behavior (DSC). The scans of a corona discharge LDPE superficial layer, stripped off the treated-polymer film and a corona discharge LDPE free of the superficial layer, are shown in Figure 4. The thermogram 2 is very similar to that of a treated sample before stripping. On the scan of the stripped layer (thermogram 1), there is a marked decrease of the fusion peak of polyethylene at 109°C. As shown in a previous work,[12] the endotherm A on the lower curve is already present on the scan of commercial untreated LDPE. This peak results from partial quenching of the sample during manufacture. The B endotherm appears after treatment. This peak suggests that the formation of more organized regions in the polymer is a result of heat evolved in the discharge cell. Moreover, some results obtained in polarized IR on PP yarns, indicate the possibility of a slight disorientation of the crystalline zones due to the heat treatment.

Influence of Solvent Treatment. The ATR spectra of a PP control sample, of a corona-treated sample and of a treated sample, immersed 2 H in water, are shown in Figure 5. After immersion, one can see a strong decrease of the bands observed after corona treatment (OH, C=O and C=C). The spectrum is quite similar to that of the control.

Figure 6 shows the surface morphology using SEM of a control PP fiber, of a corona-treated PP fiber and of the same treated fiber but immersed in water. The corona-treated fiber shows a surface with pustules which are characteristic of the effects of corona discharge. When these pustules are

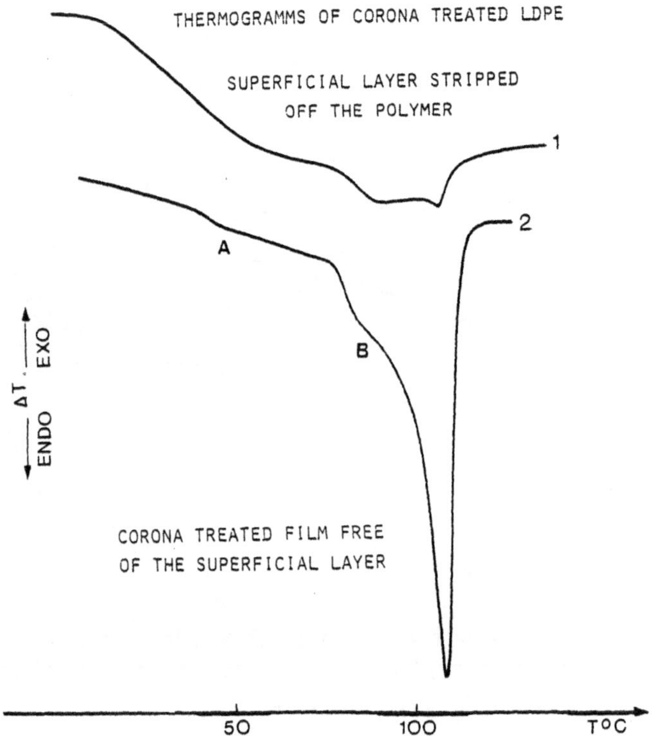

Fig. 4: Thermogramms of Corona Treated LDPE

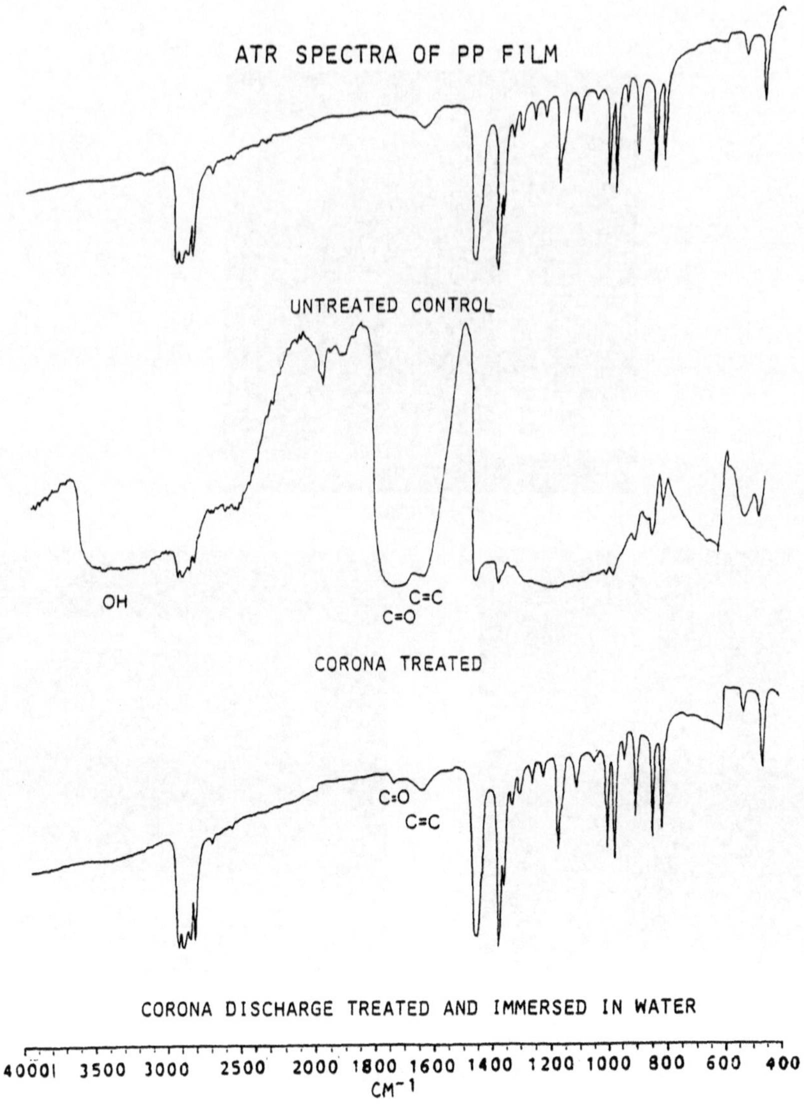

Fig. 5. ATR Spectra of PP Film

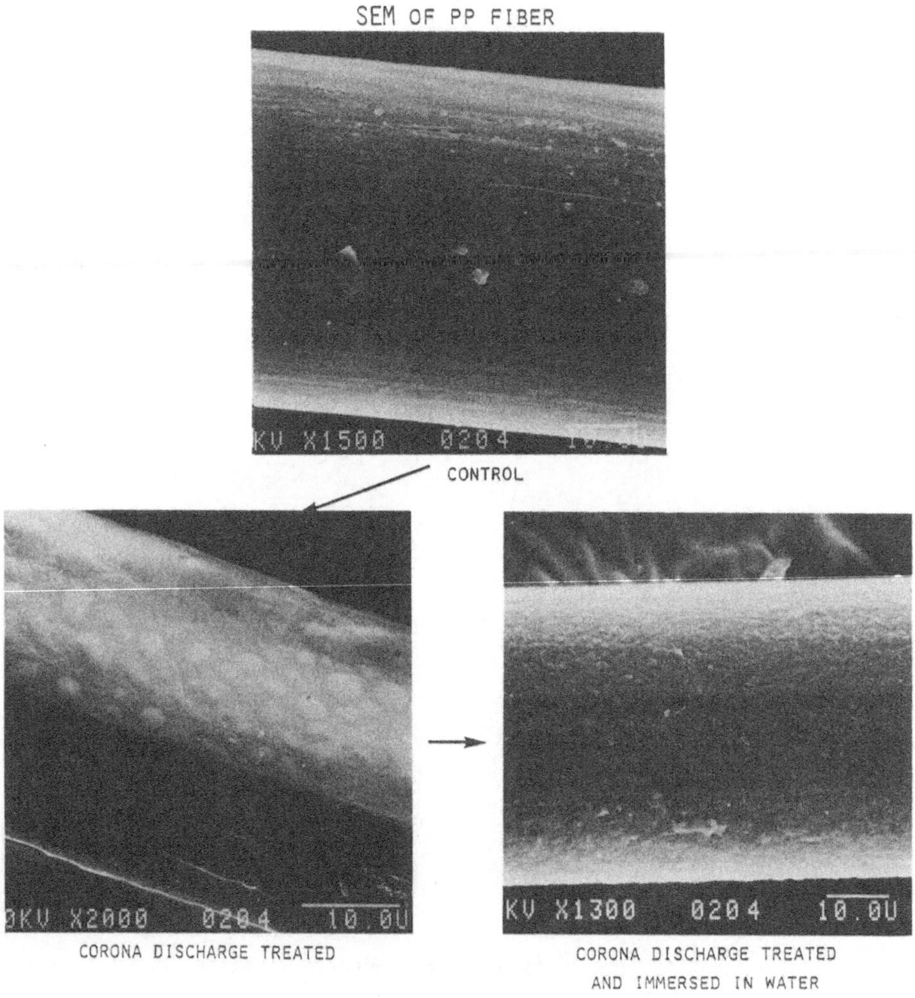

Fig. 6. SEM of PP Fiber

washed off with water, the resulting topography is much rougher than that of control fibers.

TEM of Thin Sections. The micrograph (Figure 7) is obtained by TEM and represents a transverse section of a PP film treated 90 s. on one face and 1 s. on the other face; after corona treatment, the sample was subjected to OsO_4 vapors. The black edges are probably due to a reaction between osmium and the double ethylenic bonds of the modified superficial layers. There is a correlation between the thickness of the superficial layer and the corona treatment duration. Figure 8 shows a cross section of fibers from a yarn stained by OsO_4 before (A) and after (B) corona treatment.

Evolution of the Modified Superficial Layer with the Storage Duration and Medium.

The treated material is usually stored for a few days before being used. Hence, the superficial layer behavior was studied as a function of time and relative humidity of storage. As far as the behavior of the superficial material is concerned, the role of water is a primary importance.

Evolution with Storage Duration. The amount of carbonyl groups and ethylenic bonds, as measured by ATR on a treated LDPE film (Figure 9) and on a treated PP yarn, decrease with storage duration at RH of about 45-55%.

Fig. 7. TEM of PP Film Thin Section Stained with Osmium

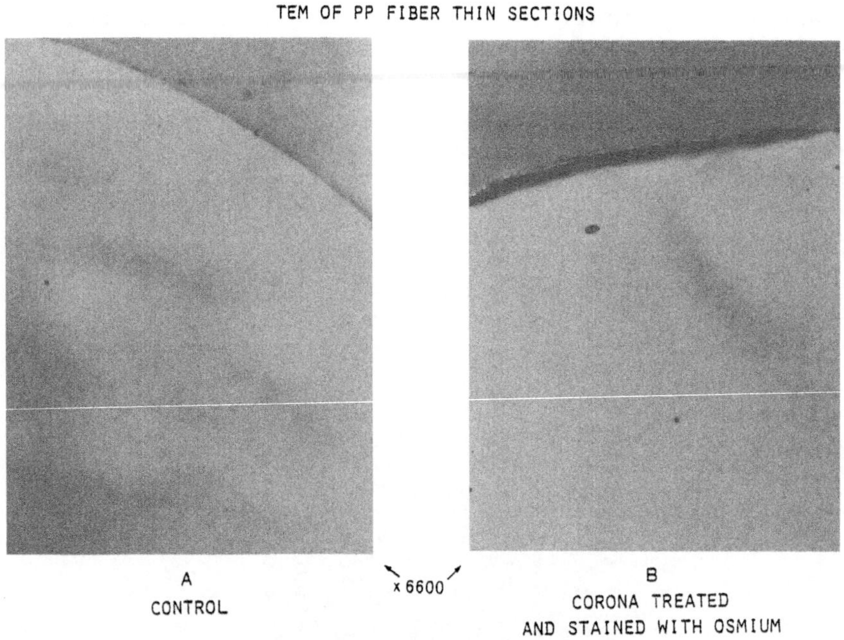

Fig. 8: TEM of PP Fiber Thin Sections

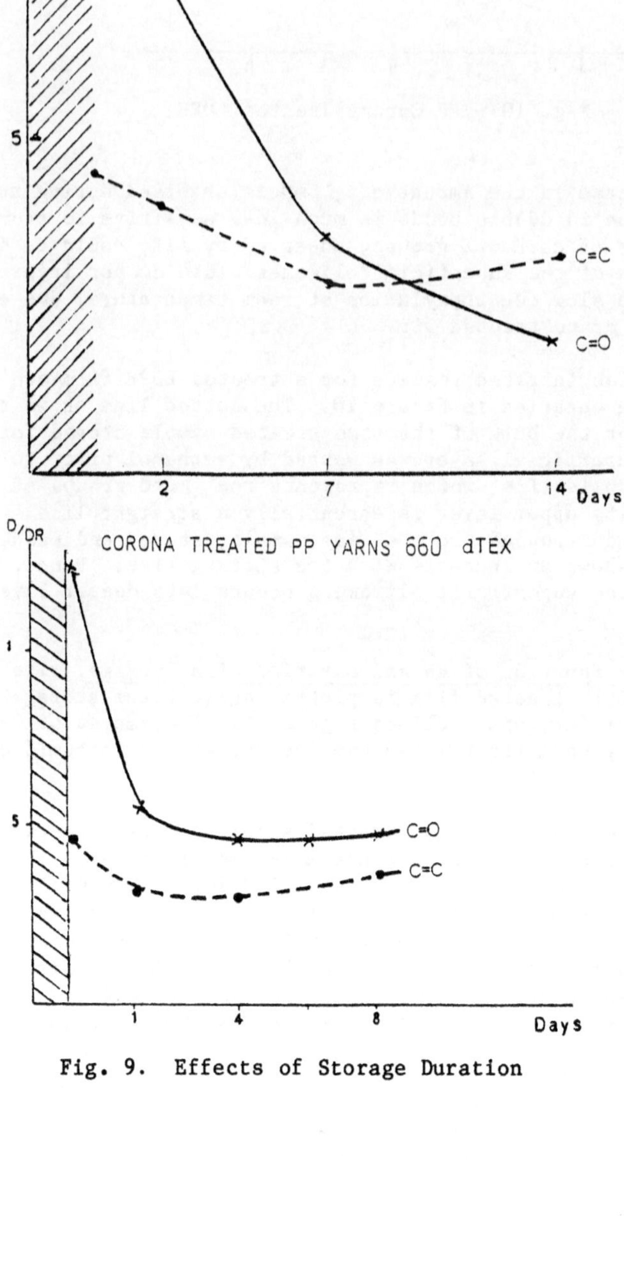

Fig. 9. Effects of Storage Duration

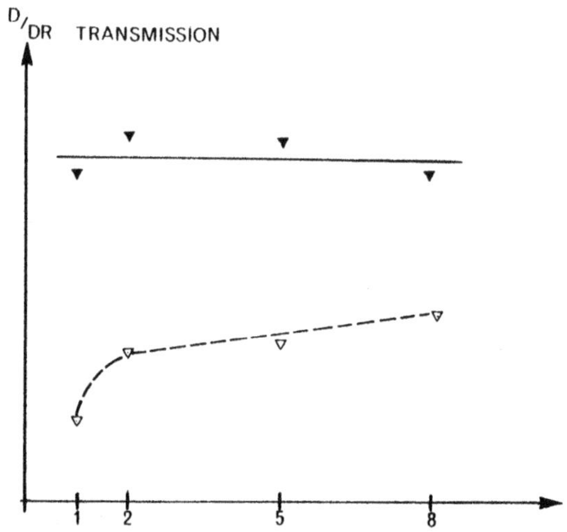

Fig. 10: PP Corona Treated LDPE

A marked decrease in the amount of >C=O is observed during the first days. The variation in double bonds is much less sensitive to storing time. The decrease of carbonyl groups, observed by AIR, could be due to a diffusion of a part of the superficial oligomers into deeper layers of the polymer and/or to a slow decarboxylation at room temperature, due either to atmospheric oxygen or to trapped air.

The transmission infrared results for a treated LDPE film are plotted against the storage duration in Figure 10. The dotted line shows the >C=O groups evolution for the bulk of the same treated sample stored for 1, 2, 5 and 8 days. The superficial layer was washed by methanol prior to each measurement. The solid line, which represents the >C=O groups of the treated film with its upper layer is essentially a straight line. The dotted line, which corresponds to the same sample, stored and washed prior each measurement, shows an increase with the storing time. Hence, a diffusion of part of the superficial oligomers occurs into deeper layers of the film.

Evolution as a Function of RH and Duration of a Storage. The >C=O groups content of LDPE treated film is plotted against the storage duration and RH of the medium (0%, 65%, 90%) on Figure 11. The amount of C=O decreases more steeply when the treated samples are stored in humid atmosphere (65% - 90%).

Autohesion Phenomena. The two treated samples surfaces were pressed for 3 min. at 80°C under 30 bars in a thermomechanical press. The stripping experiments were performed with an INSTRON apparatus (Figure 12). Because the sticking process was not optimized, the numerical values of the sticking strengths are relative values. In Figure 13, one can see the curves of bonding strength versus the joint length of heat-sealed PP film. Curve A is untreated, curve B is corona-treated and curve C is corona-treated with the superficial layer being washed by immersion in water before heat sealing.

The high increase of autohesion after corona treatment is evident from curve b as compared to that of control sample curve a. However, if the corona-treated sample is immersed, before sealing, no more autohesion is observed (curve c).

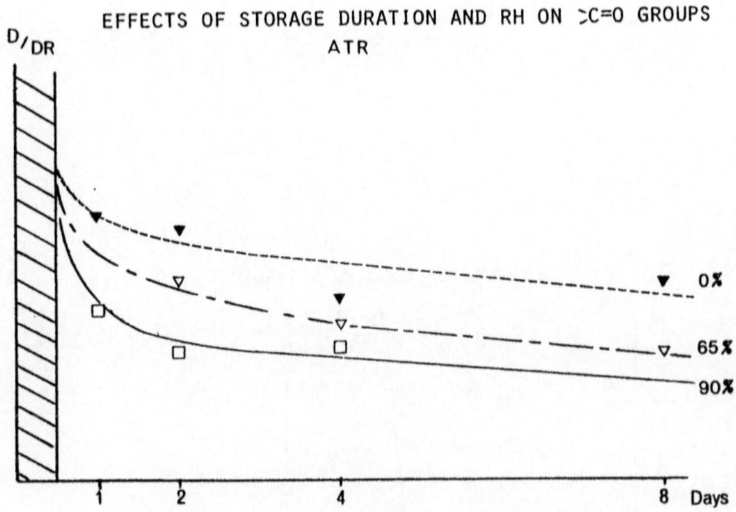

Fig. 11. Corona Treated LDPE Films

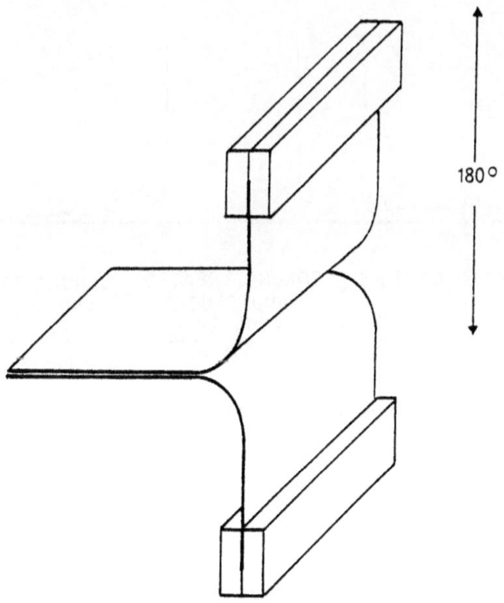

Fig. 12. INSTRON Apparatus

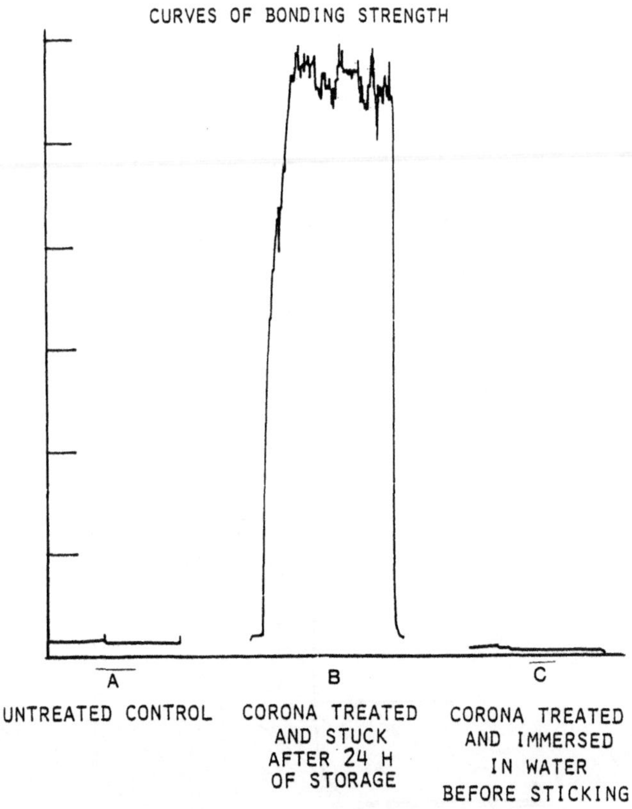

Fig. 13. Corona Discharge Treated PP Film

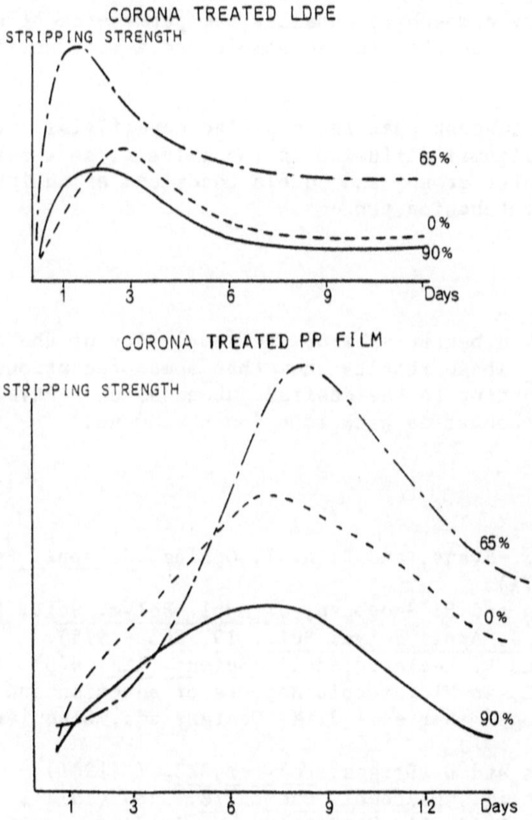

Fig. 14. Evolution of Autohesion Bonding of LDPE Treated Films as a Function of Storage Duration and Medium

Figure 14 shows how the unsticking strength of treated LDPE and PP films varies with the storage duration and medium of the treated samples before sealing. The unsticking experiments were performed 2 h after sticking. The upper graph is LDPE and the lower graph is PP. The relative humidity varies between 0% and 90%.

DISCUSSION

The data show that a corona discharge in presence of air creates an oligomeric superficial layer on polyolefin films and yarns with a number average molecular weight of about 200-250. This amorphous, waxy superficial layer is rich in double ethylenic bonds (1630 cm^{-1}) in ester type carbonyl groups (1720 cm^{-1}) and poor in olefinic moieties. It plays a very important part in the adhesion process. The appearance of this pellicle coincides with a high increase of autohesion: its disappearance by washing with water is accompanied by a drop in adhesivity. The thickness of the superficial component depends on the corona discharge duration (Figure 7).

Moreover, the graphs in the Figure 13 show that a certain amount of storage time is required between corona discharge and sticking. No autohesion is observed if sticking is attempted immediately after treatment.

The chemical modification of treated samples, especially the carbonyl groups and autohesion properties, are closely related to storage duration and medium. A dry atmosphere enhances the permanence of carbonyl groups at a superficial level on the treated sample and a more humid atmosphere aids adhesivity.

The results suggest that the modified superficial layer might act as a plasticizer by oligomer diffusion in the polymer, as a reactive product because of its polar groups and double bonds and as an interdiffusing material in the autohesion process.

CONCLUSION

In this work a better view of the complexity of the autohesion phenomena was reached. These results show that some precautions must be taken into account according to the desired autoadhesion. Physiochemical modifications were not connected with adhesion phenomena.

REFERENCES

1. C. Y. Kim, J. Evans, and D. A. I. Ooring, _J. Appl. Polym. Sci._, _15_, 1365 (1971).
2. R. H. Hansen and H. Shonborn, _J. Appl. Polym. Sci._, _B4_, 203 (1966).
3. D. K. Owen, _J. Appl. Polym. Sci._, _19_, 265 (1975).
4. M. Sotton and B. Leclercq, _Bull. Scient. ITF_, _8_(5), 10 (1979).
5. K. L. Mittal, in Microscopic Aspects of Adhesion and Lubrification Tritibology - Series 7 J. M. Georges ed., Elsevier, Amsterdam (1982).
6. D. M. Brewis and D. Briggs, _Polymer_, _22_, 7 (1981).
7. J. Yakama et al., _Makromol. Chem._ _178_, 2483 (1977).
8. F. Rouzbehi, F. Arefi, B. Catoire, M. Goldman, and J. Amouroux, _I.S.P.C._, _Eindhoven_, _2_,(7) 405 (1985).
9. B. Ranby and J. F. Rabek, _ESR Spectroscopy in Polymer Research_, Springer Verlag, Berlin Heidelberg (1977).
10. H. L. Tighzert, Thèse d'état, Lyon (1983).
11. H. L. Tighzert, T. Philibert, B. Chabert, and G. Nemoz, _Bull. Sci. ITF_, _11_(43) (1982).
12. B. Catoire et al., _Polymer_, _25_, 737 (1984).

PREDICTION OF FAILURE OF

POLYPROPYLENE FILMS

Barbara E. Linder and Robert J. Samuels

School of Chemical Engineering
Georgia Institute of Technology
Atlanta, Georgia 30332-0100

INTRODUCTION

The search for process-property relationships has occupied a central role in the development of polymer engineering. Of particular interest is the understanding, control, and prediction of such properties as modulus, yield, and failure, and the influence variables such as time, temperature, and environment have on material performance. These parameters control all deformation properties and thus determine the process, testing, and end-use applications of a material.

Many processes, testing and end-use applications involve the solid-phase deformation of polymers. Such processes as fiber and film drawing, rolling, cold forming and deep-drawing of sheets, and hydrostatic extrusion all involve solid-phase deformation. Most solid polymer testing (tensile, impact, flexing, etc.) also involves their solid-phase deformation, as does their end-use-application. Clearly a need exists for understanding, and if possible predicting the complex processes occurring.

The problem, however, is formidable. For example, a large number of variables must be considered when processing a polymer into a fiber, a film, or a molding, or when deforming the polymer during testing. A change in any one of the process or test variables will change the final product's behavior. Further, a change in fabrication process from, for example, a fiber to a film process will change the final product's form as well as its behavior. When one considers the number of variables possible within a given process, and then considers the number of possible processes, the probability of correlating these systems through process history criteria seems insurmountable.

The problem becomes even more complex when multiphase systems such as semicrystalline polymers are involved. A semicrystalline polymer such as isotactic polypropylene may be considered as an ideal composite material. The polymer is composed of two phases, one crystalline, the other noncrystalline. Both phases have the same chemical composition, differing only in their structural organization. The phases are chemically bound to each other through molecules, portions of which exist simultaneously in each phase. The molecules in both phases are anisotropic and hence orientation of either or both phases will lead to observed anisotropic bulk properties.

When a semicrystalline polymer is deformed, either during a fabrication process or some mechanical test, the molecules will respond to the temperature and strain rate viscoelastically. Since the glass transition temperature of the noncrystalline region is approximately 100°C lower than the crystal softening temperature, each phase will respond to the deformation differently. Consequently, the orientation of the two phases will differ from each other and this will result in mechanical and optical property differences. Also, because of this difference, the properties of one of the phases may control the response of the material for some temperature or stress environments, while the average of both phases may control in other environments. Clearly, semicrystalline polymers are complex materials.

Any attempt to organize this mass of environmental variables into a clear and predictable order does seem unrealistic. If this vast array of information is looked at from a different perspective, however, the problem of organization does not seem so unreasonable. If all processes are considered simply as different paths from some starting structural state of the material to some final structural state, then it immediately becomes obvious that different process paths can lead to equivalent structural states. Further, one would expect that no matter which process path the material took to reach a certain structural state, once it had reached that state its properties would be defined. The key to correlating all this information is therefore a knowledge of the structural state of the material.[1-7]

Thus, given the structural state of the polymer, it should be possible to predict the modulus, yield and failure stress, and yield and failure strain for different deformation rates and temperatures through suitable correlations. Further, if some general expression characterizing the stress-strain curve from structural parameters were available, the complete solid-phase deformation behavior could be predicted from structural state data.

In order to predict the solid-phase stress-strain behavior of a polymer to yield it is first necessary to demonstrate that structural state parameters can be used to predict not only the small deformation properties (such as modulus), but the large deformation properties (such as yield and failure) as well. That the initial structure determines the small deformation properties would be expected since little or no change in the initial structure occurs at these deformation levels. As deformation continues however, to yield and then even further to failure, the initial structure undergoes significant change. Thus, for the initial structural state of a polymer to be used to predict large deformation properties, some underlying general principal of behavior that couples the final structures to that present initially (at least at some critical points during deformation) is required.

Critical criteria for yield and failure have been developed from continuum mechanics.[8,9,10] A microscopic coupling of the initial structural state of the polymer with the macroscopic critical criteria is required however, before complete predication of the solid-phase deformation behavior of a semicrystalline polymer is feasible from its structural state parameters. A recently completed study[11] has demonstrated that the initial structural state of the polymer does indeed couple with critical yield criteria, and predictions involving processes in the high deformation region are possible. Thus, as a consequence of the above studies, the following information is now available for uniaxially oriented isotactic polypropylene film:

1. The modulus can be predicted over a wide range of temperatures

and frequencies both along and at angles to the film symmetry axis from structural state parameters.[1,12,13]

2. The yield and failure stress can be predicted at several rates and temperatures along the symmetry axis of the film,[14,15] and the yield stress can be predicted at several rates at room temperature around angles to the symmetry axis of the film[11] from structural state parameters.

The purpose of the present study is to investigate (1) whether macroscopic failure criteria can be used to describe the failure of oriented isotactic polypropylene films both along and at angles to the symmetry axis, and (2) whether the macroscopic failure criteria can be coupled with the initial microscopic orientation state of the polymer, such that the failure stress may be predicted from structural state functions directly.

EXPERIMENTAL

Material

The material of choice for the present study is isotactic polypropylene. This polymer has undergone intense examination over past years.[1,2,4-6,11-20] Thus, there is a large available reference literature on the process-structure-property behavior of the polymer. Further, whereas the response of other candidate polymers to processing can lead to various structural complications, such as gauche-trans conformational changes, preferential orientation of molecular ring structures, crystal structure transformations, etc., isotactic polypropylene is quite forgiving and does not pose such a threat.

Film Fabrication

Isotactic polypropylene films of different initial thicknesses were cast on a roll at 30°C. The film thicknesses were varied so that all films would have the same final thickness (approximately 1.5 mils.) after being drawn to different final extensions. The cast films were uniaxially drawn on rolls at 127°C. Once drawn, they were removed from the rolls, cut into strips, and allowed to relax until the measured draw ratio remained constant with time. The relaxed films were used for all measurements.

Structure Determination

The structural state parameter used in this study is the Hermans orientation function, f_p, where p is defined as C or NC referring to the crystalline or noncrystalline phase respectively.[21] The orientation function, f_p, describes the orientation of the molecular chain axis relative to the film fabrication or machine direction of the sample and is defined as:

$$f_p = (3 \langle \cos^2 \theta_p \rangle -1)/2 \qquad -1/2 \leqslant f_p \leqslant 1 \qquad (2)$$

where $\langle \cos^2 \theta_p \rangle$ is the average cosine squared value of the angle between the molecular chain axis and the reference film fabrication axis. The orientation function, f_p equals 1 when the molecular chain axis is perfectly aligned with the film fabrication axis, $f_p = 0$ when the molecular chain axis is randomly oriented with respect to the film fabrication axis and $f_p = -1/2$ when the molecular chain axis is perpendicular with respect to the film fabrication axis. When the molecular chain axis is aligned along some intermediate direction, the orientation function, f_p, will lie in the range $-1/2 < f_p < 1$.

Semicrystalline polymers are polymers composed of two phases, a hard, highly ordered, crystalline phase and a softer noncrystalline phase. The observed bulk property of a semicrystalline polymer may be controlled by either the crystalline phase, the noncrystalline phase or some average of these two phases. The crystalline orientation function may be obtained from wide angle X-ray diffraction data, while the noncrystalline orientation may be obtained from a combination of wide angle X-ray diffraction, density and birefringence methods. When the crystalline and noncrystalline orientations, as well as the volume fractions of each phase are known, the average orientation of the material may be obtained using the equation:

$$f_{AV} = V_C f_C + (1-V_C) f_{NC} \tag{2}$$

where V_C is the volume fraction of crystals. The crystalline, f_C, noncrystalline, f_{NC}, and average, f_{AV}, orientation functions are the primary structural state parameters which will be related to both the axial and off-axis failure stresses in this study.

The structural state data reported for these films[22] are the crystalline, noncrystalline, and average orientation functions, the density and the volume fraction of crystals, and are shown in Table 1. It should be noted for these films that increases in the fabrication draw ratio resulted in corresponding increases in all orientation function values of these films.

Tensile Measurements

The tensile deformation properties were obtained using an MTS model T5002 Tensile Tester. The polypropylene film samples used for these tensile tests were cut to dimensions of 0.2 inches by at least 3 inches. The long axis of the sample was cut at an angle, ϕ, to the film fabrication or symmetry axis direction in the plane of the film (Figure 1). The sample cut angles, ϕ, chosen for the 20%/minute deformation rate were 0, 1, 2, 3, 5, 10, 15, 20, 25, 30, 40, 50, 60, 70, 80, and 90 degrees while the sample cut angles for the 100%/minute deformation rate were 0, 1, 2, 3, 5, 10, 20, 30, 45, 60, 70, 80, and 90 degrees. The samples were cut using an X-ACTO knife with a Type 11 blade on a wooden cutting board. At least five samples of each orientation and sample cut angle were deformed to failure in this study.

Table 1. Morphological Data for Isotactic Polypropylene Films[a] Drawn at 127°C

Sample	Draw Fabrication Temp. (°C)	Draw Ratio	Δ_T ($\times 10^3$)	Density gm/cm^3	V_C	f_C	f_{NC}	f_{AV}
2	127	1.84	8.35	0.9111	0.681	0.371	0.052	0.269
3	127	2.70	13.40	0.9040	0.590	0.534	0.172	0.386
4	127	4.20	22.30	0.9074	0.633	0.761	0.376	0.620
5	127	5.95	26.90	0.9058	0.613	0.899	0.468	0.732

[a]The sample identification used here is consistent with the sample identifications given in [22].

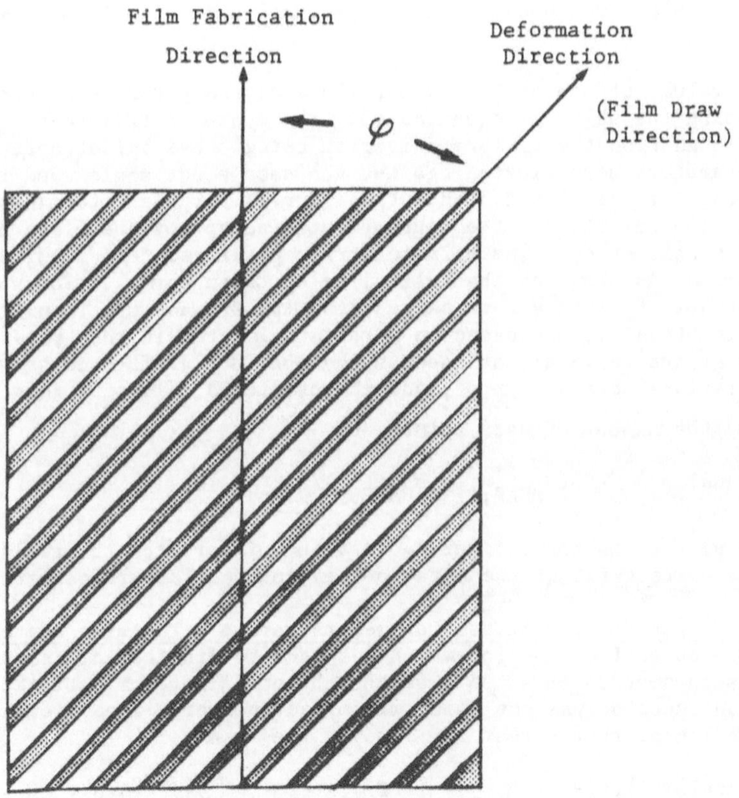

Fig. 1. Example of Template Used for Sample Preparation for Tensile Tests

Once a polymer film sample was prepared, it was placed in the grips such that the draw axis was identically the long axis of the sample. The grips being used in this study were a scale-up of grips designed for use with the Rheovibron.[23] The grips were designed to reduce both sample slippage in the clamps and poor sample alignment. They also allowed for sample mounting and alignment away from the testing equipment. A sample alignment device was constructed which insured that the long axis of the film would always be perpendicular to the edges of the grips and parallel to the motion of the crosshead of the tensile tester.[24] The sample alignment device produced samples with an initial crosshead of one inch for all measurements.

Curve Fitting

Failure Stress versus Sample Cut Angle. Several objective functions were evaluated to determine their applicability to the determination of the anisotropic material parameters from a suitable failure criterion. The objective function chosen to fit the failure stress data to the failure equation was:

$$\sum F_i = \sum(\log \sigma_{MEAS,i} - \log \sigma_{CALC,i})^2 \tag{3}$$

where $\sigma_{MEAS,i}$ was the measured failure stress and $\sigma_{CALC,i}$ was the predicted failure stress from a suitable failure criterion. This objective function would not be affected by a decade variation in the measured failure stress. The anisotropic material parameters were obtained using the subroutine ZXSSQ[25] which used a Levenberg-Marquardt algorithm to minimize equation (3).

Once values of the anisotropic failure stress parameters were obtained by minimizing the value of equation (3), the measured failure stress and that predicted from the failure criterion using these anisotropic failure stress parameters were plotted against the sample cut angle used for the deformation. Figures 2 and 3 show these curves for the deformation rates of 20% and 100%/minute for the four oriented polypropylene films being examined in this study. The failure stress predicted from a suitable failure criterion, is shown as the solid line of these curves. The values of the anisotropic failure stress parameters, A_F, B_F, and C_F, in units of Megapascals^{-2}(MPa^{-2}), discussed in more detail later in this paper, and the value of the residual, defined as the absolute difference between the measured failure stress, $\sigma_{MEAS,i}$ and the predicted failure stress, $\sigma_{CALC,i}$ divided by the number of data points, N:

$$\text{Residual} = \sum \text{ABS}(\sigma_{MEAS,i} - \sigma_{CALC,i})/N \quad (4)$$

were also plotted on these figures. A value of zero for the residual would identify a perfect fit of the data to the proposed failure criterion.

Anisotropic Failure Stress Parameters versus Structural State Parameters. The objective function which was used to fit the anisotropic failure stress parameters obtained from the chosen failure criterion to the orientation function was one which minimized the point-line distances along the perpendicular to the best fit line in the form

$$\text{Residual} = \sum(((P_i - P_{int})^2 + (f_{p,i} - f_{p,int})^2)^{1/2})/N \quad (5)$$

In this equation, P_i is the ith value of the anisotropic failure stress parameter, $f_{p,i}$ is the ith value of the orientation function being evaluated, and P_{int} and $f_{p,int}$ are the values of the anisotrophic failure stress parameter and the orientation function on the calculated curve at the point where the perpendicular to the best fit line passing through P_i and $f_{p,i}$ intersects the best fit line. In this expression, N is the number of points evaluated. The relationships which were evaluated between the anisotropic failure stress parameters and the structural state parameters were:

$$\log A_F = \text{Slope} \times f_p + \text{Intercept} \quad (6)$$

$$B_F = \text{Slope} \times f_p + \text{Intercept} \quad (7)$$

$$C_F = \text{Constant} \quad (8)$$

Statistical Analysis

In earlier work with axial failure of polypropylene films,[1,15,26] the mode of the measured failure data was used in determining a relationship between the failure and the structural state parameters. Approximately 70 data points were obtained for each oriented polymer film for the axial failure determination in the earlier work. A histogram of the failure values was then drawn for this large quantity of samples and the failure value which exhibited the highest frequency of measured values was taken as the mode.

Approximately five failure stress observations per deformation angle per initial orientation were evaluated in the current investigation. A

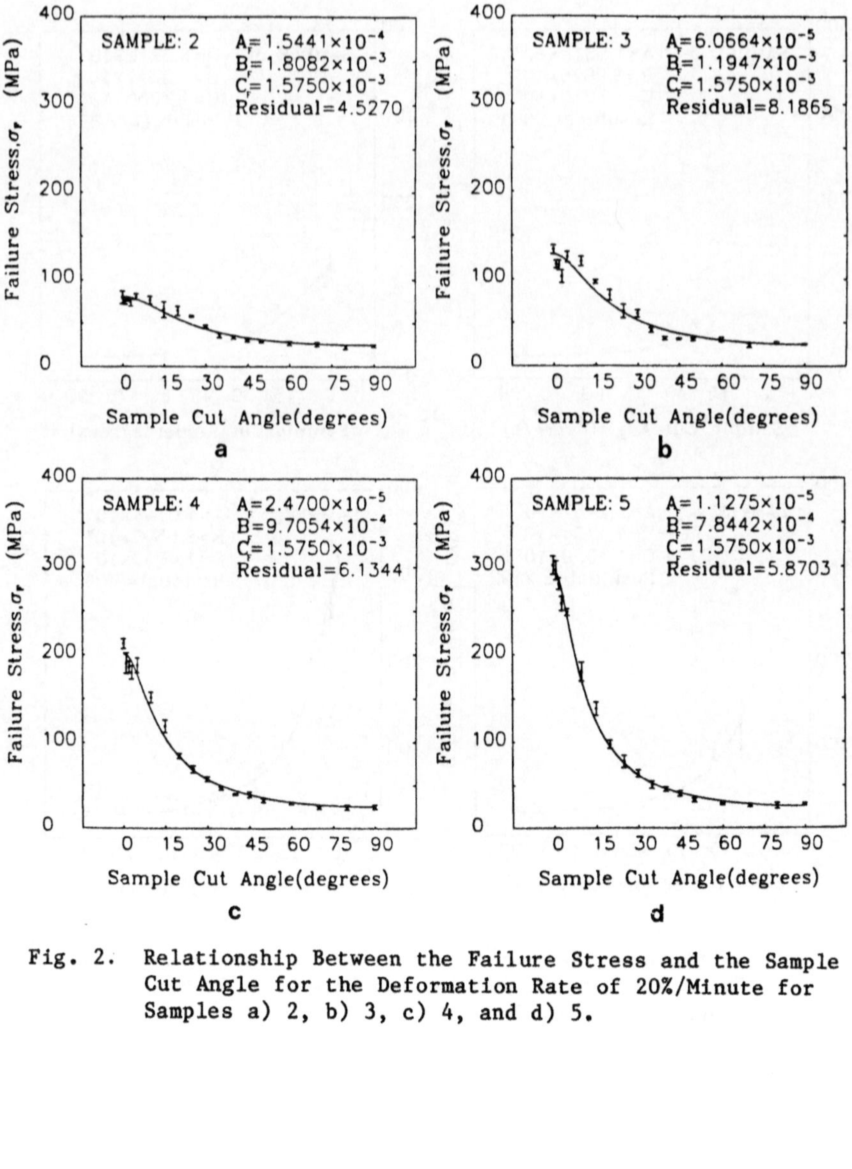

Fig. 2. Relationship Between the Failure Stress and the Sample Cut Angle for the Deformation Rate of 20%/Minute for Samples a) 2, b) 3, c) 4, and d) 5.

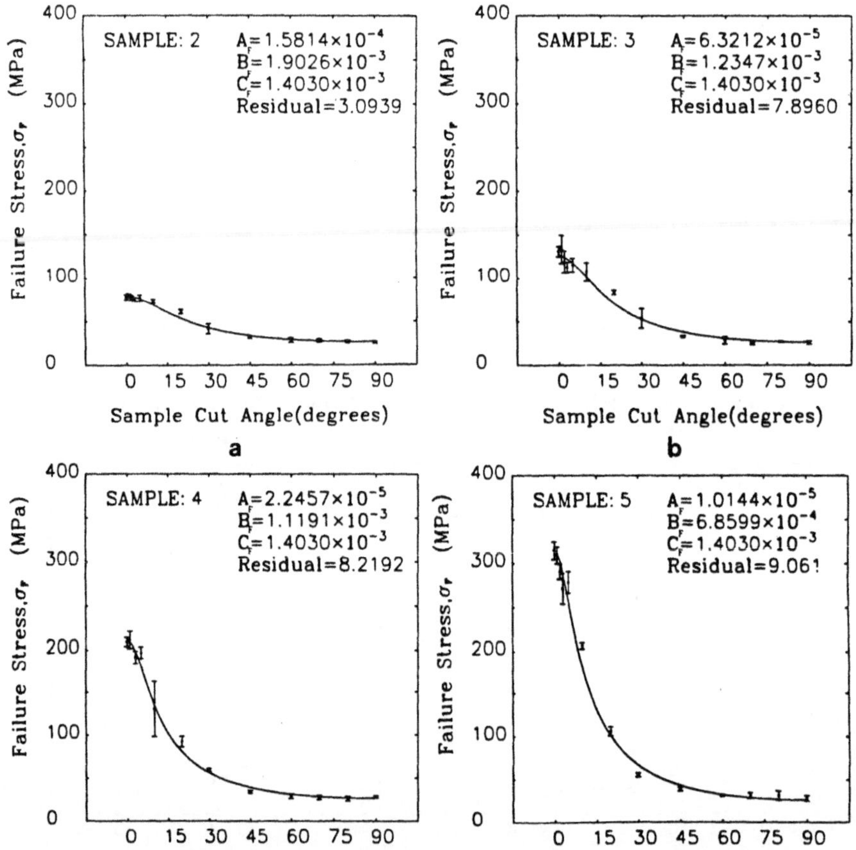

Fig. 3. Relationship Between the Failure Stress and the Sample Cut Angle for the Deformation Rate of 100%/Minute for Samples a) 2, b) 3, c) 4, and d) 5.

histogram for this small number of observations would be meaningless, therefore no mode value may be defined. The only values which may be determined from this small quantity of measured failure stress values are the median and the mean. The median is defined as the value which divides the distribution into two equal parts, while the mean is the average of all measured values. A comparison between the mode of one data set and the mean or median of another would be meaningless, therefore, either the mean or the median of all data sets should be compared. Data sets using only five data points were also available from the earlier study[26] as opposed to the 70 data points reported.[15] These five point data sets were used for comparison with the failure stress values measured in this study.

The measured data has also been found to be skewed. This means that the measured data is not symmetrically distributed about some central value. The best statistical variable to evaluate the failure stress for skewed data is the median[27] as opposed to the mean (Figure 4). Since the median failure stress may be readily determined for the data obtained in this investigation, it was used in the evaluation of both the axial and off-axis failure stresses.

RESULTS AND DISCUSSION

When a stress, σ, is applied to a cast film, the film is deformed by a given amount of strain, ε. When the polymer deformation is done at both a suitable deformation rate and temperature a stress-strain curve schematically shown in Figure 5 is obtained. Initially, a sharp rise in load with strain is observed which is linear to a strain of one half percent or less. Essentially, no structural changes occur at this low deformation. This portion of the curve corresponds to the initial elastic deformation and determines the initial modulus. As the stress increases, the slope of the stress-strain curve decreases prior to yield. This is known as the strain softening region, where the onset of plastic nonrecoverable deformation occurs. As the stress increases in the strained polymer, the yield point is reached. In this figure, the yield point is the first region in the curve where the slope equals zero. Little deformation occurs prior to yielding, therefore only minor structural changes would accompany this deformation. Following the yield point, the stress rapidly decreases with strain until the lower yield point is attained. After the lower yield

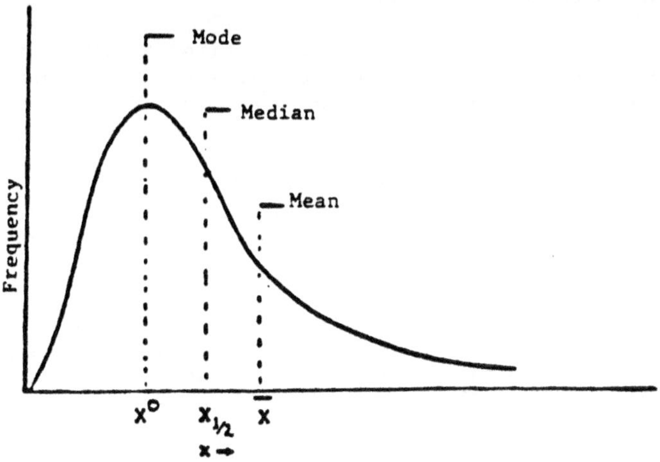

Fig. 4. Representation of the Mode, the Median and the Mean for Skewed Data

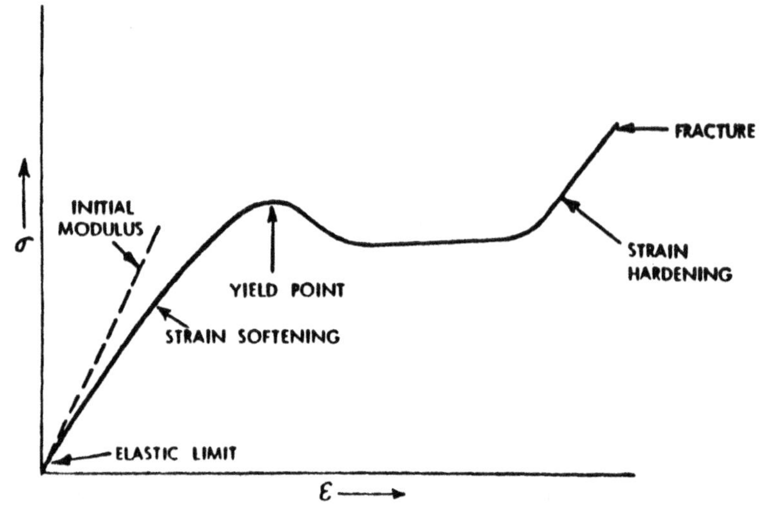

Fig. 5. Schematic Representation of a Stress-Strain Curve

point, the stress remains constant with applied strain. In this region, the sample cross-section approaches a uniform value and is known as necking. The last region shown in this stress-strain curve is the strain hardening region, where the stress again increases with strain. The final point in the stress-strain curve is the failure point, the point at which the sample breaks. The present study will concentrate on the failure behavior observed during both axial and off-axis tensile deformation of isotactic polypropylene films.

Phase Orientation and Failure Stress

Axial Deformation. Prior to any discussion of off-axis failure stress, the axial failure stress was evaluated and compared to results of an earlier axial failure study[1,15,26] of isotactic polypropylene which used films fabricated from the same film grade resin under similar processing conditions but deformed at a different fabrication draw temperature. The fabrication draw temperature for the films used in the earlier study was 110°C, as opposed to a fabrication draw temperature of 127°C for the films used in the current study.

Table 2 shows the morphological characterization data obtained for the films used in the earlier study. Both the earlier and present stress-strain measurements were made at 70°C and 50% R.H. The median value of approximately five measured failure stress data points for each oriented film, obtained at deformation rates of 20%/minute and 100%/minute from the present investigation, and 50%/minute and 100%/minute from the earlier investigation were then plotted against the initial noncrystalline, crystalline and average orientation functions as shown in Figures 6-a through 6-c.

Figure 6-a shows a linear relationship between the failure stress measured in the current study at the deformation rates of 20% per minute and 100% per minute, and that measured in the earlier study at deformation rates of 50%/minute and 100%/minute, when plotted against the initial noncrystalline orientation. This figure shows that the magnitude of the failure stress measured for the two polymer film series prepared using different fabrication conditions are the same for samples having equivalent noncrystalline orientation functions.

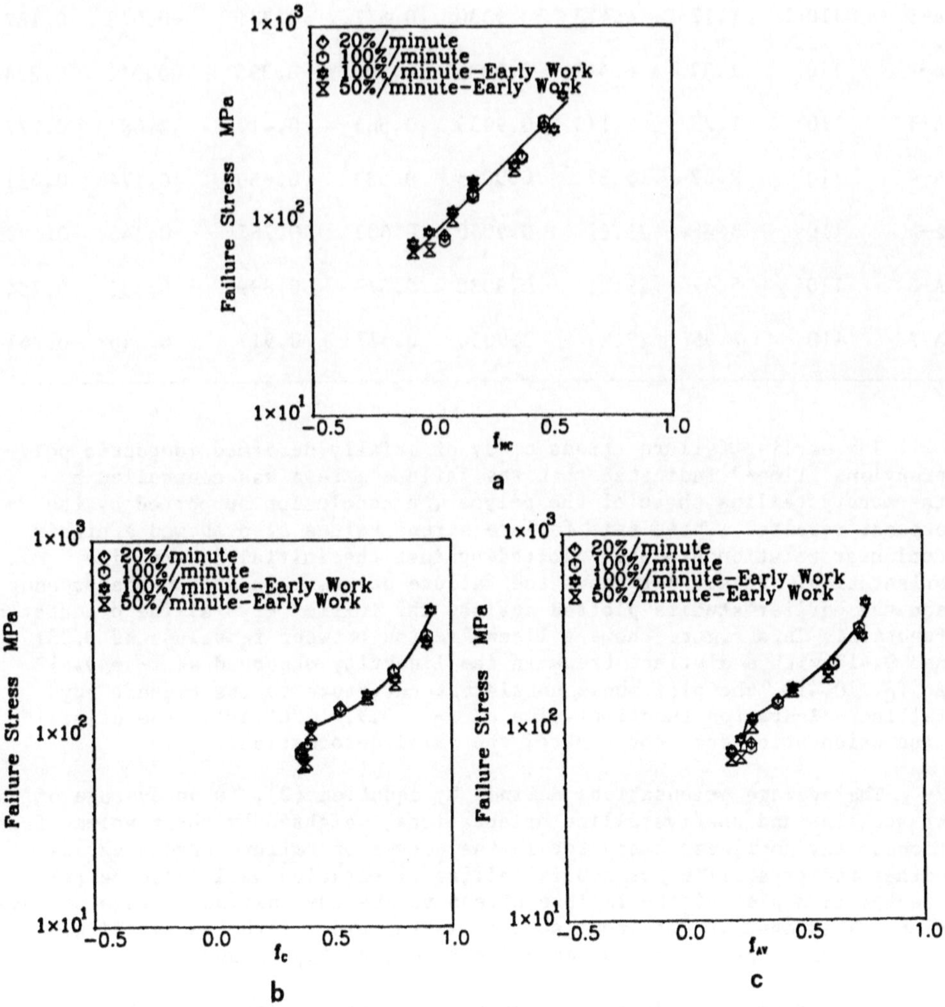

Fig. 6. Relationship Between Failure Stress at 0 degrees and the a) Noncrystalline Orientation Function, b) Crystalline Orientation Function, and c) Average Orientation Function

Table 2. Morphological Data for Isotactic Polypropylene Films[26] Drawn at 110°C

Sample	Draw Fabrication Temp. (°C)	Draw Ratio	Δ_T ($\times 10^3$)	Density gm/cm^3	V_C	f_C	f_{NC}	f_{AV}
A-1	110	1.12	4.413	0.9030	0.577	0.381	-0.078	0.187
A-2	110	1.32	6.447	0.9036	0.585	0.395	-0.011	0.226
A-3	110	1.73	9.170	0.9035	0.583	0.412	0.087	0.277
A-4	110	2.62	15.37	0.9035	0.583	0.650	0.174	0.451
A-5	110	3.46	21.67	0.9050	0.603	0.767	0.345	0.599
A-6	110	5.47	29.21	0.9030	0.577	0.899	0.556	0.754
A-7	110	7.35	29.29	0.9030	0.577	0.917	0.548	0.761

The earlier failure stress study of axially deformed isotactic polypropylene films[15] indicated that the failure stress was controlled by the noncrystalline phase of the polymer, a conclusion supported by the present results. These same failure stress values also showed a highly nonlinear relationship when plotted against the initial crystalline orientation. Figure 6-b shows the failure stress values from the present and the earlier studies plotted against the initial crystalline orientation function. This figure shows a linear region between f_C values of 0.381 and 0.412 with a distinct break in the linearity observed at $f_C = 0.412$. At $f_C > 0.412$, the plot shows consistent curvature to the highest crystalline orientation function value of $f_C = 0.917$. Clearly, the crystalline orientation does not control the axial deformation.

The average orientation, defined by equation (2), is an average of the crystalline and noncrystalline orientations, weighted by their volume fractions. Any nonlinear character in the curves of failure stress versus either the crystalline or noncrystalline orientation would also be reflected in a plot of the failure stress versus the initial average orientation. This explains the nonlinearity in the plot of the axial failure stress versus the average orientation function, f_{AV} (Figure 6-c).

The figure also shows that the failure stress measured in the current investigation is a semilogarithmic function of all three phase orientation functions, f_C, f_{NC}, and f_{AV}. The axial failure stress is linear, however, only for the equivalent noncrystalline orientation functions for both polymer film series prepared under the different fabrication conditions. The axial failure stress may therefore be considered to be controlled by the noncrystalline phase of the polymer.

Off-Axis Deformation. Relationships have been developed for isotactic polypropylene relating the axial deformation with the initial noncrystalline orientation function.[5] However, no relationship between the failure stress and the initial structural state parameters for off-axis tensile deformation have been developed. Two approaches were taken in this investigation 1) a macroscopic approach for the determination of the coefficients of a specific failure equation used to predict the failure stress from the measured data and 2) a microscopic approach, relating these coefficients to the initial structural state parameters.

Macroscopic Approach - Failure Criteria. Several failure criteria exist which predict failure for the axial and off-axis deformation of orthotropic composite materials. These include, 1) the maximum stress theory, 2) the maximum strain theory, 3) the Tsai-Hill theory, and 4) the Tsai-Wu theory.[10] Tsai[9] examined the maximum stress theory for the application of a uniaxial stress applied at an angle, ϕ, to the fiber direction of a unidirectionally reinforced composite specimen and found that the theory did not adequately represent the experimental data. He then evaluated the maximum strain theory[9] and found that this criteria predict the failure stress even more poorly than the maximum stress criterion, and he therefore continued his investigation to find a more suitable criterion.

Tsai found that the Tsai-Hill theory described the failure data for unidirectionally reinforced composites better than the other criteria. However, all three suffered some inadequacies in the prediction of measured data and the Tsai-Wu theory was developed to improve the correlation. To do this the number of terms in the predictive equation was increased. Since, upon the application of a uniaxial stress, a criterion composed of a maximum of three independent stress coefficients may be used to predict the failure stress behavior, this criterion is not appropriate for the prediction of the failure stress in the current investigation.

The third failure criteria examined by Tsai was the Tsai-Hill criterion[9] [identical in form to the Hill yield criterion[8]] for anisotrophic materials, as shown below:

$$F_F(\sigma_Y - \sigma_Z)^2 + G_F(\sigma_Z - \sigma_X)^2 + H_G(\sigma_X - \sigma_Y)^2 \tag{9}$$

$$+ 2L_F\tau_{YZ} + 2M_F\tau_{ZX} + 2N_F\tau_{XY} = 1$$

The anisotropic failure stress parameters, F_F, G_F, H_F, L_F, M_f, and N_F, are defined as

$$G_F + H_F = 1/X^2 \tag{10}$$

$$H_F + F_F = 1/Y^2 \tag{11}$$

$$F_F + G_F = 1/Z^2 \tag{12}$$

$$2F_F = 1/Y^2 + 1/Z^2 - 1/X^2 \tag{13}$$

$$2G_F = 1/Z^2 + 1/X^2 - 1/Y^2 \tag{14}$$

$$2H_F = 1/X^2 + 1/Y^2 - 1/Z^2 \tag{15}$$

$$2L_F = 1/R^2 \tag{16}$$

$$2M_F = 1/S^2 \tag{17}$$

$$2N_F = 1/T^2 \tag{18}$$

where X, Y, and Z are the tensile failure stresses along the principal axes of anisotropy and R, S, and T are the shear failure stresses with respect to those principal axes of anisotropy. In this equation, σ_X, σ_Y, and σ_Z are the applied tensile stresses and τ_{YZ}, τ_{ZX}, and τ_{XY} are the applied shear stresses in terms of the principal axes of anisotropy. Several as-

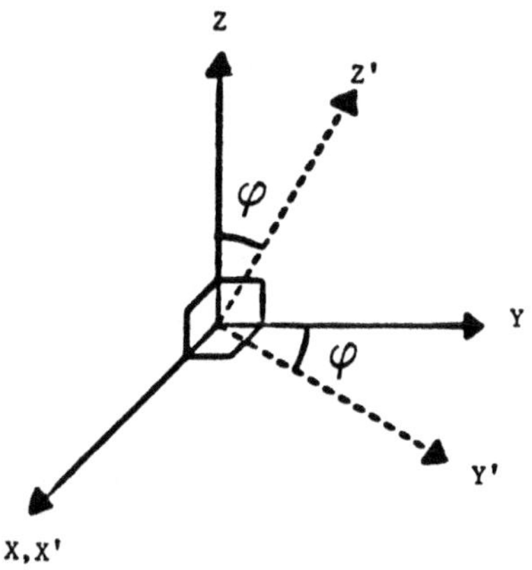

Fig. 7. Axis Definitions for the Application of a Uniaxial Stress Along the Z' Axis in the Y-Z Plane

sumptions have been made in the development of this predictive equation for failure stress of anisotropic materials. These assumptions are 1) the absolute magnitudes of the compressive and tensile failure stresses are identical, 2) the failure stress is not controlled by the hydrostatic component of stress, and 3) failure occurs at the limit of elastic behavior.

In the present study, a uniaxial tension was applied along the Z' axis in the Y-Z plane. The Y-Z plane defines the plane of the film, with the Z axis aligned parallel to the symmetry axis, and the Y axis perpendicular to the symmetry axis in the plane of the film. The X axis is the axis normal to the plane of the film. The Z' axis for this development makes an angle, ϕ, with the Z axis in the Y-X plane (Figure 7). For the application of a uniaxial stress along the Z' axis, the Tsai-Hill failure criterion may be rewritten as

$$(F_F + G_F)\cos^4 \phi + 2(L_F - F_F)\sin^2 \phi \cos^2 \phi \tag{19}$$

$$+ (H_F + F_F)\sin^4 \phi = 1/\sigma^2$$

where σ is the applied uniaxial tension along the Z' axis. As can be seen, the application of a uniaxial stress along the Z' axis, reduces the number of coefficients from six to four, those coefficients being F_F, G_F, H_F, and L_F. Equation (19) is applicable to systems which are either orthotropic, where the four constants, F_F, G_F, H_F, and L_F are unique quantities, or transversely isotropic, where the four constants may be reduced to three constants. If the system is assumed to be transversely isotropic in the X-Y plane, implying F = G, equation (19) may be simplified to:

$$2F_F\cos^4 \phi + 2(L_F - F_F)\sin^2 \phi \cos^2 + \tag{20}$$

$$+ (H_F + F_F)\sin^4 \phi = 1/\sigma^2$$

Equations (19) and (20) may be rewritten in terms of the three parameters, A_F, B_F, and C_F as

$$A_F\cos^4 \phi + B_F\sin^2 \phi \cos^2 \phi + C_F\sin^4 \phi = 1/\sigma^2 \tag{21}$$

where the three parameters are defined as:

$$A_F = (F_F + G_F) \tag{22}$$

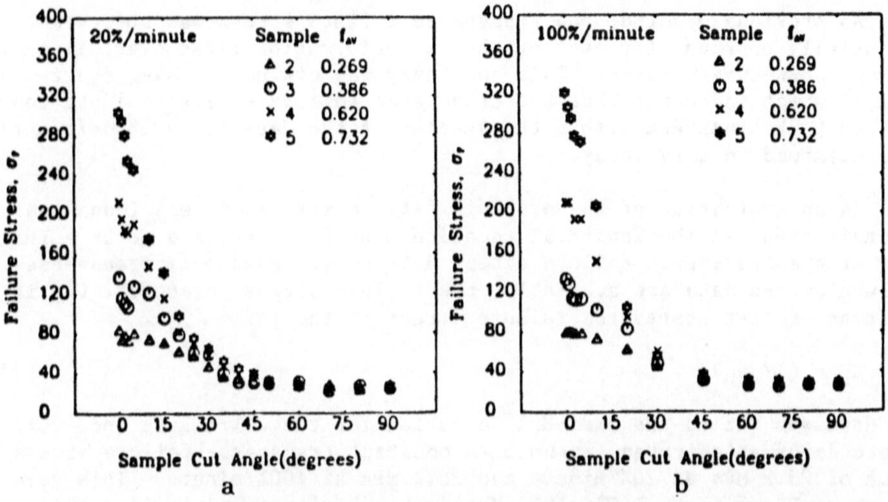

Fig. 8. Composite Failure Stress Versus Sample Cut Angle for a) 20%/Minute and b) 100%/Minute Deformation

$$B_F = 2(L_F - F_F) \tag{23}$$

$$C_F = (H_F + F_F) \tag{24}$$

for orthotropic symmetry and

$$A_F = 2\ F_F \tag{25}$$

$$B_F = 2\ (L_F - F_F) \tag{26}$$

$$C_F = (H_F + F_F) \tag{27}$$

for a material which is transversely isotropic in the X-Y plane. The values of A_F, B_F, and C_F will be identical for either a transversely isotropic system or an orthotropic system. Only identification of the magnitudes of F_F, G_F, H_F, and L_F can distinguish between these two forms of symmetry and the magnitudes of F_F, G_F, H_F, and L_F cannot uniquely be determined from tensile failure stress data alone.

The failure stress may be predicted from tensile failure stress data using an equation having no more than three independent coefficients. If an equation contains more than three independent coefficients, those constants will not uniquely be determined from the tensile failure stress data alone. Thus, the only equation appropriate for tensile data analysis is the Tsai-Hill equation as defined by equation (21); and therefore, this equation was used to describe and predict the tensile failure stress in the present study.

Relationship between Failure Stress and Sample Cut Angle. Composite plots of failure stress versus sample cut angle for the deformation rates of 20%/minute and 100%/minute are shown in Figures 8a and b. In general, the failure stress is seen to increase with increasing initial average orientation. These figures show a failure stress dependence upon sample orientation for cut angles near the axial deformation direction (0°), with this orientation dependence decreasing as the sample cut angle approaches the transverse direction of the film (90°). Along the transverse direction, there appears to be little or no dependence of the failure stress upon the initial structural state of the polymer.

As previously noted, the transverse failure stress was observed to be essentially constant for each of the two deformation rates over all initial average structural states. This constancy was evaluated using the t-test,[28] which indicated that the transverse failure stress could be considered to be constant within the scatter of the data for each deformation rate examined in this study.

In an evaluation of the off-axis failure stress, it was found that the off-axis data fit the Tsai-Hill equation when C was assumed to be a function of the transverse failure stress data only. When only transverse failure stress data are available, the failure stress parameter, C_F, is a function of that transverse failure stress in the form:

$$C_F = (1/\sigma_{90})^2 \tag{28}$$

The data was fit to the Tsai-Hill equation for the 20%/minute and 100%/minute deformation rates, assuming a constant transverse failure stress value of 25.2 MPa at 20%/minute and 26.7 MPa at 100%/minute. This gave values of C_F equal to 1.575×10^{-3} MPa^{-2} at 20%/minute and 1.403×10^{-3} MPa^{-2} at 100%/minute. Figures 2 and 3 show the failure stress versus sample cut angle for the deformation rates of 20%/minute and 100%/minute using this curve fitting scheme.

Microscopic Approach - Relationship between Anisotropic Failure Stress Parameters and Structural State Parameters. The values for A_F and B_F obtained from the best fit of the failure stress data to the Tsai-Hill equation when C_F was held constant, as shown in Figures 2 and 3, were plotted against the initial structural state parameters (Table 1). Figures 9-a and 9-b show these plots for the 20%/minute deformation rate while Figures 10-a and 10-b show the plots for the 100%/minute deformation rate.

The three solid lines on Figures 9-a and 10-a show the best fit lines predicted by equation (6), when the parameter A_F was fitted to either the crystalline, noncrystalline, or average orientation function values. Note that the solid lines predict the value of the anisotropic failure stress parameter A_F from any phase orientation function, with almost equivalent accuracy. The solid lines (Figures 9-b and 10-b) for B_F, plotted against the phase orientation functions, show the values of B_F, predicted from equation (7).

The anisotropic failure stress parameter, B_F, may be related to any of the three reported structural state parameters, with almost equivalent accuracy. Table 3 shows the tabulated slopes and intercepts obtained using the predictive equations for A_F and B_F for the deformation rates of 20%/minute and 100%/minute, using constant values of C_F equal to 1.575×10^{-3} and 1.403×10^{-3}, respectively.

Once a distinct functional relationship was determined between the three anisotropic failure stress parameters and the structural state parameters, the derived equations were used to predict the anisotropic failure stress parameters, A_F, B_F, and C_F from the structural state parameters only. The predicted anisotropic failure stress parameters were subsequently used to predict the axial and off-axis tensile failure stresses.

The anisotropic failure stress parameters were predicted from either 1) the crystalline orientation function, 2) the noncrystalline orientation function or 3) the average orientation function only, and these values of A_F, B_F, and C_F, were used to predict the value of the failure stresses using equation (21). Figures 11 and 12 show the axial and off-axis failure stresses for the four oriented polymer films when predicted from the noncrystalline orientation function only, for the deformation rates of 20%/

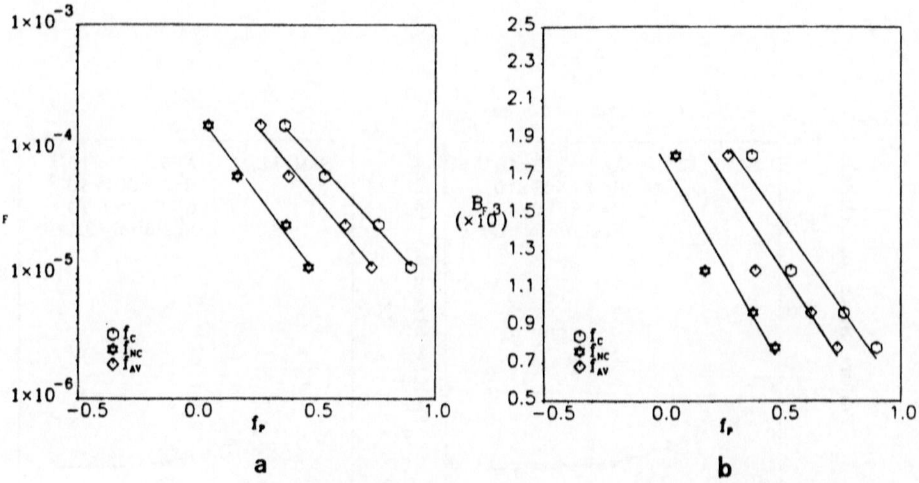

Fig. 9. Relationship Between the Anisotropic Failure Parameters, a) A_F and b) B_F and the Phase Orientation Functions for the Deformation Rate of 20%/Minute

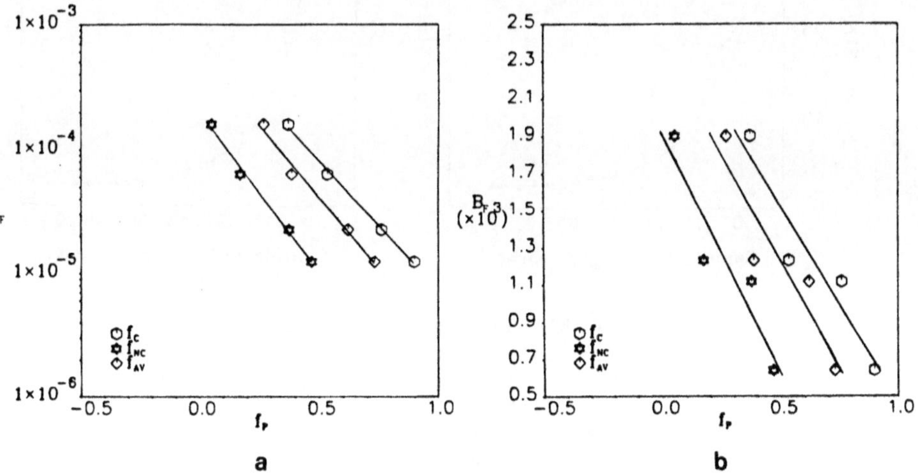

Fig. 10. Relationship Between the Anisotropic Failure Parameters, a) A_F and b) B_F and the Phase Orientation Functions for the Deformation Rate of 100% Per Minute

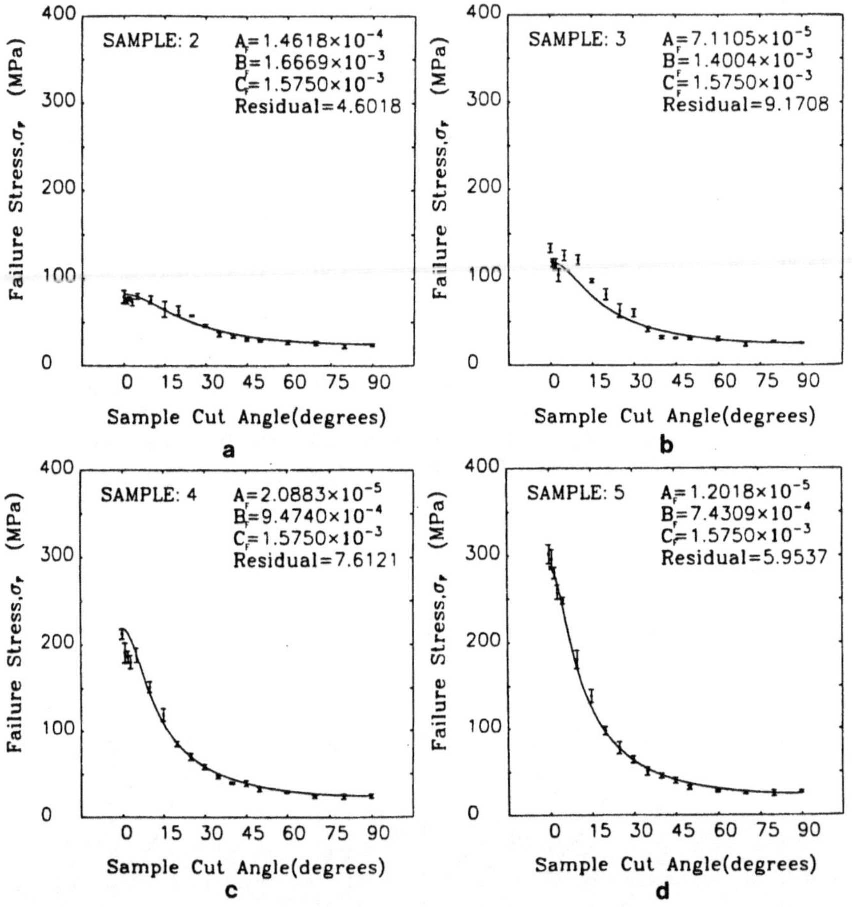

Fig. 11: Relationship Between the Off-Axis Failure Stress and the Sample Cut Angle for a Deformation at 20%/Minute. The Solid Line Shows the Failure Stress Predicted from the Noncrystalline Orientation Function Only for Samples a) 2, b) 3, c) 4, and d) 5.

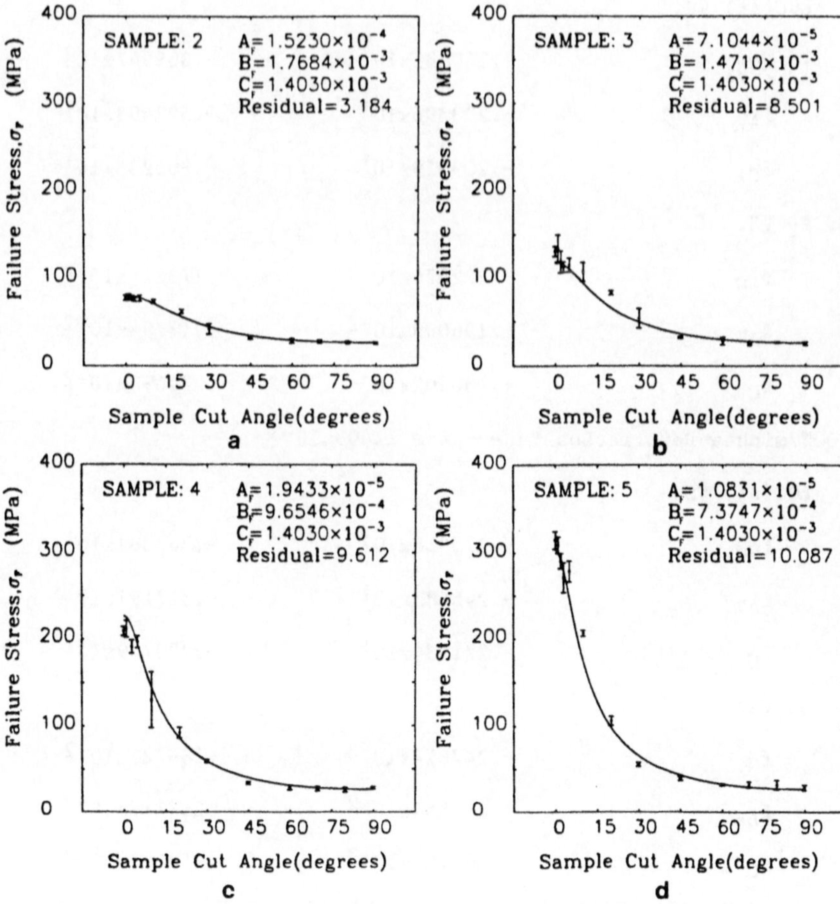

Fig. 12. Relationship Between the Off-Axis Failure Stress and the Sample Cut Angle for a Deformation at 100%/Minute. The Solid Line Shows the Failure Stress Predicted from the Noncrystalline Orientation Function Only for Samples a) 2, b) 3, c) 4, and d) 5.

Table 3. Table of Slopes and Intercepts Obtained from the Best Fit of the Anisotropic Failure Stress Parameters to the Structural State Parameters.

Orientation Function	Slope	Intercept
20%/minute Deformation Rate - $C = 1.575x10^{-3}$		
$LOG(A_F)$ VS.		
f_{NC}	$-.260831x10^{1}$	$-.369947x10^{1}$
f_{AV}	$-.233396x10^{1}$	$-.322483x10^{1}$
f_C	$-.209519x10^{1}$	$-.305235x10^{1}$
B_F VS.		
f_{NC}	$-.222076x10^{-2}$	$.178241x10^{-2}$
f_{AV}	$-.196009x10^{-2}$	$.217294x10^{-2}$
f_C	$-.180502x10^{-2}$	$.234693x10^{-2}$
100%/minute Deformation Rate - $C = 1.403x10^{-3}$		
$LOG(A_F)$ VS.		
f_{NC}	$-.275968x10^{1}$	$-.367381x10^{1}$
f_{AV}	$-.246883x10^{1}$	$-.317191x10^{1}$
f_C	$-.221934x10^{1}$	$-.298749x10^{1}$
B_F VS.		
f_{NC}	$-.247814x10^{-2}$	$.189723x10^{-2}$
f_{AV}	$-.219554x10^{-2}$	$.233719x10^{-2}$
f_C	$-.203106x10^{-2}$	$.253799x10^{-2}$

minute and 100%/minute respectively.

The noncrystalline orientation function was chosen to represent the off-axis data since the axial failure stress behavior has been found to be controlled by the noncrystalline phase of the polymer.[1,15,26] Table 4 shows the residuals obtained for the predictions of the failure stress when the anisotropic failure stress parameters are obtained from the three initial structural state parameters as compared to the initial curve fit of the data to the Tsai-Hill equation (See Figures 2 and 3). These residual values show that the failure stress may be predicted from any of the phase orientation functions. This result is simply a consequence of the limited orientation range of the present samples. The earlier axial studies, covering a wider orientation range clearly shows the control which the noncrystalline region has over failure processes in isotactic polypropylene. The values also indicate that the off-axis failure stress may be correlated

Table 4. Comparison of Residuals from Curve Fits of the Failure Stress

Sample	Initial Curve Fit (MPa)	$\sigma_F(f_C)$ (MPa)	$\sigma_F(f_{NC})$ (MPa)	$\sigma_F(f_{AV})$ (MPa)
20%/minute Deformation Rate - $C_F = 1.575\text{x}10^{-3}$ MPa^{-2}				
2	4.527	4.549	4.602	4.876
3	8.167	8.882	9.171	9.585
4	6.134	6.622	7.612	7.321
5	5.870	5.702	5.954	5.760
100%/minute Deformation Rate - $C_F = 1.403\text{x}10^{-3}$ MPa^{-2}				
2	3.094	3.100	3.184	3.529
3	7.896	8.089	8.501	9.055
4	8.219	8.075	9.612	9.144
5	9.061	9.332	10.087	9.572

to the microstructure of the polymer sample.

CONCLUSIONS

In order to predict the solid-phase stress-strain behavior of a polymer to failure, it is first necessary to demonstrate that structural state parameters can be used to predict not only the small deformation properties (such as modulus), but the large deformation properties (such as yield and failure) as well. That the initial structure determines the small deformation properties would be expected since little or no change in the initial structure occurs at these deformation levels. As deformation continues to yield, however, and then even further to failure, the initial structure undergoes significant change. Thus, for the initial structure state of a polymer to be used to predict large deformation properties, some underlying general principle of behavior that couples the final structures to that present initially (at least at some critical points during deformation) is required.

Earlier studies have shown that both the modulus[1,12,13] and yield stress[11,14] could be predicted axially and around angles from initial structural state functions. The purpose of the present study was to explore whether the same may be true for failure stress. That is, 1) whether macroscopic failure criteria can be used to describe the failure of oriented isotactic polypropylene films both along and at angles to the symmetry axis, and 2) whether the macroscopic failure criteria can be coupled with the initial microscopic orientation state of the polymer, such that the failure stress may be predicted from structural state functions directly.

The present study has shown that both the axial and off-axis failure stress of uniaxially oriented isotactic polypropylene films does correlate with the initial structural state via the Tsai-Hill failure stress parameters. Thus, the primary goals of the investigation have been fulfilled.

The absolute magnitudes of the axial failure stress values obtained for the oriented polypropylene films at the two deformation rates used in the current study, 20%/minute and 100%/minute, were found to be the same at equivalent noncrystalline structural states as the axial failure stress values from an earlier 50%/minute and 100%/minute failure stress study of oriented polypropylene films prepared using a different fabrication draw temperature. This indicates that the axial failure stress is controlled by the noncrystalline structural state of the test material and not by the fabrication process used to make the test specimen.

Finally, the above results strongly suggest that some underlying general principle of molecular deformation behavior couples the critical path determinants (modulus, yield and failure) with the initial starting structure of the material.

ACKNOWLEDGEMENTS

Financial support for this project was provided by the National Science Foundation Polymers Program, Grant No. DMR-8012692.

REFERENCES

1. R. J. Samuels, Structured Polymer Properties, Wiley-Interscience, New York (1974).
2. J. L. Kardos, J. Raisoni, S. Picarolo, and J. C. Halpin, Polymer Engineering and Science, 19, 1000 (1979).
3. A. R. Wedgewood and J. C. Seferis, Polymer Engineering and Science, 24, 328 (1984).
4. R. S. Stein and G. L. Wilkes, Structure and Properties of Oriented Polymers, I. M. Ward, Ed., John Wiley and Sons, New York, Chap. 3 (1975).
5. I. M. Ward, Mechanical Properties of Solid Polymers, Wiley-Interscience, New York (1974).
6. C. R. Desper, Characterization of Materials in Research, Ceramics and Polymers, Syracuse University Press, New York, Chap. 16 (1975).
7. R. L. McCullough, C. T. Wu, J. C. Seferis, and P. H. Lindenmeyer, Polymer Engineering and Science, 16, 371 (1976).
8. R. Hill, The Mathematical Theory of Plasticity, Clarendon Press, Oxford (1950).
9. S. W. Tsai, Strength Theories of Filamentary Structures, in: Fundamental Aspects of Fiber Reinforced Plastic Composites, R. T. Schwartz and H. S. Schwartz, Eds., Wiley, New York (1968).
10. L. R. Jones, Mechanics of Composite Materials, McGraw Hill, New York (1975).
11. B. E. Linder and R. J. Samuels, Polymer Engineering and Science, 25, 821 (1985).
12. J. C. Seferis, R. L. McCullough, and R. J. Samuels, J. Macromol. Sci.-Phys., B13(3), 357 (1977).
13. J. C. Seferis and R. J. Samuels, Polymer Engineering and Science, 19, 975 (1979).
14. R. J. Samuels, SPE ANTEC, 26, 309 (1980).
15. R. J. Samuels, Polymer Engineering and Science, 19, 66 (1979).

16. J. E. Huber and R. J. Samuels, in: Interrelations Between Processing, Structure, and Properties of Polymeric Materials, J. C. Seferis and P. S. Theocaris, Eds., Elsevier Scientific Publishing Co., New York, p. 193 (1984).
17. R. J. Samuels, Die Makromol. Chem., Suppl. 4, 241 (1981).
18. R. J. Samuels, Polymer Engineering and Science, 23, 257 (1983).
19. L. W. Collier and R. J. Samuels, SPE ANTEC, 29, 390 (1983).
20. F. M. Mirabella, J. Polymer Sci., Polymer Physics Ed., 22, 1293 (1984).
21. P. H. Hermans, "Physics and Chemistry of Cellulose Fibers with Particular Reference to Rayon," Elsevier, New York (1949).
22. R. J. Samuels, J. Macromol. Sci.-Phys., B8(1), 41 (1973).
23. A. R. Wedgewood and J. C. Seferis, Polymer, 22, 966 (1981).
24. B. E. Linder, M. S. Thesis, Georgia Institute of Technology (1984).
25. "International Math Science Library Incorporated," Edition 9, June 1, 1982, (IMSL Ed. 9) Houston, IMSL.
26. R. J. Samuels, J. Macromol. Sci.-Phys., B4, 701 (1970).
27. H. L. Pazer and L. A. Swanson, "Modern Methods for Statistical Analysis," Intext, San Francisco (1972).
28. R. V. Dilts, "Analytical Chemistry," D. Van Nostrand, New York (1974).

ULTRA-HIGH MODULUS POLYOLEFINS

FROM SINGLE CRYSTALS

T. Kanamoto, T. Hoshiba, K. Tanaka, and M. Takeda

Department of Chemistry, Science University of Tokyo
Kagurazaka, Shinjuku-ku, Tokyo 162, Japan

Roger S. Porter

Materials Research Laboratory
Department of Polymer Science and Engineering
University of Massachusetts, Amherst, MA 01003 USA

INTRODUCTION

Recently we have shown that single crystal aggregates of ultra-high molecular weight polyethylenes (UHMW-PE), and of polypropylene (UHMW-PP), grown from dilute solutions, exhibit a dramatically high extensibility compared to melt-crystallized forms. The mats of solution-grown crystals for these two polymers have been drawn up to a ratio of 250 for UHMW-PE and 66 for UHMW-PP. Such superdrawn films exhibit extremes of properties including a tensile modulus, up to 220 GPa for the former and 33 GPa for the latter polymer. These values approach the uncertain theoretical maximum modulus reported for the perfect crystal for each of the two polymers. For effective deformation, the first draw step is solid-state extrusion. The choice of original morphology for extrusion draw is of major consequence, with single crystals offering a significant advantage.

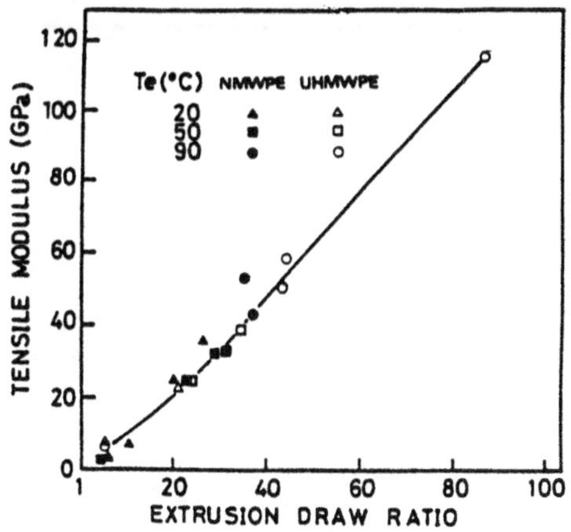

Fig. 1

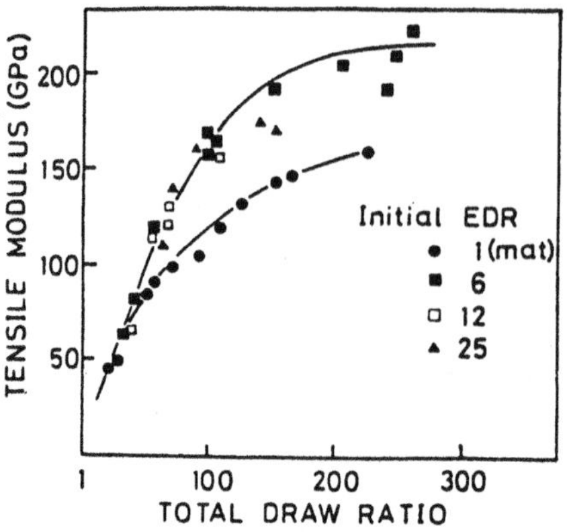

Fig. 2

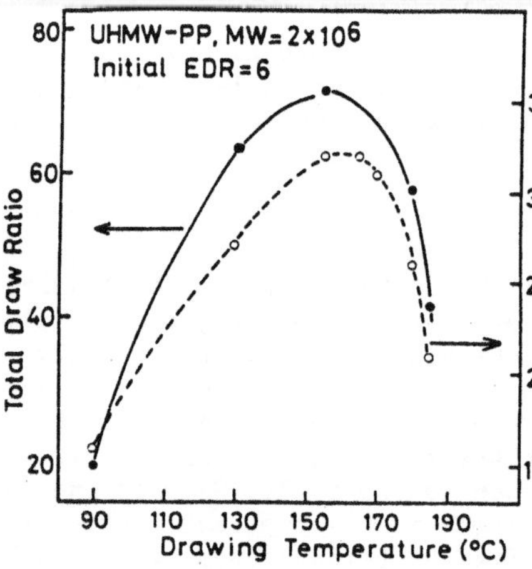

Fig. 3

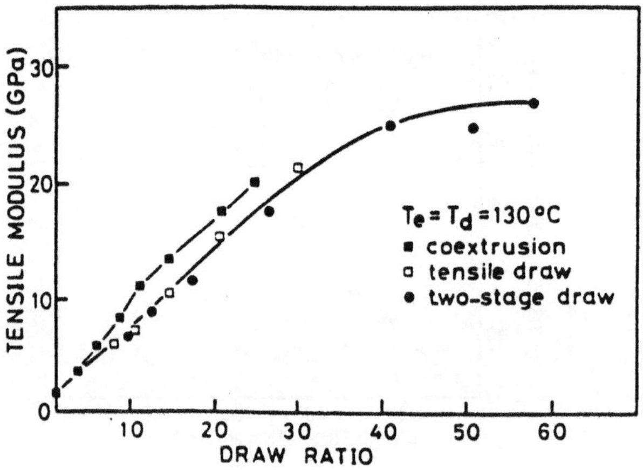

Fig. 4

Mats of solution-grown crystals formed from different and high molecular weights of polyethylene and polypropylene have been ultradrawn by three techniques; solid-state coextrusion developed by Porter et al., tensile drawing, and a combination of solid-state coextrusion followed by tensile drawing.[1-4] The effects of the drawing techniques and of polymer MW on the deformation behavior and the resultant structures and properties have been studied. Some of the results on the ultra-draw of polyethylene and on polypropylene are given in the following illustrations. The total draw ratio is the product of the initial extrusion draw ratio and the following ratio for tensile draw.

REFERENCES

1. T. Kanamoto, A. Tsuruta, K. Tanaka, M. Takeda, and R. S. Porter, Polym. J., 15, 327 (1983).
2. T. Kanamoto, A. Tsuruta, K. Tanaka, and M. Takeda, Polym. J., 16, 75 (1984).
3. J. Southern and R. S. Porter, Chem. Eng. News, p. 45 (Sept. 22, 1969).
4. P. D. Griswold, A. E. Zachariades, and R. S. Porter, Polym. Eng. Sci., 18, 861 (1978).

MONITORED PHOTOOXIDATION OF POLYPROPYLENE NONWOVENS

T. L. Vigo[1], A. Muschelewicz[2], and K.-Y. Wei[3]

USDA, ARS Textiles and Clothing Laboratory
Knoxville, TN 37916

[1]USDA, ARS, SRRC, P.O. Box 19687, New Orleans, LA 70179
[2]Dept. of Textiles, Merchandising, and Design, The University of Tennessee, Knoxville, TN 37996-1900
[3]Pall Corporation, 30 Sea Cliff Ave., Glen Cove, N.Y. 11542

INTRODUCTION

The photooxidation of polypropylene has been extensively investigated by a variety of techniques (primarily infrared spectroscopy and electron spin resonance) to elucidate the mechanisms by which this photooxidation occurs.[1] As with most polymers, hydroperoxides have been found to photosensitive effectively the degradation and cause chain scission and the formation of various decomposition products containing carbonyl and alkene functional groups.[2] Most studies however, have focused on the volatile degradation products formed from polypropylene at high temperatures (usually greater than 200°C) utilizing gas chromatography, and in some instances, mass spectrometry.[3] The studies that have addressed the polypropylene residue have been primarily concerned with functional group analysis by infrared spectroscopy and/or chemical techniques.[1] This study provides additional information on the photooxidation mechanism of polypropylene utilizing mass spectrometry to examine the nature of the volatile products liberated from the polymer residue (before and after ultraviolent exposure) at different temperatures, in conjunction with changes in various functional carbonyl groups and in the mechanical properties of the nonwoven polypropylene fabrics.

EXPERIMENTAL

Materials

Polypropylene fibers were 1.5 denier Herculon,® obtained from Hercules Fibers, and poly(methyl methacrylate) beads from Rohm and Haas, Plexiglas* V811). The following chemicals were used: o-nitrobenzaldehyde (m.p. 42-45°C, Eastman Kodak); $CoSO_4.7H_2O$, $CuSO_4.5H_2O$, $NiSO_4.7H_2O$ and $NH_2OH.HCl$

*Mention of trademarks or proprietary products does not constitute a guarantee or warranty of the product by USDA and does not imply its approval to the exclusion of other products that may be available.

(all reagent grade, Baker Chemical Co.); toluene and ethanol (ACS Certified, Fisher Scientific); chlorous acid (12 g of $NaClO_2$--J.T. Baker Reagent Grade--in 100 ml of 30% aq. HOAc--Fisher Scientific, ACS Certified); $Si(CH_3)_2Cl_2$ (Aldrich Chemical Co.); and Methylene Blue (Fisher Certified Biological Stain) buffered to a pH of 7.9 with mono- and dibasic sodium phosphate (J. T. Baker Reagent Grade).

Procedures for Preparing Films and Fibers

Acrylic films containing a solid state actinometer were prepared by stirring poly(methyl methacrylate) beads (17.8 g) in toluene (80 ml) at 25°C until a viscous solution formed, adding solid o-nitrobenzaldehyde to the solution, then stirring it for an additional hour. The solution was cast on a glass surface (previously treated with $Si(CH_3)_2Cl_2$ to facilitate removal of the film) utilizing a Gardner knife. After evaporation of the toluene for 8 h at 25°C, an acrylic film of 27 um thickness containing 0.51M o-nitrobenzaldehyde was obtained. Polypropylene nonwoven fabrics were prepared by taking carded webs (1.0 oz/yd^2) and thermally bonding them between stainless steel Ferrotype plates on a Wabash press* for 1 min at 165°C.

Photooxidation Apparatus, Calibration and Procedures

A special apparatus for exposing films and fabrics to ultraviolet light was constructed, and has been described in similar studies on the photolysis and photooxidation of Nylon 66.[4] A Hanovia mercury lamp (light source) was mounted in its reflector on a piece of wood so that its distance from the sample could be varied. Inside a wooden chamber, a water-cooled condenser served as a sample reservoir. The system was designed to run either under vacuum, in the absence of air (e.g., nitrogen could be introduced) or in the presence of air. In either instance, volatile substances could be trapped for mass spectral analysis (see schematic in Figure 1).

Calibration of the light source was achieved by first pouring a special aqueous solution of salts (36% $NiSO_4.7H_2O$-5% $CoSO_4.7H_2O$-$CuSO_4.5H_2O$) into the exterior jacket or condenser enveloping the inner quartz tube, so that the optimum bandpass filter was centered at 3200A°.[5] The lamp was

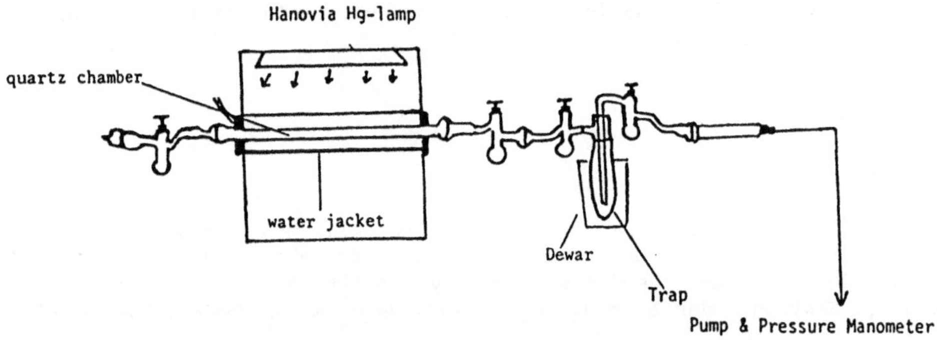

Fig. 1. Ultraviolet Exposure Apparatus for Films and Fabrics Consisting of a Mercury Vapor Lamp, Quartz Chamber and Water-Cooled Jacket, and Provision to Condense Volatile Products

placed at a distance of 6.8 cm from the center of the quartz tube and turned on for 1 h. A poly(methyl methacrylate) film containing the solid state actinometer, o-nitrobenzaldehyde, was then placed on a glass plate in the center of the quartz tube and irradiated until the substance in the film photoisomerized to o-nitrosobenzoic acid. The photoisomerization normally took about 10 min and was monitored by the disappearance of the $-NO_2$ peak at 1520 cm^{-1} in the infrared spectra of the acrylic film. This calibration method gave a light intensity value at 3200A° of 9.92×10^4 erg/cm^2 sec for the Hanovia mercury vapor high pressure lamp. The experimental value was found to be in good agreement with the light intensity value of 5.32×10^5 erg/cm^2 sec, which was obtained after 10 h irradiation as described in the literature by the lamp supplier. Figure 2 shows a schematic of the chamber used for calibration in the light source.

In the photooxidation exposures, the nonwoven polypropylene fabrics were protected from wavelengths less than 300 nm by use of a Pyrex filter. The nonwoven fabrics were exposed for 24-120 h in this chamber. Prior to irradiation, a vacuum was pulled for 24 h to seal the joints and to remove any gaseous products from previous exposures. Air was then allowed to flow through the system for an hour before the samples were actually exposed. During the photooxidation period, gaseous products were condensed and trapped in a dry ice/isopropanol bath and utilized for mass spectra and other analyses. The nonwoven polypropylene fabrics were also exposed in a Weatherometer (xenon arc light source) for 24-216 h at 45°C/54% RH--equivalent to the same number of hours of noon sunlight. Inner and outer filters used in this exposure were borosilicate and soda lime, those recommended for simulation of exposure of materials for testing their lightfastness.

Mass Spectral Analysis

Mass spectrometric analyses of the nonwoven fabrics (on condensed gases and on volatiles liberated from the solid polymer heated in the probe) were performed on an HP-5980A (Hewlett Packard) mass spectrometer operating under electron impact at 70 eV. Condensed gases were analyzed by direct introduction into the spectrometer, with the ion source heater set between 192-195°C, the gain set at 3, and the ionization potential kept constant at 70 eV. The direct probe heater was not used. For the analysis of volatiles from the solid polymer, a small piece of the fabric was placed in a capillary tube and put in a nest holder that was subsequently placed in the insertion probe. After the system was pumped down, the insertion probe was pushed into the spectrometer, the direct probe heater turned on, and the data collected. The direct probe heater warms up, commencing at 27°C, then increases to 383°C; after 4 min, the direct probe heater was turned off. A graph (Figure 3) plotting times versus probe temperatures was generated to determine the temperature of a sample in the insertion probe at a specified period of time. The average values used to generate the curve in Figure 3 were 26.9 ± 0.4°C (0 min); 74.8 ± 1.1°C (1 min); 172.0 ± 5.3°C (2 min); 302.8 ± 3.6°C (3 min) and 382.8 ± 3.0°C (4 min). The $\pm$ values are standard errors determined from an average of nine values. Mass spectra of samples obtained at probe temperatures of 100°C or less is given more attention, since minimal thermal degradation was observed in the unexposed polypropylene fibers at these temperatures.

Chemical Analyses for Various Carbonyl and Carboxyl Functional Groups in the Photooxidized Polypropylenes

The carboxyl groups were determined spectrophotometrically by measuring the amount of buffered Methylene Blue dye (buffered to a pH of 7.9 with NaH_2PO_4/Na_2HPO_4) absorbed from a solution of known concentration, adopted from the procedure used by Snyder, et al. for determining the oxidation of cellulose.[6] The aldehyde groups were also determined by this

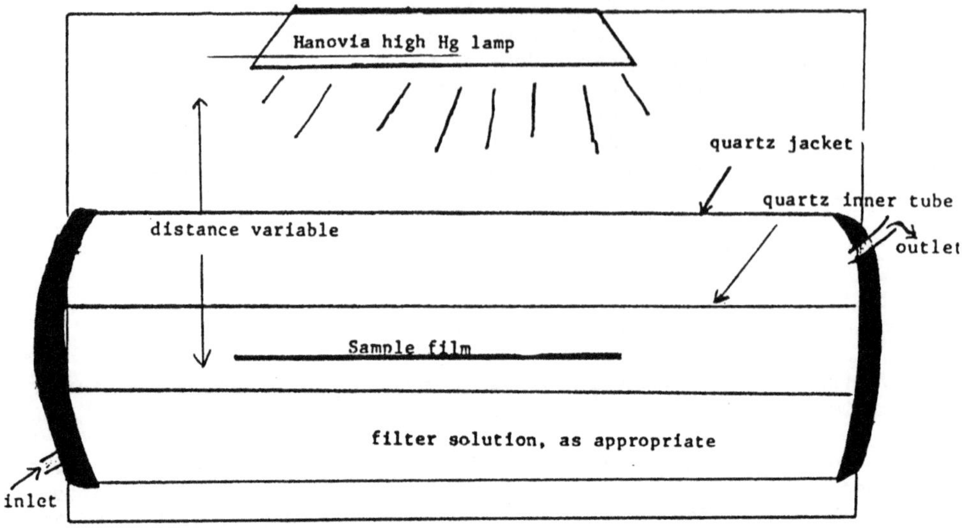

Fig. 2: Details of Irradiation Zone Showing Where Film or Fabric is Placed and Surrounded by Water-Cooled Jacket

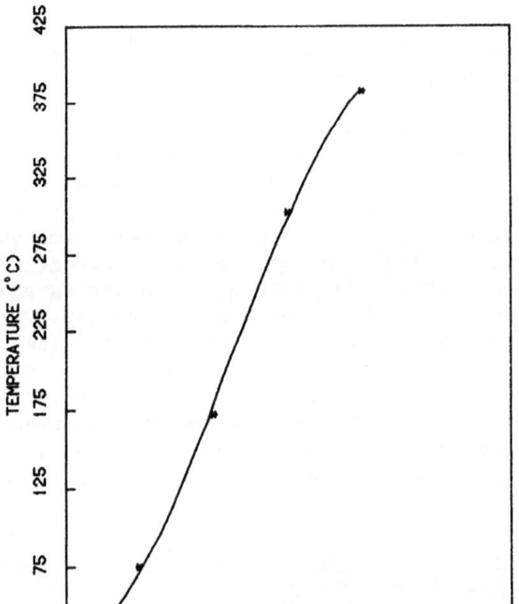

Fig. 3. Time of Sample in Mass Spectrometer Probe (min) vs. Temperature of Mass Spectrometer Probe (°C)

method by further oxidation with chlorous acid, and the increase in -COOH concentration and Methylene Blue absorption measured by difference. The total carbonyl content was determined by conversion to the oxime, and relating nitrogen content (Kjeldahl method) to carbonyl content, then determining ketone concentration by difference (from previously determined aldehyde group concentrations).

Mechanical Properties

The exposed and unexposed nonwoven fabrics were evaluated for their tensile strength, elongation, and energy-to-rupture with an Instron table model, using crosshead and chart speeds of 5.0 in/min with a Cell B full scale load (5,000 g), and a gauge length of 1 in.

RESULTS AND DISCUSSION

The analysis of nonwoven polypropylene fabrics by mass spectrometry, in conjunction with various chemical tests for aldehyde, ketone, and carboxylic acid groups, provided information about the mechanisms of photooxidation of the polypropylene before and after exposure to uv light sources. Although the modes of photochemical decomposition appear to be similar to those previously proposed using other analytical techniques, analysis of the mass spectra obtained from the uv-exposed solid samples indicate that secondary as well as tertiary hydroperoxide groups contribute to photodegradation of the polymer. This degradation also appears to occur by random chain scission as indicated by the nature of the degradation products. More detailed interpretation of the mechanisms of photooxidation follow.

The most common cited mode for chain scission of photoxidized polypropylene originates from a tertiary hydroperoxide group (initially formed by oxidation in the presence of trace amounts of impurities that act as photosensitizers). Subsequent cleavage of the tertiary hydroperoxide yields ketones[1] by one or more mechanisms:

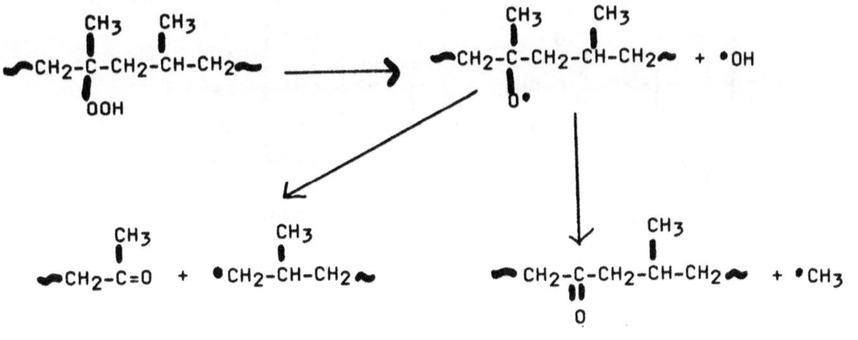

Another mode of decomposition may occur via scission of a secondary hydroperoxide group[1] to yield aldehydes and carboxylic acids:

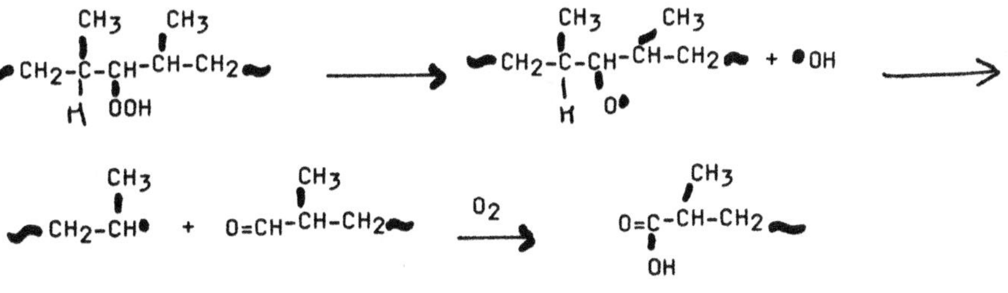

Norrish Type I and II processes, as well as other processes, can also lead to the formation of olefinic end groups on the polymer chain.[1] As noted in many previous investigations, the amount and type of degradation products vary widely with the temperature, uv exposure, mechanical influences, accessibility and crystallinity of the polypropylene, and impurities present in the polymer. Furthermore, most temperature studies have been conducted under conditions where the polypropylene is totally pyrolyzed and subsequently analyzed by mass spectrometry, infrared spectroscopy and related techniques. In our study, we first evaluated (by mass spectrometry) the condensed gases evolved from the fibers at ambient temperature during their exposure for 0-120 hr to the mercury lamp source. No discernible amounts of volatile products of any mechanistic significance were detected. Small amounts of carbon dioxide (m/z 44) and water (m/z 18) were observed, but were not considered substantial enough to use as a basis for mechanistic interpretations.

When the polypropylene fabrics previously exposed to the uv light sources were subsequently subjected to various temperatures in the mass spectral probe, they produced fragments that could be used to elucidate mechanisms by which photo- and thermal degradation of the polypropylene chain occurred. The most useful information was obtained at a probe temperature at which the unexposed polypropylene exhibited minimal thermal degradation (ca. 50°C), and the uv-exposed polypropylene exhibited substantial degradation.

The fragmentation patterns (Figures 4a-4f) of nonwoven polypropylene fabrics unexposed (0 hr) and exposed (24-216 hr) in a xenon arc source show a progressive increase in higher mass ions with increasing exposure time. Similar patterns were observed for unexposed polypropylene that was thermally degraded by increasing probe temperatures in the mass spectrometer (Figures 5a-5d), ranging from initial fragmentation at about 80°C to extensive fragmentation at 175°C. Thus, photochemical exposure lowers the temperature at which thermal degradation occurs, but the mechanisms by which chain scission occur appear to be similar, if not identical. On the basis of these fragmentation patterns, the following mass spectral assignments have been made for fragmentation of the polypropylene residue:

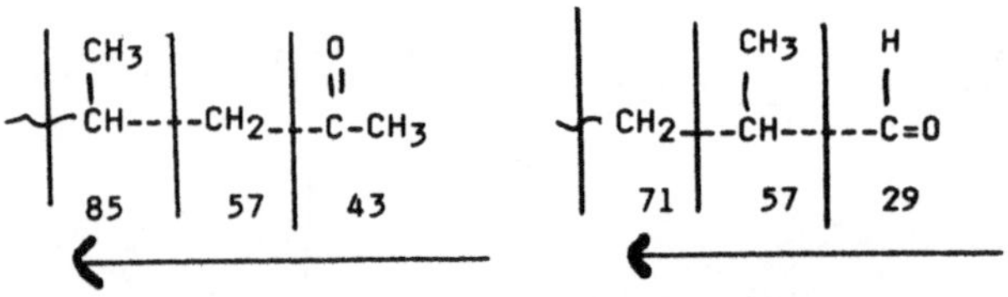

Ketone Aldehyde

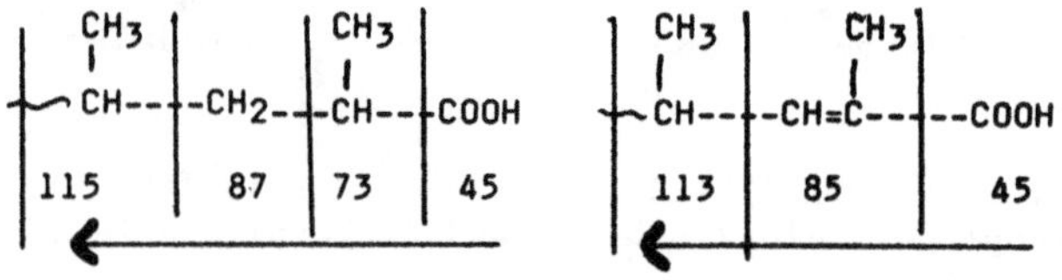

Saturated carboxylic acid Unsaturated carboxylic acid

Selected ions observed from polypropylene fabrics exposed for 24-216 h in the Weatherometer are plotted as a function of exposure time (Figure 6). The nature of the plots is consistent with the thermal fragmentation of a ketone end group (m/z 43) whose relative intensity increases with increasing exposure time. Similar trends were noted for the formation of the m/z 57 ion, which may be derived from the thermolysis of either aldehydes or ketones; the m/z 71 ion, which may be derived from the thermolysis of alde-

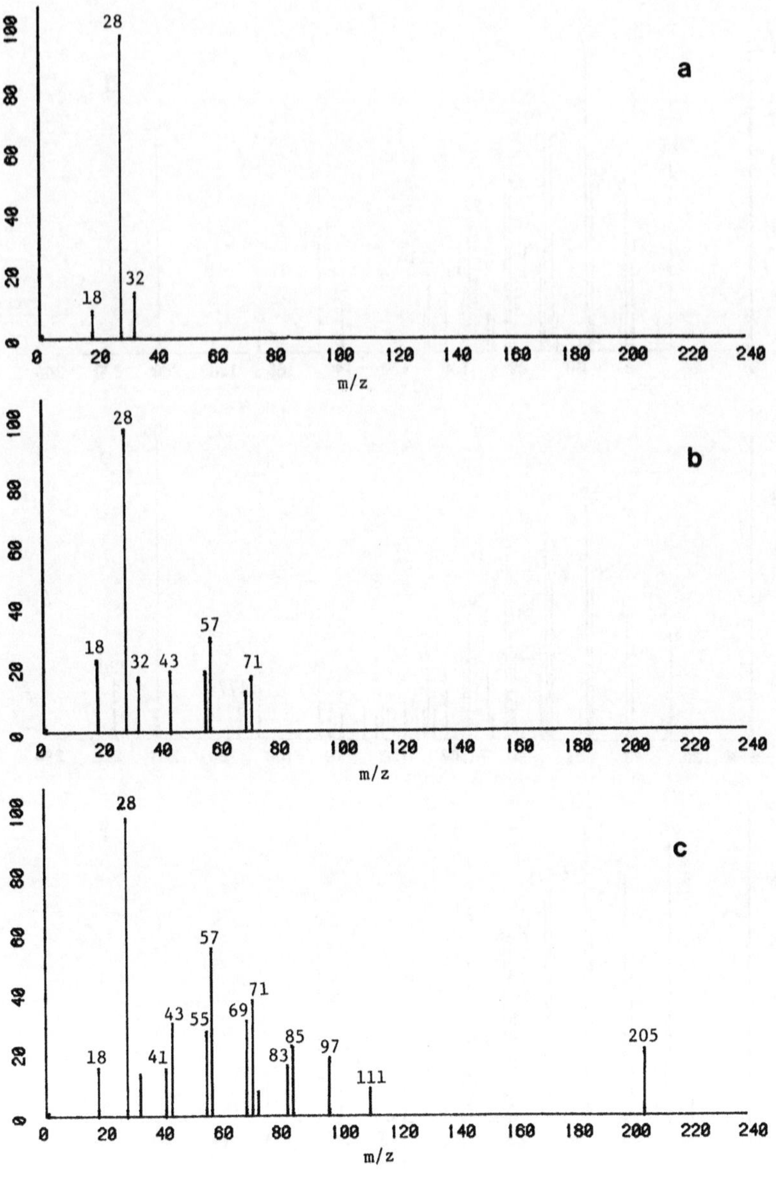

Fig. 4. Mass Spectra (70 eV) of Nonwoven Polypropylene Fabrics at a Probe Temperature of 50°C, Previously Exposed to a Xenon Arc Light Source (Weatherometer) for (a) 0, (b) 24, (c) 48 hrs

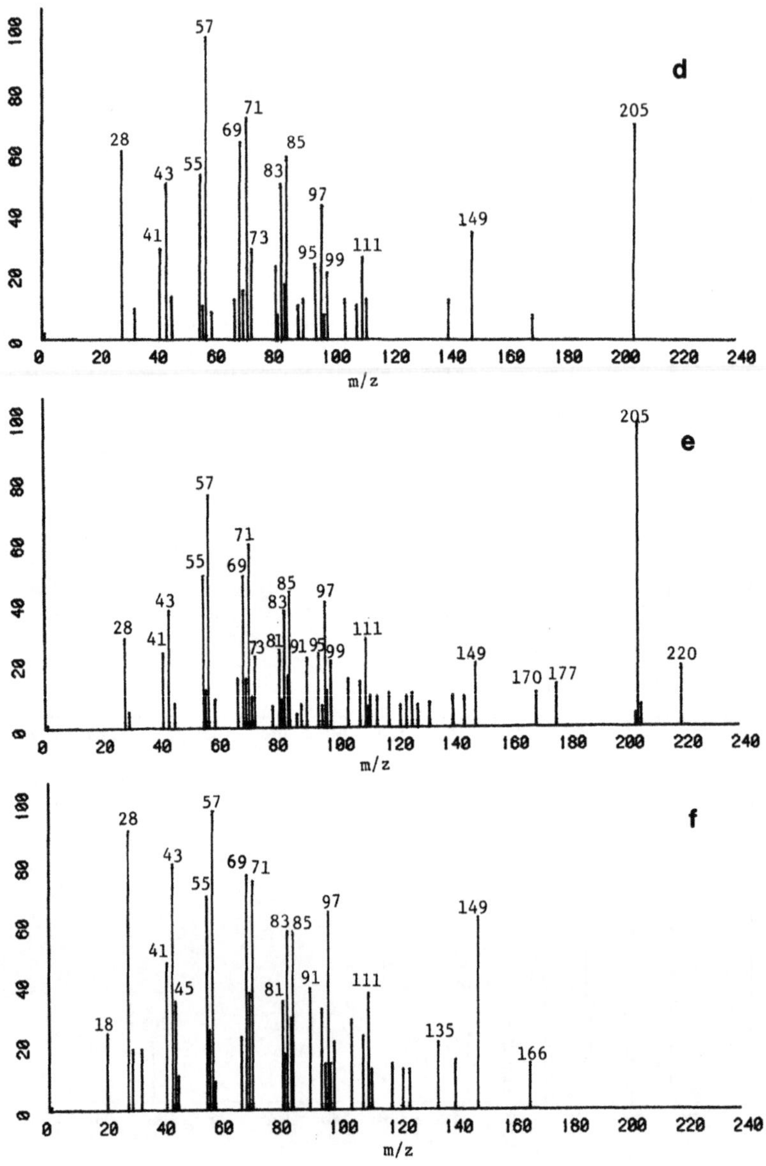

Fig. 4: Mass Spectra (70 eV) of Nonwoven Polypropylene Fabrics at a Probe Temperature of 50°C, Previously Exposed to a Xenon Arc Light Source (Weatherometer) for (d) 96, (e) 144, and (f) 216 hrs

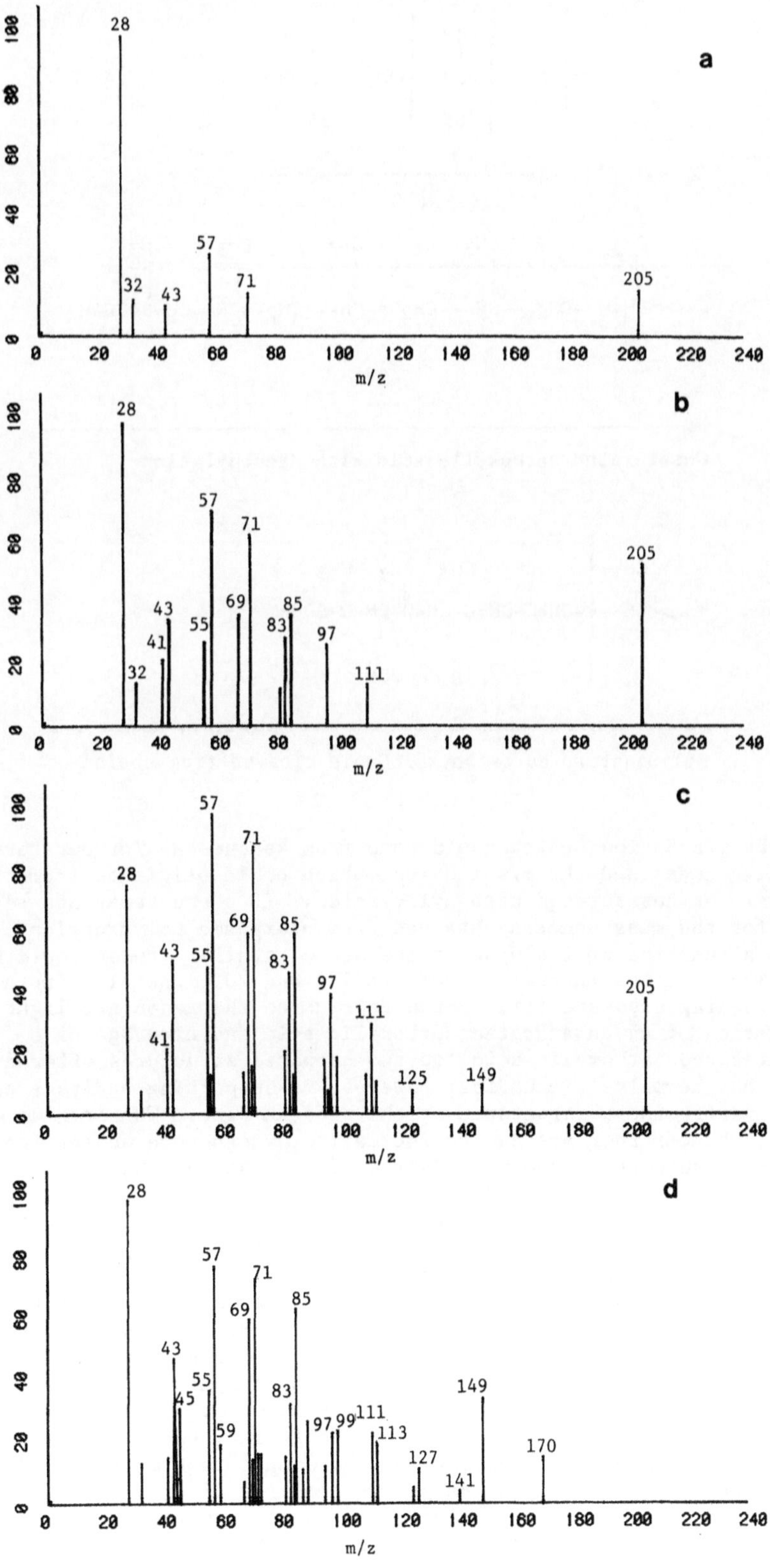

Fig. 5. Mass Spectra (70 eV) of Nonwoven Unexposed Polypropylene Fabrics at a Probe Temperature of (a) 80°C, (b) 90°C, (c) 120°, and (d) 175°C

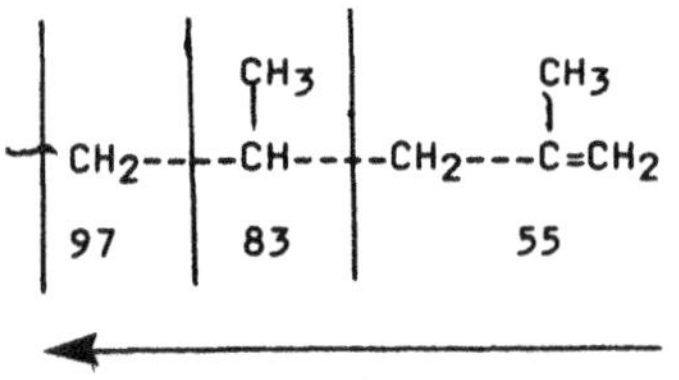

Alkene

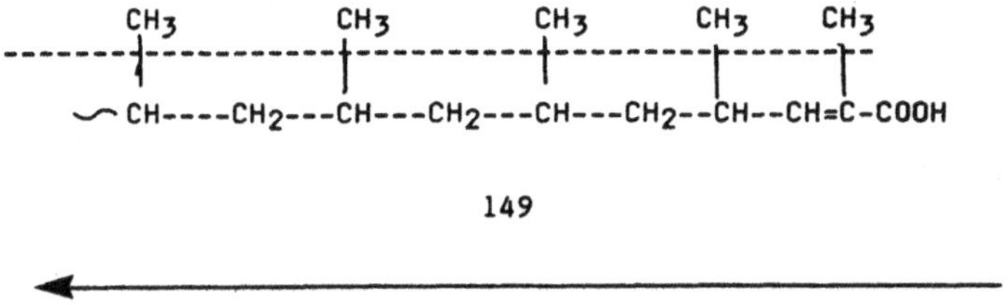

Unsaturated carboxylic acid with demethylation

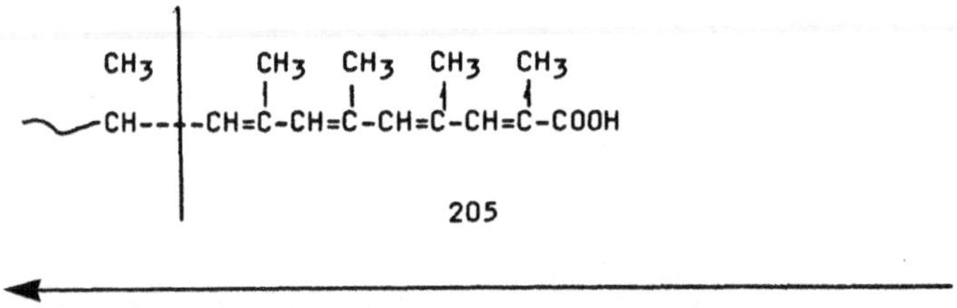

Polyunsaturated carboyxlic acid cleaved from chain

hydes; the m/z 85 ion, which could come from ketones and/or unsaturated carboxylic acids; and the m/z 113 ion, which could originate from the thermolysis of unsaturated carboxylic acids. If these same ions were plotted for the mass spectra obtained from unexposed polypropylene (Figures 5a-d) as a function of its probe temperature, similar trends would be noted. The relative intensities of m/z 149 and 205 ions also increases with increasing exposure time of the fabrics to the xenon arc light source. Demethylation of an unsaturated carboxylic acid and cleavage of a polyunsaturated carboxylic acid are two proposed structures offered to explain the thermolysis behavior; however, no supporting analyses can be given to verify these structures at the present time. The fragmentation products, whether they are due to increasing uv exposure and/or increasing temperature, support random degradation of the polymer chains.

In addition to aldehydes, ketones, and carboxylic acids, fragmentation patterns of the exposed polypropylene fabrics also are supportive of the thermal scission of alkenes to give ions at m/z 55, 83 and 97. This trend persists for unexposed polypropylene fabrics thermally stressed in the mass spectrometer direct insertion probe from temperatures of 75 to 175°C. Polypropylene fabrics that were exposed to the mercury vapor lamp have similar trends in their fragmentation patterns compared to those exposed to the xenon arc light source. However, degradation at a given temperature and exposure time is more extensive for the former type of exposure, presumably due to the closer proximity of the sample to the light. For example, when each of the fabrics was exposed for 48 hrs. and their fragmentation patterns evaluated at 50°C, the fabric exposed to the mercury lamp had high relative abundances of m/z 57 (100), m/z 71 (71) and m/z 149 (69) and moderately high relative abundances at m/z 43, 83, 85, 97, 111, and 135. In contrast, the fabric exposed to the xenon arc source for the same period of time had a m/z 57 abundance of 56, and moderate to small abundances at m/z 43, 71, no fragments at m/z 149, but a small amount (22) of m/z 205.

Chemical analyses of the exposed fabrics for various carbonyl group concentrations as a function of exposure time to the xenon arc source were also conducted to provide additional information relevant to the mechanism of polypropylene photooxidation. The results for carboxyl, aldehyde, ketone, and total carbonyl (aldehyde + ketone) are given in Table 1. The values for the carboxyl content correlate the best with the mass spectra fragmentation patterns, since the carboxyl content increases with increasing exposure time and levels off at high exposure times. The aldehyde content as well as the ketone and total carbonyl content is highest at intermediate exposure times, and then levels off in the case of the aldehyde and is negligible in the case of the ketone concentration. This disparity in agreement between the mass spectra fragments and all chemical analyses for these functional groups could be explained by differences in thermal lability of the groups. The important point is the chemical analyses indicate appreciable amounts of aldehydes, ketones, and carboxylic acids consistent with the mechanisms proposed for the random and complex fragmentation of the oxidized polypropylene in the solid state.

The tensile properties of the fabrics exposed in the Weatherometer were also determined and are depicted as a function of exposure time in Figure 7. The tensile or breaking strength, originally 49,050 mN, decreased almost linearly up to 96 h exposure, then levelled off to about 72% loss of the initial strength after 216 h exposure (13,571 mN). Similar trends were noted for loss of energy-to-rupture (originally 353,956 uJ); however, greater initial loss was observed in this property than in breaking strength after only 24 h exposure. More moderate, yet substantial losses in elongation (from an initial value of 137% to less than half that amount after 216 h exposure) also occurred. Thus, the changes in tensile properties parallel the mass spectra fragmentation patterns and to a lesser extent the chemical analyses conducted.

SUMMARY AND CONCLUSIONS

The photooxidation of nonwoven polypropylene fabrics, before and after exposure to mercury vapor and xenon arc light sources, has been investigated by mass spectral analysis of volatile products formed when the polypropylene was exposed to ultraviolet light and on its insertion in the mass spectrometer probe, by functional group analysis for carboxyl, aldehyde, ketone, and carbonyl groups, and by the tensile properties of the nonwoven fabrics. Good correlation was obtained for the analysis of aldehyde and carboxyl groups with the fragmentation patterns of the exposed fabrics by mass spectrometry, indicating random chain scission after initial formation of ketones via tertiary hydroperoxides, aldehydes after initial formation of secondary hydroperoxides, and subsequent oxidation/elimination processes to form saturated and unsaturated carboxylic acids. The fragmentation of alkenes initially formed by Norrish Type I and II processes is also indicated by the mass spectral results obtained. Polypropylene that was exposed for increasing time to either xenon arc or mercury vapor light sources had increasing mass ions when it was subject to a spectrometer probe temperature as low as 50°C. No meaningful volatile products were detected when mass spectrometry was employed to detect condensed gases emanating from polypropylene fabrics exposed for 24 to 120 h to a mercury vapor lamp. However, the mass spectra obtained from unexposed polypropylene fabrics subjected to increasing temperatures in the probe (i.e., 75-175°C) had similar, if not identical fragmentation patterns, to those of fabrics exposed and evaluated at a lower probe temperature. Loss in tensile properties generally correlated well with increasing fragmentation in

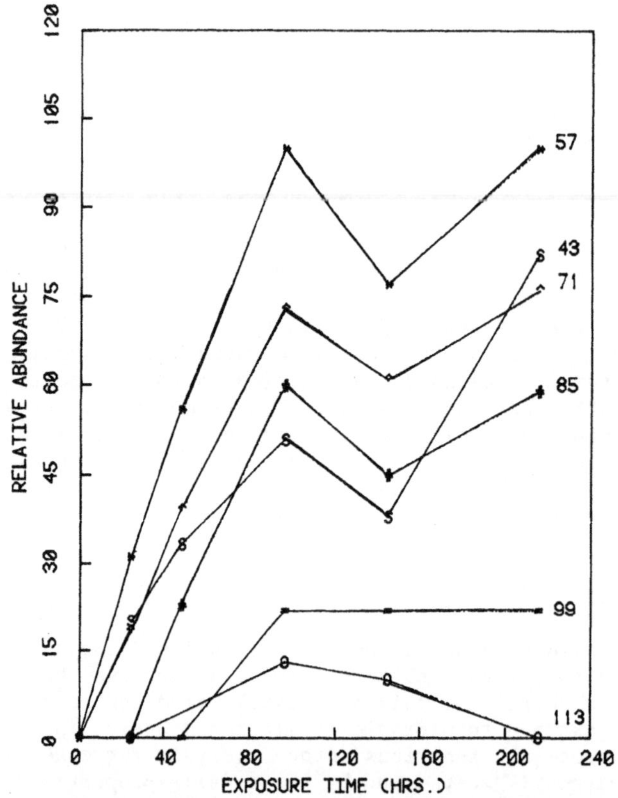

Fig. 6: Relative Abundance of m/z Fragments of Polypropylene Fabrics Exposed for 0-216 hrs. to a Xenon Arc Light Source. Legend:(S) m/z 43, (*) m/z 57, (^) m/z 71, (#) m/z 85, (=) m/z 99, and (O) m/z 113

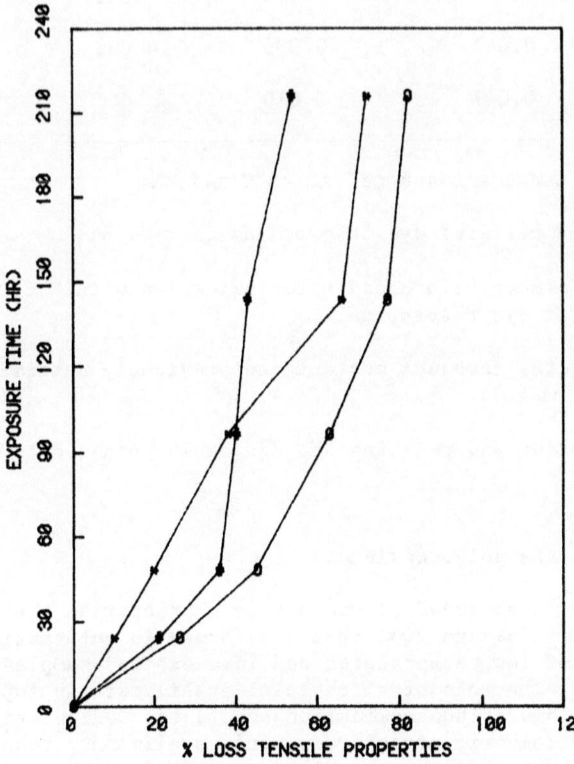

Fig. 7. Tensile Properties of Polypropylene Fabrics vs. Exposure Time in a Xenon Arc Light Source. Legend: (*) Breaking Strength, (O) Energy-to-Rupture, and (#) % Elongation-at-Break

Table 1. Functional Group Analyses of Photooxidized Polypropylene Fabric

Exposure Time (hr)[a]	mmol COOH[b]	mmol CHO[c]	mmol ketone[d]	mmol C=O[e]
24	0.020	0.013	0	0
48	0.029	0.018	0.024	0.042
96	0.040	0.010	0.015	0.025
144	0.047	0.006	0.003	0.009
216	0.047	0.010	0	0.009

[a]In Weatherometer (xenon arc source) at 45°C/54% RH.

[b]From Methylene Blue buffered dye absorbed as in ref. 6.

[c]Determined by difference before and after oxidation with $NaClO_2$, and evaluation as (b) for dye absorption.

[d]By difference of total carbonyl content and previously determined aldehyde content from (c).

[e]By conversion to oxime and relating %N (Kjeldahl) obtained to % carbonyl content.

the mass spectra of the polypropylene.

Further studies are needed to adequately characterize the fragments of higher mass ions (m/z 149 and 205) that are formed in substantial amounts in exposed samples at low temperatures and in unexposed samples at higher temperatures. Also, the role of ultraviolet stabilizers in impeding or changing the mechanisms of photooxidation should be studied utilizing the described mass spectrometric procedures, since preliminary results with Nylon 66[4] indicate that some photostabilizers slow down the formation of high mass ions, while others totally change the mechanisms by which photooxidation may occur.

REFERENCES

1. B. Ranby and J. F. Rabek, "Photodegradation, Photo-oxidation, and Photostabilization of Polymers," Wiley, New York, pp. 129-141 (1975).
2. D. J. Carlsson and D. M. Wiles, Macromolecules, 2: 597-606 (1969).
3. N. Grassie and W. B. H. Leeming, in: "Ultraviolet Induced Reactions in Polymers," S. S. Labana, Ed., ACS Symposium Series, 25: 367-390 (1976).
4. J. F. Kinstle, Y. Charit, T. L. Vigo, AZ. Muschelewicz, and K.-Y. Wei, in press.
5. P. Laporte and F. Zaraga, Appl. Optics, 20(17): 2946-2950 (1981).
6. S. L. Snyder, T. L. Vigo, and C. M. Welch, Carbohydr. Res., 34: 91-98 (1974).

ROLE OF FINISHES IN

POLYOLEFIN FIBERS

Stanley E. Ross

Emery Chemicals
P. O. Box 628
Mauldin, S.C. 29662

INTRODUCTION

Polypropylene fibers and yarns have continued to gain increasing acceptance in a wide variety of applications and uses. This is a result of a series of technological advances in polymer and fiber manufacturing processes which have taken place over a period of time. In addition, considerable effort has been devoted to the engineering of fiber properties to meet specific requirements. Among the responsible factors that play a subtle but predominant role in this successful development is that of fiber finish.

I propose to discuss the wide diversity of fiber finish requirements and to review the function of fiber finish with several examples of polypropylene manufacturing and utilization processes. I also plan to discuss some of the relevant methods used for predicting the performance of finishes, such as frictional measurements, thermal behavior evaluation, studies of chemical effect on frictional elements, and evaluation of the activity of antistats.

Last year, according to the Textile Economics Bureau, something on the order of one billion pounds of polypropylene fibers and yarns were produced in the United States (Table 1). This was divided principally among staple fibers, filament yarns, slit film yarns, and spunbonded type products. These fibers and yarns were used in the apparel, home furnishings, industrial, and automotive markets. Weaving, knitting, and nonwoven processes were used to produce end use items from these fibrous materials.

Table 1: Polyolefin Shipments (Millions of Pounds)

	1982	1983	1984
Yarn and Film Fiber	583	708	742
Staple	139	189	247
Total	722	897	989

In the home furnishings area, some of the product in which polypropylene is successfully used are upholstery fabrics, carpets, carpet backings, bedspreads, and wall coverings. Industrial applications include cordage, sewing thread and filter fabrics. A growing area for polypropylene is in nonwovens, both disposable as well as durable (Table 2). Nonwovens include latex bonded, thermally bonded, and needle punched fabrics for uses as diverse as lightweight cover stock for disposable diapers to heavy structures for civil engineering applications.

These many different types of products involve a variety of continuous filament yarns and staple fibers which have dissimilar dimensions and physical characteristics and, of course, are made by different manufacturing processes.

What, then, is the role of fiber finish in this overall picture? Fiber finish plays a very important role in:

1. assisting the fiber manufacturer to produce the product, and

2. enabling the fiber producer's customer to carry out the required yarn and fabric manufacturing processes to obtain the end product.

Without the application of fiber finish, neither of these objectives could be accomplished. As I indicated, because of the diversity of yarn and fiber physical forms, there is no single all-purpose finish. Actually, the trend has been in the opposite direction, that is, to develop specialized finishes for each different type of fiber, manufacturing process, and end use.

Let's take a brief look at the manner in which polypropylene is produced in the fiber manufacturer's plant and see how the finish fits in and what is expected of it. We will also take a look at what the fiber will be subsequently exposed to.

Commercially manufactured textile denier polypropylene is made by melt spinning (Figure 1). The polymer is melted and extruded into filaments which are cooled and solidified in an air stream. At the base of the chimney where the solidified filaments appear, they are converged and contact a finish applicator. This can be in the form of a kiss roll which is rotating in a finish trough. The amount of finish applied to the filaments can be controlled by the concentration of finish in the solution or emulsion and the total wet pick-up. Alternatively, a positive metering system can be used to pump finish to a ceramic slot in contact with the filaments.

From this point in manufacture, filaments intended for staple are combined into large tows, oriented by stretching across a heated surface, crimped, and cut into short lengths which are compatible for processing on machinery originally designed for natural fibers, principally cotton and wool. If continuous filament yarn is the desired product, the filaments are also oriented but as a discreet bundle containing a specific number of filaments, and are wound as long continuous lengths. Texturized yarns are

Table 2. Polypropylene Usage in Nonwoven Fabrics (Millions of Pounds)

1975	1982	1984
92	102	145

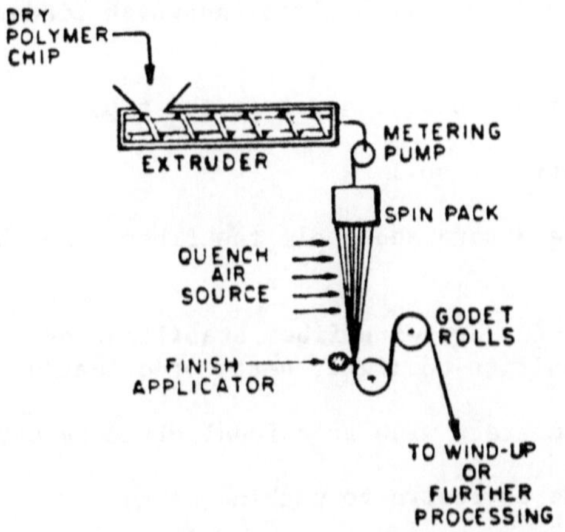

Fig. 1. Melt Spinning Process

also made as continuous filament yarns. During texturizing straight tightly packed filament bundles are converted into distorted loosely packed filament bundles. The purpose is to achieve a fabric which has a more pleasing hand, is more comfortable, better in appearance, or a yarn which gives greater cover or more resilience to a carpet.

In some cases, this original finish application carries the fibers through the whole process. In others, supplementary finishes are applied somewhere later in the process.

In general, the function of the spin finish at this point of application in the manufacturer's operation is:

1 - to provide good fiber to metal lubricity to prevent abrasion and damage to the fibers

2 - to provide the necessary fiber to fiber cohesion to produce a take-up package which is stable and also provides bundle integrity for multifilament yarns

3 - to eliminate the nuisance of static electricity by the rapid dissipation of generated charges

As you might expect, although these three functions are major considerations, they really represent only a small portion of the total fiber finish requirements. Actually, there is a large number of "musts" and "must-nots" which are demanded of fiber finishes, and I would like to enumerate some of these for your consideration. Consider the following. Fiber finishes:

1 - must have good thermal stability and not leave residues on surfaces or cause toxic fumes or undesirable odors

2 - must provide for rapid wetting of fiber surfaces to accommodate normal and high speed processes

3 - must have water solubility or be emulsifiable or solvent soluble, as the case may be, and have viscosity behavior appropriate for uniform application and removal

4 - must have good stability to storage under a variety of storage conditions

5 - must be compatible with sizes, nonwoven binders, and other fiber treatments

6 - must not cause color changes in the fiber

7 - must not attract soil

8 - must not penetrate and swell the fiber, thereby damaging fiber properties

9 - must not interfere with fiber stabilization systems or render fibers sensitive to light, heat, or oxidative degradation

10 - must not interact with frictional elements used in texturizing

11 - must not be corrosive to machine parts

Let's return to an examination of frictional properties. Fiber to fiber friction is important to the fiber producer in controlling package formation and also in filament orientation. It is important to the user in the processing steps encountered in the manufacture of spun yarns from staple. I refer to picking, carding, drawing and spinning. With continuous filmanent yarns, fiber to fiber frictional properties are important considerations in processes such as twisting, plying, braiding, sewing, and in preventing distortion in woven fabrics.

Fiber to metal friction influences can be readily observed in fiber manufacturing in yarn and tow drawing and in texturizing operations. Fiber to metal friction during carding, drafting and spinning also has a direct influence on processability and yarn quality. Friction of yarns during knitting and tufting are further examples in which there is frequent contact between yarn and metal.

Specific required values of fiber to fiber and fiber to metal friction normally cannot be assigned since the frictional requirements for all of these processes are very different and some are quite contradictory. I shall discuss this in more detail shortly.

Polypropylene, like other manmade hydrophobic fibers, is prone to develop static electricity which can interfere not only in the manufacturing process but also in subsequent textile operations.

Therefore, control of frictional and static properties is of tremendous importance in the realization of the maximum potential of polypropylene fibers and yarns. These considerations, coupled with the other attendant requirements previously discussed, represent significant demands. No single finish component is expected to satisfy all the frictional and static requirements. Frequently, a formulation consisting of several components results, each component contributing to the total performance. Further, no one combination of components or single formulation is expected to meet the many demands of the different products and processes.

A number of specialized finishes have been developed for many of the specific polypropylene fiber and yarn variants which are directed toward specific processes and end uses.

For example, let's take the case of a low denier polypropylene staple fiber which is to be carded into a web and thermally bonded for some disposable nonwoven application. This requires a formulation which in conjunction with the fiber crimp, contributes a relatively high fiber to fiber friction which is important in insuring a carded web with good cohesion, uniformity, and integrity, thereby compensating for the low stiffness of the fibers. Low fiber to metal friction is also a key factor in the processing of these staple fibers which have diameters on the order of only 15 to 20 micrometers. Static protection is also important.

Another example is a slit film or ribbon type yarn intended for woven carpet backing for tufted carpets. During its manufacture, the requirements are for good wetting of the fiber surface by the finish and moderate frictional coefficients. For tufting, however, relatively low fiber to metal friction is a very important feature because of the action of the tufting needles on the backing fabric. In this case, static protection is probably not as critical as in the previous example because of the yarn dimensions and type of process.

Low fiber to metal friction is also a desirable characteristic of a finish which has earlier been applied to a POY yarn which is to be oriented and texturized in contact with lengthy heater plates or to a flat yarn which is to be similarly texturized.

Finally, low fiber to fiber friction is a highly desirable feature of continuous filament yarns used in cordage applications which involve twisting and plying to form compact structures which have a large amount of fiber to fiber contact. High friction is undesirable since it is generally associated with low flex resistance and low energy absorption, and therefore, short life.

How do we go about selecting those finishes or components which are best suited for some specific defined requirement? Let's consider some of the more important finish characteristics which we can predict from laboratory studies.

Measurements of frictional properties can be readily carried out on yarns to which known amounts of finish have been applied under closely controlled conditions. Obviously, laboratory measurements at best only simulate actual use conditions, but we have found them to be a reasonably good predictor of behavior. One of the well-known instruments for performing frictional measurements is the Rothschild F Meter. In case of fiber to metal friction, the measurement is carried out by pulling a yarn around a circular metal pin under conditions of known pre-tension and angle of contact (Figure 2). The output tension is measured and the coefficient of friction determined from the capstan equation:

$$\frac{T_2}{T_1} = e^{ue}$$

Where T_1 and T_2 are the incoming and outgoing tensions respectively, θ the angle of contact in radians, and μ the coefficient of friction. The Rothschild instrument calculates and plots the coefficient of friction automatically. Some prefer to use the value of $T_2 - T_1$ as a measure of the frictional force since the capstan equation is not accurately obeyed by fibers and yarns.

There are a number of variables, both mechanical and physical, in addition to the pretension and angle of contact, which can influence friction results. Some of these are speed, surface roughness, surface temperature, ambient temperature and humidity, finish viscosity, uniformity of finish application, finish concentration on the fiber, and fiber size and shape.

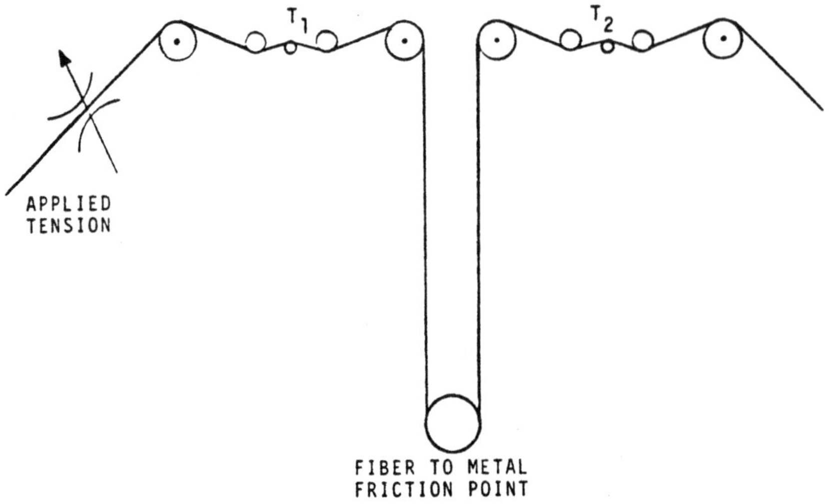

Fig. 2. Fiber to Metal Frictional Measurement

Obviously, when performing laboratory frictional experiments to compare the performance of different finishes, a compromise set of conditions is selected. Other conditions which are more appropriate to the intended manner of use are chosen as required such as perhaps, measurement of friction against a heated pin to relate to that process condition.

The fiber to fiber friction measurement is carried out in a similar way except that the yarn is twisted around itself and the force determined to pull the yarn in contact with itself (Figure 3). Again, with a knowledge of the incoming tension, the angle of wrap, and the outgoing tension, the frictional coefficient can be determined. In the case of fiber to fiber friction, we normally distinguish between static and dynamic frictional coefficients. Static friction is determined at a low speed (on the order of 1 cm/min) and dynamic friction at a higher speed. When measuring low speed friction, a stick-slip phenomenon is sometimes observed. It is this measurement which is most closely related to the "scroop" observed with staple fibers, or the cohesion of a staple fiber web as it emerges from a card, or the performance of a spin finish in yielding a yarn package which is stable and does not slough. The stick-slip phenomenon indicates that the static friction is higher than the dynamic friction.

Polypropylene fiber has a relatively high coefficient of friction against itself. This is believed to be due to plastic deformation of the surface. Application of finish normally reduces both the static and dynamic friction to a more desirable level.

During the measurement of fiber to metal friction and the passage of yarn around the metal pin, static charge is generated. The Rothschild friction meter has an electrostatic voltmeter attachment which measures the charge generated by the moving yarn. At periodic intervals, the static is discharged and allowed to rebuild. Correlation of the charge developed in this measurement with actual performance observed under various manufacturing and use conditions is generally very good provided the relative humidity is reasonably close to the test condition. An alternate method of assessing antistatic activity of a finish is to measure the time for a charge to dissipate after it has been imposed on the fiber (half-life measurement).

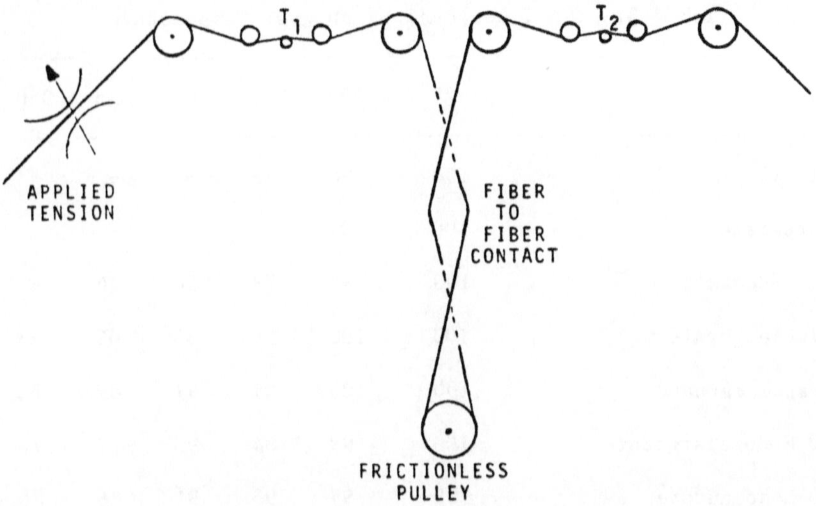

Fig. 3. Fiber to Fiber Frictional Measurement

I would like to address some of the other finish requirements which I previously listed. Let's review the volatility and thermal stability aspects.

Some of the processes for manufacturing polypropylene fiber and yarn involve contact of the fibers with heated surfaces. The surface can be in the form of heated rolls, heated plates, or crimper or texturizing chambers. It has been observed that some finishes or components of finishes tend to emit fumes as a result of volatilization in the thin film of finish on the fiber in contact with the heated surface. Further, deposition of the finish on heated surfaces can occur and on prolonged heating, residues which are insoluble or difficult to remove can be formed as a result of thermal decomposition. These residues are undesirable since they can interfere with the process.

We have used several techniques to assess these thermal properties of finishes. Thermogravimetric analysis (TGA) is one of these and has proven to be a useful tool in characterizing weight loss - temperature behavior, and gives a measure of the volatility factor. For example, a series of compounds was examined by heating 10 mg samples at 10°C per minute in the presence of nitrogen in a thermogravimetric analyzer and observing the weight loss as a function of temperature. The samples comprised mineral oil, butyl stearate, tridecyl stearate, trimethylolpropane tripelargonate, pentaerythritol tetrapelargonate, polyethylene glycol 400 monopelargonate, polyethylene glycol 400 monolaurate, and glyceryl trioleate.

From the examination of the data (Table 3), it is a relatively simple matter to obtain an indication of which finish components are the least or most volatile and therefore compatible with the temperature requirements of the process. There are, of course, the other techniques for determining volatility and smoke points which, when used in conjunction with TGA give a complete picture.

I indicated that residues remaining on surfaces can be a major source of problems by impeding the path of the yarn and by requiring more frequent cleaning cycles. One of the useful methods for determining this property is to place several grams of the finish in a dish and heat the dish to temperatures on the order of 200°C for a period of 16 hours. The residue

Table 3. TGA % Residual Weight vs. Temperature

	25°C	200	275	300	325	350	400
Mineral Oil	100	90	12	-	-	-	-
Butyl Stearate	100	94	11	-	-	-	-
Tridecyl Stearate	100	99	79	54	16	-	-
TMP Tripelargonate	100	100	96	84	63	23	-
PE Tetrapelargonate	100	100	99	97	89	72	1
Peg 400 Monopelargonate	100	99	94	90	82	70	22
Peg 400 Monolaurate	100	99	95	91	86	76	35
Glyceryl Trioleate	100	99	97	96	95	94	76

is weighed and examined for general appearance and solubility. With the more volatile materials, the ideal behavior is to leave no residue at all. With the less volatile materials, one prefers to find the finish intact, or at least not in the form of a hard to remove varnish.

Since polyurethanes are frequently used as friction discs in texturizing processes, problems of sensitivity of the polyurethane have become evident as indicated by eventual swelling and softening. This is brought about by the chemical affinity of various materials and their penetration into the polyurethane. Again, laboratory screening of finish components has well identified the offenders. The test is carried out by exposing polyurethane sections to the finish at temperatures depending upon preference, ranging from 65 to 200°C and time periods from 24 to 96 hours. Examination of the polyurethane section for dimensional change or weight gain can indicate sensitivity to the finish. It has been found that low molecular weight polyethylene glycol esters and diesters of fatty acids tend to be offenders as do ethoxylates of alcohols and lower molecular weight fatty diethanolamides. On the other hand, mineral oil, alkyl esters, alkylamine quaternaries, and sorbitan esters do not affect polyurethanes. Obviously, the offenders would be eliminated as candidates for finishes for yarns to be texturized on equipment using polyurethane parts.

What types of molecules, then, have been found to be compatible with polypropylene? A simple procedure to determine which types of finishes are unsuitable for polypropylene was described in a 1967 FMC patent (3,296,019). The test involved placing 0.2 ml of the material in question onto a sheet of polypropylene film which was secured to a clean plate. A piece of graph paper was placed between the film and the plate. If there was a change in the appearance within 24 hours, such as waviness or puckering, it was assumed that plasticization of the polyolefin had occurred and the lubricant was rejected. The test indicated that paraffinic oils and low molecular weight esters had an undesirable effect.

However, there is a wide variety of materials which pass the test and are being successfully used for the many polypropylene products. As you can well imagine there is an infinitely large number of potential combinations of components, and much of the information describing specific formulations is closely guarded. With regard to antistats, it has been found that both cationic and anionic antistats such as quaternary salts and phosphate esters work very well in eliminating static charges, and the preference of one type over another is based on considerations such as compatibility with other components of the formulation. As an aid in wetting the low surface energy polypropylene fibers, phosphate esters and ethoxylated alcohols have been found useful.

In addition to the antistats and wetting agents, components of finishes which are being successfully used commercially include uncapped and capped polyethylene glycol fatty acid esters, ethoxylated caster oils, glyceryl and sorbitan esters, ethoxylated sorbitan and sorbitol esters, diethanolamides, various higher molecular weight alkyl esters, ethoxylated alcohols and many others.

Normally, the formulated finish consists of a lubricant and an antistat and, in the case of non-water soluble lubricants, an emulsifier package. Frequently, biocides, antioxidants, anticorrosion agents, and pH control agents are also present. Given the large number of candidate materials, the development of effective finishes becomes a serious endeavor. Selection and optimization of a finish requires considerable effort and patience. The finishes developed to date have significantly contributed not only to the success of the manufacture of polypropylene fibrous materials but also to their penetration into markets formerly held by other manmade and natural fibers.

In summary, I have tried to give you an idea of the importance of fiber finish in the manufacture and utilization of polypropylene fibers. I have reviewed the many requirements of fiber finishes and some of the methods used to evaluate candidates with regard to the important properties of friction, thermal behavior, antistatic activity, and compatibility with polyurethane materials.

As new technology and higher speed processes develop, the burden will continue to fall ever more heavily on finish and its application. Finishes will continue to become more specific, and new molecules will be uncovered which will contribute unusual combinations of properties and thereby enable them to do an even better job.

THE COLORATION OF POLYPROPYLENE FIBERS

Marvin Wishman, James Leininger, and Deborah Fenton

Phillips Fibers Corporation
Greenville, South Carolina 29602

INTRODUCTION

Polypropylene, as produced, does not possess properties normally attributed to a fiber-forming polymer. Despite a relatively low melting point, it is not stable below melt temperatures and will degrade in UV light at room temperature, unless stabilizers are added. In addition, polypropylene has no dye sites and cannot be dyed by any conventional means to a colorfast product. Polypropylene properties have been progressively improved by use of an ever-growing variety of additives. Many dyeable systems have been developed and many patents issued but the problem of coloring polypropylene by making it dyeable has not been satisfactorily solved.

Each dyeable system adds to the cost of fiber making, negating one of the supposed advantages of polypropylene, compared with other synthetic fibers. As better stabilization systems became available, it became obvious that polypropylene fiber provides an opportunity to make long-life products - as long as pigments are used to color the products.

A dyeable polypropylene would have all of the problems of the dyeable synthetics viz, - nylon, polyester and acrylic. These products, if colored by dyes, could not satisfy the needs of the automotive industry and were not suitable for outdoor use. Pigmentation is required for such products. The study of pigments and their relation to other additives became more important as requirements for product stability became more severe.

The ability of hindered amine light stabilizers (HALS) to stabilize polypropylene fibers has created not only new opportunities but also new problems with regard to pigments. At one time, almost any pigment, sufficiently thermally stable to be used at melt spin temperatures, would be acceptable in terms of color stability because the fibers degraded before the pigment lost its color. With improved stabilizers, this is no longer true.

Because the variety of available stabilizers makes it possible to produce polypropylene fibers for a wide variety of end uses, there is a need for a wide variety of pigments with different cost as well as color stability characteristics. Pigments are selected based upon varied requirements, some related to the fiber-making process and some to the product. Characteristics of pigments which must be considered are (Table 1)

1. Suitability for Fiber Processing
2. Cost Effectiveness
3. Color Requirements
4. Stability Requirements of Product

SUITABILITY FOR FIBER PROCESSING

A 2 dpf polypropylene round fiber is 17.7 microns in diameter. Obviously, this establishes some filtration requirements for fiber extrusion. A pigment particle must be small enough to pass through such filtration requirements and not create problems at drawing, etc. It must also provide the color and that color must be durable. While color yield may increase as particle size is reduced, color stability is inversely proportional to particle size. Special techniques have been developed to convert pigments, as produced, into concentrates (25-50%) in which the pigment particles are not agglomerated. These concentrates must then be added by the fiber producer so that they provide color and do not harm the fiber-making process.

Conversion of pigments, as produced, to pigment concentrates is an art, and is usually done by specialists at concentrate houses rather than by the fiber producer. The fiber producer can then add the pigment in several ways, depending upon the accuracy required and the specific process being used. This procedure can be:

1. A pellet blend of natural resin and selected color concentrates can be added directly to the fiber extruder, or --

2. Powdered concentrates can be individually metered to natural resin and fed to the fiber extruder, or --

3. The required color concentrate and natural pellets can be 100% precompounded to be fed to the extruder, or --

4. A portion of the natural resin can be precompounded with color concentrates and metered into the fiber extruder along with metered natural resin.

All of these methods may work, depending on the requirements of the product being made. The success of the method depends upon the ability of the concentrate manufacturer and the skill of the fiber maker in using the concentrate.

Table 1. Characteristics of Pigments Requiring Consideration

Suitability
Cost Requirements
Color Requirements
Stability Requirements of the Product
Interaction with Other Additives

Fiber pigments must not only be of proper size to process without causing deterioration of the process or broken filaments due to particle size, but the pigment must also be stable at the temperatures at which polypropylene is spun. Pigments which are stable for several minutes at temperatures well above 260°C are required for many of these processes. The fiber producer must add stabilizers to the resin to prevent polymer degradation which would result in poor control of melt viscosity and therefore, the extrusion process itself.

Other additives are also introduced to provide final fiber properties other than color. Colorants are only part of the additive package required to produce an acceptable fiber product. Since pigment colorants, melt viscosity stabilizers, thermal and UV stabilizers and other additives are all present during melt extrusion, interaction among these additives can cause a number of fiber-making problems.

Some pigments are prodegradants during melt spinning, i.e., at the melt they cause more molecular weight change than would occur if the pigment were absent. Whether or not a particular pigment will act as a melt viscosity prodegradant depends upon the particular stabilizers used. We have found that some phthalo blue pigment concentrates cause excessive viscosity changes unless certain stabilizers are also used. It must, however, be remembered that additives remain in the fiber and affect final properties.

Melt stabilization is frequently tested by determining melt flow changes in multiple passes through an extruder. This does not necessarily correlate with what is found in fiber-making. Data obtained in actual fiber spinning trials are shown in Figure 1. Extrusion temperature and residence time are variables affecting resultant fiber melt flow. The fiber diameter may also be a variable since lower deniers have more surface area permitting more degradation. Fibers are spun to a given denier (weight in grams per 9000 meters) in order to compare effects of additives.

As indicated in Figure 1, fiber melt flow increases with extrusion residence time and temperature. The ability of stabilizers to control degradation and the role of pigments on extrusion stability are similarly determined. These spun yarns may be processed to final products to determine the effect of extrusion time and temperature on final fiber color.

COST EFFECTIVENESS OF PIGMENTS

Characterization of a single pigment may be achieved by determining the spectrophotometric reflectance curves in the visible region of fibers containing a range of concentrations of the pigment. Figure 2 shows results for pigment Red 214. The strength of the pigment is obtained by mathematical manipulation of the reflectance data.

A plot of strength vs. % pigment is shown in Figure 3 for two orange pigments, Orange 31 and Moly Orange. The color yield of Orange 1 (which is Orange 31) per unit weight is greater than for Orange 2 (which is Moly Orange) but when cost of pigment is factored in (Figure 4), Moly Orange is the more cost-effective pigment.

The modern computer makes calculations such as those required for Figures 3 and 4 relatively easy to obtain. It can also make it possible to "computer match" almost any color. If the modern spectrophotometer is attached to a computer, it can remember the reflectance curve of every colored fiber which has been determined. Thus, the samples used to determine color build-up characteristics become part of a "color library." To

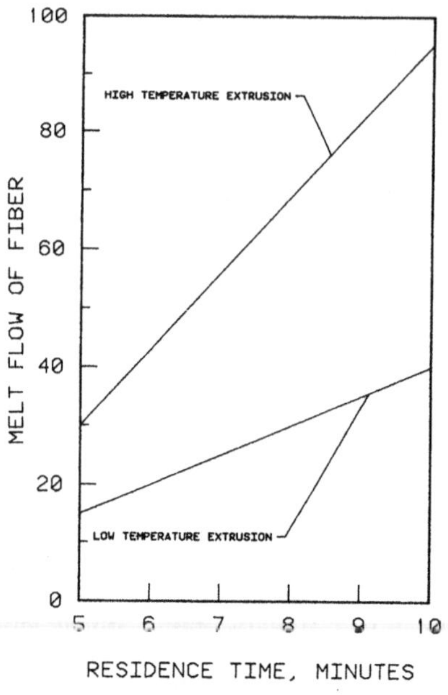
100
80
60
40
20
0
MELT FLOW OF FIBER
HIGH TEMPERATURE EXTRUSION
LOW TEMPERATURE EXTRUSION
5
6
7
8
9
10
RESIDENCE TIME, MINUTES

Fig. 1

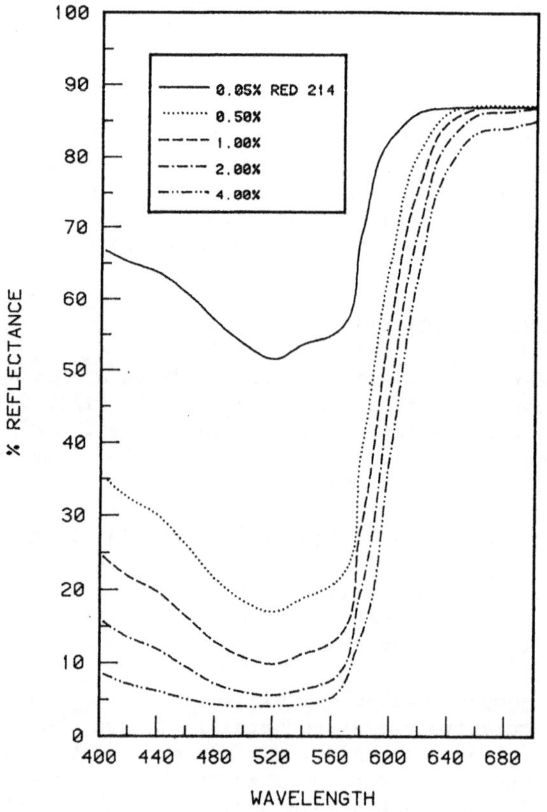
100
90
80
70
60
50
40
30
20
10
0
% REFLECTANCE
0.05% RED 214
0.50%
1.00%
2.00%
4.00%
400
440
480
520
560
600
640
680
WAVELENGTH

Fig. 2

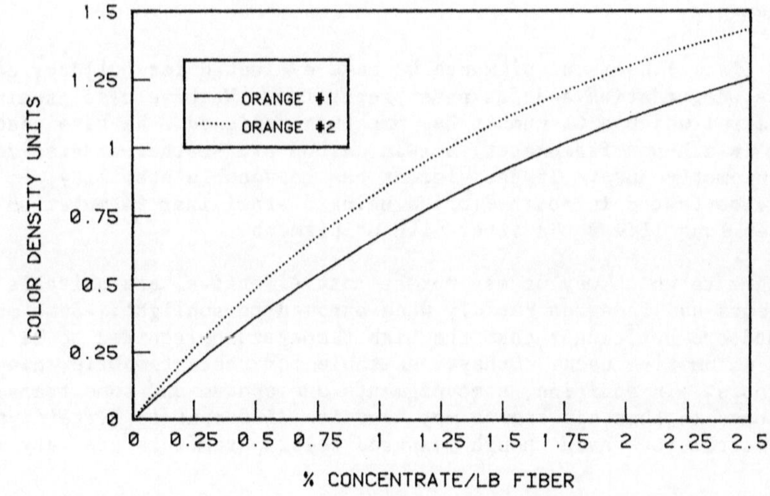

1.5
1.25
1
0.75
0.5
0.25
0
COLOR DENSITY UNITS
ORANGE #1
ORANGE #2
0
0.25
0.5
0.75
1
1.25
1.5
1.75
2
2.25
2.5
% CONCENTRATE/LB FIBER

Fig. 3

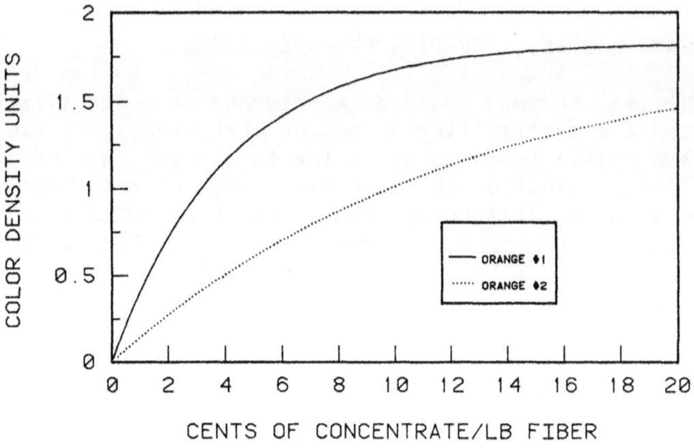

2
1.5
1
0.5
0
COLOR DENSITY UNITS
ORANGE #1
ORANGE #2
0
2
4
6
8
10
12
14
16
18
20
CENTS OF CONCENTRATE/LB FIBER

Fig. 4

match a given color, the computer simply has to search its library to find which combination of pigments and concentrations are required to reproduce the desired color. This type of computer match is demonstrated in Figure 5 where 2.4% of Yellow 83 concentrate, 0.27% of Black 7 concentrate, and 0.5% of Red 144 concentrate were selected to match a Brown with the reflectance as shown.

COLOR REQUIREMENTS

Table 2 lists 118 of the pigments we have evaluated for buildup, cost effectiveness, degradation and fastness properties. We have also examined about 50 more for which a CI number has not been assigned. We have tested all of these in a Xenon Fadeometer, a twin carbon arc Weather-O-Meter designed for automotive use. If the pigment has reasonable stability, samples are also tested in South Florida using a stabilizer formulation which provides long life for a fiber without pigment.

Some pigments which may or may not be cost effective, are suitable only for indoors and fade too rapidly when exposed to sunlight. Some may be stable outdoors but cannot take the high temperature required to be suitable for automotive uses. Others, suitable for the automobile, cannot be used outdoors. In addition, some pigments are opaque and some transparent in fibers so that appearance may be quite different in certain yarn or fabric constructions even though measured color parameters are very similar.

Thus, the color of the Brown sample in Figure 5 may have to be matched with different pigments to meet the requirements of different end uses. Indoor end uses permit a wide range of pigments so less expensive ones may be selected from the list in Table 2. The list of pigments suitable for outdoors is much smaller. For instance, Handen & Graff have listed (Table 3) eleven as suitable for high UV use. Thus, the computer must be told which pigments are applicable to a particular end use when asked to predict a composition to match a given target.

PRODUCT REQUIREMENTS & STABILIZER-PIGMENT INTERACTION

All the additives, whether added as a colorant or a stabilizer, remain in the fiber. With the availability of better stabilizers, it has been possible to achieve stability characteristics in natural (non-pigmented) fibers which were inconceivable ten years ago. Some of this improvement is due to unexpected synergistic interactions found when combining certain additives. This type of synergism is demonstrated in Table 4. The data indicate how much addition of 0.05% antioxidant will improve the thermal stability of a resultant fiber.

While the denier is different for different sets of data, dpf is not responsible for the different results. Test data indicate that long term thermal stability is virtually independent of dpf. The Ciba-Geigy data indicates that Irganox 1425 is the best of the three antioxidants when added to their resin compositions. Irganox 1425 also improves the stability of fibers but only in a predictable manner. The stability improvements, obtained with Irganox 1330 and 1010, were unexpected, i.e., they behaved synergistically with the other additives in the control.

The results in Table 5 show how difficult it is to select good pigments for a stabilizer based upon results in a different composition. Without a UV stabilizer, none of the eight pigments tested is a prodegradant. In fact, all, except Blue 29 are stabilizers. Using a nickel stabil-

Table 2. Pigments for Polypropylene Fibers

Red 5	Red 108	Red 185
Red 7	Red 112	Red 187
Red 38	Red 113	Red 190
Red 48:1	Red 119	Red 194
Red 48:2	Red 122	Red 199
Red 48:3	Red 123	Red 200
Red 53	Red 132	Red 202
Red 53:1	Red 144	Red 206
Red 57:1	Red 149	Red 207
Red 58:3	Red 151	Red 208
Red 68	Red 166	Red 214
Red 71	Red 170	Red 224
Red 88	Red 175	Red 242
Red 101	Red 177	Red 244
Red 104	Red 179	Red 246
		Red 247
Yellow 12	Yellow 93	Yellow 120
Yellow 17	Yellow 95	Yellow 138
Yellow 20	Yellow 97	Yellow 139
Yellow 24	Yellow 99	Yellow 150
Yellow 34	Yellow 108	Yellow 151
Yellow 37	Yellow 109	Yellow 154
Yellow 42	Yellow 110	Yellow 155
Yellow 53	Yellow 113	Yellow 173
Yellow 74	Yellow 116	Yellow 180
Yellow 81	Yellow 118	Yellow 181
Yellow 83	Yellow 119	
Blue 15	Green 7	Violet 19
Blue 16	Green 10	Violet 23
Blue 22	Green 17	Violet 29
Blue 28	Green 19	Violet 32
Blue 36	Green 26	
Blue 60	Green 36	
	Green 50	
Orange 13	Orange 34	Orange 46
Orange 20	Orange 36	Orange 48
Orange 23	Orange 38	Orange 49
Orange 31	Orange 43	Orange 61
		Orange 62
Black 7	Brown 11	Brown 35
Black 13	Brown 23	
Black 26	Brown 24	
Black 28	Brown 25	White 6

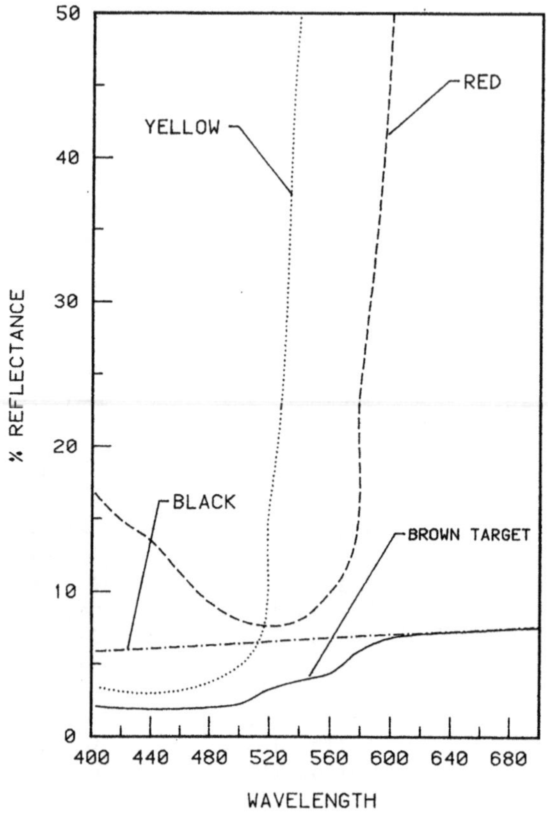

Fig. 5

Table 3. Pigments for High UV Use*

White 6	Yellow 37	Orange 20
Red 149	Red 166	Blue 15
Violet 19	Violet 23	Black 7
Green 7	Green 36	

*S. L. Handen & J. R. Graff, Fiber Producer 8: 35-36, 66-69 (1980).

Table 4. Oven-Aging of Polypropylene Fibers

Formulation	Days to Embrittlement at 110°C 3.8 dpf*	at 130°C 18 dpf
No Antioxidant Blank	1	3
0.05% Irganox 1330	2	168
0.05% Irganox 1010	5	153
0.10% Irganox 1425	9	9

*Data from Ciba-Geigy

Table 5. Interaction of Pigments with Light Stabilizers*

Pigment (0.4%)	Kilolangleys to 50% Tensile Strength (1 mil film) Unstabilized	Nickel Stabilizer	Tinuvin 770 (0.3%)
None	13	30	30
White 6	23	39	92
Red 101	15	24	-
Yellow 37	17	32	80
Orange 20	18	31	85
Blue 29	13	32	85
Yellow 93	22	31	40
Red 144	15	41	58
Blue 15	16	30	94

*P. P. Klemchuk, Polymer Photochemistry 3:1-27 (1983).

izer, which more than doubled UV stability, only Red 101 is a prodegradant. Using Tinuvin 770 at 0.3%, which improves stability of natural fiber six-fold, both Yellow 93 and Red 144 are prodegradants, a result which is not predicted by the other data.

As the amount of Tinuvin 770 is increased (Table 6), an even larger number of pigments become prodegradants. Of the 24 pigments, none of which was a prodegradant without Tinuvin 770, 13 are prodegradants when 0.75% Tinuvin 770 is used. When these same pigments were tested in the Phillips Fibers proprietary stabilization system, only 7 were not prodegradants for use in automotive fibers and/or outdoors.

Table 7 contains degradation results in a Xenon Fadeometer of a number of samples made with the same base resin but with six different HALS products. Samples were made with three different pigments at 0.75% pigment concentration. In the "No HALS" samples, the three pigments were not stabilizers as was reported in Tables 5 & 6, probably because this "No HALS" resin is quite well stabilized even without a HALS. Note that in Tables 7 & 8, the Protection Factor is determined by dividing the test sample time to degrade by "No HALS" degradation sample time. Thus, for any HALS, if the Protection Factor is smaller for the pigmented than the natural sample, the pigment is a prodegradant.

All six HALS products improved degradation stability of the natural fiber in a Xenon Fadeometer to approximately the same extent. Orange 31 was a prodegradant in every case. Blue 15 was a prodegradant only when HALS-3 was used. Red 144 was not a significant prodegradant with any of the 6 HALS products in a Xenon Fadeometer. These data do not indicate that color stability is acceptable for Red 144 using these other HALS products. The data say only that Red 144 is not a prodegradant in Xenon testing in the presence of high levels of these 6 HALS in a base resin.

The U.S. auto industry has established degradation test requirements at higher than normal testing temperatures. One such test is a twin carbon arc Weather-O-Meter test run at 190°F. Requirements for polypropylene fabrics originally were 300 hours without degradation or noticeable color change. Requirements for polypropylene fabrics have since been raised to 500 hours in similar exposure devices.

Table 8 contains test results in this type accelerated device using the same samples as those in Table 7. Orange 31 performs best in HALS 4 and 5 as it did in the Xenon Fadeometer. Blue 15 is a prodegradant in all cases. Red 144 is a troublesome prodegradant only with HALS-1. HALS-1, which behaved as well as the other five samples when tested in a Xenon Fadeometer as natural fiber (Table 7), interacted poorly with all three pigments in the high temperature Weather-O-Meter.

Of considerable interest is the performance of Red 144 with HALS-4 with regard to color stability. With other HALS products Red 144 loses much color long before degradation occurs in these accelerated devices. With HALS-4 stabilizer, Red 144 became unacceptable colorwise only shortly before degradation.

CONCLUSION

The problem of producing colored polypropylene fibers has been complicated by the development of improved stabilization compositions. Solutions to these problems are not readily predicted by theory or current knowledge of mechanism. The Edisonian approach has been used successfully in preparing many new products for more severe end uses, such as in automobiles

Table 6. Pigment - Stabilizer Interaction

Pigments (1.0%)	Degradation Protection Factors* Tinuvin 770** 0%	0.75%	Phillips Stabilization System GM Weather-O-Meter	Florida Months to Degrade
None	1.0	1.0	1.0	>24
Yellow 37	1.4	1.04	1.4	Color NG
Yellow 83	1.2	0.70	0.93	24
Yellow 93	1.6	0.65	0.73	18
Yellow 95	1.2	0.52	0.47	18
Yellow 109	1.6	0.83	0.73	15
Yellow 110	1.2	0.91	1.00	>24
Yellow 138	1.0	0.57	Color NG	18
Irg. Yellow 5768	1.7	1.52	0.67	-
Crom. Yellow 6258	1.3	1.33	0.40	-
Orange 31	1.6	0.76	0.47	12
Orange 61	1.3	1.04	0.80	>24
Irg. Orange 1147	1.8	1.33	>1.07	-
Red 48:2	1.2	0.98	Color NG	Color NG
Red 88	1.5	0.85	Color NG	15
Red 144	1.4	0.87	0.67	15
Red 149	1.2	0.98	0.73	18
Red 166	1.5	0.91	0.60	18
Red 177	2.0	1.00	0.75	15
Red 224	1.0	0.80	0.75	18
Blue 15:3	1.6	>1.30	1.07	>24
Blue 16	1.2	1.33	>1.87	>24
Blue 60	1.6	1.65	Color NG	>24
Green 7	1.8	>1.09	1.33	>24
Black 7	2.5	>1.52	>2.00	>24

*Protection Factor = Stability Pigmented Fiber/Stability Natural Fiber
**Data from Steinlin & Saar, Melliand Textilberichte II: 941 (1980)

Table 7. Degradation Protection Factor* in Xenon Fade-O-Meter

	Pigments at 0.75%			
	Natural	Orange 31	Blue 15	Red 144
No HALS	1.00	0.50	0.50	1.00
0.8% HALS-1	3.15	2.13	3.23	3.03
0.8% HALS-2	3.15	1.93	3.23	3.03
0.8% HALS-3	3.23	2.13	2.63	3.15
0.8% HALS-4	3.23	2.63	3.15	3.23
0.8% HALS-5	3.10	2.63	3.15	3.23
0.8% HALS-6	3.15	2.10	3.15	3.03

*Protection Factor = Sample Deg. Time/Time to Degrade Natural No HALS sample.

Table 8. Degradation Protection Factor* in Twin Carbon Arc Weather-O-Meter at 190°F

	Pigments at 0.75%			
	Natural	Orange 31	Blue 15	Red 144
No HALS	1.00	0.30	0.30	0.61
0.8% HALS-1	2.91	1.74	1.26	2.00
0.8% HALS-2	4.17	2.78	2.48	4.13
0.8% HALS-3	4.13	2.74	1.61	4.78
0.8% HALS-4	3.60	4.00	1.91	5.57
0.8% HALS-5	5.39	3.61	1.48	4.13
0.8% HALS-6	4.13	2.78	3.30	5.04

*Protection Factor = Sample Deg. Time/Time to Degrade Natural No HALS sample.

Table 9. Annual Production (Million Pounds)*

	Yarn & Monofilaments	Staple & Tow	Total	Total Noncellulosic	Olefin % of Total
1970	200	62	262	3,572	7.3
1972	347	70	417	5,346	7.8
1974	463	68	531	6,212	8.5
1976	516	61	577	6,615	8.7
1978	597	95	692	7,768	8.9
1980	632	116	748	7,874	9.5
1982	585	138	723	6,442	11.2
1983	709	186	895	7,527	11.9
1984	744	247	991	7,335	13.5

*(Textile Organon, January 1984)

and outdoors. The properties which can now be built into polypropylene fibers have permitted the use of polypropylene staple fibers to quadruple and polypropylene filament to grow 50% since 1976 (Table 9). Since 1978, total olefin fiber has grown 50% and increased from 8.9 to 13.5% of an otherwise stagnant noncellulosic market. As industry continues to find new uses for polypropylene's unique fiber properties, use of high quality polypropylene fibers should continue to grow.

DYEABLE FIBERS FROM BLENDS OF POLYPROPYLENE

Pushpa Bajaj, A. C. Vashista, and Amir Modjarrad

Department of Textile Technology
Indian Institute of Technology, Delhi
New Delhi 110016, India

INTRODUCTION

Unmodified polypropylene fiber is difficult to dye due to the absence of dye sites in the molecular chain. Further, the high crystallinity is a deterrent to the diffusion of dyes. Because of its hydrocarbon nature and high crystallinity, unmodified polypropylene fiber is likely to be penetrated only by oleophilic dye molecules. Since smaller amounts of oleophilic dyestuffs are available for the penetration of such a dye molecule, either the pure hydrocarbon nature of the polymer must be changed or some submicroscopic discontinuities with high interfacial energy should be created within the fiber for dye molecules.

In the past decade, considerable efforts have been expended in synthesizing the dyes, specially meant for hydrocarbon fibers, or in modifying polypropylene fiber for making it dyeable, like other conventional fibers.

Most of the PP fibers produced are colored by the process of mass coloration.[1-3] This process involves the intermixing of colorants with the polymer prior to extrusion. Organic pigments are used because of their bright, transparent and strong optical effects, while inorganic pigments offer a high degree of fastness. The following limitations restrict the applications of pigments for coloration of PP fibers.

- Some of the inorganic pigments like lead and cadmium are very toxic in nature, hence their handling during blending with polypropylene becomes difficult.
- A wide range of colors must be stored for different shades which leads to an increase in the cost.
- Reproducibility and matching of shades is difficult.
- Also, in the apparel field, the frequency of fashion change in color is quite high, and it is not possible to satisfy the customer's demand with only a score of colors.

Dyeable polypropylene fibers can be produced by (i) surface modification, (ii) attachment of dye receptive monomers through graft copolymerization, or (iii) blending with other dyeable polymers or additives before extrusion.

In view of the comparatively greater affinity of the disperse dyes for polypropylene, extensive studies have been made to produce disperse dyeable polypropylene fibers. The approach followed has been to intermix some appropriate polymeric compounds, ranging from cellulose ester to wholly synthetic polymers. Polycarbonates, polyvinyl ester, fatty acid esters, polystyrene, ABS, polyester etc. have generally been used as additives for producing dyeable PP fibers.[4-6] Ziang-Feng[7] also prepared a disperse dyeable polypropylene fiber by blending it with a copolymer (1.5-5%) of styrene and 2,2,6,6-tetramethyl-4-piperidinyl methacrylate.

Another method of producing disperse dyeable polypropylene fiber is to decrease its crystallinity value to such a level so that it may be dyed with disperse dyes similar to polyester fibers. Hedge and Hebe[8] have reported that with an increase in polystyrene content, the shrinkage of PP fibers increases i.e., the crystallinity of blend fibers decreases. White and coworkers[9] have also studied the structural changes in the polypropylene matrix due to the addition of nylon-6.

The formation of fibrils of nylon-6 in the polypropylene matrix, which have high interface energy, tends to lower the degree of crystallization of polypropylene. In addition, Halim[10] has established that due to the good adhesion between polypropylene and nylon-6, the blend can be cold drawn. In the process of cold drawing, crimping and heat-setting, it is expected that pores and cracks occur in polypropylene fiber, and these submicroscopic pores and cracks may allow the diffusion of dyes into the fibers.

Gill et al.[11] have also mentioned that during melt spinning, the crystallinity of polypropylene fiber could be lowered by using (i) high spinning temperature (ii) low quenching temperature (iii) high stretch ratio during cold drawing and (iv) low winding speed during melt spinning etc.

The purpose of this paper is to present integrated studies of polypropylene blend systems. The rheological behavior of polypropylene blended with 2-10 parts of polystyrene, polyethylene terephthalate or nylon-6 is reported in addition to the dyeability, morphological and mechanical properties of polypropylene blend fibers.

EXPERIMENTAL

Materials

Polypropylene used was Koylene-S of fibre grade supplied by Indian Petrochemicals Corporation, Ltd. Nylon-6, PET and polystyrene of the following intrinsic viscosity were supplied by different manufacturers.

	Polymer	Solvent (Temp.)	$[\eta]$	$\bar{M}_v$
1.	Polypropylene	Decalin (135°C)	0.86	72.5×10^3
2.	Polystyrene	Toluene (30°C)	0.66	22×10^4
3.	PET	TCE:Phenol 40:60	0.65	17.5×10^3

4.	Nylon-6	Formic acid (25°C)	1.1	31.4×10^3

Preparation of Blends

Blends of polypropylene with polystyrene, PET or Nylon-6, with blend compositions 2,5 and 10 wt. % polystyrene or PET, and 2,4 and 6 wt. % Nylon-6 were prepared by melt blending in a Betol single screw melt extruder (Model RM1820). The experimental details of melt blending are given below.

Screw speed	=	20 rpm
Quenching bath temp.		23°C

Temperature at different zones in the extruder.

Zone	Temperature, °C		
	PP/Polystyrene	PP/PET	PP/Nylon-6
I	220	260	225
II	230	265	230
III	230	265	235
IV	220	260	240

The blends obtained in the form of strands were cut into chips, and then dried in a vacuum oven.

Rheological Measurements

The melt flow properties of the homopolymers and blends were measured on a MCR capillary rheometer installed on Instron tester (Model 1112). The L/D ratio of the capillary used was 20. The forces corresponding to different plunger speeds, 1.25, 2.5, 5 and 10 cm/min were recorded, and from this data, the shear rates and shear stresses were calculated.

Scanning Electron Microscopy

Scanning electron micrographs of the cryogenically fractured ends of the extrudates were recorded on a Stereoscan S4-10 of Cambridge Scientific Instruments Ltd. The blends were etched with specific solvents at room temperature for 30 min to remove the minor component, i.e., PET, polystyrene or Nylon-6.

DTA measurements were done on a Stanton Redcroft DTA-671 differential thermal analyzer, using alumina as the reference material. The samples were first heated to 200°C and were kept at that temperature for 5 min so as to eliminate the effect of a previous history of crystallization. Thermograms were recorded during the cooling cycle at a constant rate of 10°C/min.

Crystallization behavior and the changes in morphological structure of blended PP in the cooling cycle was also studied by hot stage microscopy. For this purpose, Mettlar FP82 thermosystem was coupled with Mettlar FP80 to study the structural changes under a polarizing microscope.

Melt Spinning of Blends

Blended PP/N6 was melt spun through a single hole spinneret with a diameter 0.732 mm. The molten polymer coming out of the spinneret as a monofilament was passed through an air quenching duct (length 50 cm) installed just below the spinneret. The monofilament thus obtained was drawn to 6 times its length, at 80°C and the drawn fiber was wound on spools under low winding speeds of 40 m/min.

Heat-setting of PP and its blend (PP/N6) fibers was carried out at 100°C for 30 min in an air oven. For heat-setting the fibers were wound under tension on the glass bobbins.

PP/PET and PP/PS blends were melt spun in a Betol extruder (a horizontal spinning line) and the monofilament was quenched in a water bath at 5°C placed 9 in. from the spinneret. These filaments were then cold drawn to 16 cm except PP/PET (90/10) blend. The spinning temperature for PP/PS blends and PP/PET blends was 220°C and 260°C respectively.

Structure and Properties of Blend Fibers

X-ray diffraction measurements were made on Philips Norelco X-ray diffraction equipment provided with a scintillation counter and recorder. Radial scans of intensity (I) vs. diffraction angle (2θ) were recorded in the range of 8-50° of 2θ using CuK_α radiation. The orientation of the crystalline regions of the polypropylene was also measured.

Tensile properties were measured on an Instron universal tester (Model 1121) at constant strain rate 5 cm/min keeping the initial gauge length 5 cm/min.

Sonic modulus of the blend fibers was measured by a Dynamic Modulus tester PPM-5R.

Dyeing of blend fibers was carried out in a beaker dyeing machine at 100°C. Two disperse dyes with different molecular weights (Cibacet Orange 2R, mol. wt. 242 and Supracet Fast Brown, 5R, mol. wt. 369) were selected for dyeing.

The amount of the dye in the fiber was calculated by the dye stripping technique. The apparent diffusion coefficient (Da) was calculated using Hill's method,[12] which measures the dye uptake of the fiber with time intervals of 't' from an infinite dyebath.

To test light fastness, the samples and standards were mounted on a frame and partly covered with an opaque sheet, leaving the other half exposed. The samples were exposed on a Fade-O-meter for a maximum of 72 hours at 60% relative humidity and 25°C. The fading of the samples was compared from time to time with the standards.

RESULTS AND DISCUSSION

Rheological Properties of Blends

The data, at constant temperatures and varying pressures, were used to evaluate the shear stress and shear rate relationships in these blends. Variations of shear stress vs. shear rate are linear over the range of shear rates investigated (Figures 1 and 2), confirming the validity of the power-law relationship for these blends. The values of the exponent, 'n' of the power law relation

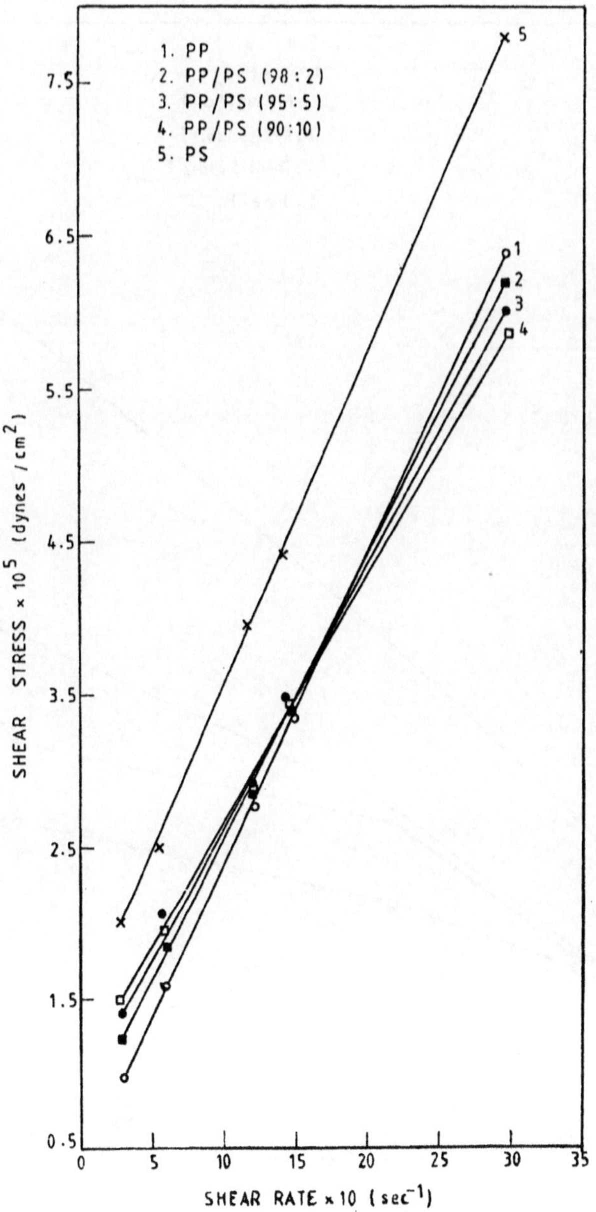
1. PP
2. PP/PS (98:2)
3. PP/PS (95:5)
4. PP/PS (90:10)
5. PS
SHEAR STRESS × 10^5 (dynes/cm^2)
SHEAR RATE × 10 (sec^-1)
7.5
6.5
5.5
4.5
3.5
2.5
1.5
0.5
0
5
10
15
20
25
30
35
1
2
3
4
5

Fig. 1

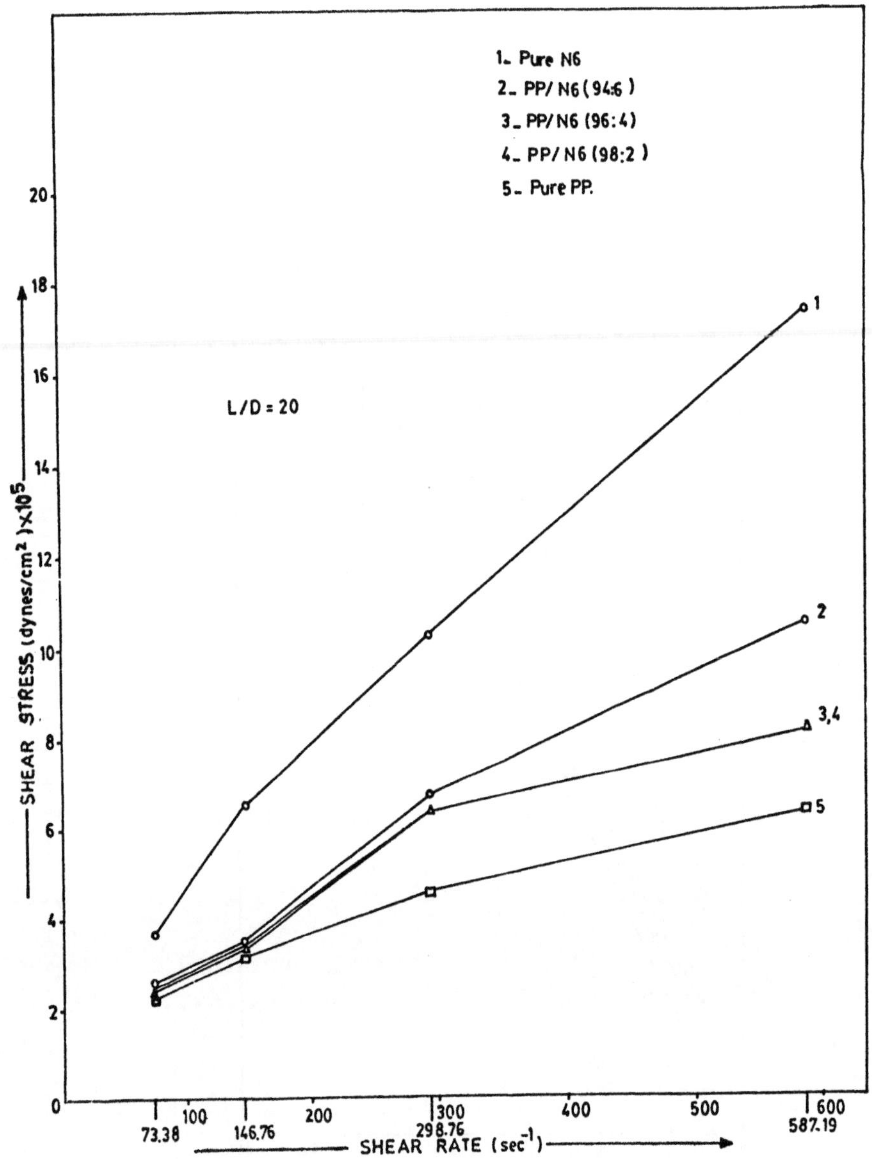
1_ Pure N6
2_ PP/N6 (94:6)
3_ PP/N6 (96:4)
4_ PP/N6 (98:2)
5_ Pure PP.
L/D = 20
SHEAR STRESS (dynes/cm²) X10⁵
SHEAR RATE (sec⁻¹)
0
2
4
6
8
10
12
14
16
18
20
100
200
300
400
500
600
73.38
146.76
298.76
587.19
1
2
3,4
5

Fig. 2

$$(\tau_w)_{app} = K\gamma^n$$

determined from these data clearly support the pseudoplastic behavior of the melt for all these samples.

Variation of apparent melt viscosity (η_{app}) with shear stress (τ_w) is shown in Figure 3. The drop in melt viscosity with increase in shear stress is due to the fact that at higher shear rates, the molecular chains tend to be aligned parallel to each other, and exude more easily through the capillary, thus a decrease in the apparent melt viscosity.

Variation of melt viscosity with blend composition is shown in Figures 4-5. Melt viscosity of PP/PS blends, at constant shear stress, is found to be nearly the same at all compositions, i.e., 98/2, 95/5 and 90/10. However, in PP/PET and PP/N6 blends, η_{app} increases with an increase in PET and N6 content. However, rheological properties of polypropylene-polystyrene blends, reported by Han, clearly indicated a reduction in melt viscosity of the blend with a minimum at a blending ratio for 50:50 by wt. of PP:PS. These differences in the rheological properties of the PS, Nylon-6 and PP are due to a combination of absolute molecular weight and mol. wt. distribution.

Thermal Behavior of PP Blends

DTA thermograms recorded for PP and its blends showed sharp endothermic melting peaks and exothermic crystallization peaks in the cooling cycle. The characteristic peaks for PP/PS blends are illustrated in Figure 6, and the data are presented in Tables 1 and 2. The T_m of PP in PP/PS blends is 167°C. Hence, there is no change in T_m by blending with polystyrene up to 10 wt. %. However, the T_m decreases in both PP/PET and PP/N6 blends.

The crystallization peaks occur in the range of 117-123° for various blends under study, while virgin PP crystallizes at 119°C. The differences in these exotherms are also found in terms of temperature of onset of crystallization, T_{onset}, initial slope of the exotherm, Si, area under the exotherm divided by the mass of PP and the width at half-height of the exotherm peak, Δw, which are known to be related to the crystallization parameters.

Increase in exothermic peak temperature, T_p in PP/PS (95:5 and (90:10) blends clearly indicates the reduced supercooling. The quantity Si which is indicative of the rate of nucleation, increases with the addition of PET or N6 as nucleating agents. However, the overall rate of crystallization is low as shown from T_c-T_p values (Table 1). Since the parameter (T_c-T_p) is a measure of the overall rate of crystallization[13], the greater the (T_c-T_p), the smaller the rate of crystallization. The quantity A/m, which is proportional to the heat of crystallization of the given sample, thus, its degree of crystallinity, decreases, with increase in the Nylon-6 or PET content, in PP blends. Δw, the exothermic peak width at half-height, increases with increase in N6 content and this clearly supports the broader crystallite size distribution.

The rate of nucleation and growth of large spherulites has been further confirmed from the thermal microscopy of PP/N6 blends. In virgin PP and Nylon-6, only a single phase was observed in the molten state, but in the blends, two clearly distinct phases were noted (Figures 7-8). One was the polypropylene matrix and the other Nylon-6 in the form of minifibers in the PP matrix. Nylon-6 remains in the form of minifibers in the molten PP until the temperature reaches the T_m of N6.

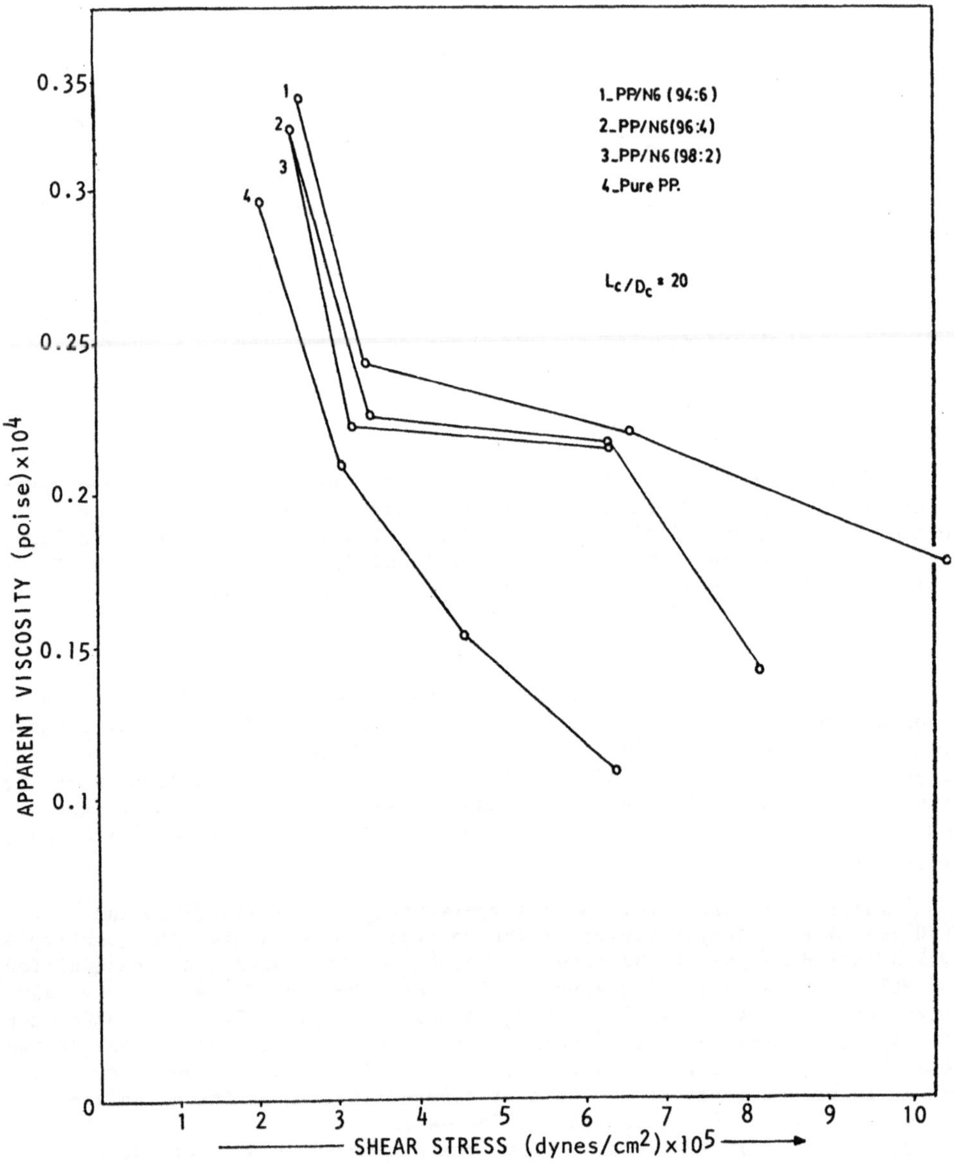

1_PP/N6 (94:6)
2_PP/N6(96:4)
3_PP/N6(98:2)
4_Pure PP.
Lc/Dc = 20
APPARENT VISCOSITY (poise)x10^4
0.35
0.3
0.25
0.2
0.15
0.1
0
1
2
3
4
5
6
7
8
9
10
SHEAR STRESS (dynes/cm^2)x10^5

Fig. 3

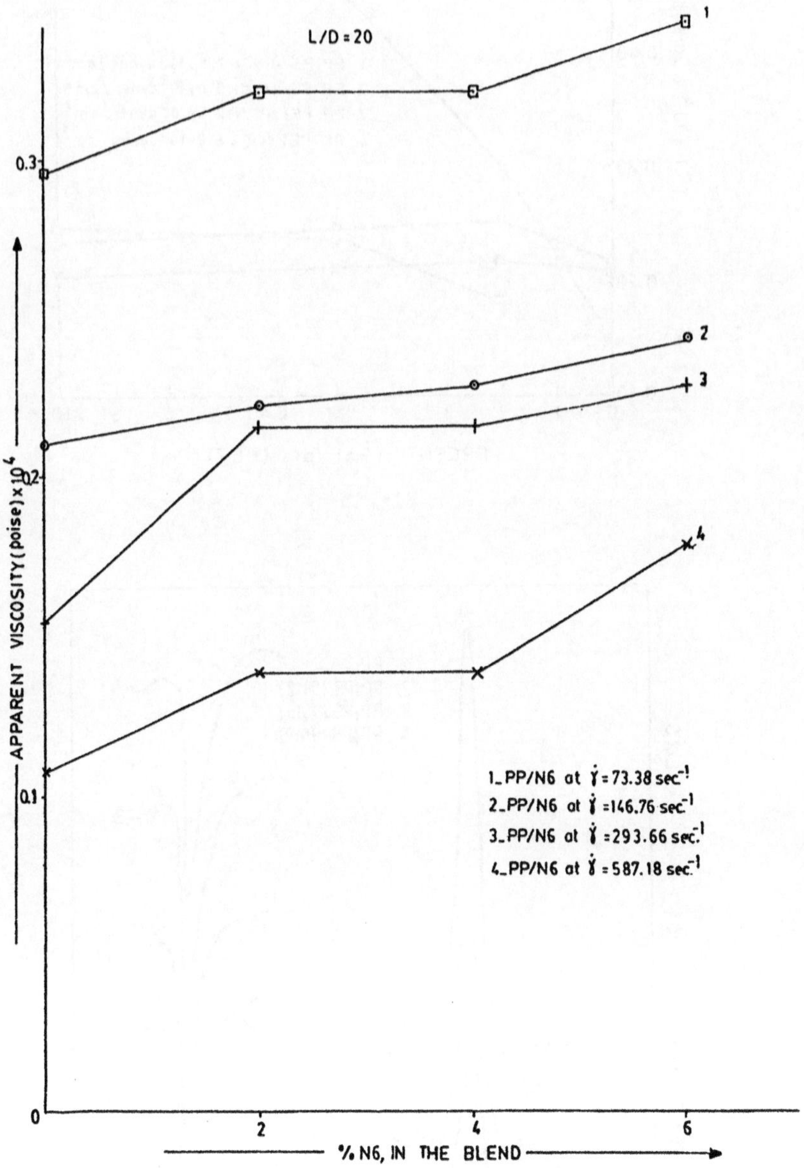
L/D = 20
1
2
3
4
0.3
0.2
0.1
0
APPARENT VISCOSITY (poise) x 10^4
2
4
6
% N6, IN THE BLEND
1_PP/N6 at γ̇ = 73.38 sec⁻¹
2_PP/N6 at γ̇ = 146.76 sec⁻¹
3_PP/N6 at γ̇ = 293.66 sec⁻¹
4_PP/N6 at γ̇ = 587.18 sec⁻¹

Fig. 4

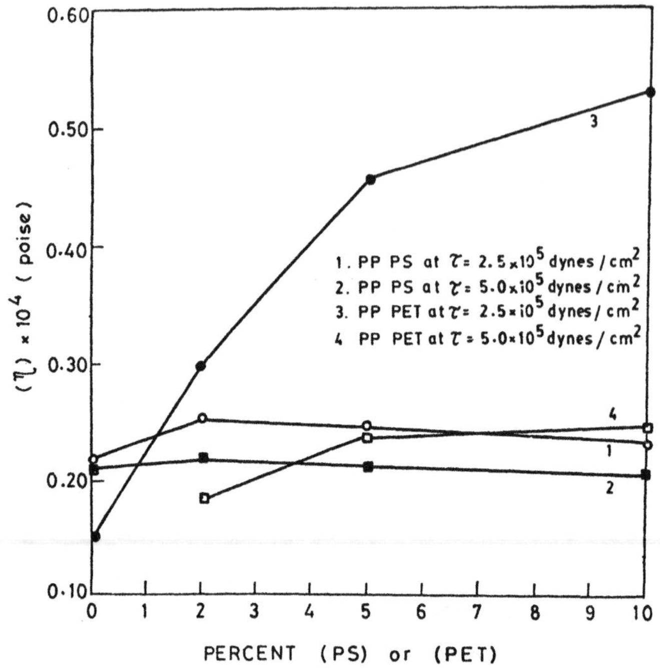
0.60
0.50
0.40
0.30
0.20
0.10
0
1
2
3
4
5
6
7
8
9
10
$(\eta) \times 10^4$ (poise)
PERCENT (PS) or (PET)
1. PP PS at $\tau = 2.5\times10^5$ dynes / cm^2
2. PP PS at $\tau = 5.0\times10^5$ dynes / cm^2
3. PP PET at $\tau = 2.5\times10^5$ dynes / cm^2
4 PP PET at $\tau = 5.0\times10^5$ dynes / cm^2

Fig. 5

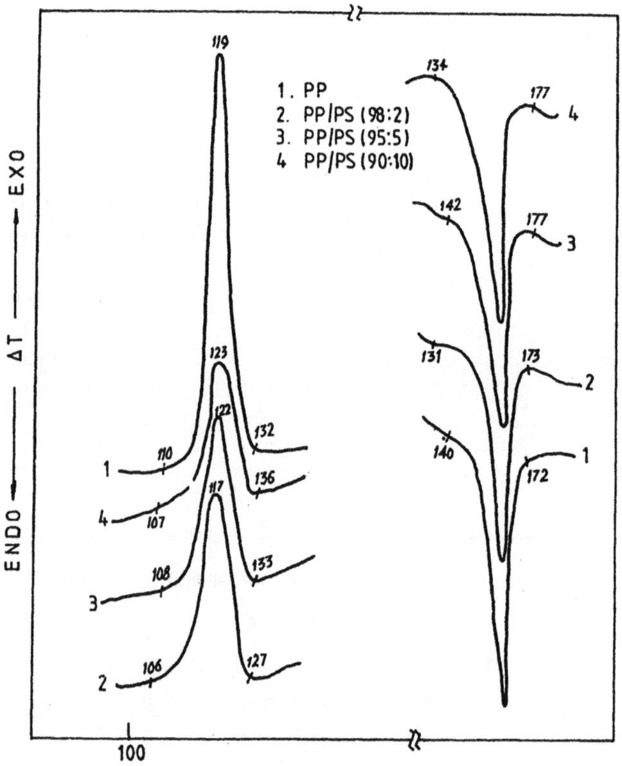
1. PP
2. PP/PS (98:2)
3. PP/PS (95:5)
4 PP/PS (90:10)
ENDO ← ΔT → EXO
TEMPERATURE , °C
100

Fig. 6

Table 1. Thermal Properties of PP/PS and PP/PET Polymer Blends

Sample Code	Endothermic peak T_i (°C)	Endothermic peak T_m (°C)	Exothermic Peak Temperature T_i (°C)	T_{pk} (°C)	T_c (°C)	$T_c - T_{pk}$ (°C)	ΔH_f cal/gm	% Crystallinity from ΔH_f Crystallinity of PP component in the blend
PP	140	167	110	119	123	4	13.3	38.0
PP/PS (98:2)	137	167	106	117	122	5	12.4	35.8
PP/PS (95:5)	142	167	108	122	128	6	12.2	35.6
PP/PS (90:10)	137	167	107	123	132	9	11.9	34.8
PP/PET (98:2)	137	165	108	120	126	6	11.6	33.6
PP/PET (95:5)	133	164	103	117	123	6	10.9	31.8
PP/PET (90:10)	138	163	103	117	126	9	10.2	30.4
PET	227	254	-	-	-	-	-	-

Table 2. DTA Data of PP/N6 Polymer Blends

Sample Code	Melting peak T_m (°C)	Crystallization peak T_c (°C)	Onset crystallization T_i (°C)	S_i (arbitrary units)	ΔW (arbitrary units)	Crystallinity, (A/m) (arbitrary units)
Pure N6	228	183	190	8.6	7.5	319.3
Pure PP	167	119	124	5.0	9.0	403.6
PP/N6 (98:2)	166	116	124	5.1	10	402.8
PP/N6 (96:4)	163	116	125	5.1	10.5	373.2
PP/N6 (94:6)	163	115	127	5.2	11.5	358.4

At 240°C, Nylon-6 droplets are seen in the PP matrix, which act as a nucleating agent for the crystallization of polypropylene. Hence, the morphology of polypropylene crystallites is much different, at the interphase, than in the bulk. Though the solubility parameter of PP and N6 are different and they form incompatible blends, there may be some mixing of two components at the interphase which influences the spherulitic growth and the rate of crystallization of polypropylene.

Structure-Properties of PP Blend Fibers

Figure 9 displays stress-strain curves of PP and PP/N6 blend fibers. The yield stress, elongation-to-break, tenacity and initial modulus decrease with increase in N6 content.

There was still some residual draw in virgin PP fibers and they could be drawn further up to draw ratio 8. However, the blend fibers could not be drawn beyond DR = 6 at 80°C. Hence, the stress-strain curves fibers drawn as DR,6 are illustrated in Figure 9. The lower drawability of blend fibers may be attributed to two phase structure and large number of small morphological units which impose more restraints on the mobility of the matrix polymer than on a small number of large units. Friedrich[14] has also shown that course spherulitic morphologies in PP exhibit the lowest values in respect of yield or fracture stress. Since the mechanical properties are a function of orientation in the fiber, low tenacity values in PP blend fibers can be correlated with the extent of molecular orientation in the crystalline phases.

The effect of the blend composition in PP/N6 filaments on a WAXS pattern is illustrated in Figure 10. There seems to be a noticeable decrease in the monoclinic form with increase in the N6 content (2 to 6 wt. %). The intensity of peaks (14.5° and 22.5°) corresponding to the planes (110) and (041) also are reduced in the blend fibers. The monoclinic α-form of PP changes its structure more towards smectic type. X-ray crystallinity and orientation in PP blend fibers also decrease with respect to the virgin PP fiber. Sonic modulus of blend fibers is also low in PP fibers. Decreased crystallization and orientation in blend fibers (Table 3) can be interpreted in terms of development of smectic paracrystalline polypropylene in the presence of small amounts of Nylon-6, PET and polystyrene and the spinline conditions.

Dyeing

The dye-uptake and diffusion coefficient (at 1 hr dyeing) values for the blend fibers, dyed with disperse dyes, Cibacet Orange 2R (mol. wt. 242) and Supracet Fast Brown 5R (mol. wt. 369) are listed in Table 4. The rate of dyeing was determined by finding the dye-uptake values at different time intervals. The time versus dye-uptake plots are given in Figures 11-12.

We found that PP fiber attains the maximum dye-uptake value in 60 minutes when dyed with Cibacet Orange 2R. However, the PP/PET fibers having 10 wt. % PET take only 40 minutes to reach the equilibrium dye-uptake value. The diffusion of dye into the fiber structure is a diffusion through a porous immobile solid.[15]

The apparent diffusion coefficient (Da) values are greater for the blend fibers for both the dyes when compared to virgin PP fibers. The diffusion process is also influenced by the size of the dye molecules. As the molecular size increases, the diffusion coefficient decreases under identical conditions.

Similar trends have been observed in PP/N6 and PP/PS blend fibers, in

Table 3. Crystallinity and Orientation Values of PP/N6 Blend Fibers

Material	Density	X-ray crystal-linity	Orien-tation index	Velocity sonic (km/sec)	Sonic modulus (g/d)
PP	0.8921	50.83	4.08	3.307	123.65
PP/N6 (98:2)	0.8930	46.75	3.86	3.031	103.83
PP/N6 (96:4)	0.8981	44.18	3.77	2.77	86.77
PP/N6 (94:6)	0.9001	40.11	3.22	2.725	83.93

Table 4. Dyeing Behavior of PP and PP/N6 Blend Fibers

			Equilibrium dye-uptake (gms/kg of fiber		Diffusion coefficient at 1 hr ($cm^2sec^{-1}x10^8$) of dyeing	
Material	Draw Ratio	Fibre dia. ($X10^{-1}cm$)	Cibacet Orange 2R (mol.wt.242)	Supracet Fast Brown 5R (mol.wt.368)	Cibacet Orange 2R	Supracet Fast Brown 5R
PP	6	1.24	0.47	0.40	1.531	0.927
PP/N6 (98:2)	6	1.22	1.0625	1.001	1.867	0.981
PP/N6 (96:4)	6	1.22	1.188	1.020	1.991	1.012
PP/N6 (94:6)	6	1.24	2.482	1.385	1.060	1.220

fact, the dye uptake values are quite high in PP/PET and PP/PS blends fibers having 10 wt. % of the second component (Figures 13-14).

REFERENCES

1. M. Ahmed, "Polypropylene Fibres-Science and Technology," Elsevier Scientific Publications Co., New York (1982).
2. J. Share, Rev. Prog. Coloration, 6, 7 (1976).
3. D. R. Bash, Proc. 2nd Intern. Conf. on PP Fibers and Textiles, PRI, London (1979).
4. W. H. Skoroszewski and B. Cope Cer. Offen. 2,354,922. cf. Chem. Abstr., 81, 79278 (1974).
5. T. Abeta and Y. Memoi, Japan Pat. 74,14,861, Chem. Abstr., 81, 171231 (1974).
6. J. Morita, Japan Kokai Pat. 76,40,110, Chem. Abstr., 86, 156915 (1977).
7. Xiang-Feng, Kao Fen Tzu Hung Hisum 5,285,290 (1979).
8. A. Judge and J. Nebe, 2nd Intern. Conf. on PP Fibers, PRI, London (1979).
9. J. L. White, J. E. Spruiell, and B. C. Goswami, J. Appl. Polym. Sci., 28, 2011 (1983).
10. M. A. Halim, Ph.D. Thesis on "X-ray diffraction studies of PP, polyamide fibres," University of Manchester (1980).
11. R. A. Gill and C. Benzomin, Plastics and Rubber Processing, (March 1980).
12. R. J. Harwood, R. McGregor, and R. H. Peters, J. Soc. Dyers and Color, 88, 216 (1972).
13. M. N. Beck and H. D. Ledbetter, J. Appl. Polym. Sci., 9, 2131 (1965).
14. K. Friedrich, Prog. Colloid Polym. Sci., 66, 299 (1979).
15. S. Rosenbaum, Textile Res. J., 34, 291 (1964).

POLYOLEFINS - WHAT NEXT?

H. F. Mark

Polytechnic University
Brooklyn, New York

INTRODUCTION

In this Symposium on "Advances in Polyolefins" a galaxy of distinguished experts have presented an impressive and stimulating account of the recent developments in Science and Engineering of this important class of polymeric materials. Professor Raymond B. Seymour, the organizer of the Symposium has kindly asked me to add a "Concluding Chapter" to it. This chapter could have been a "Summary" or a "Forecast." The other chapters represent a closely knit and complete survey of the "Advances" - therefore, a "Summary" would not make much sense.

Concerning a "Forecast": When President Hoover was asked in 1930 what he thinks about the development of the economy in the USA, he said: "The economy in this country may develop favorably or it may not" and when G. B. Shaw in 1923 was asked to foretell the future success of his play, "Saint Joan," he is said to have answered: "Forecasting is always difficult but it gets even more risky when the future is involved." This is, apparently, the verdict of experts in Politics and in the Arts but in Science and Engineering we are in a somewhat safer position because we have a large and well documented material on the development of the present "State of the Art" and instead of forecasting we are capable to "extrapolate" on something which is based on existing knowledge of supply and clearly visible demands.

FUNDAMENTAL ASPECTS

It is now almost 50 years since, in 1937, ICI announced the first synthesis of what is now known as LDPE, at high pressures and with free radical initiators; it is about 34 years that Phillips Petroleum followed up with what is now called HDPE using an activated chromium-catalyst at medium pressures and it is 32 years since, in 1954, Karl Ziegler and Giulio Natta revolutionized the field by the use of bimetallic complex coordination catalysts.

Yet, even today, the exact mechanisms of initiation, propagation, transfer and termination are still a matter of discussion. This is clearly shown in a book of Y. V. Kissin, which appeared in 1985 on the "Isospecific Polymerization of Olefins with Heterogeneous Ziegler-Natta Catalysts.[1]

The papers of the Symposium have convincingly shown that most of the progress during the last decades came from the discovery of new catalytic systems. Several "generations" of catalysts - four or five - have been developed; they started from the simple, classical titanium-aluminum compositions, went to systems at the surface of carriers, were "activated" with magnesium chloride and similar compounds, "stabilized" with electron donors and "modified" with various other organic molecules. The results are impressive:

1. The yields went up to 10^5 - 10^6 parts of polymer per part of catalyst,

2. The stereoregulation increased to 99.5% and more,

3. The average molecular weights can be controlled over a wide range and

4. Even the molecular weight distribution can be affected to a certain degree - narrow or broad.

Is there anything left? Yes, there is.

a. Not all demanded favorable influences are displayed by one particular catalyst: if the yield is very high, the stereoregulation is not at its best; this means that compromises are necessary.

b. It would be nice if all catalysts could be made less sensitive against impurities (poisons) such as oxygen, moisture, acetylenic species and, quite general, to any polar or strongly polarizable organic compound.

c. Since, presently, the trend turns sharply to copolymers we should have a much better control of the relative reactively ratios between ethylene and the various comonomers. For the different species of VLDPE and LLDPE, catalysts are needed which introduce the presently used comonomers - 1-butene, 1-hexene and 1-octene - more favorably into the growing chain. In the future, similar demands may arise for other olefinic comonomers, like pentenes, cyclopentadienes and vinyladamantene.[2]

d. It would be very desirable to have catalysts which present the synthesis of segmented copolymers (stereo block copolymers) of ethylene with all comonomers - including propylene.

This is a long, but probably as yet incomplete list of demands which shows that there is much work ahead to establish a better understanding of the basic catalytic actions which may help us to contribute to the solution of some of the still existing problems.

Besides the synthesis there exist also demands in such aftertreatments as endgroup capping, crosslinking together with thermal and chemical stabilization.

Engineering Aspects

A particularly useful - almost indispensible - source of up-to-date information in the entire field of Polyolefins are the "digests of Polymer Developments" by Norbert Platzer.[3]

It can be seen that, in the future, as especially important engineering capability is to "swing," i.e. to operate one and the same plant equipment with appropriate changes of ingredients and processing parameters for the commercial production of a variety of polyolefins, for instance VLDPE,

LLDPE, and HDPE - covering a density range from 0.900 to 0.955 - are all copolymers of ethylene with other alpha olefins and may be made by the Unipol Process of UCC or by one of the slurry or solution processes by Dow, DuPont or BP Chemicals. It has been found that by the proper selection of the catalyst and the monomer feed, all of them may be produced in the same plant equipment. Table 1 shows a few selected "swing" capabilities which already exist.[4] It is very probable that in the near future, much more swing capacity of this type will come into operation.

A similarity expansive situation prevails in the production of copolymers of ethylene with polar monomers.[5] An important representative of this class is the copolymer of ethylene and vinyl acetate - EVA - which is made by high pressure copolymerization (800-2,000 bars) in tubular reactors utilizing free radical initiation. If the acetyl groups are removed by hydrolysis - a copolymer of ethylene and vinyl alcohol - EVAL - is obtained. With the aid of a similar process - free radical initiation at high pressures - copolymers of ethylene with acrylic monomers - acids, esters and nitriles - are prepared and it is obviously advantageous to design plant equipment that may "swing" from one comonomer to several others.

The interest in these compositions is presently sparked by their exceptional characteristics as oxygen barriers for the use of plastics in the food packaging field. It has been found that EVAL alone or in blends or grafts with other compositions is an excellent oxygen barrier as long as it is "bone" dry.[6] As a result many multilayer packaging films and containers are now being designed with dry EVAL as the barrier layer and with several types of adhesives and isotactic polypropylene as the backbone for rigidity and toughness.

Table 1. Polyethylene and Capacity in Million lbs.

Producer	Total Capacity	Swing Potential
UCC	2115	1620
Dow	2010	965
USI	1705	100
Chevron	1475	575
Phillips	1375	300
Exxon	1145	850
Mobil	950	950

Altogether it appears to be evident that there are still many future engineering innovations which will give the polyolefins the place which they deserve in plastics technology.

REFERENCES

1. Y. V. Kissin, "Isospecific Polymerization of Olefins with Heterogeneous Ziegler-Natta Catalysts," Volume 9 of the series "Polymers: Properties and Applications," Springer-Verlag, New York (1985).
2. Compare e.g. Kazuto Fominaga and Masami Haga in "Chemical Economy and Engineering Review," Vol. 17, pp. 23-29 (1985).
3. "Digest of Polymer Development," published by Springborn Laboratories, Inc., Enfield, Connecticut 06082, Quarterly Reports, last issue (January 1986).
4. See f.i., "Modern Plastics," Vol. 63, pp. 54-57 (March 1986).
5. Compare e.g. pages 77 to 87 of the "Platzer Digest," (January 1986).
6. See f.i. "Modern Plastics," Vol. 63, pp. 58-61 (March 1986).

AUTHOR INDEX

SUBJECT INDEX

GPSR Compliance

The European Union's (EU) General Product Safety Regulation (GPSR) is a set of rules that requires consumer products to be safe and our obligations to ensure this.

If you have any concerns about our products, you can contact us on ProductSafety@springernature.com

In case Publisher is established outside the EU, the EU authorized representative is:

Springer Nature Customer Service Center GmbH
Europaplatz 3
69115 Heidelberg, Germany

Batch number: 10762209

Printed by Printforce, the Netherlands